Safety and quality issues in fish processing

Related titles from Woodhead's food science, technology and nutrition list:

Fish drying and smoking (ISBN: 1 56676 668 0)

This book brings together the work of an international group of fish technologists, food scientists, microbiologists, chemists and engineers. It will provide readers with an understanding of the physical, chemical and biological factors influencing the functionality of the product at every stage from harvest to consumption.

Seafood safety, processing and biotechnology (ISBN: 1 56676 573 0)

The 25 reports in this book were prepared by international food scientists specialising in seafood. Topics include seafood quality, toxicity, analytical techniques, HACCP, modelling, seafood microbiology, and new food and non-food uses of seafood.

Food processing technology: principles and practice Second edition (ISBN: 1 85573 533 4)

The first edition of *Food processing technology* was quickly adopted as the standard text by many food science and technology courses. The publication of this completely revised new edition is set to confirm the position of this textbook as the best single-volume introduction to food manufacturing technologies available. New chapters include computer control of processing, novel 'minimal' technologies including processing using high pressures or pulsed electric fields, ohmic heating and an extended chapter on modified atmosphere packaging.

Details of these books and a complete list of Woodhead's food science, technology and nutrition titles can be obtained by:

- visiting our web site at www.woodhead-publishing.com
- contacting Customer Services (email: sales@woodhead-publishing.com; fax: +44 (0) 1223 893694; tel.: +44 (0) 1223 891358 ext. 30; address: Woodhead Publishing Limited, Abington Hall, Abington, Cambridge CB1 6AH, England)

If you would like to receive information on forthcoming titles in this area, please send your address details to: Francis Dodds (address, tel. and fax as above; e-mail: francisd@woodhead-publishing.com). Please confirm which subject areas you are interested in.

Safety and quality issues in fish processing

Edited by
H. Allan Bremner

CRC Press
Boca Raton Boston New York Washington, DC

WOODHEAD PUBLISHING LIMITED
Cambridge England

Published by Woodhead Publishing Limited
Abington Hall, Abington
Cambridge CB1 6AH
England
www.woodhead-publishing.com

Published in North America by CRC Press LLC
2000 Corporate Blvd, NW
Boca Raton FL 33431
USA

First published 2002, Woodhead Publishing Limited and CRC Press LLC
© 2002, Woodhead Publishing Limited
Reprinted 2005
The authors have asserted their moral rights.

British Library Cataloguing in Publication Data
A catalogue record for this book is available from the British Library.

Library of Congress Cataloging in Publication Data
A catalog record for this book is available from the Library of Congress.

Woodhead Publishing Limited ISBN 1 85573 552 0
CRC Press ISBN 0-8493-1540-9

Cover design by The ColourStudio
Project managed by Macfarlane Production Services, Markyate, Hertfordshire
(e-mail: macfarl@aol.com)
Typeset by MHL Typesetting Limited, Coventry, Warwickshire
Printed and bound by Replika Press Pvt. Ltd., India

Contents

Contributors

Chapters 1 and 10

Professor H. Allan Bremner
Allan Bremner and Associates
21 Carrock Court
Mount Coolum
Queensland 4573
Australia

Tel/Fax: +61(0)7 5446 2560
E-mail: bremqual1@optus.net.com.au

Formerly at
Danish Institute for Fisheries
    Research
Department of Seafood Research
Building 221 Søltofts Plads
Technical University of Denmark
DK-2800 Kgs. Lyngby
Denmark

E-mail: hab@dfu.min.dk

Chapter 2

Donn R. Ward
Professor and Associate Head
N.C. State University
College of Agriculture and Life
    Sciences
Department of Food Science
Box 7624
Raleigh NC 27695
USA

Tel: +1 919 515 2951
Fax: +1 919 515 4694
E-mail: Donn_Ward@ncsu.edu

Chapter 3

Ms Sirilak Suwanrangsi
Special Exchange Projects Officer
    (Thailand)
Fish, Seafood and Production
    Division
Canadian Food Inspection Agency
59 Camerot Drive
Nepean

Ontario
Canada K1A 0H9

Tel: +1 613 225 2342 (ext. 4541)
Fax: +1 613 228 6648
E-mail: suwanrangs@inspection.gc.ca

Chapter 4

Lahsen Ababouch
Chief
FIIU
F-608
FAO
Rome
Italy

Tel: +39 06 5705 4057
Fax: +39 06 5705 5188
E-mail: Lahsen.Ababouch@fao.org

Chapter 5

L. Nilsson and L. Gram
Danish Institute for Fisheries
   Research
Department of Seafood Research
Søltofts Plads,
Building 221
Technical University of Denmark
DK-2800 Kgs. Lyngby
Denmark

Tel: +45 45 25 49 21
Fax: +45 45 88 47 74
E-mail: lni@dfu.min.dk

Chapter 6

S. Yamada and E. Zychlinsky
Hitachi Chemical Diagnostics, Inc.
630 Clyde Court

Mountain View
CA 94043
USA

E-mail: yamada17@earthlink.net

H. Nolte
Department of Internal Medicine
Asthma and Allergy Unit
University of Copenhagen
Denmark

Chapter 7

J. Oehlenschläger
Institute for Fishery Technology and
   Fish Quality
Federal Research Centre for Fisheries
Palmaille 9
D-22767 Hamburg
Germany

Tel: +49 40 38905 151
Fax: +49 40 38905 262
E-mail: oehlenschlaeger.ibt@bfa-
   fisch.de

Chapter 8

K. D. Murrell
WHO/FAO Collaborating Centre
   for Emerging Parasitic Zoonoses
Danish Centre for Experimental
   Parasitology
The Royal Veterinary and
   Agricultural University
Dyrlaegevej 100
DK-1870 Frederiksberg C
Denmark

Tel: +45 35 28 27 75
Fax: +45 35 28 27 74
E-mail: kdm@kvl.dk

Chapter 9

G. Palleschi, D. Moscone, L. Micheli
and D. Botta
Dipartimento di Scienze e
Technologie Chimiche
Università di Roma 'Tor Vergata'
Via della Ricerca Scientifica
00133 Roma
Italy

Tel: +39 06 72594337
Fax: +39 06 72594328
E-mail: Giuseppe.Palleschi@uni
roma2.it

Chapter 11

Dr Alexandra Barbosa
ICBAS – Insituto de Ciências
Biomédicas de Abel Salazar
Largo Prof. Abel Salazar, 2
4099-003 Porto
Portugal

Tel: +351 222 062 272
Fax: +351 220 622 232
E-mail: octopus@icbas.up.pt

Professor H. Allan Bremner
Allan Bremner and Associates
Formerly at
Danish Institute for Fisheries
Research
Department of Seafood Research
Building 221 Søltofts Plads
Technical University of Denmark
DK-2800 Kgs. Lyngby
Denmark

E-mail: hab@dfu.min.dk

Professor Paulo Vaz-Pires
ICBAS – Insituto de Ciências
Biomédicas de Abel Salazar
Largo Prof. Abel Salazar, 2
4099-003 Porto
Portugal

Tel: +351 222 062 272
Fax: +351 222 062 232
E-mail: vazpires@icbas.up.pt

Chapter 12

Paw Dalgaard
Danish Institute for Fisheries
Research (DIFRES)
Department of Seafood Research
Ministry of Food, Agriculture and
Fisheries
DTU
Building 221
DK-2800 Kgs. Lyngby
Denmark

Tel: +45 45 25 25 66
E-mail: pad@dfu.min.dk

Chapter 13

Norman F. Haard
Institute of Marine Resources
Department of Food Science and
Technology
University of California
Davis
CA 95616
USA

Tel: +1 530 752 2507
Fax: +1 530 752 4759
E-mail: nfhaard@ucdavis.edu

Chapter 14

I.P. Ashton
Unilever R&D
Colworth House
Sharnbrook
Beds MK44 1LQ
England

E-mail: ian.ashton@unilever.com

Chapter 15

M. Frederiksen
Danish Institute for Fisheries
   Research (DIFRES)
Department of Seafood Research
Building 221 Søltofts Plads
Technical University of Denmark
DK-2800 Kgs. Lyngby
Denmark

Tel: +45 45 88 33 22
Fax: +45 45 88 47 74
E-mail: maf@dfu.min.dk

Chapter 16

M. Gudmundsson and H. Hafsteinsson
Technological Institute of Iceland
   (MATRA)
Keldnaholt
IS-112 Reykjavik
Iceland

Tel: +354 570 71 00
Fax: +354 570 71 11
E-mail: magnusg@iti.is
E-mail: hannes.hafsteinsson@iti.is

Chapter 17

George M. Hall
Department of Chemical Engineering
Loughborough University
Leicestershire LE11 3TU
England

Tel: +44 (0) 1509 222 517
Fax: +44 (0) 1509 223 923
E-mail: G.M.Hall@lboro.ac.uk

Chapter 18

Peter E. Doe
School of Engineering
University of Tasmania
GPO Box 252-65
Hobart
Australia 7001

Tel: +61 3 6226 2129
Fax: +61 3 6226 7863
E-mail: Peter.Doe@utas.edu.au

Chapter 19

E. Martinsdóttir
Project Manager
Research and Development Division
Icelandic Fisheries Laboratories
P.O. Box 1405
IS-121 Reykjavik
Iceland

Tel: +354 5620240
Fax: +354 5620740
E-mail: emilia@rf.is

Chapter 20

N. Hedges
Unilever R&D
Colworth House
Sharnbrook
Bedfordshire MK44 1LQ
England

Tel: +44 (0) 1234 781781
E-mail: nick.hedges@unilever.com

Chapter 21

H. Rehbein
Institute for Fishery Technology and
    Fish Quality
Federal Research Centre for Fisheries
Palmaille 9
D-22767 Hamburg
Germany

Tel: +49 40 38905 167
Fax: +49 40 38905 262
E-mail: rehbein.ibt@bfa-fisch.de

Chapter 22

Asbjorn Gildberg
Norwegian Institute of Fisheries and
    Aquaculture Research Ltd.
N-9005 Tromsø
Norway

Tel: +47 77 62 90 00
Fax: +47 77 62 91 00
E-mail: asbjorn.gildberg@fiskforsk.
    norut.no

Chapter 23

Dr Carmen G. Soleto and Dr Ricardo
    I. Peréz-Martín
Instituto de Investigaciones Marinas
Eduardo Cabello 6
36208 Vigo
Spain

Tel: +34 986 214471
Fax: +34 986 292762
E-mail: carmen@iim.csic.es

Chapter 24

Bo M. Jørgensen
Danish Institute for Fisheries
    Research (DIFRES)
Department of Seafood Research
Ministry of Food, Agriculture and
    Fisheries
DTU
Building 221
2800 Kgs. Lyngby
Denmark

Tel: +45 45252566
E-mail: boj@dfu.min.dk

1

Introduction

H. Allan Bremner, Allan Bremner and Associates, Mount Coolum

Fish is an exceptionally important component of the human diet and an enormous industry exists to provide a huge variety of consumer products in which fish is a major component. These offerings range from whole fish, large and small, to pieces of fish such as cuts and fillets, to canned fish in a multitude of forms, to dried and cured products, to fish oils and extracts, to frozen portions and complete meals through to reformed and gelled products. The list is enormous, the variety even within one product type is extensive and the range of species used as food runs well into the thousands. Each of these variations and combinations presents a huge matrix of possibilities, opportunities and problems. Over the last 80 or so years, fish technologists and scientists have been endeavouring to draw some general rules from observation and experimentation on fish and fish products to control and predict their properties under a vast variety of circumstances. The two main driving themes for these efforts have been in safety and quality – expressed mostly in terms of measurable properties.

This volume picks up these driving themes to cover major issues in safety and quality that are not only important topics of investigation relevant to industry today but that will continue to be important into the future. Each author is an expert in their own particular field and they have summed up the situation to provide a current benchmark of existing knowledge. In addition they have pointed to solutions to problems, where they exist, and have also indicated current gaps in the knowledge base and described research and investigations required to capitalise and expand on this base. In many instances they have described how new understandings, approaches and technologies will have impact and thus effect change in the way operations are carried out to provide better, safer and more stable products with greater surety than previously. It has also been important to describe how one area may relate to another, for example how improvements in analytical techniques have increased understanding of composition, properties, nutritional attributes and of contamination and that this information is relevant to safety and quality considerations.

The volume is organised into three major sections concerned with 'Ensuring safe products', 'Analysing quality attributes' and, 'Improving quality within the supply chain'. In the first part, relating to safety, the chapters deal with the over-riding issue of ensuring that the fish products are safe for human consumption. The volume is not concerned *per se* with safety of the processes themselves with regard to premises, vessels, installations, machinery and personnel except where this impinges on the product, or the perception of the product, but does include reference to factors such as allergies in process staff. Safety in the context of this volume means freedom from pathogenic organisms at infective levels including parasites. It also covers contaminants such as heavy metals and other residues, allergens and toxins. These are viewed from the perspectives of detection, identification, quantitation, evaluation and implication. As such this includes aspects of processing, safety management and risk assessment. Risk assessment and control is covered by outlining and emphasising the value of the HACCP approach and by providing examples of how this is done in practice to establish conditions to minimise risk and to ensure a safe product.

Although this volume has the word quality in the title, the intent is that the word is not used vaguely as a 'catch all' term and endeavour has been made throughout each contribution to try to be specific and to be exact wherever possible. Thus the section on analysis of quality includes discussion on the use of this term and of the commonly used terms freshness and shelf-life to set the scene for chapters on the major causative factors of change in properties of fish products in all forms whether raw, stored, part-processed or finished product. These major factors are covered in chapters dealing with modelling of the effects of the extrinsic bacterial agents involved in spoilage, elaboration of the roles of the intrinsic enzymes, and the processes of oxidation all of which affect one or more properties.

The third section on improvements starts with a fresh look at managing quality along the whole supply chain and then includes quality management of stored fish and of frozen fish and the factors that affect shelf-life. Correct identification of species is included here as it is an important part of business and regulatory practice but it also relates to safety and to analytical improvements. The newer non-thermal technologies using high pressures are summarised and an up-to-date understanding of the ancient, but incidental, practice of using lactobacilli as a preservation technique and the equally ancient, but more deliberate, technique of drying fish to preserve them is covered. The final chapter deals extensively with more efficient utilisation and contains a wealth of ideas on this aspect.

The volume is aimed at several levels as it contains information that is both current and very relevant to future practices. Each chapter is extensively referenced with key information. The book is aimed at being a substantial addition to industry, institutional, research and personal libraries. It will be invaluable for industry technologists, consultants, researchers, graduate and post-graduate students and for government authorities involved in regulation or inspection and control.

Part I

Ensuring safe products

2

HACCP in the fisheries industry

D. R. Ward, North Carolina State University, Raleigh

2.1 Introduction

Over the past decade, the Hazard Analysis and Critical Control Point (HACCP) system has become internationally recognized as the system of choice, with respect to the prevention and control of food safety hazards. In some respects, the evolution of HACCP from concept to an international standard has been relatively rapid. Credit for the development of HACCP is traditionally given to the 1971 Food Protection Conference (APHA, 1972), with the first industry application by The Pillsbury Company for astronaut feeding during the inception of the NASA manned space program. The basic concepts of HACCP, however, are found in the Hazard Opportunity Studies (HAZOP) which have been employed by the chemical and engineering industries for hazard controls dating back to the mid-1930s (Mayes and Kilsby, 1989). After HACCP's introduction, the low-acid canned food industry and the US Food and Drug Administration (FDA) quickly picked up on the preventive controls and documentation aspects of HACCP. Other segments of the food processing industry voluntarily introduced HACCP, or elements of HACCP, into their food safety control systems. However, it was not until 1985 that HACCP moved, in any meaningful way, into the national spotlight. In that year, the National Academy of Science (NAS, 1985) indicated that while HACCP had worked well for the low-acid canned food industry, it had not been successfully transferred to other food commodities. The implication being that processors of other food commodities should use HACCP. They also pointed out that HACCP must be an industry-driven program, with the role of the regulatory agency being that of approval of the processing plant's basic plan design, on-site verification, and inspector training.

It is important to recognize that HACCP has been in a constant state of evolution since its introduction. Principles have been added and renumbered; application guidelines, prerequisite programs and decision trees have been developed; and definitions revised and added. HACCP, which began as a voluntary program, is now mandatory for various products and processes by both the FDA and the United States Department of Agriculture (USDA). As a voluntary program HACCP evolved based largely on the experiences of food processors who were trying to incorporate this science-based system into diverse and complex processing environments. Now that HACCP is becoming the regulatory system of choice, both nationally and internationally, policy issues will be likely to shape its evolution more than science.

Although the principles of HACCP have not changed since FDA first mandated its implementation by the seafood processing industry (December 17, 1997), there have been changes in FDA's expectations. When first introduced, FDA's expectations were very rudimentary: existence of a written HACCP plan (if required), critical control point monitoring records, and sanitation monitoring records of eight required areas (although not part of HACCP, sanitation monitoring is part of the HACCP regulation). It is important to note that some products and processes are not associated with any readily identifiable food safety hazard. Therefore, in these specific situations, a HACCP plan is not necessary. However, if a plant does not have a HACCP plan, due to the fact that a significant safety hazard could not be identified, the person responsible for making that determination should be prepared to demonstrate that they have conducted a thorough hazard analysis. FDA's rules do not require a written hazard analysis, nor do they require predetermined corrective actions or that verification records be listed in the plan. These specific peculiarities are not the result of FDA's lack of interest in these HACCP components; they are the consequence of FDA being the first agency to develop a HACCP rule for review by the Office of Management and Budget (OMB). During this review, OMB's staffers did not understand the seven HACCP principles as comprising a unified food safety system. Consequently, they eliminated such requirements as a written hazard analysis and predetermined corrective actions. Nonetheless, if a plant opts not to have a written hazard analysis, or chooses not to share that analysis with the inspector, and then claims its product does not pose a significant food safety hazard, it should be prepared for intense questioning by the inspector. Irrespective of whether a plant needs to have a written HACCP plan or not, all seafood processing operations must maintain sanitation records.

Some consumer groups have been highly critical of FDA's expectations of the seafood industry, particularly during the first few years of the HACCP regulations. Such criticism was both unfortunate and unnecessary. While the HACCP concept is very simple (i.e., identify the food safety hazards, control those hazards, and provide relevant documentation), development and implementation of an actual plan can be very difficult, especially for an industry that was unaccustomed to a high level of regulatory structure. FDA's strategy provided industry with the opportunity to grow into the new HACCP-regulatory environment. Also, it allowed time for FDA inspectors to become accustomed to their role in this new regulatory climate. While inspectors have always had major responsibilities for evaluating food-processing facilities, HACCP expanded those responsibilities to include not only the processing environment but also a higher level of accountability for reviewing the adequacy of the actual manufacturing processes. It is important to understand that the products produced during this time were no less safe than they were before HACCP, and for many plants that had good HACCP plans that were appropriately implemented, the risks associated with food safety hazards were reduced.

2.2   HACCP principles

In essence, HACCP is a two-part system. The first part focuses on defining the nature of the product being produced and developing a flow diagram which details each operational step in the process. Understanding the nature of the product is essential to determining the potential food safety hazards. Important aspects to know include the intended use (i.e., raw ready-to-eat; raw ready-to-cook; cooked ready-to-eat); method of distribution and marketing (i.e., refrigerated, frozen, etc.); and the intended consumer (i.e., infants, elderly, general population). The significance of the intended consumer is often a confusing point. This does not mean that food should be safer for one segment of the population than another. It does recognize, however, that some segments of the population are more vulnerable than other segments. For example, we know that infants are more vulnerable due to the fact that they have developing immune systems, and we know that the elderly are more vulnerable because their immune system may be in decline. With this knowledge in mind, if a food manufacturer was targeting either of these populations, then its HACCP plan should reflect the increased risk through tighter critical limits, and/or increased monitoring frequency, and/or enhanced verification schedules.

The second part of HACCP consists of applying the seven principles. The following is a brief review of the HACCP principles as developed by the National Advisory Committee on Microbiological Criteria for Foods (NACMCF, 1998).

2.2.1   Conduct a hazard analysis (Principle 1)

The purpose of the hazard analysis is to develop a list of hazards that are of such significance that they are reasonably likely to cause illness or injury if not effectively controlled. Consequently, in the context of HACCP, the word 'hazard' is always limited to safety.

In HACCP there is no hierarchy among the principles in terms of importance. All seven principles are equally important and must ultimately be integrated into an overall plan. However, in the opinion of the author, if the principles were ranked, the hazard analysis principle would be among the most consequential, simply because all hazards must be correctly identified and characterized (reasonably likely to occur or not reasonably likely to occur) or the ensuing plan will be flawed, and some potentially relevant hazard could exist without a preventive control.

2.2.2   Determine the critical control points (Principle 2)

A critical control point (CCP) is defined as a step at which control can be applied and is essential to prevent or eliminate a food safety hazard, or reduce it to an acceptable level. Examples of CCPs could include thermal processing, chilling, testing ingredients for chemical residues, product formulation control, and testing product for metal contamination.

2.2.3   Establish critical limits (Principle 3)

A critical limit is a parameter, established at CCPs, which targets conditions essential for the production of safe food. It can be a maximum and/or minimum value to which a biological, chemical or physical parameter must be controlled at a CCP to prevent, eliminate, or reduce to an acceptable level the occurrence of a food safety hazard. Failure to achieve the critical limit means that the CCP is not in control and the food being produced must be considered unsafe.

2.2.4   Establish monitoring procedures (Principle 4)

Monitoring is a planned sequence of observations or measurements to assess whether a CCP is under control and to produce an accurate record for future use in verification. Monitoring at CCP is done to determine whether or not the critical limit(s), established for each CCP, is being met. Monitoring serves three main purposes: first, it is essential to food safety management in that it facilitates tracking of an operation. If monitoring indicates that there is a trend toward loss of control, then action can be taken to bring the process back into control before a deviation from a critical limit occurs. Second, monitoring is used to determine when there is a loss of control and a critical limit deviation occurs at a CCP, i.e., exceeding or not meeting a critical limit. When a deviation occurs, an appropriate corrective action must be taken. Third, it provides written documentation for use in verification.

2.2.5   Establish corrective actions (Principle 5)

When there is a deviation from established critical limits, corrective action is necessary. As recommended by the NACMCF (1998), corrective actions are predetermined components of a written HACCP plan. However, while FDA's rules (1995) require that corrective actions be taken, they are not required to be predetermined. Corrective actions include the following elements: (a) determine and correct the cause of noncompliance; (b) determine the disposition of non-compliant product; and (c) record the corrective actions that have been taken.

2.2.6 Establish verification procedures (Principle 6)

Verification is defined as those activities, other than monitoring, that determine the validity of the HACCP plan. Of the seven principles, this one inevitably proves to be the most challenging for trainers to teach and for students to understand. Perhaps part of this lies in the fact that in the evolution of HACCP this principle was the last to be developed and as a consequence it attempts to deal with several problematic issues that had become evident. The activities (other than monitoring) that determine the validity of the plan include:

1. Evaluating whether the facility's HACCP system is functioning according to the written plan.
2. Determining (initial validation) if the plan is scientifically and technically sound, that all hazards have been identified, and that if the HACCP plan is properly implemented these hazards would be effectively controlled. This includes determining if CCPs have been properly identified and that the critical limits are scientifically valid for hazards being controlled. Equipment calibration is also part of validation.
3. A subsequent validation (sometimes referred to as revalidation) is necessary if there is an unexplained system failure; a significant product, process, or packaging change occurs; or a new hazard is recognized.
4. A periodic comprehensive verification should be conducted, even if there have been no substantive changes to the plan. FDA requires that seafood processors review their plan on an annual basis.

2.2.7 Establish record-keeping and documentation procedures (Principle 7)

Generally, records maintained for the HACCP system should include:

1. Summary of the hazard analysis, including rationale for determining hazards and control measures. (FDA does not require a written hazard analysis).
2. The HACCP plan:
   a. Listing of the HACCP team and assigned responsibilities (not required by FDA)
   b. Description of the food, its distribution, intended use, and consumer (not required by FDA)
   c. Verified flow diagram (not required by FDA)
   d. HACCP Plan Summary Table:
      i. steps in the process that are CCPs.
      ii. the hazard(s) of concern
      iii. critical limits
      iv. monitoring (procedures and frequency)

      v. corrective action
      vi. verification procedures and schedule (procedures and frequency)
      vii. Recording-keeping procedures

2.3   Hazards

In the context of HACCP, a hazard is defined as a biological, chemical, or physical agent that is reasonably likely to cause illness or injury in the absence of its control (NACMCF, 1998). As stated earlier, the proper identification of all hazards that are reasonably likely to occur is essential to the development of a HACCP plan. When discussing hazards associated with seafood one often finds that the issue can quickly become very complex. Because the word 'seafood' represents literally hundreds of commercially available species, each with a potentially unique species-related hazard, processors must be vigilant. The food safety hazards associated with seafoods may be different due to inherent differences in physiology, environment, post-harvest handling, and method of preparation. In the US FDA has published the third edition of The *Fish and Fisheries Products Hazards & Controls Guidance.* The purpose of the 'guide' is to assist processors to identify hazards and to formulate control strategies. The guide is divided into three sections; (i) Potential Vertebrate Species Related Hazards, (ii) Potential Invertebrate Species Related Hazards, and (iii) Potential Process Related Hazards.

2.3.1   Species-related hazards

In the guide, the division between vertebrate species (fish with backbones) and invertebrate species (fish without backbones) was done as a means to enhance the user's ability to find and retrieve information and as a means of facilitating hazard identification. Table 2.1 illustrates the classification of potential hazards in vertebrate species according to the guide (FDA, 2001). The concern associated with biological hazards is limited to parasites, while potential chemical hazards are natural toxins (e.g., ciguatera, amnesic shellfish poisoning, gemplotoxin, tetrodotoxin), histamine (scombrotoxin), chemical (e.g., environmental and pesticides), and drugs (aquacultured species).

Table 2.1   Potential vertebrate species-related hazards

Market names	Latin names	Hazards				
		Biological	Chemical			
		Parasites	Natural toxins[1]	Histamine	Chemical	Drugs

[1] This hazard applies only if the product is market uneviscerated.

Table 2.2   Potential invertebrate species-related hazards

Market names	Latin names	Hazards				
		Biological		Chemical		
		Pathogens	Parasites	Natural toxins	Chemical	Drugs

Table 2.2 illustrates the classification of potential hazards in invertebrate species according to the guide (FDA, 2001). Inasmuch as invertebrate species include molluscan shellfish, which are often eaten raw, the presence of human pathogens becomes a relevant biological hazard, in addition to parasites. Since invertebrates are not associated with histamine poisoning, the hazard is not listed as a potential concern.

2.3.2   Process-related hazards

Table 2.3 illustrates the categories relevant to determining the process-related seafood hazards in the guide (FDA, 2001). Obviously, the categories are not intended to be an exhaustive list of all possible hazards but merely an effort to help processors think through the hazard analysis step as it pertains to process-related hazards. The increasing complexity potentially introduced by processing is very apparent.

2.4   Developing and implementing HACCP plans

Developing a good HACCP plan has proven to be a challenging experience for many seafood processors. As a university faculty member with extension responsibilities, the author has had the opportunity to provide HACCP training to many seafood processors, as well as regulatory personnel, and on many occasions has assisted with developing HACCP plans. The nature of a typical three-day HACCP training session is such that most attendees leave with enough understanding of the HACCP concept to do a reasonably good job of writing a plan for their specific products and processes. The fact that Seafood HACCP Alliance's training incorporates a number of different models representing a cross-section of products, processes, and hazards, has proven to be very beneficial. Usually processors can find a model, that relates well to their own product and/or process (e.g., molluscan shellfish, cooked ready-to-eat, smoked fish, and fresh ready-to-cook). These models are invaluable because they allow processors to see HACCP principles in a specific example as opposed to an abstract discussion. Moreover, prior to seeing a model, many processors seem to have the mistaken notion that a HACCP plan was some sort of huge document. Typically, they are both surprised and delighted to discover that the plan for even the most complex model is merely a few pages. In fact, as a HACCP

Table 2.3 Potential process-related hazards

Finished product food	Package type	Method of distribution and storage	Hazards									
			Biological							Chemical	Physical	
			Pathogen growth temperature abuse	C. botulinum growth	Toxin formation, inadequate drying	S. aureus toxin batter	Pathogen survival through cooking	Pathogen survival through pasteurization	Pathogen contamination after pasteurization	Allergens/additives	Metal inclusion	Glass inclusion

trainer, before I start teaching Principle 1, I will have the class turn to Principle 7 and the completed two-page plan for the 'cooked shrimp' model. This is the model that we build as we work our way through the seven principles. Using this approach helps take the mystery out of the expected outcome, and helps ease the mind of the more nervous individuals in the class. It also helps to clarify, by example, exactly what it is that we is attempting to develop.

Developing a HACCP plan is a big step, but the biggest step is implementation of the plan. The plan merely details what processors said they are going to do, the trick now becomes to deliver on what the plan promises. For most processors, even small processors, implementation of the plan involves the participation of other plant employees. These employees are typically involved in some aspect of monitoring and therefore they have responsibility for documenting monitoring results. Early experience with the seafood HACCP regulation showed that it was not uncommon to find that plant employees were not adequately trained in their specific HACCP function and therefore they did not realize the importance of monitoring at the frequency specified in the plan (e.g., once every two hours). Additionally, many employees were not aware of the importance of documenting their monitoring activity at the time the activity was performed. Given the thousands of seafood processors in the US, just about anything that could be a problem with implementing HACCP was a problem for someone, somewhere. That said, HACCP implementation was not a problem for all seafood processors. Many processors clearly demonstrated that they understood the concepts undergirding HACCP as well as the mandated regulatory requirements imposed by FDA.

Paradoxically, one of the primary factors associated with the challenge of HACCP implementation is also one of HACCP's primary benefits. HACCP requires structure. Processors unaccustomed to operating within a structured program that requires specific activities to be performed and documented on a proscribed schedule find HACCP implementation a frustrating experience. More often than not, these processors will acknowledge that they recognize the value of the structure created by HACCP, they just have difficulty in getting used to the adjustments necessary to make the structure work.

2.5   Sanitation standard operating procedures (SSOPs)

Typically programs such as sanitation, good manufacturing practices (GMPs), employee training, recall plan, etc., are considered, in the context of HACCP, as prerequisite programs. Meaning that while these programs are not actually part of HACCP, in order for HACCP to perform properly as a preventive food safety control program, the prerequisite programs must be in place and functioning. In the US, food processors have long been required to operate under GMP regulations, which include sanitation. The problem has been, however, that the GMPs are vaguely worded and do not require routine monitoring or record keeping. Consequently, in order to put more teeth into the agency's ability to force better compliance with existing sanitation requirements, within the same regulation that mandated HACCP (FDA, 1995), FDA also mandated monitoring and documentation of eight key sanitation areas. In order to justify including the sanitation requirement in the proposed HACCP rule (FDA, 1994), FDA reported sanitation deficiency data from its Establishment Inspection Reports (EIR) from 1988–90. The following data are from 715 EIRs covering 561 facilities:

- 23% of the receiving area facilities were not clean/orderly or in good repair
- 26% of the facilities lacked effective insect and rodent control measures in the receiving area
- 16% failed to handle ice in a sanitary manner and to protect it properly
- 35% do not adequately clean or sanitize
- 21% had processing equipment that was not constructed so that it was easily cleaned and sanitized
- 18% of the processing equipment was not made from suitable materials
- 15% had hand sanitizers that were not kept at proper concentrations
- 18% failed to have hand sanitizers in the processing area
- 33% had processing areas that were not maintained in a clean and sanitary manner
- 42% had processing areas with exterior openings that were not sealed/covered properly to prevent the entrance of pests or insects
- 16% had waste materials not being collected/covered in suitable containers or not being disposed of properly
- 23% handled finished product in a manner that did not preclude contamination
- 22% documented employees were not taking necessary precautions to avoid food contamination.

It is very easy to read data such as described above and to make sweeping assumptions about an entire industry. While there is no excuse for any of the deficiencies mentioned, it is exceedingly difficult to operate day after day without a lapse somewhere. The expectation for implementing sanitation standard operating procedures (SSOPs) is not that deficiencies will be eliminated but that they will be minimized, and when observed that they will be corrected in a timely manner.

With the above deficiencies in mind, FDA mandated that 'each processor shall monitor conditions and practices during processing with sufficient frequency to ensure, at a minimum, conformance with the conditions and practices specified in the GMPs that are both appropriate to the plant and the food being processed and relate to the following eight key areas:'

1. Water
2. Food contact surfaces
3. Cross-contamination
4. Hand washing and sanitizing and toilet facilities
5. Protection from adulteration

6. Labelling and storage of toxicants
7. Employee health
8. Exclusion of pests

There are those who contend that in the absence of a HACCP program, adherence to these sanitation requirements would go a long way toward minimizing the food safety risk associated with seafood consumption. Given that both HACCP and SSOPs are required, the combined effect should be a significant reduction in risk.

As was the case with the introduction of HACCP, in the early stages of the SSOP implementation period, FDA inspectors found that many processors were not complying with the requirements of the regulation. No doubt there were many factors associated with the lack of compliance. In the opinion of the author, one of the factors had to do with training. Although the HACCP training developed by the Seafood HACCP Alliance included a section on the new sanitation requirements, initially those of us writing and editing the Alliance's HACCP training manual did not fully appreciate the training needs associated with sanitation. Consequently, the early training sessions did not address the new sanitation requirements sufficiently to impress on processors FDA's expectations. To the Alliance's credit, when the need became apparent, it developed a separate training manual and a one-day training program to assist processors to:

*   understand why sanitation is essential to the success of a HACCP program
*   understand the requirements of the sanitation component of this regulation
*   provide examples of possible checklists for documentation of sanitation monitoring.

2.6   The new millennium

In the introduction to this chapter it was suggested that FDA's initial expectations of industry, with respect to HACCP, were rather rudimentary. Over the last few years, FDA has demonstrated that it is raising the bar. It has begun to be more resolute in its evaluation of preventive controls for such hazards as *Clostridium botulinum* as well as *Listeria monocytogenes* for products and in processes where such hazards are reasonably likely to occur. The agency has expanded its expectations for the control of pathogens from the harvest area (molluscan shellfish) to include both *Vibrio parahaemolyticus* and *Vibrio vulnificus*. Furthermore, FDA has been taking more aggressive action in those instances where processors have demonstrated an inability to comply with the requirements of the regulation.

If the axiom 'change is constant' is accurate, then HACCP will change. The question then becomes 'who or what will drive the change?' Now that HACCP is a regulatory system, change will in large measure come from the experiences of the regulatory agencies. Historically, these agencies have been slow to embrace innovation and change. Until the advent of HACCP 'floors, walls, and ceilings,' 'command and control,' 'appearance and odor,' or 'gotcha' characterized the inspection philosophy of USDA or FDA for many years. Under HACCP, the regulatory agencies do not have to change in response to a newly identified hazard or processing technology, the burden has shifted to industry. Some could argue that the same paralysis that gripped the agencies will affect industry. Possible, but not likely. Some companies, that were not very innovative prior to HACCP, will continue to be satisfied to leave well enough alone, but most operations will continue to embrace innovation as long as it enhances profitability. Moreover, they will quickly learn to evaluate any potential new innovation in the context of its impact on the current HACCP system.

2.7   Conclusion

For most of its existence HACCP has been largely a voluntary, industry-driven program. Its evolution has been the result of needs and opportunities identified by industry practitioners. Now that HACCP has become a regulatory tool, industry may view its primary challenge merely to comply with HACCP and cease to believe that they have a role in the evolution of HACCP. If this happens, HACCP will no longer be 'industry driven', but driven by regulatory processes. Exactly how this will impact the continuing evolution of HACCP remains unclear. No doubt future changes in HACCP will be driven more by regulatory policies, which may or may not be based on science. Whatever happens one thing is clear, while the history of HACCP has been relatively brief, it has been interesting. Moreover, the opportunity continues to exist for HACCP to have a profound impact on seafood safety.

2.8   References

APHA (1972) *Proc. National Conference on Food Protection.* American Public Health Association. Food and Drug Administration, Washington, DC.

FDA (1994) *Federal Register*, Vol. 59, No. 19, p. 4187. Proposal to Establish Procedures for the Safe Processing and Importing of Fish and Fishery Products: Proposed Rule. Food and Drug Administration, Washington DC.

FDA (1995) *Federal Register*, Vol. 60, No. 242, p. 65096. Procedure for the Safe and Sanitary Processing and Importing of Fish and Fishery Products: Final Rule. Food and Drug Administration, Washington DC.

FDA (2001). *Fish and Fisheries Products Hazards & Controls Guidance: Third Edition.* Department of Health and Human Services, Public Health Service, Food and Drug Administration Center of Food Safety and Applied Nutrition, Office of Seafoods, Washington, DC.

MAYES, T. and KILSBY, D.C. (1989) The use of HAZOP hazard analysis to identify critical control points for microbiological safety of food. *Food Quality and Preference* 1(2), 53.

NACMCF (1998) Hazard analysis and critical control point principles and applications guidelines. *J. Food Protection* 61(6), 762–75.

NAS (1985) *An evaluation of the Role of Microbiological Criteria for Foods and Food Ingredients.* National Academy of Sciences, National Academy Press, Washington, DC.

3

HACCP in practice: the Thai fisheries industry

S. Suwanrangsi, Thai Department of Fisheries, Bangkok

3.1 Introduction

Seafood safety is undergoing a period of unprecedented change, fuelled at the domestic level by increasing consumer concern over foodborne hazards, and at international level by demands for effective food hygiene and food safety control systems across national boundaries. HACCP implementation in the fishery industry worldwide in particular is becoming more widespread in response to these pressures, with importing countries requiring implementation by overseas suppliers. HACCP is recognized internationally as the best tool to control food safety, though there are certain limitations in applying it at the early stage of the production chain. More recently, risk assessment has emerged as an important new technique in assessing microbiological and other hazards. The fishery industry needs to prepare for safety and quality control systems based on risk assessment.

Since 1993 Thailand has emerged as a major exporter of fishery products. The export value of these products increased from 2.3 billion US$ in 1990 to 5 billion US$ in 1997. The economic crisis that hit Asia in the middle of 1997 reduced this total to 4.2 billion US$ in 2000. Japan has traditionally been the main importer of Thai fishery products. In 1998, the USA became the largest importer of Thai fishery products for the first time, importing 31% of total Thai production. Other major importers of Thai fishery products are the European Union (EU), Australia, Canada and Asian countries. China has emerged as a significant importing country in the past few years, with significant potential for the future. The types of fish product exported in 1999 were (by volume): canned tuna and other canned seafood (45%), frozen shrimp (12%), frozen fish products (18%), frozen cephalopods (8%), frozen molluscs (4%) and fresh fish (17%).

3.2   The development of HACCP systems in Thailand

In recent years the global community has been seeking a common approach to maximizing the quality and safety of all food products. This approach includes the use of Hazard Analysis and Critical Control Point (HACCP) systems as a mean of assuring proper food handling, processing and retail sale to consumers. The use of HACCP systems in the fishery industry is now on a global scale. Since it first emerged, the concept has increased in importance, partly through its endorsement by Codex Alimentarius, and partly through its adoption by the European Union and USA as a requirement for the import of high-risk food products such as fish. Currently over forty countries have announced HACCP initiatives for the control of fish production, processing and distribution.

The Thai Department of Fisheries (DOF) implemented HACCP based fish inspection programs in 1991 in response to the development of the first voluntary HACCP systems in the industry. The program involved reviewing existing inspection methods, new training for inspectors and developing new HACCP-based audit procedures. Since 1996, HACCP based quality systems have been mandatory for the export fish processing sector. The DOF requires that approved businesses must have a HACCP program implemented, documented and verified by the Department.

In response to the growing use of HACCP systems, the DOF has increased its support to industry. The most important early activity for successful implementation of HACCP from 1991–97 was training on the principles and application of HACCP for the fish processing industry. However, since 1998 the focus has been much more on audit of established HACCP systems and the development of guidelines and other supporting information. Training on HACCP principles and methodology has been delegated by the DOF to other food institutes and academic institutions. To meet the needs of industry for guidance in preparing a documented HACCP plan and prerequisite program, as well as implementing and maintaining the program, DOF has laid down and published:

- a guide on basic HACCP methodology, updated regularly to meet changing international guidelines and importing countries' regulatory requirements
- guidance on the particular regulatory requirements of importing countries
- a handbook on HACCP documentation
- guidelines on assessment processes
- guidelines on particular hazards and critical limits
- generic HACCP plans for major commodities, through workshops and working groups within industry.

Various government agencies, universities and private HACCP consulting firms currently offer training on HACCP. DOF also provides training on an *ad-hoc* basis for industry. DOF also conducts surveillance studies on raw materials, water used in processing and other research to support hazard identification. Close monitoring of industry performance in HACCP implementation has been carried out by inspection of facilities, concentrating on the effectiveness of critical control points, record review and verification procedures within the business's HACCP system. Implementation by the industry is classified into three stages, depending on how advanced a business is in HACCP implementation:

1. The initial planning stage.
2. The development stage.
3. The post-implementation phase (where the business has a fully implemented HACCP system).

It is accepted that HACCP implementation may be a gradual process with businesses moving gradually from one stage to another. Other food industries such as milk, meat, poultry, vegetables and fruit products have started implementing HACCP systems. HACCP training and audits for these sectors are carried out by government agencies and institutes such as the Department of Livestock Development, the Thai Industrial Standard Institute and the National Food Institute.

3.3   HACCP methodology

The Department of Fisheries HACCP Program is focused on product safety. Other quality related elements, and especially hygiene and Good Manufacturing Practices (GMP), are to be met by a prerequisite program (PRP). The processor must have basic sanitation, hygiene control and GMPs, through a PRP, as the foundation for the HACCP system. In most cases much of what is required for a PRP is already being done in response to the business's own needs and the requirements of its retail customers, though development of individual elements within a prerequisite program may well have been piecemeal. A systematic review helps to identify gaps and create a more coherent and clearly documented system. Experience suggests that an effective prerequisite program makes HACCP planning much easier, removing many hazards and potential control points which might otherwise produce an over-complex, unfocused and unwieldy HACCP plan. HACCP planning can then concentrate on relatively few hazards and critical control points.

Each processing establishment must develop a HACCP plan appropriate to their raw materials and processing practice. The processor must identify potential hazards associated with the products and processing environment. Once the hazards are identified, critical control points can be determined using a decision-tree approach or expert or experienced judgement, keeping them to the minimum required to control product safety. Using this approach, confusion between critical control points (CCP) and control points (CP) for product quality and regulatory requirements is greatly reduced. For each critical control point the company must establish a critical limit, set up and validate a monitoring system, and decide on appropriate corrective action.

The Department of Fisheries requires that HACCP plans and prerequisite programs be documented. A system of record keeping must be established to record monitoring activities, instances of non-compliance found, corrective action taken, and verification and internal audit activities, including any modifications of the program. Guidelines for program development and documentation are also provided and handbooks are available to the fisheries industry in the local language. Key documents to be maintained are listed in the appendix at the end of this chapter. Businesses are also responsible for having personnel training to carry out their duties within the HACCP system. In many cases the individual plant has its own laboratory capable of performing microbiological and contaminant testing. Some are well equipped with instrumental techniques such as HPLC, especially those dealing with aquaculture shrimp production. If a laboratory is not available in-house, the business is responsible for setting up a monitoring program which makes use of third-party testing.

The Fish Inspection and Quality Control Division of the DOF will assess the HACCP programs of the processing plants in three ways:

1. By verifying the design and appropriateness of the documented HACCP program against the product and processing requirements of the plant.
2. By conducting independent audits of prerequisite programs and HACCP plans, with the frequency of audit based on history of compliance and performance of individual establishment.
3. By collecting samples of products, input materials, whether raw materials or processing materials such as water or ice, to ensure that the HACCP system process is effective and products are safe.

These will determine how effectively the plant HACCP program is operating and this, in turn, will determine the frequency of the regular inspection and audit of the plant and of the products exported. The role of DOF in providing guidance during inspection is to:

- explain clearly health and safety standards, regulation guidelines or requirements used as references to their inspections
- confirm the business's understanding of HACCP principles
- encourage the application of all seven principles of HACCP
- conduct assessments according to good audit practices
- provide a clear explanation of the assessment process
- explain any non-compliance, using objective evidence, but not how to correct the non-compliance.

HACCP audit policy and procedures have been developed for the field inspectors. Training on HACCP verification and HACCP system audit are provided, together with harmonization meetings among regional inspectors to ensure consistency. Inspectors are also given ISO 9000 Lead Assessor Training to build up their general audit skills.

3.4   Common problems in HACCP implementation

There are a number of common problems in the effective implementation of HACCP systems. A number of these relate to preparation for a HACCP program. These include:

- differing HACCP models
- differing regulatory requirements
- the scope of HACCP implementation
- resource requirements
- inadequate prerequisite programs.

There are also a number of problems in designing and setting up the HACCP system, including:

- information about hazards and effective hazard analysis
- establishing critical control points and limits
- documentation.

Finally, businesses encounter difficulties in maintaining and improving their HACCP systems in such areas as:

- audit procedures
- measuring the success of HACCP implementation.

The following sections look in turn at these problems.

3.4.1   Differing HACCP models

HACCP systems differ greatly. The EU and the United States Food and Drug Administration (FDA) seafood HACCP programs concentrate on food safety. Other models, such as those developed by the US National Marine Fisheries Services and the Canadian Quality Management Program, for example, also include quality issues. The Australian government, for example, has developed SQF 2000, a system combining HACCP and selected elements of ISO 9000. The Codex Committee on Fish and Fishery Products is currently revising the Code of Practice for Fish and Fishery Products to incorporate both HACCP principles and essential quality issues such as composition and labelling. The Codex model proposes the use of Defect Action Points (DAPs) based in part on Critical Control Points (CCPs) in HACCP systems.

The current range of models can be very confusing, even for experts, let alone the small-scale processor, and puts a greater responsibility on regulatory authorities such as the DOF to provide clear guidance and training. In developing countries, the fisheries industry is dominated by small primary processors. Resources are a major issue, both in terms of the finance and expertise available for HACCP implementation. Guidance on HACCP systems needs to take account of these limitations, and there is a greater need for government support.

3.4.2   Meeting differing regulatory requirements and standards

A particular problem facing businesses exporting to a range of markets is that of differing regulatory requirements and competing HACCP models from differing national administrations. An additional problem is a constant stream of changes and additions as differing national regulations evolve. This problem of multiple and moving standards places a particular responsibility on government agencies such as the DOF. The DOF works closely with the industry in defining the common minimum requirements for a HACCP system which will meet the main requirements of a company's export markets. Some companies have then developed a basic HACCP plan, which, once implemented, has then gradually expanded to meet particular national statutory requirements where necessary. Regular review, often in consultation with the Department of Fisheries, is required to keep up with developments in differing national jurisdictions.

3.4.3   The scope of HACCP implementation

To date HACCP systems have only been implemented on any scale by the manufacturing sector within the fishery industry worldwide. In developing countries HACCP systems are still limited mainly to the export sector. To be truly effective, HACCP implementation must start at the aquaculture/harvesting stage and continue through processing, distribution and retail handling to educating consumers in handling food products. There have been some initiatives to address HACCP implementation within aquaculture, but these are limited. The lack of HACCP implementation in other parts of the supply chain has retarded the development of HACCP systems by local processors.

3.4.4   Resource requirements

Businesses often underestimate the resources required in HACCP implementation in such areas as upgrading prerequisite systems, setting up and maintaining documentation, and training. In particular, companies implementing HACCP systems in practice have often underestimated the amount of training required. These commitments include:

- Sanitation and hygiene training for line personnel in building effective prerequisite systems.
- In-depth training for the HACCP team leader. Companies implementing HACCP systems successfully were those willing to send team leaders to accredited courses, so that they could be sure of having someone familiar with all the stages of HACCP implementation.
- Training for HACCP teams, often through a one-day course on HACCP principles and the role of HACCP teams.
- Audit training to equip personnel to carry out internal audits.
- Training in monitoring and verification skills for CCP staff.

3.4.5   Inadequate prerequisite programs

It has been estimated that over 70% of fish products traded internationally have quality defects ranging from decomposition, contamination with pathogens or foreign bodies, to discrepancies in stated weight and incorrect labelling. This failure illustrates the scale of the problems faced in effective implementation of HACCP systems. Given the involvement of small-scale producers in developing countries in the supply chain, there are major weaknesses in basic good manufacturing and hygiene practice.

HACCP can be effective only if it is based on a solid foundation of Good Manufacturing Practices (GMP), Sanitation Standard Operating Procedures (SSOPs), and clear Standard Operating Procedures (SOPs). In practice, for many of those involved in the industry, the first priority needs to be sound prerequisite systems. Once these are functioning effectively, experience with fish processors in Thailand suggests that HACCP systems become much simpler to design and manage, both for large and small businesses. The DOF provides training, guidelines and documentation on developing effective prerequisite programs, especially for small-scale processors.

3.4.6   Information about hazards and effective hazard analysis

Each species of fish, molluscan shellfish or crustacea may have quite different hazards which also vary from one country or region to the next. The development of aquaculture has resolved those problems to some extent by creating a more controlled environment for fish breeding before harvesting. However, aquaculture species have distinct hazards of their own. The DOF has published a *Fishery Products Hazards and Control Guide* to assist industry with hazard identification for key products originating in Thailand. The Guide explains hazards related to individual species and processes as well as recommended control measures.

Hazard analysis has proved one of the major challenges for many businesses. Whilst, partly as a result of the work of agencies such as the DOF, many of the hazards affecting fish products are now well documented, a significant number of businesses remain ignorant of the full range of hazards and their potential impact on their products. They have been used for 'reactive' rather than proactive hazard analysis, dealing with individual hazards as they arose, particularly in response to customer concerns over final product quality. In many cases businesses lack in-house microbiological expertise. The Thai Fisheries Department staff has a key role to play in assisting businesses start the process of hazard analysis. They have developed a simple hazard table, dividing hazards into biological, chemical or physical hazards, further subdivided by species. This framework has provided a starting point for the HACCP team within the business. Fisheries Department staff also provide information on key information sources to consult, together with training materials on some of the main hazards, and initial advice based on experience with similar businesses.

3.4.7 Establishing CCPs and critical limits

Because they confused quality and safety issues, and failed to define the scope of the HACCP study, a number of businesses produced initial HACCP plans which incorporated a maze of Critical Control Points (CCPs) which could not be monitored adequately and which resulted in a mountain of records. Experience of HACCP plans across the fish-processing sector suggests that, once prerequisite systems were operating effectively, and once quality issues were separated out, most HACCP systems were relatively simple with far fewer CCPs than many original HACCP plans.

Businesses setting up CCPs sometimes found difficulty in establishing critical limits, given different recommendations in the scientific literature or conflicting requirements from differing regulatory authorities. The validation of critical limits was another area of difficulty, especially for small-scale processors who could not run their own challenge tests. The Department of Fisheries has been able to provide some support in the use of government or university research laboratories to run challenge tests. However, there remain gaps in appropriate guidance on validation procedures for critical limits, and this is an area where the Department itself is still developing the appropriate technical knowledge and skills.

3.4.8 Documentation

Creating and managing HACCP documentation is an obvious area where businesses face difficulties, usually from creating too much documentation of poor quality. Many of the problems are the result of confusion and poor individual document design. Experience suggests that it is important to keep documentation simple. CCP monitoring procedures should, for example, deal with a few essential issues such as:

- the purpose of the CCP
- the critical limit to be monitored
- how it is to be monitored
- the person responsible for monitoring
- where results are to be recorded
- how frequently results are to be recorded
- what to do if the critical limit is exceeded.

3.4.9 Audit procedures

Since many companies only started to implement HACCP systems at the end of the 1990s, many still have not yet had the opportunity to conduct their first full internal audit. In some cases, firms have been slow to do so because they are unfamiliar with the concept. Many tend to delay auditing because they are concentrating on day-to-day production issues and ensuring that their HACCP systems are operating smoothly. The Department of Fisheries has encouraged all businesses implementing HACCP systems to undertake an annual internal audit and, providing support is offered, most firms try to do so. A few firms are most resistant to internal auditing altogether.

Many businesses lack the resources to fund an external audit, and must rely on appointing an internal audit team. Because auditing is still not widespread, support from the Department of Fisheries has proved essential for many companies. The Department is able to recommend accredited courses for internal auditing. Given the pressures in dealing with day-to-day issues, it is easy for audits to slip and, therefore, particularly important to set a deadline for training auditors, planning and conducting the audit, and providing a written report.

3.4.10 Measuring the success of HACCP implementation

HACCP systems are implemented to ensure that safe food is produced. For a more integrated HACCP-based system, quality and wholesomeness of products may also be indicators of effectiveness. However, it can be very difficult to verify these. Most regulatory agencies used performance-based analysis, to determine compliance and effectiveness of system. Once the company adheres to the procedures of the validated HACCP system, meets hygiene conditions, complies with regulations concerned, and product history indicates either no problems or that problems have been corrected satisfactorily, it is deemed that the HACCP system of such a processor is effective. However, these criteria may not truly indicate the effectiveness of a HACCP system. Should there be other ways used, for example, target levels of pathogen reduction in particular products or achievement of a food safety objective (FSO) to evaluate HACCP effectiveness? These issues have yet to be worked out by regulatory agencies.

3.5   Future trends

HACCP systems demand a business and staff at all levels willing to take more responsibility for and control of the business's operations, rather than relying on others to identify problems and improvements, whether they be customers, regulators or more senior managers. The success of HACCP implementation in the Thai fisheries industry has also been dependent on close cooperation between industry and government, and the willingness of the latter to play a proactive role in HACCP implementation.

There are a number of improvements that need to be made; these include the following. HACCP systems are still concentrated in processing. So far they have not been extended further along the supply chain, particularly in primary production, marine capture and aquaculture. Many people have still argued for the extension of HACCP principles into such areas as retail sales and consumer handling of fish products. HACCP take-up amongst smaller businesses remains poor. Government needs to keep abreast of changes in HACCP regulations and standards internationally, and there is a need for improved practical guidelines for hazard analysis. Inspectors need to gain more expertise and experience in effective inspection techniques, translating HACCP theory into effective practice. As an example, inspectors need to be able to use generic HACCP models flexibly, accepting variations which are appropriate to a particular business. There needs to be more research in such areas as the validation of critical limits.

Harmonization of HACCP and regulatory verification at international level will help make HACCP implementation more effective and consistent within and between countries. Codex Alimentarius and other international organizations need to speed up this process and provide a forum to discuss and resolve the differences that hinder the development and implementation of HACCP.

3.6   Sources of further information and advice

There are excellent sources of information that can be referred to, including the following. Much of the information can be accessed from the internet.

HACCP principles and application to fish and fishery products
Canadian Food Inspection Agency 1997. Re-engineer Your Quality Management Program Plan: A Manual for Fish Processors. Canadian Food Inspection Agency, Ontario. 57 p.

Codex Alimentarius. 1999. Vol. B (Recommended International Code of Practice for Fish and Fishery Products.) FAO Documents Office. Joint FAO/WHO Food Standards Programme, Rome. 157 p.

Dillon, Mike and Chris Griffith. 1996. How to HACCP: An illustrated guide. 2nd edn, M.D. Associates, North East Lincolnshire, UK. 120 p.

Dillon, Mike and Chris Griffith. 1997. How to Audit: Verifying Food Control Systems. M.D. Associates, North East Lincolnshire, UK. 215 p.

European Commission. 1997. European Community Legislation: Fishery Products. DG VI-Unit VI/B/II.2. Veterinary and Zootechnical Legislation. Brussels, Belgium.

FAO. 1995. The use of hazard analysis critical control point (HACCP) principles in Food Control: Report of an FAO Expert Technical Meeting 12–16 December 1994, Vancouver, Canada. FAO, Rome.

FAO. 1998. Food Quality and Safety Systems: A training manual on food hygiene and the Hazard Analysis and Critical Control Point (HACCP) system. FAO, Rome. 232 p.

ILSI Europe. 1997. A Simple Guide to Understanding and Applying the HACCP Concept. 2nd edn, International Life Science Institute Europe, Brussels. 13 p.

Leaper, S. 1997. HACCP: A Practical Guide. 2nd edn, Campden & Chorleywood Food Research Association, Gloucestershire, UK. 51 p.

Mayes, T. and Sara Mortimore. 2001. Making the Most of HACCP Learning from other's experience. Woodhead Publishing Limited. UK. 272 p.

Mortimore, Sara and Carol Wallace. 1998. HACCP A practical approach. 2nd edn, A Chapman & Hall Food Science Book, Maryland. 403 p.

The National Seafood HACCP Alliance for Training and Education. 1997. Hazard Analysis and Critical Control Point Training Curriculum. 3rd edn, Sea Grant, North Carolina. 276 p.

Standard Committee on Agriculture and Resource Management. A Guide to the Implementation and Auditing of HACCP. 1997. CSIRO Publishing PO Box 1139. Collingwood Australia.

Suwanrangsi, S. 2001. HACCP Implementation: Experiences of Developing Countries. US Seafood News Volume 9, Issue 3.

Suwanrangsi, S. 2001. HACCP Implementation in the Thai Fisheries Industry. In: Mayes T. and Mortimore, S. (eds). Making the Most of HACCP. Woodhead Publishing Limited. Cambridge. 183–200.

US Food and Drug Administration. 1991. Current Good Manufacturing Practice in Manufacturing, Packing, or Holding Human Food, 21 CFR 110 ch. 1. In Code of Federal Regulations (CFR). DHHS/FDA, Rockville, Md.

US Food and Drug Administration. 1995. Procedures for the safe and sanitary processing and importing of fish and fishery products. Federal Register. 60(242): 65096–202.

US Food and Drug Administration. 1996. Food and Drug Administration seafood HACCP program, 21 CFR 123. In Code of Federal Regulations (CFR). FDA/Office of Seafood, Washington, DC.

US Food and Drug Administration. 1996. FDA Law, Regulations and Guidelines for Fishery Products. In Sanitation Standards for Trade in Fishing Products: US. Part I. APEC/NFS/DFH, Chile.

Hazards related to fishery products

ASEAN Canada Fisheries Post Harvest Technology Project, Phase II. 1995. Canned Tuna Quality Management Manual. Marine Fisheries Research Department, Southeast Asian Fisheries Development Center. Singapore. 203 p.

ASEAN Canada Fisheries Post Harvest Technology Project, Phase II. 1997. Quality Management for Aquacultured Shrimp. Marine Fisheries Research Department, Southeast Asian Fisheries Development Center. Singapore. 129 p.

Huss, H.H. 1994. Assurance of Seafood Quality: FAO Fisheries Technical Paper 334. FAO, Rome. 169 p.

Huss, H.H. 1995. Quality Changes in Fresh Fish: FAO Fisheries Technical Paper 348. FAO, Rome. 195 p.

US Food and Drug Administration. 1998. Fish & Fisheries Products Hazards & Controls Guide, 2nd edn, DHHS/PHS/FDA/CFSAN/Office of Seafood, Washington, DC. 276 p.

Compendium of Fish and Fishery Product Processing Methods, Hazards 1999. and Controlshttp: *http://seafood.ucdavis.edu/haccp/compendium/compend/htm*

WHO. 1997. Food safety issues associated with products from aquaculture: report of a joint FAO/NACA/WHO study group. Joint FAO/NACA/WHO Study Group on Food Safety Issues Associated with Products from Aquaculture, Bangkok, Thailand. (WHO technical report series: 883) 68 p.

HACCP implementation in the fisheries industry
Cato, James C. 1998. Seafood Safety, Economics of Hazard Analysis and Critical Control Point (HACCP) programmes: FAO Fisheries Technical Paper 381. FAO, Rome. 70 p.
Martin, Roy E., Robert L. Collette and Joseph W. Slavin (eds). 1997. Fish Inspection, Quality Control, and HACCP: A Global Focus. Technomic Publishing Company, Inc., Pennsylvania. 802 p.
Southeast Asian Fisheries Development Center. 2001. Proceedings of the 1st regional Workshop on Application of HACCP in the Fish Processing Industry in Southeast Asia. Marine Fisheries Research Department, SEAFDEC. Singapore. 124 p.

Websites
USFDA-HACCP: *http://www.cfsan.fda.gov/~comm/haccprel.html*
Seafood National Information Center: *http://seafood.ucdavis.edu/home.htm*
WHO-Food safety: *http://www.who.int/peh/food.htm*
FAO-HACCP: *http://www.fao.org/waicent/search/default.asp*
CODEX Alimentarius Standard: *http://www.codexalimentarius.net/STANDARD/ standard.htm*

Appendix: Documented HACCP-based Quality Program

Background information
1. Company information: name, address, Phone and Fax. number, person responsible for HACCP plan management, and top management commitment. Top management shall be committed to:
   - the development of a HACCP program and implementation
   - reassessing and modifying the HACCP program.
2. Organizational chart includes people involved in HACCP plan management or details relating to HACCP team and management of system.
3. Organization's responsibilities stated in job description of key personnel involved in HACCP program management.
4. Recall (product identity and traceability).
5. Employee qualification and training.

Prerequisite program
Sanitation Standard Operating Procedures describing procedures taken to control critical areas in sanitation of processing plant.

1. Construction and layout; equipment and maintenance of hand-washing facilities, hand sanitizing facilities, and toilets.
2. Water and ice quality.
3. Condition and cleanliness of food contact surfaces.
4. Control of contamination of food, food contact surface and packaging.
5. Maintenance of hand washing, hand sanitizing, and toilet facilities.
6. Prevention of cross-contamination of food, food packaging material, and other food contact surfaces.
7. Use, types and labelling of chemicals.
8. Exclusion of pests from the food plant.
9. Control of employee health conditions that could result in the micro-biological contamination of food, food packaging materials, and food contact surfaces.

HACCP plan
HACCP plan consists of

1. Product description stated details related to characteristics of products necessary for analysis of hazard.
2. Process flow diagram listed important steps in production of products.
3. HACCP principal elements
    (a) hazard analysis
    (b) CCP determination
    (c) critical limits and validation
    (d) monitoring
    (e) corrective actions
    (f) records
    (g) verification.

Standard operating procedures (SOPs as reference in individual HACCP plans as controls to be taken at each step). This may include
SOP 1. histamine formation
SOP 2. pathogenic growth
SOP 3. container integrity (for canned, control atmosphere or vacuum pack products)
SOP 4. calibration of monitoring equipment
SOP 5. labelling.

Quality system audit procedures
Periodic review of overall system

1. Internal audit
2. Third-party audit
3. Regulatory audit.

4

HACCP in the fish canning industry

L. Ababouch, FAO, Rome

4.1 Introduction

Since its discovery by the Frenchman Nicholas Appert at the beginning of the nineteenth century, the method of preserving food using heat has become a fairly well established process. It is sometimes referred to as 'appertization'. For many decades, this process was based on a trial and error approach until the discovery of microbiology and heat transfer principles at the end of the nineteenth century, which enabled the development of its scientific foundations.

Nowadays, consumers enjoy billions of fish cans worldwide, representing over 14 million tons or 12.5% of the world catch destined for human consumption (FAO, 2000). Also, the canning process can be highly automated, especially in developed countries, or a labor-intensive process as in many developing countries such as Thailand, Philippines, Brazil or Morocco, where this industry provides significant employment opportunities. This presents the canning industry with specific safety and quality challenges that require innovative and preventative approaches to maintain or improve its safety record. This is more so because some poisoning outbreaks due to consumption of canned fish, although very rare, can result in severe health hazards, damage the company responsible or undermine a whole industry. For example, the 1982 outbreak of botulism that caused the death of one person in Belgium following the consumption of contaminated canned salmon led to the examination of the entire 1980 and 1981 production of the Alaskan canned salmon industry and a series of recalls involving over 50 million cans worldwide (Thompson, 1982).

This chapter provides an up-to-date description of the safety and quality challenges facing the fish industry and ways and means to address these challenges, particularly through the implementation of HACCP principles.

4.2   The canning process, safety and spoilage

HACCP principles were embraced by the fish canning industry some 30 years ago ahead of other segments of the food industry, as the sole and most cost-effective way to ensure product safety and quality. Before explaining the underlying reasons for this, it is worth clarifying some misconceptions regarding food preservation using heat. Viable micro-organisms can be recovered from safe and stable canned foods. End product sampling and analysis is unreliable and provides very little assurance that a canned food lot is safe or will not spoil. Therefore, a reliable safety and quality assurance program should systematically integrate (i) the quality of raw material; (ii) hygiene and sanitation; (iii) proper thermal processing (including design, application and monitoring); (iv) proper can seaming.

4.2.1   Food safety considerations

Canning of food for preservation requires the use of an hermetically sealed container which is impermeable to liquids, gases and micro-organisms, and the use of a heat process sufficient to inactivate micro-organisms capable of proliferating under normal non-refrigerated conditions of storage and distribution (Smelt and Mossel, 1982; Larousse, 1991). Any canned food that meets these two requirements is considered 'commercially sterile'. In addition, the US Food and Drug Administration (FDA), under Regulation 21CFR Part 113.3 (e), defines 'commercial sterility' as the condition achieved by the control of a_w and application of heat which renders the food free of micro-organisms capable of reproducing under normal non-refrigerated conditions of storage and distribution.

Commercial sterility is different from 'absolute sterility'. The latter means total absence of viable micro-organisms, whereas viable micro-organisms can be recovered from commercially sterile canned fish if (Ababouch, 1999):

- the micro-organism is an obligate thermophilic spore-forming bacterium but the normal storage temperature is below the thermophilic range ($< 40°C$)
- the micro-organism is acid tolerant and the food pH is within the high acidity range (pH < 4.6)
- the canning process uses a combination of heat and food water activity (a_w) reduction.

Thus, the storage temperature of the canned food, its pH and water activity (a_w) all play a crucial role, along with the chosen thermal process, to ensure safety and shelf stability of the final product. In fact, the three parameters (pH, T and a_w) affect directly the severity of the thermal process to be used for food preservation and the commercial sterility of the heat stabilized product. Likewise, they affect the subsequent growth of pathogenic and spoilage bacteria that survive the heat treatment (ICMSF, 1981).

Canned seafoods are characterized by a pH > 4.6 and $a_w > 0.98$. Foods with a pH greater than 4.6 are called 'low acid canned foods' (LACF) for which the micro-organism of major concern is *Clostridium botulinum*. Some strains of *C.*

botulinum produce spores that are the most heat resistant of all pathogenic micro-organisms. Consequently, the fish canning industry must rely on thermal processes sufficient to ensure the lowest probability of survival of *C. botulinum* spores so as to present no significant health risk to consumers. Experience has shown that the minimum heat process necessary to preserve a LACF should enable the reduction of the most heat resistant *C. botulinum* spores to 10^{-12} of its initial count. This is known as the *botulinum* cook or the $12D$ concept (Stumbo, 1973; Pflug, 1980). D is the thermal reduction time or the time necessary to inactivate 90% of a given microbial population by heating at a constant temperature. Stumbo reported that it is probably safe to assume that on average, resistant *C. botulinum* spores contaminate foods at a rate of no more than one spore per container. Thus, a thermal process based on the $12D$ concept should achieve a probability of survival of one spore in one of one trillion containers. In other words, the probability of one container being non-sterile is equal to 10^{-12}, i.e., one can in one trillion cans is not sterile.

Because of this very low target probability of survival of *C. botulinum* spores in thermally preserved products, sampling and examining end products is not reliable in ensuring product safety. Indeed, it is impossible to verify in a production lot that the probability for any one container to be non-sterile is $\leq 10^{-12}$. Table 4.1 shows that the probability of finding at least one container that is not sterile in a random sample of size N is a function of N and of the percentage of non-sterile containers in the lot of processed containers. For instance, if this percentage is 0.01% (10^{-4}), the probability of finding one non-sterile container in a sample of 10,000 containers is only 0.63. This probability,

Table 4.1 Probability of finding at least one non-sterile container in a sample of size N (Pflug, 1980)

% of non-sterile containers in lot	Number of units in random sample						
	10	20	50	100	500	1000	10,000
0.001	0.000100	0.000200	0.000500	0.001000	0.004988	0.009885	0.095164
0.002	0.000200	0.000400	0.001000	0.001998	0.009950	0.019802	0.181271
0.005	0.000500	0.001000	0.002497	0.004988	0.024691	0.048772	0.393477
0.01	0.001000	0.001998	0.004988	0.009951	0.048773	0.095167	0.632139
0.02	0.001998	0.003992	0.009951	0.019803	0.095172	0.181286	0.864692
0.05	0.004989	0.009953	0.024696	0.048782	0.221248	0.393545	0.993270
0.1	0.009955	0.019811	0.048794	0.095208	0.393621	0.632305	0.999955
0.2	0.018921	0.039249	0.095253	0.181433	0.632489	0.864935	1.000000
0.5	0.048890	0.095390	0.221687	0.394230	0.918428	0.993346	1.000000
1.0	0.095618	0.182093	0.394994	0.633968	0.993430	0.999957	1.000000
2.0	0.182927	0.332392	0.635830	0.867380	0.999959	1.000000	1.000000
5.0	0.401263	0.641514	0.923055	0.994079	1.000000	1.000000	1.000000
10.0	0.651322	0.878423	0.994846	0.999973	1.000000	1.000000	1.000000

based on a Poisson Distribution, is very low (0.095) for a percentage of non-sterile containers equal to 0.001% (10^{-5}), and almost nil for a percentage equal to 0.0001% (10^{-6}) or less. In light of these data, it is legitimate to question the soundness of end product sampling and analysis as requested by the European Union for example (EEC, 1991). Not only is it not reliable, but most worrying is the fact that it provides very false security when the results of the control indicate that samples are commercially sterile.

4.2.2   Spoilage considerations

In addition to pathogenic bacteria, the fish canning industry must define the maximum tolerable level for the survival of spoilage bacterial spores or the commercial viability of a canning operation can be threatened. In this respect, a $5D$ process is considered adequate for mesophilic non-pathogenic spores, of which those of *C. sporogenes* PA 3679 are the most heat resistant.

In practice, canned seafoods are processed beyond the minimum *botulinum* cook because of the occurrence of thermophilic spore formers of greater heat resistance. The accepted rate of non-pathogenic thermophilic spore survival is 10^{-2} to 10^{-3}/container ($2D$ or $3D$ process). These higher risks of spoilage due to thermophilic spores are considered acceptable because thermophiles are not pathogenic and given reasonable storage temperature ($< 35°C$), the survivors do not germinate. Presently, the fish canning industry uses thermal processes that vary from $F_0 = 5$ minutes to $F_0 = 20$ minutes depending on the type of product and the technology used. F_0 is the sterilization process equivalent time, defined as the number of equivalent minutes at $T = 121.1°C$ delivered to a food container calculated using a Z value (the temperature increase required for a tenfold decrease in the D value) of $10°C$. Lower F_0s yield microbially safe and shelf stable products without undue impairment of flavor, consistency, color or nutrient content.

4.3   The regulatory context

The above-mentioned theoretical considerations demonstrate beyond doubt that end product sampling and analysis is not reliable to ensure the safety of canned fish. In addition, the severe outbreaks of botulism incriminating canned foods in the 1960s led food control authorities and the food canning industry in the early 1970s to embrace safety and quality approaches embodied in the Code of Practice for Good Manufacturing Practices (GMP) and in the HACCP principles. Food control authorities have enacted regulations that require mandatory application of these approaches (EEC, 1994; FDA, 1997). International organizations, such as the FAO/WHO *Codex Alimentarius* Commission (CAC) supported this approach that required:

• canned food products to be prepared/processed in certified canneries. The certification process requires that the plant meets minimal requirements in terms of layout, design and construction, equipment, personnel hygiene and qualifications and plant sanitation

- the canning industry to be responsible for developing and implementing a HACCP-based in-plant quality control program
- the food control authority to be responsible for certifying canneries, approving and monitoring in-plant HACCP-based programs.

In this respect, the most comprehensive regulations are those of the US FDA which stem from the National Canners Association – FDA Better Process Control (BPC) Plan in 1971. The plan was drawn as a GMP regulation (21 CFR Part 108 titled *Emergency Permit Control* and Part 113 titled *Thermally processed low acid foods packaged in hermetically sealed containers* which became effective in January 1973. A few years later, safety concerns regarding the hazard of botulism in heat-sensitive low-acid foods that are acidified to permit less severe thermal processing requirements led the FDA Commissioner to add a separate GMP regulation (Part 114 for *acidified foods*) which became effective in May 1979.

The US BPC plan places the responsibility for production of safe food products on individual food industry employees. The plan requires that operators of thermal processing and packaging systems work under the supervision of a person who has attended and completed a prescribed course of instruction at a school approved by the FDA commissioner. The BPC schools represent a co-operative venture between the universities, FDA and industry personnel. More recently, the FDA Commissioner has authorized the holding of BPC schools in foreign countries in Africa, Latin Amercia and South East Asia. In parallel, the canning industry, through its professional associations in Europe and America conducts research to establish reliable heat processes and container closure evaluation schemes and advise the industry regarding technological developments and their quality and safety implications.

4.4   Hazards in fish canning

Canned fish enjoys an excellent product safety record. The main biological hazards that need to be prevented using a HACCP plan are botulism, histamine poisoning and *staphylococcus* enterotoxin poisoning (Ababouch, 1995; Huss, 1994). Chemical hazards due to the contamination of fish, especially with heavy metals such as mercury, lead or tin are of lesser concern to the fish canning industry. Contamination with mercury, lead and cadmium occurs in the aquatic environment and shall be dealt with by the appropriate food control authorities through a national monitoring program of the aquatic environment. Several comprehensive publications regarding these hazards, their causative agents and ecology are available (Huss, 1994; FDA, 2001; Jouve, 1996). The salient information is reported here as a background for the application of HACCP in the fish canning industry.

4.4.1   Histamine poisoning

Histamine poisoning is a chemical intoxication occurring from a few minutes to several hours following the ingestion of foods that contain unusually high levels of histamine (Taylor, 1986). Histamine is produced in foods by the decarboxylation of histidine (Fig. 4.1). This reaction is catalysed by the enzyme histidine decarboxylase found in some bacterial species. These include various species of Enterobacteriaceae, *Clostridium, Vibrio, Photobacterium* and *Lactobacillus*. However, all the strains of certain species such as *Morganella morganii* are capable of histidine decarboxylase activity, whereas the enzyme is present in only a few strains of other species such as *Klebsiella pneumoniae* and *Lactobacillus buchneri*. The presence of most of these bacteria is often the result of unacceptable food handling and hygiene practices.

Fish in general, and canned fish in particular, have been involved in an overwhelming majority of the incidents of histamine poisoning. This is because fish species, such as tuna, mackerel, sardines, saury, seerfish, and mahi-mahi, contain large amounts of free histidine in their muscle tissues, which serves as a substrate for histidine decarboxylase. In addition, proteolysis, either autolytic or bacterial, may play a role in the release of histidine from tissue proteins. In fact, histamine poisoning has historically been referred to as scombrotoxin poisoning because of the frequent association of the illness with the consumption of spoiled scombroid fish such as tuna and mackerel.

Despite the compelling evidence which points to histamine as the causative agent in many food poisoning incidents, it has been virtually impossible to reproduce the illness in oral challenge studies with human volunteers. The paradox between the lack of toxicity of pure histamine taken by human volunteers and the apparent toxicity of even smaller doses of histamine in spoiled canned fish has been attributed to the possible occurrence of histamine toxicity potentiators in the spoiled fish. These potentiators would act to reduce the threshold dose of histamine needed to precipitate histamine poisoning symptoms in humans challenged orally. Food-borne substances that have been suggested as potentiators of histamine poisoning comprise trimethylamine, trimethylamine oxide, agmatine, putrescine, cadaverine, anserine, spermine and spermidine. Other potentiators include some pharmacological diamine oxidase blockers, heavy intake of alcohol and certain disease (liver cirrhosis, upper gut bleeding, bacterial overgrowth in the intestines) (Fig. 4.2).

Potentiation of histamine toxicity is likely by inhibition of histamine-metabolizing enzymes present in the intestinal tract, namely diamine oxidase

Fig. 4.1   Formation of histamine.

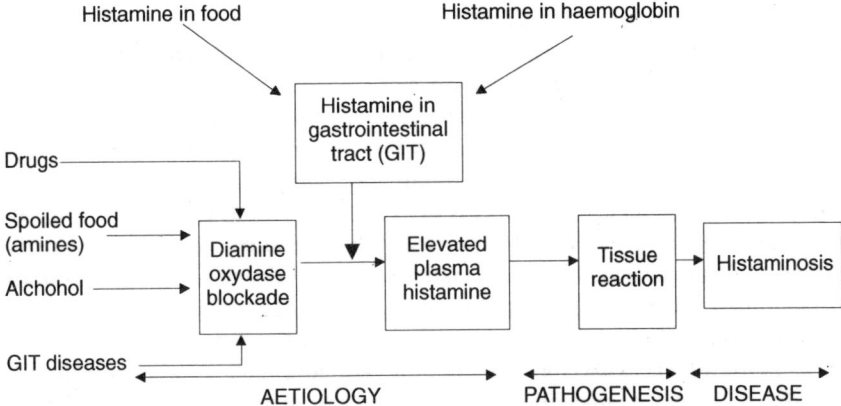

Fig. 4.2 The disease concept of food-induced histaminosis.

(DAO) and histamine N-methyltransferase (HMT). HMT is very selective for histamine while DAO also oxidises other diamines such as putrescine. In the absence of potentiators, these enzymes metabolise histamine and thus prevent its absorption into the circulation.

Prevention of histamine accumulation in foods, especially fish destined for canning, relies mostly on cooling the fish as soon as possible after the catch. Because most of the histamine-producing bacteria are mesophiles, fish storage at temperatures < 5°C, coupled with implementation of good hygienic practices, are sufficient to control histamine accumulation. In the case of small pelagic species such as sardines, mackerel and anchovies, which are often caught in large quantities where fish icing is not practical, the fish should be refrigerated quickly using either refrigerated seawater or chilled seawater. This is not always the case in many countries and fish canners rely on histamine analyses of received fish before agreeing to process the fish further.

4.4.2 Botulism
Botulism is a neuroparalytic disease caused by the consumption of foods that contain neurotoxins produced by toxigenic strains of *C. botulinum*. Botulism resulting from the consumption of fish has been recorded in Russian literature since 1818 and was first reported in the USA in 1899 (Bryan, 1986).

The close association of type E botulism with fish products has long been recognized. All known cases of type E botulism until 1980, except three, have been caused by marine or freshwater fish products or sea mammals (Huss, 1981). However, fish may be carriers of other toxin types as well. Indeed, several investigators reported that type E botulism accounted for only 20–70% of the botulism outbreaks caused by marine products, whereas type A accounted for 21–44% and type B accounted for 9–80% (Huss, 1981). Canned fish account for a very small percentage of botulism outbreaks traced to fishery products. Of the 165 outbreaks documented by Huss (1981) from the USA, Canada, Japan, USSR, and Scandinavia, only five outbreaks (3%) were traced to canned fish. The two major causes of survival and toxin production by *C. botulinum* in canned seafoods are underprocessing and post-process contamination (Pflug, 1980).

Underprocessing can result from a failure in the design of the sterilization process or from a failure in delivering the scheduled process because of equipment or human error in process delivery. To prevent underprocessing, all parameters that affect a sterilizing process's design and delivery must be clearly defined and monitored. In several countries, seafood-canning companies must register their establishments with the appropriate regulatory authority and must file their scheduled process. The regulation identifies several factors, which form a useful checklist for canners who are formulating new canned fish scheduled processes, amending existing ones or wishing to review their control procedures. Filing requires the provision of information on container dimension, target F_0 value, process temperature, process time, product initial temperature, product fill weight, product consistency, liquids to solids ratio and particle size, packing style, container stacking patterns in retort or retort baskets, number of baskets/retort, retort operation (e.g. venting and/or condensate removal) and cooling method (Warne, 1988).

Post-process contamination results from poor can seam, contaminated cooling water and can-handling damage. It is a rare event which can occur randomly with a very low probability (3.8×10^{-12}) that compares well with the risk associated with the minimum acceptable thermal process of low-acid canned foods. To prevent post-process contamination, appropriate measures should be taken to control the double seam operation, to disinfect cooling water and to handle properly sterilized cans.

4.4.3   Other hazards

Other hazards rarely associated with canned fish are food-borne illnesses due to post-process contamination, mostly *Staphylococcus* enterotoxin poisoning (SEP) (Smelt and Mossel, 1982) or the risk of poisoning with heavy metals such as mercury, lead or tin (Sainclivier, 1983).

In the early 1960s, several outbreaks of SEP, in which canned fish were suspected occurred in France and prompted research to assess the validity of thermal processes used in the fish canning industry (Thuillot *et al.*, 1964, 1968). These authors concluded that *S. aureus* cells could not survive the thermal processes commercially used and ascribed the presence of these organisms in canned sardines to post-process contamination. Other investigation suspected under-processing as the main reason for the presence of viable *S. aureus* organisms in the incriminated canned fish, possibly because of their protection from the deleterious effect of heat, by the oil in which the canned fish were packed (Ababouch, 1992). Post-process contamination of canned seafoods can be prevented by applying the same rules recommended to prevent post-process contamination with *C. botulinum.*

Regarding chemical hazards, toxic chemicals may be found in tissues of fish and other predatory species as a result of the concentration of these chemicals in the higher trophic levels of the food chain. This phenomenon is called biomagnification. The chemicals can also be found in the fish tissues as a result of bioaccumulation, where increasing concentrations of chemicals in the body tissues accumulate over the life span of the individual. Thus a large (i.e. older) fish will contain more of the chemicals than a small (i.e. younger) fish of the same species (Huss, 1994).

The 1953 methyl mercury poisoning incident that occurred in Minamata, Japan resulted in 44 deaths and 111 cases of severe disturbance of hearing and vision and brain damage (Sainclivier, 1983). This incident prompted several studies to assess the level of heavy metals in the oceans. These investigations have concluded that the level of mercury in the oceans has not changed significantly from its natural level of 0.01–0.02 ppb in seawater (Huss, 1988). Also, an FAO/WHO expert group has established a 'Provisional tolerable weekly intake' of 0.2 mg methyl mercury per person. This prompted several countries to impose tolerances varying from 0.5 to 1 ppm of mercury in fish. (Sainclivier, 1983; EEC, 1991).

Other chemicals of potential public health interest in canned seafoods are lead and tin (EEC, 1991). Lead is present in cans sealed by lateral soldering. Although very remote, the possibility exists for lead to diffuse from the lateral seam, under local acidic conditions, and contaminate the food. Tin is not considered as dangerous as mercury or lead. Tin contaminates canned foods following the corrosion of tin cans. Thus, lacquering the cans for better protection against corrosion prevents leaking of tin into the foods. Furthermore, the danger of food contamination by tin is decreasing as the canning industry is using more and more aluminium alloys, tin-free steel and other tin-free materials to produce cans (Warne, 1988; Larousse, 1991).

4.5 Spoilage of canned fish

Spoilage in canned fish is usually indicated by leakage, a swelling of the container or an abnormal odor or appearance of its contents. In some cases, the presence of microbially induced toxins, capable of food poisoning, is not accompanied by any external or internal visible signs of spoilage. Four causes can lead to the spoilage of canned foods:

1. pre-spoilage or incipient spoilage that takes place before the product is thermally processed
2. underprocessing
3. thermophilic spoilage
4. post-process spoilage.

4.5.1   Pre-spoilage

Pre-spoilage or incipient spoilage, takes place before the product or the ingredients are thermally processed. It may be caused by microbial or enzymatic action resulting in gas accumulation, development of off-odors and the presence of excessive numbers of dead microbial cells in the end product. If the responsible micro-organisms are pathogenic such as *Staphylococcus aureus,* histamine-producing bacteria, they produce thermostable toxins which will not be significantly affected by the thermal process and will cause food poisoning.

In fish received at the processing factory, microbial counts of 10^7 per gram or above are not uncommon. Fish habitat and feed and vegetable growing field soil are the main sources of organisms found on or in raw products destined for canning. Additional contamination can come from surfaces in contact with the food during harvesting and transportation, from washing water and handling practices and from the ingredients (sugar, salt, syrups, starch, spices, etc.). The type and number of these micro-organisms will be greatly affected by the different operations that fish will undergo before they are thermally processed. Fish destined for canning are washed to remove blood, slime and other foreign materials. Washing will often eliminate up to 90% of the food surface microbial load. However, only water with acceptable microbiological quality should be used. Chlorination of the washing water, to levels of 1–4 ppm of residual chlorine, is very useful in this respect (Ababouch, 1999). Preparatory operations such as beheading and evisceration expose food tissues and accelerate microbial growth and food spoilage if delays occur, specially under high temperatures conducive to bacterial growth.

The cooking of fish significantly inactivates heat-sensitive micro-organisms, namely vegetative bacteria, sensitive spores such as type E *C. botulinum* spores, yeasts and moulds. However, heat-resistant organisms, especially spores of *C. botulinum* types A and B and thermophilic sporeformers survive cooking. Thus, extreme care should be exercised to implement proper hygienic practices and avoid delays, otherwise contamination by hygiene-related organisms such as *S. aureus* may lead to spoilage and the accumulation of thermostable toxins that will not be inactivated during thermal processing.

4.5.2   Underprocessing

Underprocessing, also referred to as understerilization, means that the product did not receive sufficient heat treatment to become commercially sterile. It is often indicated by the survival of bacterial spores exclusively, because these will resist the insufficient heat treatment, whilst vegetative bacteria, yeasts and moulds will not. The causes, consequences and methods of prevention of underprocessing have been presented in section 4.4.

4.5.3   Thermophilic spoilage

Thermophilic spoilage occurs when the time-temperature conditions are conducive to the growth of thermophilic bacteria. These are not pathogenic organisms. They occur naturally in soil and sediments and their spores are frequently isolated in low numbers from commercially sterile food products. Their number can increase if ingredients such as starch, sugar or spices are used in the product and are excessively loaded with these spores.

Thermophiles can grow in canned foods if cooling hot retorted cans takes place at ambient temperature or if finished products are stored at temperatures above 40°C. In the first case, cooling is significantly slow and the temperature of the cans will be in the range 40°C–75°C for periods long enough to promote growth of thermophiles. This is, however, rare nowadays because most of the retorts are equipped to provide for rapid cooling of the cans.

Prevention of thermophilic spoilage can be achieved by cooling the retorted cans rapidly to reach a temperature < 40°C and storing finished products at less than 35°C to inhibit the growth of any surviving thermophiles. Also, thorough washing of the raw material to remove sediments and prevention of recontamination with soil or other sources of thermophiles during the preparatory operations is of paramount importance. When ingredients such as sugar, starch and spices are used, processors should exercise great care to ensure that these ingredients do not contain excessive numbers of thermophilic spores.

4.5.4 Post-process contamination

Post-process contamination or leaker spoilage takes place when microbial contaminants leak into the can after heat sterilization due to a failure of the container to maintain an hermetic seal. It is undoubtedly of greater economic importance as it accounts for 60–80% of the spoilage of canned foods.

The micro-organisms involved in leaker spoilage can be any type found on can-handling equipment, in cooling water or on the skin of can handlers. These include bacterial cocci, short and long rods, yeasts and moulds, aerobic sporeformers or, more likely, a mixture of many of these organisms. Post-process contamination can also result in outbreaks of botulism or *Staphylococcus* enterotoxin poisoning. Leaker spoilage is often associated with the integrity of the can seams, the presence of bacterial contaminants in the cooling water or on wet can runways and abusive can-handling procedures after heat processing. Cooling water can be the primary source of organisms responsible for leaker spoilage.

Although very rare nowadays, defective tin plate, can manufacturing defects and mishandling empty cans will affect the can integrity. Can manufacturing defects of interest are defective side seams, over or under-flanging which interfere with double seam formation and defective double seams resulting from faulty seam operation or end compound distribution. The most common type of damage from improper handling of empty cans is bent flanges or cable cuts. The latter occurs when cans are held back while the conveyor cable continues to run.

Filling operations can impact directly on the quality of the seam. Overfilling, particularly a cold product, and the subsequent expansion of the product during processing can cause end distortion and seam damage. Also, if the filler leaves product hanging over the can flange, it may interfere with can seaming and result in leakage, especially with fibrous products such as leafy vegetables and meat and fish products containing pieces of bones.

Double seam deficiencies can be transient or permanent. Transient leaks are reversible in that a leakage path opens but subsequently closes leaving no detectable deformation at the point of the leak. A permanent leak path through the double seam may exist due to an improper seam construction, the use of non-leak-resistant compounds and improper side-seam soldering and side-seam tightness.

Handling practices of the filled containers can impact dramatically on the hermetic seal. Blows due to can dropping in retort crates without cushioning the fall or to cans rolling and striking solid surfaces such as another can or the bar of an elevator can lead to a leak. The effects of repeated blows on the seam are cumulative and may lead to contamination similar to that resulting from a single violent blow. Seam deformations may also result from mechanical impacts, abrupt pressure changes as occur when retorted cans are suddenly exposed to atmospheric pressure for cooling. During thermal processing, the can contents expand considerably and may result in permanent distortion of the can ends, unless adequate headspace under vacuum and counter-pressure during cooling of large size cans is provided.

To minimize post-process contamination, it is necessary to ensure good controls over empty can inspection and handling systems, can seam integrity, adequate chlorination of cooling water and minimal can abuse during in-plant can handling, transportation and distribution. As can integrity is critical, can seams and seaming machines should be inspected as frequently as feasible, at least every 30 minutes for a visual inspection of a seamed can and every four hours for a thorough seam teardown and examination. Cooling water should be chlorinated to a chlorine level of 2–5 ppm, to allow for a content of no less than 1 ppm after cooling. Cooled wet cans should be dried in a restricted-access area and must not be handled until they are dry. Care should be exercised thereafter to minimize abuse leading to dents and leakage.

4.5.5   Other causes of spoilage
Canned foods can also spoil from non-microbial causes. Internal can corrosion can lead to the accumulation of hydrogen which relieves vacuum and swells the can making it unmarketable. Externally, corrosion often causes pinholes that allow micro-organisms to penetrate the can and spoil its contents.

The production of canned food salads containing fish and other vegetables such as corn, peas or green beans requires proper handling to avoid objectionable food color defects. Indeed, contamination of food with metals such as copper or iron before it is placed in the can or a reaction between the food and the container can lead to objectionable food color defects. These include blue-greying of corn and blackening of peas, corn, shrimps and other fish meats. This is often the result of protein sulphur compounds breaking up under high temperature during blanching or cooking, and their combining with iron forming black iron sulphide. The use of enamel lined cans for these products eliminates this problem.

4.6   The application of GMP in the fish canning industry

For the reasons detailed previously, many countries have enacted GMP regulations that require the fish canning industry to meet minimal requirements in terms of plant layout, design and construction, equipment, personnel hygiene and qualifications and plant sanitation. These regulations are internationally harmonized in the revised version of the Codex Alimentarius Code of Practice – General Principles of food hygiene (CAC/RCP1-1969, Rev.3, 1997) (FAO, 1997). Presently, the Codex Committee on Fish and Fishery Products (CCFFP) is adapting these principles to the fish industry and the draft Code of Practice (presently at step 3 of the CCFFP adoption procedure) defines clearly the GMP requirements and the modalities for HACCP implementation to prevent both safety problems and quality defects in fish and fishery products. The reader is encouraged to consult the draft Code of practice for fish and fishery products and follow its developments during the coming session of the CCFFP in May 2002.

4.7   The application of HACCP in the fish canning industry

As stated earlier, the food canning industry is the oldest segment of the food industry that has adopted GMP and HACCP principles since the early 1970s. Consequently, this industry has accumulated enough experience in this area and generic models of HACCP are available through various media and publications. These models can be used given proper adaptation to the particular handling and processing conditions of a given cannery. Most importantly, the success of implementing an adapted HACCP plan rests with the level of awareness of the company decision makers, the qualification of the supervisors in charge of CCPs, the monitoring of these CCPs and the degree of freedom the supervisors have to implement agreed upon corrective actions rapidly. These are the ingredients that provide for the cost-effectiveness and success of a HACCP program or its failure.

What follows is the safety and quality approach that was implemented successfully in the fish canning industry of Morocco (sardines and mackerel) and Senegal (tuna) where the author worked extensively. The companies that implemented this approach enjoy worldwide exports. They meet the GMP requirements of the EU and FDA and their staff underwent successful HACCP and BPC school training in Morocco during the period 1994–1998.

Before HACCP implementation, the canning facilities are upgraded to meet the sanitary requirements in terms of layout, equipment, personnel hygiene, water supply quality and cleaning and sanitation. Good handling practices, including safe handling of empty and filled cans, maintenance of retorts and seaming machines, instrument calibration, are all part of the safety and quality assurance programs.

The following annexes are attached to the HACCP manual to describe the appropriate preventative and monitoring procedures to be implemented:

Annex 1: Standard sanitation operating procedures SSOP (layout, circulation of personnel and products, personnel hygiene, pest control, water supply) and control of SSOP (control of personnel hygiene, of cleaning and sanitation, of water quality and water chlorine level).
Annex 2: Standard handling, icing and transportation of fresh fish.
Annex 3: Standard sterilization procedure.
Annex 4: Measurement of fish temperature.
Annex 5: Sensory evaluation of fresh fish.
Annex 6: Determination of total volatile bases (TVB).
Annex 7: Determination of histamine.
Annex 8: Verification of the container closure.
Annex 9: Determination of heat penetration and temperature distribution in the retorts.

The following sections contain details of the HACCP components.

4.7.1   HACCP team

In general, the HACCP team in these companies included the QC manager, the production manager, the hygiene manager and an outside technical advisor as seen fit. Because of the large number (up to 600) of workers, often female, the hygiene manager is often a female.

The duties of each HACCP team member are explicitly identified for the hygiene manager in the plan and are carried out under the supervision of the QC manager. The latter validates all actions necessary for implementing the HACCP plan. If needed, the hygiene manager will refer to the general manager for the implementation of cumbersome and costly actions, to whom the different options and solutions will be presented without any compromise on safety. If necessary, the technical adviser is consulted to provide scientific and technical advice as seen fit. Depending upon their respective duties, the HACCP team members received the necessary training in HACCP, hygiene and sanitation, laboratory techniques or retort and seaming control.

4.7.2   Description of the products and their intended use

The companies selected manufacture up to 30 different products, using sardines, mackerel or tuna. Table 4.2 describes the contents, packaging and distribution conditions of some of these products as they appear in the HACCP plans. The canned fish products are used for human consumption by the general public. They are eaten without any further cooking as an appetizer, as such, or after

Table 4.2 Example of a description of canned fish products manufactured in Morocco

Product	Contents	Packing materials and format	Shelf life
1. Canned sardines in vegetable oil	Beheaded tail-off sardines: 75% soya oil: 24% salt: 1% pH = 6.2–6.5. water activity a_w = 0.98	(1) Tin format 1/6 P 30 2 pieces, simple or easy open lid. (2) Aluminum alloy format 1/6 P 30 easy open lid. (3) ½ H 40. (4) 1/6 P 30 DAS R 26	5 years at ambient temperature
2. Canned sardines *au naturel*	Beheaded and tail-off sardines: 75% water: 24% salt: 1% pH = 6.2–6.5. water activity a_w = 0.98	Tin or aluminum alloy (1) format 1/6 P 30 2 pieces simple lid; (2) 1/6 P 30 ES easy open lid. All cans are individually packed in paper holsters.	3 years at ambient temperature
3. Canned sardines in olive oil	Beheaded tail-off sardines: 75% Olive oil: 24% salt: 1% pH = 6.2–6.5. water activity a_w = 0.98	Tin or aluminum alloy (1) format 1/6 P 30 2 pieces simple lid; (2) 1/6 P 30 2 pieces easy open lid. Some cans are individually packed in paper holsters.	5 years at ambient temperature
4. Skinless boneless canned sardines in olive oil	Skinless boneless sardines: 75% Olive oil: 24% salt: 1% pH = 6.2–6.5. water activity a_w = 0.98	Tin or aluminum alloy (1) format 1/6 P 30 2 pieces simple lid; (2) 1/6 P 30 2 pieces easy open lid. (3) 1/6 P 22 2 pieces easy open lid. All cans are individually packed in paper holsters.	5 years at ambient temperature
5. Skinless boneless canned sardines *au naturel*	Skinless boneless sardines: 75% Water: 24% salt: 1% pH = 6.2–6.5. water activity a_w = 0.98	Tin or aluminum alloy (1) format 1/6 P 30 2 pieces simple lid; (2) 1/6 P 30 2 pieces easy open lid. (3) 1/4 P 22, 2 pieces easy open lid. All cans are individually packed in paper holsters.	3 years at ambient temperature
6. Skinless boneless canned sardines in tomato sauce	Skinless boneless sardines: 75% Water + tomato paste: 22%; Soya oil: 2%; salt: 1% pH = 5.9-6.2. water activity a_w = 0.98	Tin (1) format 1/6 P 30 2 pieces simple lid; (2) 1/6 P 22 2 pieces easy open lid. All cans are individually packed in paper holsters.	3 years at ambient temperature
7. Mackerel fillets canned in soya oil	Mackerel filets: 75% soya oil: 24% salt: 1% pH = 6.2–6.5. water activity a_w = 0.98	Tin (1) format 1/6 P 30 2 pieces simple lid; (2) 1/6 P 30 2 pieces easy open lid. All cans are individually packed in paper holsters.	5 years at ambient temperature
8. Mackerel fillets canned *au naturel*	Mackerel filets: 75% Water: 24% salt: 1% pH = 6.2–6.5. water activity a_w = 0.98	Tin or aluminum alloy (1) format 1/6 P 30 2 pieces easy open lid. All cans are individually packed in paper holsters.	3 years at ambient temperature
9. Mackerel fillets canned in tomato sauce	Mackerel filets: 75% Tomato paste: 22%; Soya oil: 2%; salt: 1% pH = 5.8–6.1. water activity a_w = 0.98	Tin or aluminum alloy (1) format 1/6 P 30 2 pieces easy open lid. All cans are individually packed in paper holsters.	3 years at ambient temperature

mixing with other food or salads. A large proportion of the production is exported worldwide, mainly to Europe and the USA.

4.7.3   Flow diagrams

Despite the variety of products, the production flow diagrams are similar and can be summarized as presented in Fig. 4.3. The HACCP plan contains a figure (not provided here for confidentiality) representing the plant layout and the circulation of personnel and products. Table 4.3 presents examples of thermal processing technical information needed in the GMP manual. It is related to

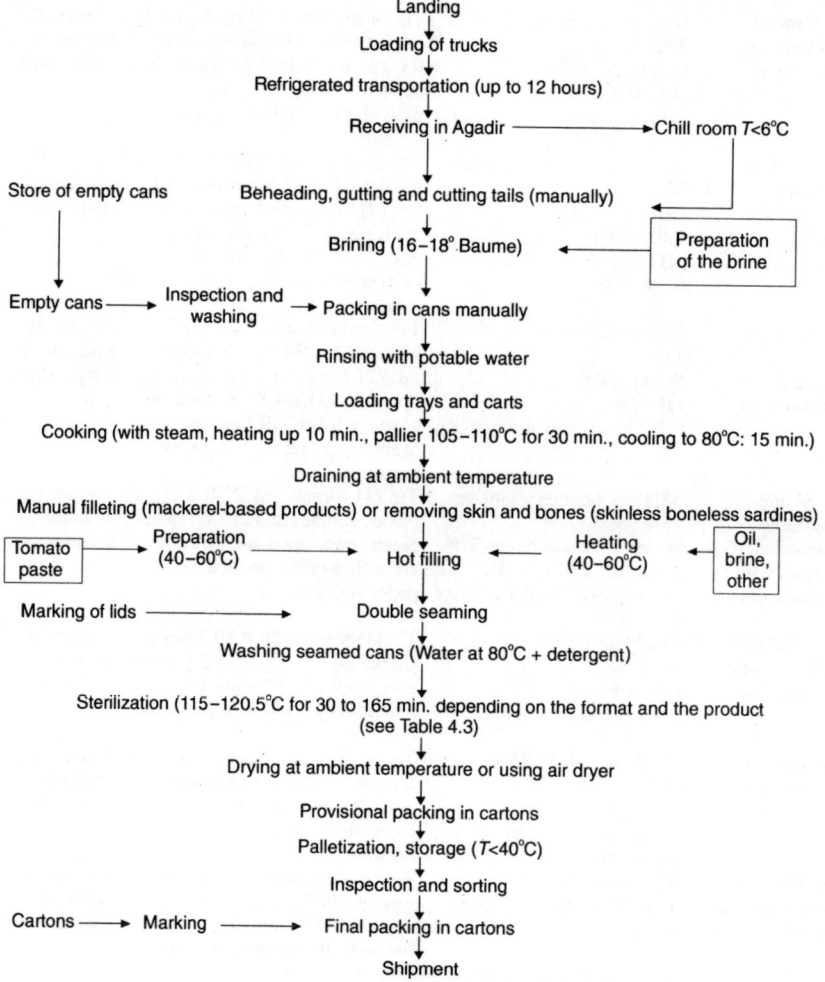

Fig. 4.3   Example of a flow diagram for canning sardines or mackerel in Agadir, Morocco.

Table 4.3 Technical parameters of sterilization in a fish cannery in Morocco

Technical parameters of the Steriflow retorts	Thermal processing target values
Retort type	Steriflow using overheated water
Minimal F_0	7 to 14 minutes, generally > 10 minutes
Can format	¼ P22 (115 g), 1/6 P 25 (125 g); 1/6 P 30 (125 g); ½ P oval (375 g); ½ H 40 (365 g); ½ HL (425 g); ½ B (425 g).
Sterilization temperature	122.5°C, with an overshooting at 123.5°C.
Heating duration	23–55 minutes at 122.5°C
Minimal initial temperature	30°C
Filling method	Manual fish packing and overfilling with liquid
Ratio solid/liquid	75% fish; 25% liquid (oil, water, tomato sauce)
Stacking of the cans in the retort basket	In bulk
Number of baskets per retort	4 Baskets
Sterilization system	Water is overheated to a pressure of 6 bars
Cooling method	The heating water is cooled through a heat exchanger and recycled to be used for cooling

Technical parameters of the vertical steam retorts	Thermal processing target values
Retort type	Vertical retorts using steam
Minimal F_0	7–14 minutes, generally > 10 minutes
Can format	1/6 P 30 DAS (125g); ½ HL (425g)
Sterilization temperature	115°C.
Heating duration	55–85 minutes at 115°C
Minimal initial temperature	30°C
Filling method	Manual fish packing and overfilling with liquid
Ratio solid/liquid	75% fish; 25% liquid (oil, water, tomato sauce)
Stacking of the cans in the retort basket	In bulk
Number of baskets per retort	1 Basket
Sterilization system	Introduction of steam through the by-pass. Venting for at least 10 minutes at 105C. Close the by-pass. Open the regulating valve as temperature reaches 115°C. This is start time.
Cooling method	Using cold water chlorinated at 2–5 ppm of active chlorine

vertical retorts that operate under pressure in steam and Steriflow retorts using cascading high-pressure water. The latter, which represent the latest technology in thermal processing, recycle the water that was used for sterilization. After being cooled through the Steriflow heat exchanger, this water is used to cool the cans. This eliminates the problem of post-process contamination from water and allows substantial savings in water consumption.

4.7.4   Hazard analysis
All potential hazards that can compromise product safety have been studied and analyzed. In order to achieve this, a HACCP team uses published scientific data, its own experience, the feed-back of its clients and that of the regulatory and control agencies, the expertise of its technical adviser and the quality-related technical specifications of its clients. All the available information was used to assess the potential of each hazard, its cause, severity and likelihood and the preventative measure(s) implemented to prevent it.

The potential causes of hazards were either a contamination (from fish, personnel, water, ice, equipment or personnel) or survival (to sanitation, cooking and sterilization) of hazardous micro-organisms, the production or persistence of

Table 4.4   Hazard analysis applied to canned fish

Hazard	Severity	Risk	Preventative measure(s)
Botulism because of insufficient thermal processing or because of post-process contamination during cooling	+++++	+	• Proper sterilization • Proper training of personnel in charge of sterilization • Proper chlorinating of cooling water.
Histamine poisoning because of contaminated raw fish or histamine accumulation during preparation.	++	+++	• Training of purchase supervisor in proper freshness assessment • Proper icing and refrigeration • Control of histamine level at receiving when in doubt.
High levels of heavy metals in canned fish	++++	+	• Good knowledge of fishing zones • Ensure the purchase of fish caught only in pollution-free areas
Post-process contamination with pathogens or toxic materials because of bad container closure	++++	+	• Training of container closure supervisor • Maintenance of seaming equipment
Staphylococcal poisoning because of bad handling of wet and hot freshly sterilized cans	+++	+	• Air drying of wet cans • Storage of wet cans in restricted-access area
Spoiled canned fish because of spoiled (high TVN) raw fish or of spoilage during processing	+++*	+	• Refrigerated fish transportation • Trained staff on sensory evaluation • TVN analysis if in doubt • SSOP
Thermophilic spoilage because of slow cooling or long storage at $T > 40°C$	++*	+	• Rapid cooling to reach $T < 40°C$ in less than 1 hour • Storage of finished products at $T < 40°C$

+: Very low, ++: low, +++: average, ++++: high, +++++: very high.
*: The severity and risk here refer to economic impact of spoilage.

toxic chemicals (such as histamine, staphylococcal enterotoxins, *botulinum* toxins). For each cause of a hazard, the most appropriate preventative measure was identified.

The complete hazard analysis is presented in Table 4.4. The preventative measures adopted are described in detail as annexes and identify the staff in charge of their application. The persistence of heavy metals, such as mercury, in raw and canned fish was not considered a hazard because the national monitoring plan since 1989 indicated levels far below 0.5 ppm in Moroccan canned fish. However, canning companies are requested to exercise care in this regard before processing fish caught in areas different from the traditional ones or in case of any alert given by the Moroccan Inspection Agency which is carrying out a monthly surveillance program.

Table 4.5 summarizes the ensuing HACCP steps including the identification of critical control point(s) (CCP), where preventative measures are implemented to prevent each hazard, the identification of critical limits, monitoring procedures and corrective actions to apply when monitoring reveals loss of control. In general each step is explained in detail in the HACCP manual. The identification of CCPs used the decision tree that was developed by the *Codex Alimentarius.*

4.7.5   Record-keeping

Forms are used to record the results of each monitoring activity and any corrective action implemented, whether it is relevant to GMP, SSOP or HACCP. For proper communication between the HACCP team members, the record-keeping forms should be designed to be as simple as possible, yet to clearly identify who is responsible for the implementation of preventative, monitoring and corrective actions and who should validate these actions or be informed of their respective outcome.

4.7.6   Verification procedure

The HACCP plans are presently verified annually. During the first three to four years of their implementation, they were verified every six months. The verification process involves:

- Evaluation of all the inspection data obtained from the laboratory of the food control authority. This laboratory carries out chemical analyses (TVB and histamine) of commercial sterility and mercury on each lot of finished product before shipment. All these data are statistically analyzed to assess the quality level of production over the year. Any quality problem detected by these analyses will be immediately addressed by the QC manager to identify why the HACCP system did not operate properly to prevent this problem.
- Evaluation of all the monitoring data collected by the company during the implementation of HACCP. The deviation cases will be given careful

Table 4.5 Example of HACCP plan for the production of canned sardines and mackerel in Morocco

Critical control point	Significant hazard	Preventative (measure(s))	Critical limit	What to monitor	How to monitor	Monitoring frequency	Who monitors	Corrective action(s)	Records	Verification
1. Receiving raw fish	Spoiled fish or fish with high histamine levels	Proper Fish icing and handling, after landing, refrigerated transportation	Fish $T < 6°C$ TVB <250 ppm Histamine < 50 ppm	Fish temperature TVB Histamine	Temperature measurement Chemical method Spectro-fluorimetry	Every received lot If freshness doubtful Every received lot	Reception supervisor QC manager QC manager	Fish sorting according to freshness. If needed, determination of TVB Reject if TVB > 250 ppm Reject if histamine > 50 ppm	Fish receiving (form 1+ 4) used for freshness, TVB and histamine	Daily record review Daily record review
2. Container closure	Post-process contamination	Training of the container closure supervisor Maintenance of closure equipment	No leaking containers	Visible defects	Visual inspection	Every 30 minutes	Container closure inspector	Isolate cans closed since last control and discard any leaking ones.	Container integrity records (forms 2+4)	Daily record review
				Integrity defects	Double seam analysis	Every 2 hours	Inspector	Isolate cans closed since last control and discard any leaking ones	Container integrity records (forms 2+4)	Daily record review
3. Sterilization	Survival of spores, especially those of *Cl. Botulinum*	Regular maintenance of the retorts. Training of the retort operating supervisor	$F_0 = 7$–14 minutes Sterilization parameters (Table 4.2)	Heat penetration & temp. distribution T, time, P	Ellab logging data equipment Recording	Twice a year and as needed Every retort cycle	QC manager Retort operator	Identify cause of deficiency and repair the retort Repair retort	Automatic recordings Sterilization (forms 3+4)	Review at every run Daily record review
4. Cooling sterile cans	Microbial contamination Thermophilic spoilage	Chlorination of cooling water. Cooling time < 1 hour	Residual free chlorine 1 ppm Duration of cooling < 1 hour	Chlorine level Duration of cooling	Chemical analysis Timing	Every day Every retort cycle	QC manager Retort operator	Identify lot and resterilize Isolate lot and check for thermophilic spoilage	Cooling water record Sterilization (forms 3+4)	Review at every run Review at every run
5. Storage of the finished product	Thermophilic spoilage	Storage at $T <$ 40°C	$T < 40°C$	Record temperature	Calibrated Thermometer	Daily	Storage supervisor	Isolate lot and check for thermophilic spoilage	Storage temp. form (5 + 4)	Daily in the summer

attention to assess why they were not prevented through the HACCP system applied. The appropriate revision will be introduced accordingly.

- Evaluation of the feed-back information from the clients.
- Annual audit by the technical advisor to assess the actual validity of the preventative, monitoring and corrective procedures implemented. This audit will encompass evaluation of the HACCP records, the equipment maintenance exercise, the calibration of the monitoring instruments and procedures. An audit report will be issued and submitted to management, followed by a meeting with the HACCP team, in group and individually, to assess new revisions to the HACCP plans, to introduce new procedures, equipment, standards, technology development and client's requirements. This opportunity is also used to include new developments in the field and newer requirements from the clients.

4.8   Future trends

The fish canning industry is undergoing product and technological development to meet consumer demand for minimally processed and diversified products and to integrate environmental (energy saving, clean technologies) as well globalization considerations. Nonetheless, HACCP will remain for many years to come at the center of safety and quality assurance in this industry mainly because of its reliability, cost-effectiveness and ability to integrate easily new developments.

4.9   Sources of further information and advice

Nowadays, with the developments in information technology, the best sources of information and advice are accessible via the internet. Here are selected sites concerned with the quality and safety of canned fish.

http://www.cfia-acia.agr.can
http://www.fao.org
http://ww.who.org
http://www.europa.eu.int/comm/food
http://ww.cfsan.fda.gov
http://www.foodhaccp.com

4.10   References and further reading

ABABOUCH, L.H. 1991. Histamine food poisoning. An update. *Fish Tech News* (FAO). 11 (1): 3–5, 9.
ABABOUCH, L.H. 1992. Safety of canned seafoods. In *Quality Assurance in the*

Fish Industry. Huss, H.H. ed. Elsevier Science Publishers. Amsterdam, pp. 259–67.

ABABOUCH, L.H. 1995. *Assurance de la qualité dans l'industrie halieutique.* Actes Editions. IAV Hassan II. Rabat. Morocco. 214 pages.

ABABOUCH, L.H. 1999. Heat treatment of foods. Spoilage problems associated with canning. Pages 1016–23. R. Robinson, C. Blatt and P. Patel (editors). *Encyclopedia in Food Microbiology.* Academic Press Limited. London.

BRYAN, F.L. 1986. Seafood-transmitted infections and intoxications in recent years. In *Seafood quality determination.* Kramer, D.E. and Liston, J. eds. Elsevier Science Publishers. Amsterdam.

DATTA, A.K. 1992. Thermal processing: food canning. In *Encyclopedia of Food Science and Technology* Hui, Y.H. (Editor in Chief). Volume 1, pp. 260–3. Volume 4, pp. 2561–5. Wiley Interscience Publication. New York.

EEC, 1991. Council Directive 91/493/EEC of 22 July 1991 laying down the health conditions for the production and the placing on the market of fishery products. *Official Journal of the European Communities.* No. L 268: 15–32.

EEC, 1994. Commission Decision 94/356/EEC of 20 May 1994 laying down the modalities for the application of own checks for fish and fishery products. *Official Journal of the European Communities.* No. L 156: 50–7.

FAO, 1997. Recommended International Code of Practice-General Principles of Food Hygiene CAC/RCP 1 – 1969, Rev. 3. Rome.

FAO, 2000. *The State of Fisheries and Aquaculture* (SOFIA). FAO. Rome.

FDA, 1997. Title 21 of the Code of Federal Regulations, Parts 123 and 1240. Volume 60, No. 242. Pages 65095-202.

FDA, 2001. *Fish and fishery products hazards and control guidance.* Third edition. Center for Food Safety and Applied Nutrition. 326 pages. USFDA, Washington DC.

GAVIN, A. and WEDDIG, L.M. 1995. *Canned foods. Principles of thermal process control, acidification and container closure evaluation.* Sixth edition. The Food Processors Institute. Washington DC.

HUSS, H.H. 1981. *Clostridium botulinum* type E and botulism. Technological Laboratory. Ministry of Fisheries. Technical University. Lyngby. 2800. Denmark.

HUSS, H.H. 1988. *Fresh fish quality and quality changes.* FAO Fisheries Series No. 29. FAO. Rome.

HUSS, H.H. 1994. Assurance of seafood quality. FAO Fisheries Technical Papers No. 334. 169 pages. Rome.

ICMSF, 1981. *Microbial ecology of foods. Volume 1: Factors affecting the growth of micro-organisms.* Academic Press. New York.

JOUVE, J.L. 1996. *La qualité microbiologique des aliments. Maîtrise et critères.* 563 pages. Polytechnica. Paris.

LAROUSSE, J. 1991. Coordonateur. *La conserve appertisée. Aspects scientifiques, techniques et économiques.* Techniques et Documentation. Lavoisier. Paris.

PFLUG, I.J. 1980. *Syllabus for an introductory course in the microbiology and engineering of sterilization processes*. Environmental sterilization Services. Saint Paul. Minnesota.

SAINCLIVIER, M. 1983. *L'industrie alimentaire halieutique. Volume 1: Le poisson matière première*. Editions Sciences Agronomiques. Rennes. France.

SMELT, J.P.P.M. and MOSSEL, D.A.A. 1982. Application of thermal processes in the food industry. In *Principles and practice of disinfection, preservation and sterilisation*. Russell, A.D., Hugo W.B. and Ayliffe, G.A.J. (eds). Blackwell Scientific Publishers. London.

STUMBO, C.R. 1973. *Thermobacteriology in food processing*. Academic Press, London.

TAYLOR, S.L. 1986. Histamine food poisoning: toxicology and clinical aspects. CRC *Critical Reviews in Toxicology*. 17: 91–117.

THOMPSON, R.C. 1982. The tin of salmon had but a tiny hole. *FDA Consumer*. June 1982: 7–9.

THUILLOT, M.L., BROSSARD, J., THOMAS, G. and CHEFTEL, H. 1964. Contribution à l'étude de la stérilisation des conserves de sardines *Clupea pilchardus* à l'huile en boîtes de grand format. *Annales de l'Institut Pasteur (Lille)*. 15: 145–55.

THUILLOT, M.L., BROSSARD, J., THOMAS, G. and CHEFTEL, H. 1968. A propos de la thermorésistance de S. aureus dans l'hule. *Annales de l'Institut Pasteur (Lille)*. 19: 153–57.

WARNE, D. 1988. Manual on fish canning. FAO technical paper 285. Rome.

5

Improving the control of pathogens in fish products

L. Nilsson and L. Gram, Danish Institute for Fisheries Research, Lyngby

5.1 Introduction

Food-borne diseases are of major concern to consumers, producers and authorities alike. Despite an increased awareness, the number of cases and outbreaks does not appear to be decreasing (Todd, 1997). Worldwide, the number of food-borne diseases is recognized as being significantly underestimated, since reporting is limited (Todd, 1997; Gram and Huss, 2000). Surveillance of food-borne diseases and identification of etiological agents has for many years been done in countries such as the UK, US and Holland and statistics on food-borne diseases thus rely heavily on data from these areas. Many foods are implicated in food-borne disease outbreaks (Olsen et al., 2000; Gillespie et al., 2001). Seafoods rank third on the list of products which have caused food-borne disease in the United States between 1983 and 1992 (Lipp and Rose, 1997). Combining statistics from the US and UK showed that between 10 and 25% of food-borne outbreaks could be traced to seafood products (Liston, 1990, Bryan 1988, Wallace et al., 1999, Gillespie et al., 2001, Olsen et al., 2000). However, seafoods accounted for only 3.6% and 10% of the number of cases of food-borne illness in the US from 1978–1987 (Liston, 1990) and in New York from 1980–1994 (Wallace et al., 1999), respectively. Seafood-borne disease may be caused by a variety of agents, including aquatic toxins, biogenic amines, bacteria, virus and parasites (Gram and Huss, 2000).

Shellfish, in particular molluscan shellfish, is the seafood product most frequently implicated in seafood-borne disease (Liston, 1990) and consumption of shellfish accounted for 25% of seafood outbreaks in the US from 1993 to 1997 (Table 5.1). Outbreaks connected with shellfish are mostly caused by enteric viruses (Glatzer, 1998; Shieh et al., 2000, Lipp and Rose, 1997). In the

Table 5.1 Foodborne disease outbreaks caused by fish and fish products in the US from 1993 to 1997 (Olsen *et al.*, 2000)

Pathogenic agent	No. of outbreaks (% of total) associated with			
	Shellfish[1]		Fish	
Bacteria	5	11%	2	1%
Biotoxins	3	6%	55	39%
Histamine	2	4%	66	47%
Parasites	0	0%	0	0%
Virus	11	23%	1	<1%
Known	21	45%	124	89%
Unknown	26	55%	16	11%
Total	47	100%	140	100%

[1] Both molluscan shellfish and crustaceans

US 23% of outbreaks connected with shellfish were caused by viruses (Table 5.1). Molluscan shellfish are filter-feeding organisms which can accumulate human viruses, pathogens and biotoxins from marine algae to levels considerable greater than those in the overlying water (Rippey, 1994; Metealf *et al.*, 1979; DePaola *et al.*, 1994). Since molluscan shellfish often are eaten raw or only steamed before consumption (Table 5.3) they are a common cause of seafood-borne disease (Gram and Huss, 2000).

Scombroid poisoning, which is caused by biogenic amines (scombrotoxins), is the most important cause of illness associated with fish products (Lipp and Rose, 1997). Scombrotoxin accounted for 47% of seafood-associated outbreaks in which an etiologic agent could be identified in the US during 1993 to 1997 (Table 5.1). Almost as important as a cause of disease from fish are aquatic biotoxins, in particular ciguatera poisoning (Olsen *et al.*, 2000). In addition several bacteria, in particular *Vibrio* spp., are known to cause food-borne diseases in seafood (Feldhusen, 2000; Food and Drug Administration, 1994).

5.2 Microbial health hazards in fish products.

Pathogenic bacteria which are indigenous to the marine or estuarine environment (Table 5.2) are naturally present on live fish and crustaceans. These bacteria are mostly found in low numbers in live fish with the exception of marine vibrios, and the risk of causing disease is insignificant unless growth of these organisms occurs during storage (Huss, 1997; Gram and Huss, 2000; Feldhusen, 2000). Marine vibrios, such as *V. parahaemolyticus* and *V. vulnificus*, may be found in high numbers in shellfish and in shellfish-eating fish from tropical waters and during the summer months in temperate zones (Reilly and Käferstein, 1997; DePaola *et al.*, 1994; Motes *et al.*, 1998; Ruple and Cook, 1992). Indigenous pathogens may cause serious illness and death in

Table 5.2 Major food poisoning organisms associated with seafood

Organism causing disease	MID of toxin or live cells	Primary habitat
Bacteria of aquatic origin		
Clostridium botulinum type E	$0.1–1$ μg toxin	Ubiquitous in aquatic environment, soil, ocean sediment, intestinal tract of fish, surface of fish
Marine *Vibrio spp.*		
V. cholerae	10^8 cfu/g	Estuarine and coastal warm
V. parahaemolyticus	$10^5–10^6$ cfu/g	waters (>15°C), intestines of
V. vulnificus	Unknown	shellfish-eating fish and tract of oysters
Histamine producing bacteria	>100 mg histamine/100 g	Members of *Enterobacteriaceae* from the aquatic environment
Dinoflagellates; maybe bacteria associated with the algae	Paralytic shellfish poisoning (PSP) toxin	Aquatic environment, accumulated in bivalves (e.g. mussels, oysters)
Bacteria from the general environment		
Listeria monocytogenes	unknown–10^8 cfu/g	Widespread in nature, soil, foilage, faeces, seafood processing environments
C .botulinum (mesophilic)		Widespread in soil
Bacteria from the human/animal reservoir		
Shigella spp.	$10^2–10^5$ cfu/g	Faecal polluted coastal regions or ponds; cause faecal contamination of seafood
Salmonella spp.	$10–10^6$	
Escherichia coli	$10–10^8$ cfu/g	
Staphylococcus aureus	$0.14–0.19$ μg toxin/kg bodyweight	Pond water, human carrier (cause postharvest contamination)
Viruses		
Hepatitis A	Living virus can infect	Faecal polluted water,
Norwalk virus	humans	accumulation in shellfish
Algae		
dinoflagellates	E.g. ciguatoxins, PSP, ASP, DSP, NSP toxins	Open waters, marine tropical waters; accumulation in shellfish (e.g. mussels, oysters)
Parasites	Some living parasites can infect humans	Fish and shellfish

persons (e.g. botulism and vibriosis), but the number of people reported to be ill is very low (Wallace *et al.*, 1999; Feldhusen, 2000). Over a 15-year period 1980–1994 in New York only 37 people were hospitalized due to consumption of seafood with a confirmed pathogen; 13 of these illnesses were caused by *Salmonella*, 13 by scombrotoxin, four by *C. botulinum*, three by *Vibrio* and three by other agents. However, three deaths were reported due to *C. botulinum* and one due to *V. vulnificus* (Wallace *et al.*, 1999).

Bacteria from the general environment, for example, mesophilic *Clostridium botulinum* (types A and B) and *Listeria monocytogenes*, easily contaminate fish and shellfish. The prevalence of *L. monocytogenes* is high in fish products (Hartemink and Georgsson, 1991; Hudson *et al.*, 1992), especially in cold-smoked fish and other lightly preserved fish products (Ben Embarek, 1994; Jørgensen and Huss, 1998; Rørvik *et al.*, 1997; Farber, 1991). *L. monocytogenes* has been identified as the causative agent in three sporadic cases of seafood-borne listeriosis (Brett *et al.*, 1998; Ericsson *et al.*, 1997; Miettinen *et al.*, 1999) and has been suspected to be the causative agent in at least two other outbreaks involving seafood (Lennon *et al.*, 1984; Riedo *et al.*, 1990).

A few non-indigenous bacteria for example, *Salmonella* spp., *Shigella* spp. and *E. coli* are associated with faecal contamination of seafood (Table 5.2). Contaminations occurring from faecal polluted water and also errors in food handling such as cross-contamination or poor hygiene have been contributing factors in some outbreaks (Wallace *et al.*, 1999). Further, contamination with *Staphyloccocus aureus* may also occur during processing, in particular when manual handling is involved. This can be a risk in cooked products, such as shrimps and crustaceans, where staphylococci may grow when the normal spoilage flora is inactivated (Table 5.2).

One of the most common causes of seafood-borne disease from fresh fish and semi-processed fish products is scombroid (histamine) poisoning (Table 5.1) which is a chemical intoxication due to consumption of biogenic amines. The disease is similar to a mild hypersensitivity reaction and, typically, passes quickly. Amines such as histamine are produced by bacterial decarboxylation of amino acids such as histidine (Lehane and Olley, 2000). Scombroid fish, i.e., fish with high levels of free histidine, are most often implicated (e.g. tuna and mackerel), although non-scombroid fish may also be involved (e.g. mahi-mahi, sardines, herring and bluefish). Growth of histamine-producing bacteria is accelerated by temperature abuse, and outbreaks typically occur due to lack of, or improper, chill storage of fresh-caught fish on boats, in markets, in processing plants or in homes (Bryan, 1988). Also, seafood-related outbreaks typically occur more frequently in the summer months (Gillespie *et al.*, 2001). Most bacteria-producing biogenic amines belong to the family *Enterobactericea* or the genera *Clostridium* and *Lactobacillus* (Lehane and Olley, 2000). Sufficient levels of histamine may be produced without the product being sensorially unacceptable. Since histamine is heat stable subsequent cooking or processing of spoiled fish does not affect the histamine content significantly (Lehane and Olley, 2000). Biogenic amines may also be formed in lower amounts during chill storage by psychrotolerant bacteria such as *Photobacterium phosphoreum* or some lactobacilli (Jørgensen *et al.*, 2000).

Viruses are the most common illness associated with shellfish, and Hepatitis A is one of the most serious illnesses associated with shellfish (Rippey, 1994). A large outbreak of hepatitis A occurred in Shanghai with 292,301 cases and 32 deaths (Halliday, 1991). The virus was transmitted through raw clams contaminated with sewage water (only 3.5% with hepatitis A had cooked their clams before consumption). However, most illnesses are caused by gastro-intestinal viruses (Rippey, 1994; Wallace *et al.*, 1999; Jay, 1996) in which Norwalk or Norwalk-like viral agents generally are suspected as the etiologic agent. However, these viral pathogens are only rarely identified since methods for isolation, culturing and identifying them are limited or do not exist (Rippey, 1994; Wallace *et al.*, 1999). The faecal-oral route is the mode of transmission, and polluted water is an obvious source.

Bivalves may become toxic to man after feeding on toxigenic dinoflagellates (algae). All filter-feeding molluscs are able to accumulate toxin-producing algae and it is not possible by human intervention to avoid this process happening (Liston, 1990). However, public awareness, coastal engineering and classification of waters may help to protect the public against marine toxins (Wallace *et al.*, 1999). Marine toxins, such as ciguatera, may also accumulate in tropical reef fish and may cause very serious illness leading to death, and this depends on the specific biotoxin.

Several parasites found in fish are capable of infecting humans as non-traditional hosts. Transmission is generally due to consumption of raw or undercooked infected fish, and by cooking, freezing or curing in salt/acid this transmission can be avoided. The main microbial hazards associated with different seafood product groups are described by Huss (1995) and summarized in Table 5.3.

5.3   Traditional preservation strategies

Traditionally preservation technologies mostly act by inhibiting the growth of microorganisms in food rather than by inactivating the cells (Gould, 1996). Heat is the only traditional preservation technology that has a lethal effect on microorganisms although new emerging technologies with lethal effect on microorganisms are being introduced in the food industry (see Section 5.4). Traditional preservation of fish products relies heavily on chill and frozen storage, reduction in water activity by addition of NaCl, removal of oxygen and air by vacuum- and modified atmosphere packaging, heat treatment, smoking and addition of a few inorganic and organic preservatives (nitrite, sorbate, benzoate).

Chill storage of fish products is essential to prevent growth and toxin production of food-borne pathogens (Table 5.3). However, refrigerated temperatures between 4–8°C are not sufficient to prevent growth and/or production

Table 5.3 Risk categories of seafood products with different microbial ecology (modified from Huss, 1995)

Product	Preservation methods	General risk category	Main health hazards	
			agent	releasing factor
Raw or partially cooked shellfish	No or light heat treatment, refrigeration	High	Virus Biotoxins *Vibrio* spp. *Salmonella typhi*	Filter feeding nature Temperature abuse, Inefficient preservation stategy
Fresh/frozen fish	Refrigeration, MAP packing, frozen	Low[1]	Histamine producers Marine toxins Parasites *C. botulinum* and indigenous pathogens	Temperature abuse, uncooked before consumption
Lightly preserved fish product	< 6% NaCl (wps), refrigeration (4–8°C), (sorbate, benzoate, nitrite, smoke)	High	*L. monocytogenes* Indigenous pathogens	Poor GMP, ineffecient preservation Temperature abuse during storage
Semi-preserved fish	> 6% NaCl (wps) pH < 5; temp. <10°C (sorbate, benzoate, nitrate)	Low	*C. botulinum* Histamine producers	Poor raw material, NaCl < 10% wps, pH > 5 Poor raw material
Minimally processed seafood ('sous vide')	Mild cooking under vacuum Chill storage	High	*C. botulinum* *L. monocytogenes* *S. aureus* *E. coli,* *Salmonella* spp. *Vibrio* spp.	Insufficient to eliminate *C. botulinum* post-process contamination
Pasteurized fish (e.g. hot-smoked fish)	Brined or dry-salted, heat treatment (77.2–98.8°C/1 min.), chill storage	High	*C. botulinum* *L. monocytogenes* *S. aureus,* *E.coli,* *Salmonella* spp.	>5C, <3% NaCl wps Post-process contamination Temperature abuse Post-process contamination
Sterilized (canned) fish	Heat treatment	Low	Histamine-producing bacteria *C. botulinum*	Poor raw material Post-process contamination, underprocessing

Note
1. Scombroid (histamine) poisoning is very common (Table 5.1) but typically a very mild disease and the risk is therefore rated as 'low'.

of toxin of the psychrotrophic pathogens *Listeria monocytogenes* and *Clostridium botulinum*, and additional preservation methods may be needed to ensure the safety of refrigerated processed foods of extended durability (REPFED, Huss, 1997). In lightly preserved fish products, such as cold-smoked fish, addition of moderate NaCl levels (< 6% NaCl in the water phase), smoking and refrigeration are the only preservation parameters used to prevent the growth of pathogens (Table 5.3). Sorbate, benzoate and nitrite may be added to some lightly preserved fish products, for example, brined shrimps. However, the processing and preservation methods used for lightly preserved fish products are not sufficient to inactivate and prevent the growth of *L. monocytogenes* (no critical control point exists in the HACCP plan), which is halo- and psychrotolerant. Since these products are consumed without further cooking it may represent a health risk factor for especially susceptible persons.

5.4   New preservation strategies

Mild preservation strategies are becoming more important in modern food industries in response to consumer demands for high quality foods that are less heavily processed and preserved (e.g., 'sous vide' foods that are mildly heated and stored at chill temperature). As a consequence, traditional ways to control safety hazards in foods are replaced by milder preservation methods such as reduced use of NaCl and chemical additives in combination with refrigeration. Many of these 'ready-to eat' and novel food products represent new food systems with respect to health risks. Growth of food-borne, psychrotrophic pathogens such as *Listeria monocytogenes*, *Clostridium botulinum* and *Yersinia enterocolitica* is not prevented during refrigeration and a number of food-borne illnesses, caused by these psychrotrophic pathogens has questioned the safety of minimally processed foods such as lightly preserved fish products and emphasized the need for improving the existing preservation methods. A wide range of natural antimicrobial systems (Table 5.4) from microorganisms (e.g. bacteriocins), plants (antimicrobial compounds from herbs and spices) and animals (e.g. lysozyme, protamine, lactoperoxidase, lactoferrins) have been characterized and examined for their potential use in food preservation. Further several non-thermal food processing techniques (e.g. high hydrostatic pressure and electroporation) have been developed to meet consumers' demand for high quality foods. The major advantages of these techniques are that they cause only minimal damage to the flavour and texture, but have the same antimicrobial effect as pasteurization or sterilization, and result in limited use of additives (Gould, 1996; Gould, 2000). Also insight into the synergistic action of preservation technologies when used in combination may assist in the development of effective mild preservation strategies.

Table 5.4 New and emerging natural preservation technologies and the synergistic impact of novel and traditional preservation techniques

Technology	Example	Antimicrobial action and spectrum	Synergistic action	
			Technology	Reference
Bacteriocins	Nisin	Disrupt the cytoplasmic membrane	CO_2	Nilsson et al., 2000
			Low temperature	Nilsson et al., 2000
		Mainly Gram-positive bacteria but also some Gram-negative bacteria via chelating agents, hydrostatic pressure or injury which may cause increased susceptibility to the action of nisin		Rogers and Montville, 1994
			Low pH	Parente et al., 1998
			Chelating agent (e.g. EDTA)	Stevens et al., 1991, 1992 Kalchayanand et al., 1992
			EDTA + lysozyme	Gill and Holley, 2000
			Leucocin	Parente et al., 1998
			Pediocin	Hanlin et al., 1993
			Lysozyme	Monticello, 1989 Nattress et al., 2001 Chung and Hancock, 2000
			Lactate	Nykänen et al., 2000
Enzymes	Lysozyme	Degrades the bacterial cell wall	Low temperature	Johansen et al., 1994
			EDTA	Samuelson et al., 1985
		Mainly Gram positive bacteria	Osmotic shock	Carminati et al., 1985
			Transferrins	Johnson, 1989
	Glucose oxidase	Oxidizes glucose in the presence of O_2 and forms H_2O_2 Bacteria in general		
	Lacto-peroxidase	Oxidation of thiocyanate by H_2O_2 to form hypothiocyanate Bacteria in general	H_2O_2	Wolfson and Sumner, 1993
			Nisin	Boussouel et al., 1999
Iron chelators	Lactoferrin	Binding of iron Bacteria in general	High pressure	Masschalck et al., 2001

Table 5.4   Continued

Technology	Example	Antimicrobial action and spectrum	Synergistic action	
			Technology	Reference
Herbs and spices	Garlic	a) Fermentation of garlic to lactic acid b) Allicin, ajoene and isothiocyanates Many microorganisms		
Ultra high pressure (UHP)		Sublethal injury of cells	Nisin	Ter Steeg et al., 1999 Masschalck et al., 2001
		Vegetative cells	Lysozyme Nisin + lysozyme	Hauben et al., 1996 Masschalck et al., 2000 Hauben et al., 1996
			CO_2 + low temperature Acid	Amanatidou et al., 2000 Wuytack and Michiels, 2001
Pulsed high electric field (PHEF)	Pasteuriza- tion effect Vegetative cells			
Ultra- sonication		Pasteurization or sterilization effect at e.g. 10°C, when used in combination with slight overpressure. Spores and vegetative cells	Heat + slight overpressure ('mano- thermosonica- tion')	Sala et al., 1995

5.5   Biological preservation

In particular, there has been an interest in the use of biological preservation as a 'natural' preservation method to enhance the safety and/or extend the shelf-life of foods. Biological preservation includes the use of living microorganisms and/ or antimicrobial compounds of plants, animal or microbial origin without changing the sensory quality of the product. The antimicrobial action of lactic acid bacteria (LAB) and their very sporadic involvement in disease makes them attractive candidates in biological preservation of foods as shown in Table 5.5. In this chapter biological preservation will be discussed with focus on so-called protective cultures and/or their metabolites, notably bacteriocins.

Table 5.5 Mode of action, antimicrobial spectrum and sensory characteristics of antimicrobial compounds produced by lactic acid bacteria

Antimicrobial compound	Mode of action on sensitive cells	Antimicrobial spectrum	Sensory characteristics
Organic acids	Membrane disruption, inhibition of metabolic reactions, stress on intracellular pH homeostatis, accumulation of toxic anions	Bacterial and fungal cells	Acids, astringent, sour, pickly and off-flavours
H_2O_2	Oxidation of lipids, increased membrane permeability	Bacteria and fungal cells	Discoloring and rancidity
H_2O_2 in lactoperoxidase system	Oxidation of essential protein and enzyme sulfydryl groups	Bacteria and fungal cells	Discoloring and rancidity
Bacteriocins	Disruption of cell membrane, PMF and ion gradient	Close related strains which include several pathogenic bacteria (e.g. *L. monocytogenes*)	No
Reuterin or 3-hydroxypropion-aldehyde	Inhibition of sulfhydryl enzymes involved in DNA synthesis	Broad-spectrum	No
CO_2	Creating anaerobic conditions	Aerobic bacteria	Weak acid flavour
Diacetyl	Interferes with arginine utilization	Bacteria (mainly Gram-negative) and fungal cells	Butter and cheese flavour

5.5.1 Use of LAB as protective cultures

In biological preservation, microorganisms are applied for their antimicrobial action whereas microorganisms used as starter cultures for food fermentation primarily are added to bring about beneficial sensory changes of the product. Most work on biopreservation has focused on inhibition of food-borne pathogenic organisms but some studies have also evaluated biological preservation as a means of preventing spoilage (Einarsson and Lauzon 1995).

Many studies have focused on LAB as potential candidates for protective cultures in biopreservation because they

- are generally recognized as safe (GRAS) and are assumed not to pose any health risk to man
- are generally considered as 'food-grade' organisms
- produce several antimicrobial metabolites during growth
- have a good image among consumers as probiotic culture that may provide health benefits to man
- have been used for centuries in food fermentation and therefore their physiology and interaction in foods has been studied intensively.

LAB inhibit other microorganisms through a range of mechanisms (Table 5.5). Their ability to produce lactic acid and acetic acid are the most important inhibitory property in fermented foods. In non-fermented foods, their production of bacteriocins or competition for nutrients are important properties. Selection of LAB protective cultures for foods has mostly been based on the ability of the culture to produce bacteriocins. Bacteriocins are peptides or proteins with a static or lytic effect typically against closely related microorganisms. They do not affect the sensory quality of the food negatively and are often active against both spoilage and pathogenic bacteria. It is generally assumed that bacteriocins do not pose any health risk to humans (Cleveland *et al.*, 2001), since the peptide- or protein-like characteristics of bacteriocins are inactivated by enzymes in the stomach and gut, which prevent any antimicrobial action in the gut.

Several investigations have shown a successful use of bacteriocinogenic protective LAB in food systems, however, only a few have been conducted in fish products (Table 5.6). In fish products protective cultures have mostly been used to inhibit the growth of *L. monocytogenes* in lightly preserved fish products, since there is a clear need for development of improved preservation techniques for this product category (Huss *et al.*, 1995; Ben Embarek, 1994; Nilsson, 1999). Studies have been carried out to evaluate the potential application of *Carnobacterium piscicola* as bioprotective cultures to control the growth of *L. monocytogenes* in cold-smoked salmon (Nilsson *et al.*, 1999; Duffes *et al.*, 1999). The background is that these cultures are often isolated as the dominant species on cold-smoked salmon (Paludan-Müller *et al.*, 1998; Leroi *et al.*, 1996) and are known to exhibit antilisterial activity (Ahn and Stiles, 1990; Buchanan and Klawitter, 1992; Stoffels *et al.*, 1992; Nilsson *et al.*, 1999). Further, it has been shown that *C. piscicola* in some cases prolongs the shelf-life of the products (Leroi *et al.*, 1996). These characteristics make *C. piscicola* a promising candidate for biopreservation of cold-smoked salmon. Nilsson *et al.*, (1999) demonstrated a successful use of selected *Carnobacterium piscicola* strains (a bacteriocin-producing and non-bacteriocin-producing strain) to control the growth of *L. monocytogenes* in vacuum-packed cold-smoked salmon (Fig. 5.1). Duffes *et al.*, (1999) obtained a bacteriostatic or bacteriocidal antilisterial action in cold-smoked salmon when *C. piscicola* strains were used as protective cultures. However, growth of *L. monocytogenes* was not affected by a non-

Table 5.6 Control of undesirable bacteria in different fish products by the addition of bacteriocinogenic lactic acid bacteria

Specific food	Bacteriocinogenic LAB	Target organisms	Reference
Brined shrimps	*Lb. saké* LKE5	*L. monocytogenes*	Jeppesen and Huss, 1993
Iced shrimps	*Lb. sake* subsp. *sake*	*V. parahaemolyticus*	Moon *et al.*, 1982
Cold-smoked salmon	*Lb. lactis* subsp. *lactis*	*L. monocytogenes*	Wessels and Huss, 1996
	C. piscicola A9b	*L. monocytogenes*	Nilsson *et al.*, 1999
	C. piscicola V1 and *C. divergens* V41	*L. monocytogenes*	Duffes *et al.*, 1999
	C. piscicola	Spoilage flora	Leroi *et al.*, 1996
Fermented minced fish	*Ped. pentosaceus* and *Lb. plantarum*	*V. parahaemolyticus* *Cl. perfringens*	Aryanta *et al.*, 1991
Cod fillets	*Lb. plantarum*	Psychrotrophic bacteria	Rose *et al.*, 1988
Fresh crab meat	*Lb. delbrueckii* subsp. *bulgaricus*	Psychrotrophic bacteria	Gilliland and Speck, 1975

bacteriocin-producing strain of *C. piscicola*. Both studies showed that high inoculum levels of *C. piscicola* does not accelerate the spoilage process of cold-smoked salmon (Nilsson *et al.*, 1999; Duffes *et al.*, 1999). Further, no production of biogenic amines was detected except for low amounts of tyramine (Duffes *et al.*, 1999).

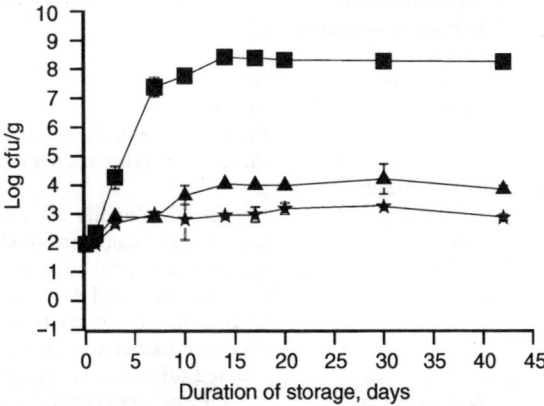

Fig. 5.1 Growth of *L. monocytogenes* on vacuum-packed cold-smoked salmon stored at 5°C. ■, *L. monocytogenes* alone, ▲, *L. monocytogenes* as a culture with *C. pisicola* A9b (bacteriocin-producing strain), and ★, *L. monocytogenes* as co-culture with *C. pisicola* A10a (non-bacteriocin producing strain). The error bars indicate ±standard deveiations of the means (after Nilsson *et al.*, 1999).

Bacteriocin-mediated inhibition of histamine formation has been reported in cheese by bacteriocin-producing lactic acid bacteria (Joosten and Nunez, 1996). To our knowledge similar investigations have never been done in fish products. However, the applicability of bacteriocin-producing protective cultures is still being questioned for several reasons. The antimicrobial efficiency of bacteriocins is negatively affected by several intrinsic food factors (Table 5.7; Stiles, 1996; Jung et al., 1992; Daeschel, 1993; Chung et al., 1989; Gänzle et al., 1999). Also, bacteriocin production may be reduced or inhibited under food-relevant conditions (Table 5.7; Nilsson et al., 2002; Himelbloom et al., 2001; Hugas et al., 1996; Leroy and de Vuyst, 1999). Further, the organisms to be inhibited may become resistant to the bacteriocins (Table 5.7; Crandall and Montville 1998; Harris et al., 1991; Ming and Daeschel, 1993; Mazzotta and Montville, 1997; Nilsson et al., 2000). Therefore, it could be of interest to focus more on non-bacteriocin-producing organisms as bioprotective cultures. It must also be emphasized that several studies have demonstrated antimicrobial properties of LAB that do not produce bacteriocins. Thus Nilsson et al. (1999) demonstrated that growth of Listeria monocytogenes in cold-smoked salmon

Table 5.7   Desirable characteristics of bacteriocinogenic protective culture and factors influencing these properties

Characteristic	Effect	Factors influencing inhibitory activity
Grow fast in the food environment	Limits presence of nutrients and thereby discourages colonization of undesirable microorganisms	The natural environment of the culture reflects the ability to grow in different foods. Dependent on several intrinsic food factors, e.g. presence of nutrients and additives, and storage temperature
Production of bacteriocin(s)	Enhances inhibition of specific undesirable microorganisms	pH NaCl Nitrite Temperature Presence of induction factors Presence of other bacteria
Production of stable bacteriocin activity	Avoids batch variation during storage	pH Change in solubility and charge of bacteriocins binding to food components (proteins and lipids) inactivation by other food additives and proteolytic enzymes Change in the sensitivity of target organism (development of resistance) Limited diffusion through the product
Production of reliable bacteriocin activity	Avoids batch variation	Parameters important for production, e.g. NaCl levels, temperature, pH Cell number of bacteriocin-producing strain, since production generally is cell-density dependent

could effectively be inhibited by a *Carnobacterium piscicola* that did not produce bacteriocins. This inhibition may involve depletion of essential nutrients for microbial growth (Buchanan and Bagi, 1997; Degnan *et al.*, 1992; Nilsson *et al.*, 1999; Lyver *et al.*, 1998; Leroi *et al.*, 1996). Degnan *et al.* (1992) suggested that the inhibitory effect of *Pediococcus acidilactici* against *L. monocytogenes* was caused by competitive antagonism. By investigating the interaction between the protective culture *C. piscicola* and *L. monocytogenes* it was suggested that inhibition of the pathogen was partially attributed to nutrient depletion (Buchanan and Bagi, 1997). Similar suggestions were made by Nilsson *et al.* (1999) when studying the interaction between another strain of *C. piscicola* and *L. monocytogenes*. In a study by Lyver *et al.* (1998) toxin production and growth of *Cl. botulinum* type E was inhibited due to competitive inhibition by *Bacillus* species in cooked surimi nuggets. It has also been shown that a competitive natural flora of fresh meat, which was dominated by Gram-negative bacteria, sensitized *L. monocytogenes* to acid when present in the same niche (Samelis *et al.*, 2001).

Protective cultures should be used only as an additional 'barrier' in a multicomponent food preservation system (Stiles, 1996). A combination of preservation treatments (hurdle-technology, Leistner, 1995) allows the required level of protection to be achieved while at the same time the intrinsic preservation of food is reduced due to an additional or synergistic effect on microbial growth.

5.5.2 Bacteriocins

Bacteriocins are defined as peptides or proteins which are produced and excreted by bacteria that exhibit bactericidal activity towards closely related species (Tagg *et al.*, 1976). Several bacteriocins have been characterized within the last decades and their potential use in biopreservation of foods including fish products (Table 5.8) has been investigated. According to Klaenhammer (1993) bacteriocins of LAB can be grouped into four major classes in which class I and II are most frequently produced: class I, lantibiotics, which contain lanthione rings that are produced post-translationally (e.g. nisin and subtilin); class II, small, heat-stable, non-lantibiotic peptides having an N-terminal YGNGV consensus sequence (e.g. pediocin and carnocin); class III, larger (> 30 kDa), heat-labile proteins (e.g. helveticin); and IV, complex bacteriocins which contain glyco- and/or lipid moieties in addition to protein (e.g. pediocin SJ-1). Today, nisin is the only bacteriocin approved as a food preservative in more than 50 countries. However, the EU directive on nisin has restricted the use of nisin in some dairy products (cheese, cream, puddings, milk), vegetable and canned foods. Nisin acts by forming pores in the cytoplasmic membrane of sensitive cells (pore size: 0.2–1.2 nm) and this induces the efflux of small hydrophilic compounds such as ATP, ADP, monovalent cations and amino acids (Abee *et al.*, 1994; Winkowski *et al.*, 1996). This results in dissipation of the membrane potential ($\Delta\Psi$) and ionic gradient (ΔpH) across the membrane and subsequently,

Table 5.8 Control of *Listeria monocytogenes* in different fish products by the addition of bacteriocins from lactic acid bacteria

Specific food	Bacteriocin	Reference
Cold-smoked salmon	Nisin	Nilsson *et al.*, 1997
	Divercin	Duffes *et al.*, 1999
	Carnocin	Duffes *et al.*, 1999
	Carnocin	Nilsson *et al.*, unpubl. data
Cold-pack lobster meat	Nisin	Budu-Amoako *et al.*, 1999
Blue crab	Nisin	Degnan *et al.*, 1994
Cold-smoked trout	Nisin	Nykänen *et al.*, 2000

destruction of the energy metabolism leading to cell death (Bruno *et al.*, 1992; Moll *et al.*, 1997). Apart from nisin, the most promising bacteriocins used for preservation purposes are those belonging to the class IIa (pediocin-like bacteriocins) due to their biological activity, which is generally higher than that of other bacteriocins and highly active against *Listeria* strains, and due also to their physicochemical properties (e.g. heat-stable) (Ennahar *et al.*, 2000). Also, those bacteriocins are produced by organisms for example, *Carnobacteria*, *Lactobacillus*, *Leuconostoc*, *Pediococcus* and *Enterococcus faecium* that belong to the natural LAB microflora of many foods. Class IIa bacteriocins act by increasing the permeability of the membrane of their target cells. The antimicrobial action is enhanced by binding of bacteriocin to a specific receptor in the membrane of the target cells although an antimicrobial effect is also obtained by non-specific binding (Chen *et al.*, 1997).

In general, bacteriocins of LAB exhibit strong antimicrobial activity against several food-borne pathogenic bacteria including *Listeria monocytogenes*, *Staphylococcus aureus* and *Clostridium botulinum*. The inhibitory spectrum of bacteriocins is generally restricted to Gram-positive bacteria. However, disruption of the outer membrane of Gram-negative bacteria (e.g. *Salmonella* and *E. coli*) via chelating agents (e.g. EDTA), hydrostatic pressure or injury may cause increased susceptibility to the action of bacteriocins (Stevens *et al.*, 1991 and 1992; Kalchayanand *et al.*, 1992). The antibacterial effect of bacteriocins, such as nisin, are known to act synergistically in combination with other food preservation techniques, for example, CO_2 packaging, low temperature, low pH, chelating agents, other bacteriocins like leucocin and pediocin, and lysozyme (Table 5.4). Preservation of cold-smoked salmon with a combination of nisin and CO_2 atmosphere packaging was found to be effective in controlling the growth of *L. monocytogenes* (Nilsson *et al.*, 1997). It was suggested that the synergistic action of CO_2 and nisin occurs at the cytoplasmic membrane (Nilsson *et al.*, 2000). CO_2 destabilized the cytoplasmic membrane of *L. monocytogenes* which was demonstrated by an increased efflux of carboxyfluorescein from *L. monocytogenes* derived liposomes in presence of CO_2. Moreover, CO_2 enhanced the nisin-induced CF leakage from *L. monocytogenes* derived liposomes (Nilsson *et al.*, 2000). In a study by Nykänen *et al.* (2000) a synergistic antilisterial action was demonstration between nisin and lactate on cold-smoked rainbow trout. It was concluded that by using a combination of lactate and nisin it may be possible to produce more safe cold-processed fish products with respect to *L. monocytogenes* and higher quality products (Nykänen *et al.*, 2000).

Bacteriocin produced by *C. piscicola* (class IIa bacteriocins) has been used to eliminate and prevent growth of *L. monocytogenes* in cold-smoked salmon (unpublished data by Nilsson; Duffes *et al.*, 1999). Crude extracts of bacteriocin produced by *C. piscicola* were bacteriocidal at 4°C and 8°C, and growth was delayed or inhibited at 8°C and 4°C, respectively (Duffes *et al.*, 1999). However, it was shown that bacteriocin produced by *C. piscicola* A9b cannot be used alone to prevent growth of *L. monocytogenes* in cold-smoked salmon juice, due to development of resistant *L. monocytogens* cells. However, when bacteriocin is used in combination with the producer strain the biopreservative technique successfully eliminated and subsequently prevented outgrowth of resistant *L. monocytogenes* cells (unpubl. data by Nilsson).

Bacteriocins do not work efficiently when used as 'stand-alone' preservatives in refrigerated foods that contain no other barriers to microbial growth (Montville *et al.*, 1995). However bacteriocins can provide synergism when used in a multiple barrier preservation system as shown in Table 5.4 and therefore add an extra degree of protection to foods.

5.5.3 Antimicrobial enzymes derived from animals

Several antimicrobial animal-derived enzymes have been isolated, for example, lysozyme, lactoperoxidase, lactoferrin and protamine.

Lysozyme is a naturally occurring lytic enzyme that is produced in many animals (found in milk and eggs) and man. The enzyme catalyzes the hydrolysis of peptidoglycan which is the major component of the cell wall of Gram-negative and particular Gram-positive bacteria. It is known to inhibit some Gram-positive bacteria, but since the outer membrane of Gram-negative bacteria prevents access to the underlying peptidoglycan it is ineffective against Gram-negative bacteria when used alone (Gould, 1996). However, the susceptibility of Gram-negative bacteria to lysozyme is enhanced by disruption of the outer membrane which may be obtained by, for example, detergents, chelators and high pressure. Lysozyme has been used commercially to prevent late blowing in semi-hard cheeses, which is caused primarily by *Clostridium tyroburicum* (Bester and Lombard, 1990; Cunningham *et al.*, 1991). Synergy between lysozyme and other antimicrobial systems, for example, nisin, EDTA, high pressure, osmotic shock, transferrins and low temperature have been demonstrated (Table 5.4). In a study by Gill and Holley (2000) it was demonstrated that the combined use of lysozyme, nisin and EDTA added to ham and bologna before cooking, restricted the growth of some spoilage (*Brochothrix thermospacta, Lb. curvatus* and *Leu. mesenteroides*) and some pathogenic bacteria (*E. coli* and *L. monocytogenes*). A synergistic action between lysozyme and nisin has been demonstrated in a few studies (Chung and Hancock, 2000; Nattress et al., 2001) for example, when used on fresh pork (Nattress et al., 2001) against spoilage bacteria. It was found that the more lysozyme added in the mixture, the longer the antimicrobial effects were efficacious (Nattress et al., 2001).

Lactoperoxidase and lactoferrin are widespread in nature (milk, tears, saliva and many other types of body secretions of mammals). These antimicrobial systems should not invoke toxicological concern due to their natural occurrence in foods (Yamauchi et al., 2000). Lactoperoxidase catalyses the oxidation of thiocyanate by H_2O_2 to form hypothiocyanate which interacts with bacterial membrane proteins (Kamau et al., 1990; Reiter and Harnulv, 1984). This three-component system has effectively been used in milk stored at abuse temperature, since lactoperoxidase and thiocyanate are generally present in sufficient amounts to obtain an antimicrobial action. However, hydrogen peroxide has to be added to activate the three-component system. Further preservation of several milk products, for example, cottage cheese, mozzarella and yoghurt have been suggested (Ekstrand, 1994). The lactoperoxidase system has shown bacteriocidal and bacteriostatic activity against several food-borne pathogenic bacteria, for example, Staphylococcus spp., E. coli, L. monocytogenes, Salmonella spp. and Bacillus cereus.

Lactoferrin has bacteriostatic properties against several pathogenic bacteria, for example, Salmonella spp., L. monocytogenes and E. coli, fungi, protozoa and viruses, which is due to its iron-binding capacity, resulting in iron depletion of the bacterial growth medium (Masschalck et al., 2001; Ellison, 1994; Van der Strate et al., 2001). Further, lactoferrin is believed to bind to the outer membrane of Gram-negative bacteria, due to its polycationic nature, resulting in destabilization of the membrane and additional releasing of lipopolysaccharides (Ellison, 1994). A synergistic interaction of lactoferrin under high pressure has been demonstrated on several pathogenic bacteria, for example, Salmonella typhimurium, E. coli and Staph. aureus (Masschalck et al., 2001).

Protamine is a cationic antimicrobial peptide, which can be extracted from the sperm cells of vertebrates including fish. Protamine has shown antimicrobial action against a wide range of Gram-positive and Gram-negative food-borne pathogenic and spoilage bacteria (Johansen et al., 1995; Truelstrup Hansen et al., 2001). Protamine acts by solubilizing the cell envelope to cause leakage of intracellular components (Johansen et al., 1997), however, there are no reports of its successful use in fish or other food products.

5.5.4   Antimicrobial compounds derived from plants (herbs and spices)
In addition to contributing flavour to foods many spices, herbs and essentials oils reduce the proliferation of microorganisms due to antifungal and antibacterial activity (Deans and Ritchie, 1987). However, the concentration needed to obtain an antimicrobial effect often exceeds sensory acceptable levels and consequently use is generally restricted to spicy products. In the western countries the addition

of spices to fish products is known in marinated fish products and 'gravad' (sugar-salted) fish. Recent studies have shown that the specific spoilage organism of CO_2 packed fish, *Photobacterium phosphoreum*, can be inhibited by oregano extract (Mejlholm and Dalgaard, 2002). The antimicrobial activity of oregano is due to carvacrol, which is the major component of the essential oil fraction of oregano and thyme (Ultee *et al.*, 1999).

In Asian and Southeast Asian countries spices such as garlic, ginger, chilli and pepper are commonly used in preparation of traditional lightly fermented fish products (Paludan-Müller *et al.*, 1999). In *som-fak*, which is a Thai fermented fish product composed of minced fish fillet, salt (2–5%) and ground boiled rice (2–12%), minced garlic is added in a final concentration of approximately 4% (Paludan-Müller *et al.*, 1999). Garlic is added as a flavouring agent, but may also act as an antimicrobial agent due to the inhibitory activity of garlic against particular Gram-negative bacteria (Feldberg *et al.*, 1988; Beuchat, 1994) or by stimulating the growth of LAB by garlic-fermentation (Nes and Skjelkvåle, 1982; Zaika and Kissinger, 1984; Paludan-Müller *et al.*, 1999). Garlic contains approximately 30% of fructo-oligosaccharides, mainly in the form of inulin (Van Loo *et al.*, 1995) which is fermented by several LAB (Paludan-Müller *et al.*, 1999). Allicin and its degradation products (e.g. ajoene) have been isolated as the major antimicrobial compound in garlic, which is known to inhibit growth of a wide range of food-borne pathogenic bacteria, for example, *Cl. botulinum*, *Salmonella* spp., *E. coli* and *Staph. aureus* (Cowan, 1999; Naganawa *et al.*, 1996), probably due to inactivation of sulphydryl containing enzymes (Feldberg *et al.*, 1988), inhibition of RNA synthesis (Feldberg *et al.*, 1988) or inhibition of lipid synthesis (Hughes and Lawson, 1991).

The antimicrobial effects of clove, cinnamon, cardamom, turmeric and pepper have been investigated on histamine production and histidine decarboxylase activity of *Morganella morganii* (a main histamine-producing bacteria in fish) in mackerel at a level of 3% (Shakila *et al.*, 1996). Clove and cinnamon showed a significant inhibitory effect on histamine and other biogenic amines (putrescine and tyramine) (Shakila *et al.*, 1996).

Mehiawah, which is a fish sauce product from the Middle East (Bahrain) pose only a few health risks for consumers, although the product may be stored at ambient temperature over many months (Gasaluck *et al.*, 1996; Ijong and Ohta, 1996). The inhibitory effect of the ingredients added during processing showed that a salt content (between 3.5–4.2%) in combination with spices (black pepper, coriander, cumin and fennel seeds), wheat and lemon eliminates the pathogens *E. coli*, *Salmonella typhi* and *Staph. aureus,* and *V. parahaemolyticus* was eliminated after 21 days of storage at 25°C (Al-Jedah *et al.*, 2000). In the control sample without addition of spices, wheat or lemon, the pathogens, except for *Staph. aureus* were still present after 28 days, with *E. coli* having reached approximately 2×10^7 cfu/ml (Al-Jedah *et al.*, 2000).

Microbial spoilage limits the shelf-life of salt cured fish, mainly due to red halophilic bacteria (Prasad and Seenayya, 2000; Prasad and Rao, 1995). The treatment of 20% natural solar salt brines with clove oil at 0.02 and 0.5% (v/v) concentrations, resulted in the complete elimination of *Halomonas* spp. and red halophilic bacteria after 30 s and 1 min. exposure, respectively (Prasad and Seenayya, 2000).

5.6   Use of lactic acid bacteria for food fermentation

Lactic acid bacteria (LAB) have traditionally been used as starter cultures in various fermented dairy, meat, vegetable and beverage products. Conversion of carbohydrates to organic acids, primarily lactic acid, has been the main reason for the use of LAB in food fermentation. Starter cultures are used to bring about beneficial sensory changes of the food and as a secondary effect the product is preserved due to the reduction of pH in the milieu. Antimicrobial metabolites other than organic acids may be produced during fermentation and these may enhance the antimicrobial action of LAB and thereby improve the safety and shelf life of the product.

LAB have not been used commercially to a great extent in the seafood industry, except for trials on silage fermentation for animal feed (Hammoumi et al.,1998; Dapkevicius *et al.*, 2000). Microbiological fermented fish products are common in South-East Asia (Adams et al.,1987) and the fermentation process depends on the natural microflora present in the processing environment and raw materials (Paludan-Müller *et al.*, 1999). LAB with antibacterial properties do occur in fermented fish products and it was suggested that the antibacterial activity of LAB may be essential to ensure the safety of fermented fish products (Østergaard *et al.*, 1998). In high-salted Thai-fermented fish products growth of LAB is inhibited (Paludan-Müller *et al.*, 2002) allowing growth of *Staphylococcus* spp. (Paludan-Müller *et al.*, 2002; Tanasupawat *et al.*, 1991). Further, a high prevalence of *L. monocytogenes* (12%) was also found in fermented fish from Malaysia (Hassan *et al.*, 2001).

5.7   Non-thermal food processing techniques

High hydrostatic pressure (HP), pulsed high electric field (PHEF or 'electroporation') and ultrasonication are promising food preservation techniques that prevent destruction of flavour, texture, colour and nutrient content as known for traditional pasteurization and sterilization techniques. HP processing has been used for preservation of several foods (Cheftel and Culioli, 1997) including salmon products (Ohshima *et al.*, 1993). Although the mechanism of inactivation by HP is not clear several products preserved by using this technique have been introduced to the Japanese market, for example, strawberry-, kiwi-, and apple-jam, fruit-juices and rice cakes (Spilimbergo *et al.*, 2002).

In vitro model studies have shown an enhanced antibacterial effect of combining HP and gases (Spilimbergo *et al.*, 2002). HP processing at low temperatures combined with modified atmosphere packaging (50% O_2 and 50% CO_2) was used to extend the shelf life and safety of fresh salmon by reduction of spoilage bacteria (LAB and *Shewanella putrefaciens*) and reduction of pathogens (*L. monocytogenes* and *S. typhimurium*) (Amanatidou *et al.*, 2000). The mode of action was probably due to intracellular formation of reactive oxygen species as well as to phase transition phenomena (Amanatidou *et al.*, 2000).

Enhanced inactivation of *Bacillus subtilis* spores was also provided by the combined use of HP in acid environment (Wuytack and Michiels, 2001). Treatment of *B. cereus* spores with nisin and/or pulsed-electric-field (PEF) did not result in inactivation of spores or increased heat sensitivity as a result of sublethal damage (Pol *et al.*, 2001). However germinating spores were sensitive to PEF treatment (Pol *et al.*, 2001).

It has been demonstrated that *E. coli* can develop resistance to high pressure inactivation by spontaneous mutation (Hauben *et al.*, 1996). However, application of hurdle technology has shown an enhanced inactivation of *E. coli* pressure-resistant mutants by high hydrostatic pressure in the presence of nisin and lysozyme (Masschalck *et al.*, 2000). In agreement with these data Hauben *et al.*, (1996) showed that high pressure disrupts the permeability of the *E. coli* outer membrane for water-soluble proteins like lysozyme and nisin.

Irradiation has been tested extensively over a 40-year period and the antimicrobial effects of ionizing radiation are well documented (Patterson and Loaharanu, 2000). Its use for specific purposes has been approved in more than 35 countries and it is extensively used as a decontamination procedure for spices. In total, use of irradiation has been approved for some types of fish products in 13 countries and low-dose irradiation (1.0-2.0 kGy) of shellfish may be an efficient way of removing *Vibrio* spp. (Molins *et al.*, 2001). Low doses of gamma irradiation, today referred to as cold-pasteurization, has been tested in several foods. As with other inactivation procedures, the killing kinetics of cold-pasteurization are affected by the parameters of the food product (dry matter, lipid, etc.) and procedures for, for example, inactivation of *Listeria monocytogenes* must be based on individual product formulation (Sommers and Thayer, 2000).

5.8   Conclusion and future trends

During the last decades, research into preservation principles of (micro)-biological origin has flourished. In particular, use of LAB and/or their bacteriocins is an area of immense interest. LAB are being used commercially today in fermentation of foods. Therefore selection and application of naturally occurring LAB as bioprotective cultures seems logical since it offers an additional 'barrier' to the preservation system and thereby minimizes the risk associated with the growth of undesirable pathogenic and/or spoilage bacteria. When relying on bacteriocinogenic LAB, it is necessary to understand the relationship between bacteriocin production and factors affecting production and efficiency since several intrinsic food factors negatively affect bacteriocin activity. The role of food-relevant factors on the efficiency of protective cultures may have been neglected and is an important area for future research. It has been pointed out that bacteriocinogenic LAB and bacteriocins should not be used alone in preservation of foods but should be used as part of a multicomponent food preservation system. Several studies have demonstrated that non-bacteriocinogenic LAB seem to be as effective in inhibiting growth as bacteriocin producing LAB. Use of such strains would overcome, for example, resistance problems and investigating the mechanisms of action and the robustness of the inhibition deserves further studies.

The combination of preservation strategies is attractive since it may reduce the intrinsic preservation of foods and further minimize problems with bacterial adaptation to stresses known as resistance. However little is known about the interactive effects of combining preservation technologies and therefore research in this area is needed to optimize such strategies.

It has been suggested that LAB can be used to create novel foods in the fish industry (Gelman et al., 2001; Glatman et al., 2000; Morzel et al.,1997). These products should be launched as safe, healthy products with all the nutritional advantages of fish (Glatman et al., 2000; Gelman et al., 2001). Overall, there is no doubt that LAB can and should be used as preserving agents, in particular against selected pathogenic organisms, in, for example, ready-to-eat seafood products.

5.9   References

ABEE T. F., ROMBOUTS M., HUGENHOLTZ J., GUIHARD G. and LETELLIER L., Mode of action of nisin Z against *Listeria monocytogenes* Scott A grown at high and low temperatures. *Appl. Environ. Microbiol.* 1994; 60:1962–8.

ADAMS M. R., COOKE R. D. and TWEEDY D. R., Fermentation parameters involved in the production of lactic acid preserved fishglucose substrates. *Int. J. Food Sci.* Technol. 1987; 22:105–14.

AHN C. and STILES M. E., Plasmid-associated bacteriocin production by a strain of *Carnobacterium piscicola* from meat. *Appl. Environ. Microbiol.* 1990; 56:2503–10.

AL-JEDAH J. H., ALI M. Z. and ROBINSON R. K., The inhibitory action of spices against pathogens that might be capable of growth in a fish sause (mehiawah) from the Middle east. *Int. J. Food Microbiol.* 2000; 57:129-133.

AMANATIDOU A., SCHLÜTER O., LEMKAU K., GORRIS L. G. M., SMID E. J. and KNORR D., Effect of combined application of high pressure treatment and modified atmospheres on the shelf life of fresh Atlantic salmon. *Innov. Food Sci. Emerg. Technol.* 2000; 1:87–98.

ARYANTA R. W., FLEET G. H. and BUCKLE K. A., The occurrence and growth of microorganisms during the fermentation of fish sausage. *Int. J. Food Microbiol.* 1991; 13:143–56.

BEN EMBAREK P. K., Presence, detection and growth of *Listeria monocytogenes* in seafoods: a review. *Int. J. Food Microbiol.* 1994; 23:17–34.

BESTER B. H. and LOMBARD S. H., Influence of lysozyme on selected bacteria associated with Gouda cheese. *J. Food Prot.* 1990; 53:306–11.

BEUCHAT L. R., Antimicrobial properties of spices and their essential oils. In V. M. Dillon and R. G. Board (eds.), *Natural antimicrobial systems and food preservation.* CAB International, Wallingford, Oxon, 1994.

BOUSOUEL N., MATHIEU F., BENOIT V., LINDER M., REVOL-JUNELLES A. M. and MILLIÉRE J. B., Response surface methodology, an approach to predict the effects of a lactoperoxidase system, nisin, alone or in combination, on *Listeria monocytogenes* in skim milk. *J. Appl. Microbiol.* 1999; 86:642–52.

BRETT M. S. Y., SHORT P. and MCLAUCHLIN J., A small outbreak of listeriosis associated with smoked mussels. *Int. J. Food Microbiol.* 1998; 43:223–9.

BRUNO M. E. C., KAISER A. and MONTVILLE T. J., Depletion of proton motive force by nisin in *Listeria monocytogenes* cells. *Appl. Environ. Microbiol.* 1992; 58:2255–9.

BRYAN F. L., Risks associated with vehicles of food-borne pathogens and toxins. *J. Food Prot.* 1988; 51:498–508.

BUCHANAN R. L. and BAGI L. K., Microbial competition: Effect of culture conditions on the suppression of *Listeria monocytogenes* Scott A by *Carnobacterium pis*cicola. *J. Food Prot.* 1997; 60:254–61.

BUCHANAN R. L. and KLAWITTER L. A., Characterization of a lactic acid bacterium, *Carnobacterium piscicola* LK5, with activity against *Listeria monocytogenes* at refrigeration temperatures. *J. Food Safety* 1992; 12:199–217.

BUDU-AMOAKO E., ABLETT R. F., HARRIS J. and DELVES-BROUGHTON J., Combined effect of nisin and moderate heat on destruction of *Listeria monocytogenes* in cold-pack lobster meat. *J. Food Prot.* 1999; 62:46–50.

CARMINATI D., NEVIANTI E. and MUCHETTI G., Activity of lysozyme on vegetative cells of *Clostridium tyrobutyricum. Latte* 1985; 10:194–8.

CHEFTEL J. C. and CULIOLI J., Effects of high pressure on meat: a review. *Meat Sci.* 1997; 46:211–36.

CHEN Y., SHAPIRA R., EISENSTEIN M. and MONTVILLE T. J., Functional characterization of pediocin PA-1 binding to liposomes in the absence of a protein receptor and its relationship to a predicted tertiary structure. *Appl. Environ. Microbiol.* 1997; 63:524–31.

CHUNG K. T., DICKSON J. S. and CROUSE J. D., Effect of nisin on growth of bacteria attached to meat. *Appl. Environ. Microbiol.* 1989; 55:1329–33.

CHUNG W. and HANCOCK R. E., Action of lysozyme and nisin mixtures against lactic acid bacteria. *Int. J. Food Microbiol.* 2000; 60: 25–32.

CLEVELAND J., MONTVILLE T. J., NES I. F. and CHIKINDAS M. L., Bacteriocins: safe, natural antimicrobials for food preservation. *Int. J. Food Microbiol.* 2001;

71:1-20.

COWAN M. M., Plant products as antimicrobial agents. *Clin. Microbiol. Rev.* 1999; 12:564–82.

CRANDALL A. D. and MONTVILLE T. J., Nisin resistance in *Listeria monocytogenes* ATCC 700302 is a complex phenotype. *Appl. Environ. Microbiol.* 1998; 64:231–7.

CUNNINGHAM, F. E., PROCTOR V. A. and GOETSCH S. J., Eggwhite lysozyme as a food preservative: an overview. *World's Poult. Sci. J.* 1991; 47:141–63.

DAPKEVICIUS M. L. N., NOUT M. J. R., ROMBOUTS F. M., HOUBEN J. H. and WYMENGA W., Biogenic amine formation and degradation by potential fish silage starter microorganisms. *Int. J. Food Microbiol.* 2000; 57:107–14.

DAESCHEL M. A., Application and interactions of bacteriocins from lactic acid bacteria in foods and beverages. In Hoover, D. G. and L. R. Steenson (eds), *Bacteriocins of lactic acid bacteria.* Academic Press Inc. San Diego, pp. 63–91, 1993.

DEANS S. G. and RITCHIE G., Antibacterial properties of plant essential oils. *Int. J. Food Microbiol.* 1987; 5:165–80.

DEGNAN A. J., KASPAR C. W., OTWELL W. S., TAMPLIN M. L. and LUCHANSKI J. B., Evaluation of lactic acid bacterium fermentation products and food-grade chemicals to control *Listeria monocytogenes* in blue crab (*Callinectes sapidus*) meat. *Appl. Environ. Microbiol.* 1994; 60:3198–203.

DEGNAN A. J., YOUSEF A. E. and LUCHANSKY J. B., Use of *Pediococcus acidilactici* to control *Listeria monocytogenes* in temperature-abused vacuum packaged wieners. *J. Food Prot.* 1992; 55:98–103.

DEPAOLA A., CAPERS G. M. and ALEXANDER D., Densities of *Vibrio vulnificus* in the intestines of fish from the U.S. Gulf Coast. *Appl. Environ. Microbiol.* 1994; 60:984–8.

DUFFES F., CORRE C., LEROI F., DOUSSET X. and BOYAVAL P., Inhibition of *Listeria monocytogenes* by in situ produced and semipurified bacteriocins on *Carnobacterium* spp. on vacuum-packed, refrigerated cold-smoked salmon. *J. Food Prot.* 1999; 62:1394–1403.

EINARSSON H. and LAUZON H. L., Biopreservation of brined shrimp (*Pandalus borealis*) by bacteriocins from lactic acid bacteria. *Appl. Environ. Microbiol.* 1995; 61:669–76.

EKSTRAND B., Lactoperoidase and lactoferrin. In V. M. Dillon and R. G. Board (eds.), *Natural antimicrobial systems and food preservation.* Biddles Ltd. Guildford, pp. 15–63, 1994.

ELLISON R. T., The effects of lactoferrin on Gram-negative bacteria. In T. W. Hutchens, B. Lönnerdal and S. Rumball (eds), Lactoferrin – Structure and Function. Plenum Press, New York, pp. 71–87, 1994.

ENNAHAR S., SASHIHARA T., SONOMOTO K. and ISHIZAKI A., Class IIa bacteriocins: biosynthesis, structure and activity. *FEMS Microbiol. Rev.* 2000; 24:85–106.

ERICSSON H., EKLOW A., DANIELSSON-THAM M. L., LONCAREVIC S., MENTZING L. O., PERSSON I., UNNERSTAD H. and THAM W., An outbreak of listeriosis suspected to have been caused by rainbow trout. *J. Clin. Microbiol.* 1997; 35:2904–7.

FARBER J. M., *Listeria monocytogenes* in fish products. *J. Food Prot.* 1991; 54:922–4, 934.

FELDBERG R. S., CHANG S. C., KOTIK A. N., NADLER M., NEUWIRTH Z., SUNDSTROM D. C. and THOMPSON N. H., In vitro mechanisms of inhibition of bacterial cell growth by allicin. *Antimicrob. Agents Chemother.* 1988, 32:1763–8.

FELDHUSEN F., The role of seafood in bacterial food-borne diseases. *Microbes and Infect.* 2000; 2:1651–60.

FOOD AND DRUG ADMINISTRATION, Proposal to establish procedures for the safe processing and importing of fish and fishery products; proposed rules. *Federal Register* 1994; 59:4142–214.

GÄNZLE M. G., WEBER S. and HAMMES W. P., Effect of ecological factors on the inhibitory spectrum and activity of bacteriocins. *Int. J. Food Microbiol.* 1999; 46:207–17.

GASALUCK P., YOKOYAMA K., KIMURA T. and SUGHARA I., Some chemical and microbiological properties of Thai fish sauce and paste. *J. Antibact. Antifung. Agents* 1996; 24:385–90.

GELMAN A., DRABKIN V. and GLATMAN L., Evaluation of lactic acid bacteria, isolated from lightly preserved fish products, as starter cultures for new fish-based food products. *Innovative Food Sci. Emerging Technol.* 2001; 1:219–26.

GILL A. O. and HOLLEY R. A., Inhibition of bacterial growth on ham and bologna by lysozyme, nisin and EDTA. *Food Res. Int.* 2000; 33:83–90.

GILLESPIE I. A., ADAK G. K., ÓBRIAN S. J., BRETT M. M. and BOLTON F. J., General outbreaks of infectious intestinal disease associated with fish and shellfish, England and Wales, 1992–1999. *Comm. Disease Publ. Health*, 2001; 4:117–23.

GILLILAND S. E. and SPECK M. L., Inhibition of psychrotrophic bacteria by lactobacilli and pediococci in non-fermented refrigerated foods. *J. Food Sci.* 1975; 40:903–5.

GLATMAN L., DRABKIN V. and GELMAN A., Using lactic acid bacteria for developing novel fish food products. *J. Sci. Food Agric.* 2000; 80:375–80.

GLATZER M. B., Shellfish-borne disease outbreaks in the US, 1992–1998. Internal technical report. U. S. Food and Drug Administration. South-east Regional Office, Atlanta, GA., 1998.

GOULD G. W., Industry perspectives on the use of natural antimicrobials and inhibitors for food applications. *J. Food Prot.* Supp. 1996; 1:82–6.

GOULD G. W., Preservation: past, present and future. *British Med. Bulletin*, 2000; 56:84–96.

GRAM L. and HUSS H. H., Fresh and processed fish and shellfish. Irradiation. In Lund, B.M., T.C. Baird-Parker and G.W. Gould (eds), *The Microbiological Safety and Quality of Foods*. Aspen Publishers. pp. 472–506, 2000.

HALLIDAY M. L., An epidemic of hepatitis A attributed to the ingestion of raw clams in Shanghai, China. *J. Infect. Dis.* 1991; 164:852–9.

HAMMOUMI A., FAID M., EL YACJIOUI M. and AMAROUCH H., Characterization of fermented fish waste used in feeding trials with boilers. *Process Biochemistry.* 1998; 33:423–7.

HANLIN M. B., KALCHAYANAND N., RAY P. and RAY B., Bacteriocins of lactic acid bacteria in combination have greater antibacterial activity. *J. Food Prot.* 1993; 56:252–5.

HARRIS L. J., FLEMMING H. P. and KLAENHAMMER T. R., Sensitivity and resistance of *Listeria monocytogenes* ATCC 19115, Scott A and UAL500 to nisin. *J. Food Prot.* 1991; 54:836–40.

HARTEMINK R. and GEORGSSON F., Incidence of *Listeria* species in seafood and seafood salads. *Int. J. Food Microbiol.* 1991; 12:189–96.

HASSAN Z., PURWATI E., RADU S., RAHIM R. A. and RUSUL G., Prevalence of *Listeria* spp. and *Listeria monocytogenes* in meat and fermented fish in Malaysia. *The Southest Asian J. Trop. Med. Publ. Health.* 2001; 32:402–7.

HAUBEN K. J. A., WUYTACK E. Y., SOONTJENS C. C. F. and MICHIELS C. W., High-pressure transient sensitization of *Escherichia coli* to lysozyme and nisin by disruption of outer-membrane permeability. *J. Food Prot.* 1996; 59:350–5.

HIMELBLOOM B., NILSSON L. and GRAM L., Factors affecting production of an antilisterial bacteriocin by *Carnobacterium piscicola* A9b in laboratory media and model fish systems. *J. Appl. Microbiol.* 2001; 91:506–13.

HUDSON J. A., MOTT S. J., DELACY K. M. and EDRIDGE A. L., Incidence and coincidence of *Listeria* spp., motile aeromonads and *Yersinia enterocolitica* on ready-to-eat fleshfoods. *Int. J. Food Microbiol.* 1992; 16:99–108.

HUGAS M., NEUMAYER B., PAGES F., GARRIGA M. and HAMMES W. P., Antimicrobial activity of bacteriocin-producing cultures in meat products. *Fleischwirtschaft* 1996; 76:649–52.

HUGHES B. G. and LAWSON L. D., Antimicrobial effects of *Allium, sativum* L. (garlic), *Allium ampeloprasum* L. (elephant garlic) and *Allium cepa* L. (onion), garlic compounds and commercial garlic supplement products. *Phytother. Res.* 1991; 5:154–8.

HUSS H. H., Quality Assurance in the Fish Industry. FAO Fish Tech. Pap. No. 334., Food and Agricultural Organization, Rome, Italy, 1995.

HUSS H. H., Control of indigenous pathogenic bacteria in seafood. *Food Control.* 1997; 8:91–8.

HUSS H. H., BEN EMBAREK P. K. and FROM JEPPESEN V., Control of biological hazards in cold smoked salmon production. *Food Control* 1995; 6:335–40.

IJONG F. G. and OHTA Y., Physicochemical and microbiological changes associated with bakasang processing. *J. Sci. Food Agri.* 1996; 71:69–74.

JAY J. M., Viruses and other food-borne biohazards. In *Modern food microbiology* (ed. Jay, J. M.), 5th edition. Chapman Hall, New York, pp. 612–626, 1996.

JEPPESEN V. F. and HUSS H. H., Antagonistic activity of two strains of lactic acid bacteria against *Listeria monocytogenes* and *Yersinia enterocolitica* in a model fish product at 5C. *Int. J. Food Microbiol.* 1993; 19:179–86.

JOHANSEN C., GILL T. and GRAM L., Antibacterial effect of protamine assayed by impedimetry. *J. Appl. Bacteriol.* 1995; 78:297–303

JOHANSEN C., VERHEUL A., GRAM L., GILL T. and ABEE T., Protamine-induced permeabilization of cell envelopes of Gram-positive and Gram-negative bacteria. *Appl. Environ. Microbiol.* 1997; 63:1155–9

JOHNSON E. A., The potential application of antimicrobial proteins in food preservation. *J. Dairy Sci.* 1989; 72:123–4.

JOOSTEN H. M. L. J. and NUNEZ M., Prevention of histamine formation in cheese by bacteriocin-producing lactic acid bacteria. *Appl. Environ. Microbiol.* 1996; 62:1178–81.

JØRGENSEN L. V. and HUSS H. H., Short communication. Prevalence and growth of *Listeria monocytogenes* in naturally contaminated seafood. *Int. J. Food Microbiol.* 1998; 42:127–31.

JØRGENSEN L. V., HUSS H. H. and DALGAARD P., The effect of biogenic amine production by single bacterial cultures and metabiosis on cold-smoked salmon. J. Appl. Microbiol. 2000; 89:920–34.

JUNG D. S., BODYFELT F. W. and DAESCHEL M. A., Influence of fat and emulsifiers on the efficacy of nisin in inhibiting *Listeria monocytogenes* in fluid milk. *J. Dairy Sci.* 1992; 75:387–93.

KALCHAYANAND N., HANLIN M. B. and RAY B., Sublethal injury makes Gram-negative and resistant Gram-positive bacteria sensitive to the bacteriocins, pediocin AcH and nisin. *Lett. Appl. Microbiol.* 1992; 15:239–43.

KAMAU D. N., DOORES S. and PRUITT K. M., Antibacterial activity of the lacto-peroxidase system against *Listeria monocytogenes* and *Staphylococcus aureus* in milk. *J. Food Prot.* 1990; 53:1010–14.

KLAENHAMMER T. R., Genetics of bacteriocins produced by lactic acid bacteria. *FEMS Microbiol.* Rev. 1993; 12:39–86.

LEHANE L. and OLLEY J., Histamine fish poisoning revisited. *Int. J. Food Microbiol.* 2000; 58:1–37.

LEISTNER L., Principles and applications of hurdle technolog, ch. 1. In G. W. Gould (ed.), *New methods of food preservation.* Blackie Academic and Professional, Glasgow, 1995.

LENNON D., LEWIS B., MANTELL C., BECROFT D., DOVE B., FARMER K., TONKIN S.,YEATES N., STAMP R. and MICKLESON K., Epidemic perinatal listeriosis. *Pediatr. Infect. Dis.* 1984; 3:30–4.

LEROI F., ARBEY N., JOFFRAUD J. J. and CHEVALIER F., Effect of inoculation with lactic acid bacteria on extending the shelf-life of vacuum-packed cold-smoked salmon. *Int. J. Food Sci. Tech.* 1996; 31:497–504.

LEROY F. and DE VUYST L., The presence of salt and a curing agent reduces bacteriocin production by *Lactobacillus sakei* CTC 494, a potential starter culture for sausage fermentation. *Appl. Environ. Microbiol.* 1999; 65:5350–6.

LIPP E. K. and ROSE J. B., The role of seafood in foodborne diseases in the United States of America. *Revue Scientifique et Technique* (International Office of Epizootics) 1997; 16:620–40.

LISTON J., Microbial hazards of seafood consumption. *Food Technol.* 1990; 44:56–62.

LYVER A., SMITH J. P., AUSTIN J. and BLANCHFILED B., Competitive inhibition of *Clostridium botulinum* type E by *Bacillus* species in a value-added seafood product packaged under a modified atmosphere. *Food Res. Int.* 1998; 31:311–19.

MASSCHALK B., GARCÍA-GRAELLS C., VAN HAVER E. and MICHIELS C. W., Inactivation of high pressure resistant *Escherichia coli* by lysozyme and nisin under high pressure. *Innov. Food Sci. Emerg. Technol.* 2000; 1:39–47.

MASSCHALK B., VAN HOUDT R. and MICHIELS C. W., High pressure increases bacteriocidal activity and spectrum of lactoferrin, lactoferricin and nisin. *Int. J. Food Microbiol.* 2001; 64:325–32.

MAZOTTA A. S. and MONTVILLE T. J., Nisin induces changes in membrane fatty acid composition of *Listeria monocytogens* nisin-resistant strains at 10°C and 30°C. *J. Appl. Microbiol.* 1997; 82:32.38.

MEJLHOLM O. and DALGAARD P., Antimicrobial effect of essential oils on the seafood spoilage microorganism Photobacterium phosphoreum in liquid media and in fish products. *Lett. Appl. Microbiol.* 2002; 34:27–31.

METEALF T. G., MULLIN B., ECKERSON D. and LARKIN E. P., Bioaccumulation and depuration of enteroviruses by soft-shelled clam, *Mya arenaria. Appl. Environ. Microbiol.* 1979; 38:275–82.

MIETTINEN M. K., SIITONEN A., HEISKANAN P., HAAJANEN H., BJÖRKROTH K. J. and KORKEALA H. J., Molecular epidemiology of an outbreak of febrile gastroenteritis caused by *Listeria monocyotgenes* in cold-smoked rainbow trout. J. Clin. Microbiol. 1999; 37:2358–60.

MING X. and DAESCHEL M. A., Nisin resistance of food-borne bacteria and the specific resistance responses of *Listeria monocytogenes* Scott A. *J. Food Protect.* 1993; 56:944–48.

MOLINS R. A., MOTARJEMI Y. and KÄFERSTEIN F.K., Irradiation: a critical control point in ensuring the microbiological safety of raw foods. *Food Control.* 2001; 12:347–56.

MOLL G. N., CLARK J., CHAN W. C., BYCROFT B. W., ROBERTS G. C. K., KONINGS W. N. and DRIESSEN A. J. M., Role of transmembrane pH gradient and membrane binding in nisin pore formation. *J. Bacteriol.* 1997; 179:135–40.

MONTICELLO D. J., Control of microbial growth with nisin/lysozyme formulations. *Eur. Pat. Appl.* 89123445.2., 1989.

MONTVILLE T. J., WINKOWSKI K. and LUDESCHER R. D., Models and mechanisms for bacteriocin action and application. *Int. Dairy J.* 1995; 5:797–814.

MOON N. J., BEUCHAT L. R., KINKAID D. T. and HAYS E. R., Evaluation of lactic acid bacteria for extending the shelf life of shrimp. *J. Food Sci.* 1982; 47:897–900.

MORZEL M., FRANSEN N. G. and ARENDT E. K., Defined starter cultures used for fermentation of salmon fillets. *J. Food Sci.* 1997; 62:1214–17, 1230.

MOTES M. L., DE PAULA A., COOK D. W., VEAZEY J. E., HUNSUCKER J. C., GARTHRIGHT W. E., BLODGETT R. J. and CHIRTEL S. J., Influence of water temperature and salinity on *Vibrio vulnificus* in Northern Gulf and Atlantic coast oysters (*Crassostrea virginica*). *Appl. Environ. Microbiol.* 1998; 64:1459–65.

NAGANAWA R., IWATA N., ISHIKAWA K., FUKUDA H., FUJINO T. and SUZUKI A., Inhibition of microbial growth by ajoene, a sulfur-containing compound derived from garlic. *Appl. Environ. Microbiol.* 1996, 62:4238–42.

NATTRESS F. M., YOST C. K. and BAKER L. P., Evaluation of the ability of lysozyme and nisin to control meat spoilage bacteria. *Int. J. Food Microbiol.* 2001; 70:111–19.

NES I. F. and SKJELKVÅLE R., Effect of natural spices and oleoresins on *Lactobacillus plantarum* in the fermentation of dry sausage. *J. Food Sci.* 1982; 47:1618–25.

NILSSON L. Control of *Listeria monocytogenes* in cold-smoked salmon by biopreservation. Ph. D. thesis, Danish Institute for Fisheries Research, Department of seafood Research, Lyngby, Denmark, 1999.

NILSSON L., CHEN Y., CHIKINDAS M. L., HUSS H. H., GRAM L. and MONTVILLE T. J., Carbon dioxide and nisin act synergistically on *Listeria monocytogenes*. *Appl. Environ. Microbiol.* 2000; 66:769–74.

NILSSON L., GRAM L. and HUSS H. H., Growth control of *Listeria monocytogenes* on cold-smoked salmon using a competitive lactic acid bacteria flora. *J. Food Prot.* 1999; 62:336–42.

NILSSON L., HUSS H. H. and GRAM L., Inhibition of *Listeria monocytogenes* on cold-smoked salmon by nisin and carbon dioxide atmosphere. *Int. J. Food Microbiol.* 1997; 38:217–27

NILSSON L., NIELSEN M. K., NG Y. and GRAM L., Role of acetate in production of an autoinducible class IIa bacteriocin in *Carnobacterium piscicola* A9b. *Appl. Environ. Microbiol.* 2002; 68:2251–2260.

NYKÄNEN A., WECKMAN K. and LAPVETELÄINEN A., Synergistic inhibition of *Listeria monocytogenes* on rainbow trout by nisin and sodium lactate. *Int. J. Food Microbiol.* 2000; 61:63–72.

OHSHIMA T., USHIO H. and KOIZUMI C., High pressure processing of fish and fish products. *Trends Food Sci. Technol.* 1993; 4:370–5.

OLSEN S. J., MACKINNON L. C., GOULDING J. S., BEAN N. H. and SLUTSKER L., Surveillance for food-borne-disease outbreaks United States, 1993-1997. *Morb. Mort. Weekly Rep.* 2000; 49:1–64

ØSTERGAARD A., EMBAREK P. K. B., WEDELL-NEERGAARD C., HUSS H. H. and GRAM L., Characterization of anti-listerial lactic acid bacteria isolated from Thai fermented fish products. *Food Microbiol.* 1998; 15:223–33.

PALUDAN-MÜLLER C., DALGAARD P., HUSS H. H. and GRAM L., Evaluation of the role of *Carnobacterium piscicola* in spoilage of vacuum- and modified atmosphere-packed cold-smoked salmon stored at 5C. *Int. J. Food Microbiol.* 1998; 39:155–66.

PALUDAN-MÜLLER C., HUSS H. H. and GRAM L., Characterization of lactic acid bacteria isolated from a Thai low-salt fermented fish product and the role of garlic as substrate for fermentation. *Int. J. Food Microbiol.* 1999; 46:219–29.

PALUDAN-MÜLLER C., MADSEN M., SOPHANADORA P., GRAM L., and MÖLLER P. L., Fermentation and microflora of *plaa-som*, a Thai fermented fish product prepared with different salt concentrations. *Int. J. Food Microbiol.* 2002; 73(1):61–70.

PARENTE E., GIGLIO M. A., RICCIARDI A. and CLEMENTI F., The combined effect of nisin, leucocin F10, pH, NaCl and EDTA on the survival of *Listeria monocytogenes* in broth. *Int. J. Food Microbiol.* 1998 40:65–75.

PATTERSON M. F. and LOAHARANU P., Irradiation. In Lund, B.M., T.C. Baird-Parker and G.W. Gould (eds), *The Microbiological Safety and Quality of Foods.* Aspen Publishers. pp. 65–88, 2000.

POL I. E., VAN ARENDONK W. G. C., MASYWIJK H. C., KROMMER J., SMID E. J. and MOEZELAAR R., Sensitivities of germinating spores and carvacrol-adapted vegetative cells and spores of *Bacillus cereus* to nisin and pulsed-electric-field treatment. *Appl. Environ. Microbiol.* 2001; 67:1693–9.

PRASAD M. M. and RAO C. C. P. Storage studies on dry salt cured fish with special reference to red discolouration. *Fish Technology* 1995; 31: 162–6.

PRASAD M. M. and SEENAYYA G., Effect of spices on the growth of red halophilic cocci isolated from salt cured fish and solar salt. *Food Research Int.* 2000; 33:793–98.

REILLY A. and KÄFERSTEIN F., Food safety hazards and the application of the principles of hazards analysis and critical control point (HACCP) system for their control in aquaculture production. *Aquac. Res.* 1997; 28:735–52.

REITER B. and HARNULV G., Lactoperoxidase natural antimicrobial system: natural occurrence, biological functions and practical applications. *J. Food Prot.* 1984; 47:724–32.

RIEDO F. X., PINNER R. W., TASCA M., CARTTER M. L., GRAVES L. M., RREAVES M. W., PLIKAYTIS B. D. and BROOME C. V., A point source food-borne listeriosis outbreak: documentated incubation period and possible mild illness. Abstract 972 in Program Abstract, 30th Int. Conf. Antimicrobial Agents and Chemotherapy, Atlanta, GA, p. 248, 1990.

RIPPEY S. R., Infectious diseases associated with molluscan shellfish consumption. *Clin. Microbiol. Rev.* 1994; 7:419–25.

ROGERS A. M. and MONTVILLE T. J., Quantification of factors which influence nisin's inhibition of *Clostridium botulinum* 56A in a model food system. *J. Food Sci.* 1994; 59:663–8.

ROSE S. M., BARTLETT F. M. and BLIGH E. G., The effect of lactic acid starter cultures on fish spoilage bacteria. *Proceedings of the 13th Annual Conference of the Tropical and Subtropical Fisheries Technological Society of the Americas*, Florida sea Grant program, 9–11 Nov. 1987, Orlando, Florida, USA, pp. 596–603, 1988.

RUPLE A. D. and COOK D. W., *Vibrio vulnificus* and indicator bacteria in shellstock and commercially processed oysters from the Gulf coast. *J. Food Prot.* 1992; 55:667–71.

RØRVIK L. M., KJERVE E., KNUDSEN B. R. and YNDESTAD M., Risk factors contamination of smoked salmon with *Listeria monocytogenes* during processing. *Int. J. Food Microbiol.* 1997; 37:215–19.

SALA F. J., BURGOS J., CONDON S., LOPEZ P. and RASO J., Effect of heat and ultrasound on microorganisms and enzymes. In G. W. Gould (editor), *New methods of Food Preservation*, Blackie Academic and Professional, Glasgow, pp. 176–204, 1995.

SAMELIS J., SOFOS J. N., KENDALL P. A. and SMITH G. C., Influence of the natural microbial flora on the acid tolerance response of *Listeria monocytogenes* in a model system of fresh meat decontamination fluids. *Appl. Environ. Microbiol.* 2001; 67:2410–20.

SAMUELSON K. J., RUPNOW J. H. and FRONING G. W., The effect of lysozyme and ethylenediaminetetraacetic acid on *Salmonella* on broiler parts. *Poultry Sci.* 1985; 64:1488–90.

SHAKILA R. J., VASUNDHARA T. S. and RAO D. V., Inhibitory effect of spices on *in vitro* histamine production and histidine decarboxylase activity of *Morganella morganii* and on the biogenic amine formation in mackerel stored at 30 degrees C. *Zeitschrift für Lebensmittel-Untersuchung und Forschung*, 1996; 203:71–6.

SHIEH Y., MONROE S. S., FANKHAUSER R. L., LANGLOIS G. W., BURKHARDT W. III and BARIC R. S., Detection of norwalk-like virus in shellfish implicated in illness. *J. Infect. Dis.* suppl. 2000; 181:360–66.

SOMMERS C. H. and THAYER D. W., Survival of surface-inoculated *Listeria monocytogenes* on commerically available frankfurters following gamma irradiation. *J. Food Safety.* 2000; 20:127–37.

SPILIMBURGO S., ELVASORE N. and BERTUCCO A., Microbial inactivation by high-pressure. *J. Supercri. Fluids.* 2002; 22:55–63.

STEVENS K. A., KLAPES N. A., SHELDON B. W. and KLAENHAMMER T. R., Antimicrobial action of nisin against *Salmonella typhimurium* lipopolysaccharide mutants. *Appl. Environ. Microbiol.* 1992; 58:1786–8.

STEVENS K. A., SHELDON B. W., KLAPES N. A. and KLAENHAMMER T. R., Nisin treatment for inactivation of *Salmonella* species and other Gram-negative bacteria. *Appl. Environ. Microbiol.* 1991; 57:3613–15.

STILES M. E., *Biopreservation by lactic acid bacteria*. Antonie van Leeuwenhoek. 1996; 70:331–45.

STOFFELS G., NES I. F. and GUÐMUNDSDÓTTIR Á., Isolation and properties of a bacteriocin-producing *Carnobacterium piscicola* isolated from fish. *J. Appl. Bacteriol. 1992; 73:309–16.*

TAGG J. R., DAJANI A. S. and WANNAMAKER L. W., Bacteriocins of Gram-positive bacteria. *Bacteriol. Rev.* 1976; 40:722–56.

TANASUPAWAT S., HASHIMOTO Y., EZAKI T., KOZAKI M. and KOMAGATA K., Identification of *Staphylococcus carnosus* strains from fermented fish and soy sauce mash. *J. Gen. Microbiol.* 1991; 37:479–94.

TER STEEG P. F., HELLEMONS J. C. and KOK A. E., Synergistic actions of nisin, sublethal ultrahigh pressure, and reduced temperature on bacteria and yeast. *Appl. Microbiol.* 1999; 65:4148–54.

TODD E. C. D., Epidemiology of food-borne diseases: a worldwide review. *Rapp. trimest. statist. sanit. mond, 1997; 50:30–50.*

TRUELSTRUP HANSEN L., AUSTIN J. W. and GILL T. A., Antibacterial effect of protamine in combination with EDTA and refrigeration. *Int. J. Food Microbiol.* 2001; 66:149–61.

ULTEE A., KETS E. P. W. and SMID E. J., Mechanisms of action of carvacrol on the food-borne pathogen *Bacillus cereus*. *Appl. Environ. Microbiol.* 1999; 65:4606–10.

VAN DER STRATE B. W. A., BELJAARS L., MOLEMA G., HARMSEN M. C. and MEIJER D. K. F., Antiviral activities of lactoferrin. *Antiviral Res.* 2001; 52:225–39.

VAN LOO J., COUSSEMENT P., DE LEENHEER L., HOEBREGS H. and SMITS G., On the presence of inulin and oligofructose as natural ingredients in the western diet. *Crit. Rev. Food Sci. Technol.* 1995; 35:525–52.

WALLACE B. J., GUZEWICH J. J., CAMBRIDGE M., ALTEKRUSE S. and MORSE D. L., Seafood-associated disease outbreaks in New York, 1980-1994. *Am. J. Prev. Med.* 1999; 17:48–54.

WESSELS S. and HUSS H. H., Suitability of *Lactococcus lactis* subsp. *lactis* ATCC 11454 as a protective culture for lightly preserved fish products. *Food Microbiol.* 1996; 13:323–32.

WINKOWSKI K., LUDESCHER R. D. and MONTVILLE T. J., Psysiochemical characterization of the nisin-membrane interaction with liposomes derived from *Listeria monocytogenes*. *Appl. Environ. Microbiol.* 1996; 62:323–27.

WOLFSON L. M. and SUMNER S. S., Antibacterial activity of the lactoperoxidase system: a review. *J. Food Prot.*, 1993; 56:887–92.

WUYTACK E. Y. and MICHIELS C. W., A study on the effects of high pressure and heat on *Bacillus subtilis* spores at low pH. *Int. J. Food Microbiol.* 2001; 64:333–41.

YAMAUCHI K., TOIDA T., NISHIMURA S., NAGANO E., HUSUOKA O., TERAGUCHI S., HAYASAWA H., SHIMIMURA S. and TOMITA M., 13-week oral repeated administration toxicity study of bovine lactoferrin in rats. *Food Chem. Toxicol.* 2000; 38:503–12.

ZAIKA L. L. and KISSINGER J. C., Fermentation enhancement by spices: identification of active component. *J. Food Sci.* 1984; 49:5–9.

6

Identifying allergens in fish

S. Yamada and E. Zychlinsky, Hitachi Chemical Diagnostics, Inc. Mountain View and H. Nolte, University of Copenhagen

6.1 Introduction: the pattern of fish allergy

Food allergy is an immunoglobulin E (IgE)-mediated reaction of the body's immune system to a food or food component (food allergen). The normal individual usually develops tolerance to ingested food proteins, which are harmless. However, in a few individuals the same food proteins can provoke an allergic reaction. The food allergens are antigens, which are proteins or glycoproteins capable of stimulating the body's immune system to produce antigen-specific immunoglobulins. Patients with fish allergy will produce IgE specific antibodies against specific amino acid sequences found in the protein structures of fish antigens. It is not well understood why some subjects will develop these antibodies against fish antigens and others will not. The prevalence of food allergy in the general population has been estimated to be about 2–8%. Fish allergy is one of the most common forms of food allergy along with egg, milk, and peanut allergy.

High levels of serum specific IgE can indicate the presence of an allergic disease. The food allergens are capable of cross-linking IgE antibodies on the surface of mast cells in tissues or in basophil leukocytes in the blood. When two IgE molecules are cross-linked the cells will release substances, such as histamine, which will cause an allergic reaction, sometimes very severe. This reaction is often termed hypersensitivity; more specifically, 'type one hypersensitivity'. This reaction should not be confused with 'hypersensitivity dermatitis' also termed contact dermatitis. In contact dermatitis the reaction is T-lymphocyte mediated and will appear 24 to 48 hours after the initial contact with the antigen. No immunoglobulin antibodies are involved. Contact dermatitis to fish antigen is an occupational hazard typically reported in the fish industry.[1]

The evaluation of patients suffering from adverse reactions after the ingestion of food can be difficult if the source of the reaction is non-allergenic. If no immunological mechanism is responsible, the reaction is termed food intolerance. Food intolerance is often confused with food allergy due to the similar reactions after the ingestion of food. For instance, a patient might have abdominal symptoms after the ingestion of milk. This reaction is often caused by lactose intolerance rather than an allergy to milk. The same is true for fish such as tuna and mackerel where scombroid fish poisoning occurs due to improper refrigeration of the fish. An accurate diagnosis between food intolerance and food allergy is very important in treating patients.

Common symptoms of this reaction include rapid onset of breathing difficulties, runny nose, sneezing, oral itching, vomiting, nausea, abdominal cramps, and diarrhea. In some cases, consumption of allergens can cause a life-threatening reaction called anaphylaxis. Signs of anaphylaxis include swelling of the mouth, severe sneezing, shortness of breath, and shock.[2] The most common food allergens are allergenic proteins from shellfish, milk, fish, egg, and peanut. It is believed that 8% of the children in the United States are affected by food allergy and up to 2% of the adults.[2] In recent years, a trend has been seen to substitute seafood and fish from pork and beef as important sources of nutrition in an effort to 'eat healthier'. Fish ingestion reduces coronary disease,[3,4] however, fish have potent allergens that may cause type one hypersensitivity in both pediatric and adult populations.

Allergy to fish is most prevalent in populations where fish consumption is high. In Norway, the prevalence of fish allergy is 1/1000 for the general population.[5] One study indicated that 30% of children with food allergy had symptoms when ingesting fish.[6] Allergenic activity resides in the meat (muscle), but recently, concern has been raised as to whether products such as fish gelatin, made from skin and bones, also contain allergenic activity in the form of collagens. Another concern is the aerosolization of seafood particles. Aerosolization is the dispersion of tiny particles of solid or liquid that are suspended in air or gas, which may cause sensitization through inhalation. This is often reported in the seafood industry where contact dermatitis and occupational asthma are direct results of exposure to seafood during processing. Jeebhay et al.[7] stated that occupational contact dermatitis ranges from 3–11% and occupational asthma ranges from 7–36%. Goetz and Whisman[8] showed similar SDS-PAGE profiles from shrimp boiling water and the distillate (steam collected over boiling shrimp), which contained the heat stable allergen, tropomyosin. This was also true when they studied scallops. Due to increased production and consumption of seafood, seafood allergens aerosolized during preparation are a source of potential respiratory and contact allergy.

Several fish, such as cod, are known to cause allergic responses in individuals. Allergy to cod has been extensively studied by various people. The dominant major allergen in codfish, gad c 1, is a glycoprotein with a molecular weight of 12 kDa and is a Ca^{2+} binding parvalbumin. Although other molecular weight proteins have been described as allergens in cod and other fish species, it seems that there is strong correspondence between IgE reactivity to gad c 1 and clinical reactivity to cod and other species of fish. The substantial clinical cross-reactivity between cod and other fish species is paralleled by an IgE cross-sensitization between many fish species. A corresponding protein, parvalbumin from carp, has been thoroughly examined.[9] This protein is distributed in muscle and fiber cells, and by inhibition experiments, the protein can inhibit 80–90% of the total IgE-binding activity to extracts of cod, tuna fish, and salmon. In a study done by de Martino et al.,[10] children with allergy to cod demonstrated positive skin prick test results to other fish species. The children were not clinically sensitized to all species of fish, indicating the possibility of tolerance to other species of fish. Pascual et al.[11] concluded that there are several common antigens among different fish species. McCants et al.[12] found cross-reactivity between different species of fish by the use of skin prick testing. A significant percentage of patients were positive to all species of fish studied, even if that patient never ingested that species. A variety of species share allergenic determinants, but patients can have species-specific reactivity. It is not well understood why patients may have severe reactions to one fish species and then tolerate the ingestion of other fish species without allergic reactions although they have developed IgE reactivity as determined by skin tests or in-vitro specific IgE measurements to such fish. In spite of the well-established allergenicity of fish, surprisingly, few studies have been performed describing the fish allergens and clinical relevance.

In a study by Yamada et al.,[13] two species of tuna, *Thunnus alalunga* (Albacore tuna) and *Thunnus albacares* (Yellowfin tuna) were characterized by sodium dodecyl sulfate-polyacrylamide gel electrophoresis (SDS-PAGE). Atopic patients, sensitized to tuna, were immunoblotted to characterize the IgE-binding components of both types of tuna. The word 'atopic' usually refers to a tendency to have allergies through a hereditary predisposition, however some patients may be the only one in their family to have allergic symptoms. These blotting studies, along with the antigens' IgE binding properties enabled the researchers to define proteins specific to each species. The biologic activities of the extracts were evaluated by performing a histamine release assay. Raw, freshly cooked, and canned tuna were also extracted and their protein profiles were characterized by SDS-PAGE and immunoblot. Using the methods described below, characterization of allergenic extracts led to the observation of tuna specific allergenic proteins and differences in patient responses to the different tuna extracts.

6.2 Materials and methodology for identifying allergens: the case of tuna

6.2.1 Tuna fish extracts

Freeze-dried defatted Albacore tuna was purchased from Biopol Laboratory, Inc. (Spokane, WA, USA) and freeze-dried defatted Yellowfin tuna was obtained from Crystal Laboratory (Luther, OK, USA). Fresh raw tuna was obtained from Mitsuwa (Sunnyvale, CA, USA). Star-Kist's Chunk Light Tuna in spring water was used for the canned tuna extract.

6.2.2 Patient serum samples

Eight tuna positive sera were obtained from Plasma Lab International (Everett, WA, USA) and tested using two *in-vitro* diagnostic allergy tests; CLA Allergy Test (Hitachi Chemical Diagnostics, Inc. Mountain View, CA, USA) and ImmunoCAP® (Pharmacia & Upjohn Diagnostics Kalamazoo, MI, USA). Histamine release testing was done by The Reference Laboratory (Copenhagen, Denmark). A non-allergenic patient sample with negative specific IgE and histamine release results was used as a control.

6.2.3 Histamine release assay

The induction of histamine release in patients suffering from type one hypersensitivity is caused by the stimulation of mast cells and basophils through antigen exposure. The histamine release assay measures the amount of histamine released in patient serum.

6.2.4 SDS-PAGE

Tuna extracts at a 0.5 mg/mL protein concentration in NuPAGE® LDS Sample Buffer with NuPAGE® Reducing Agent were heated for ten minutes at 70°C and 25 μL were separated on a 4–12% Bis-Tris Gel using the NuPAGE® MOPS SDS Running Buffer.[14] The gel was run at a constant voltage of 200 volts for 50 minutes. Pre-stained molecular weight standards (Invitrogen, CA, USA) were used to determine the molecular weight of the proteins.

6.2.5 Immunoblot assay

Extracts were separated as described above. The proteins were electrophoretically transferred onto a 0.2 μm nitrocellulose membrane (Biorad, CA, USA) using 30 volts for one hour. NuPAGE® Transfer Buffer with 10% Methanol was used. The membrane was incubated with the patient serum overnight and horseradish peroxidase-labeled goat anti-human IgE was added. Metal Enhanced DAB (Pierce, IL, USA) was used to detect bound IgE. The antigen-IgE complex was detected by the formation of brown colored precipitate on the membrane.

6.2.6 IgE inhibition study

The following method was used to inhibit patient serum with tuna extracts as stated in a previous publication.[13] Three milliliters of patient (c) serum were incubated with Albacore and Yellowfin extracts (0.5 mg protein/mL) overnight at room temperature. An immunoblot with both extracts was done with the inhibited serum. A control with non-inhibited serum was run. Patient (c) and (h) were inhibited with Albacore and Yellowfin at the same concentration and tested using ImmunoCAP.

6.3   Analyzing results

Tuna extracts were analyzed by SDS-PAGE. Figure 6.1 shows the coomassie stained gel of different tuna extracts. Lanes (a) and (b) show the protein profile

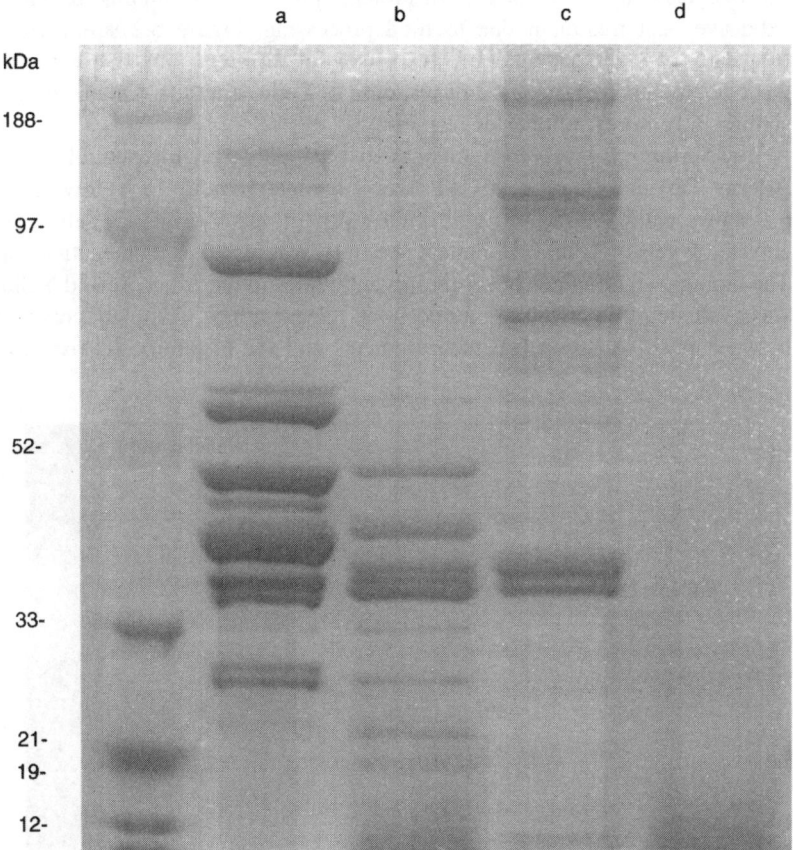

Fig. 6.1 Protein profile of Albacore, Yellowfin, fresh cooked tuna, and canned tuna extracts separated using a 4–12% polyacrylamide gel and stained with a coomassie blue staining reagent. (a) Albacore (12.5μg), (b) Yellowfin (12.5μg), (c) fresh cooked tuna (12.5μg), (d) canned tuna (12.5μg) gel stained with a coomassie stain. Molecular Colored Standards are in the first column; myosin (188 kDa), phosphorylase B (97 kDa), glutamic dehydrogenase (52 kDa), carbonic anhydrase (33 kDa), myoglobin blue (21 kDa), myoglobin red (19 kDa), and lysozyme (12 kDa).

of Albacore and Yellowfin, respectively. In this gel system, Albacore showed more proteins with molecular weights in the range of 52 kDa to 188 kDa compared to Yellowfin. At ~21 kDa, and ~46 kDa, the Yellowfin extract has proteins that do not appear in the Albacore extract. Not only is the species of tuna important in understanding allergies to tuna, but also the preparation of the fish extracts. Fresh cooked tuna (lane c) and canned tuna (lane d) protein profiles were different from Albacore and Yellowfin. Cooked tuna formed high molecular weight protein complexes that are not seen in the defatted raw material of either species of tuna. Canned tuna was previously thought to have low allergenicity due to the long food process of canning.[15] Some patients had positivie IgE immunoblots with the cooked and canned extracts (results not shown). This suggests that certain proteins preserve IgE-binding activity after extensive heat treatment due to food processing. Figure 6.2 summarizes the frequency of eight patients' IgE responses for different tuna fish extracts. All positive patients had specific IgE response to Yellowfin tuna. Canned tuna had a significantly lower number of responses.

The Serum IgE levels from tuna-positive patients were measured by the CLA Allergy Test and Pharmacia CAP assays for allergenicity to Yellowfin. Table 6.1 shows patient sera (a-h) with high levels of specific IgE to Yellowfin and elevated levels of Total IgE. Patient serum (i) was tested as the negative control. The biologic significance of serum IgE antibodies to the Albacore and Yellowfin was evaluated by performing a histamine release assay. Complete concordance between positive serum IgE measurements and the histamine release test was

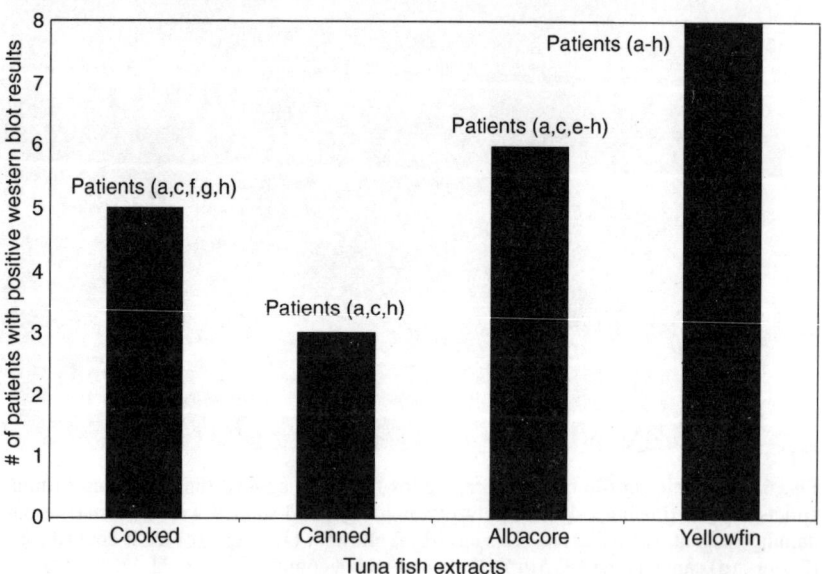

Fig. 6.2   Frequency of patients' positive IgE response to different tuna fish extracts using Western Blot.

Table 6.1   In-vitro characterization of tuna fish positive sera

Patient	Total IgE IU/mL	CAP IgE Class (IU/mL) Yellowfin	CLA IgE Class (LU)[†] Yellowfin	Histamine release Class (ng/mL histamine)* Albacore	Yellowfin
a	1545	2 (1.12)	3 (213)	1 (56)	2 (99)
b	1230	3 (3.38)	3 (195)	2 (102)	2 (147)
c	550	3 (11.30)	4 (300)	0 (22)	2 (61)
d	846	3 (9.39)	4 (300)	1 (94)	2 (106)
e	10269	3 (3.52)	4 (300)	0 (9)	2 (48)
f	3216	2 (2.69)	4 (212)	1 (50)	1 (54)
g	5310	4 (26.40)	4 (300)	2 (108)	2 (73)
h	274	3 (11.00)	4 (300)	2 (181)	3 (120)
i (control)	2	0 (<0.10)	0 (0)	0 (12)	0 (26)

*ng/mL histamine reflects the value obtained by calculating the area under the curve of six extract dilutions.
[†] LU is an arbitrary luminometer value.

noted for the Yellowfin extract. Patients (c and e) did not release histamine to Albacore. None of the tuna fish extracts induced significant histamine release from basophils passively sensitized with serum from a non-allergic patient, indicating lack of toxicity or interfering histamine content of the tuna fish preparations.

By means of immunoblotting (results not shown), patient (c) inhibited with Albacore extract and blotted against both species of tuna, showed a significant reduction in band intensity with a protein ~32 kDa in both species, but no reduction in band intensity was seen with the protein at ~46 kDa. The same patient was inhibited with Yellowfin extract and no protein bands could be detected. This suggests that the ~46 kDa protein is not present in the Albacore extract, and is specific to Yellowfin. Similar results were seen using inhibition studies with the ImmunoCAP tests for Yellowfin (see Table 6.2).

The results suggest that the allergenic components in Albacore and Yellowfin are different, as well as in different preparations of tuna. Inhibition studies

Table 6.2   ImmunoCAP inhibition results

Test condition*	IU/mL	CAP Class	% Inhibition
Patient (c) + PBS (control)	10.80	3	
Patient (c) + albacore	4.72	3	56.20%
Patient (c) + yellowfin	1.67	2	84.50%
Patient (h) + PBS (control)	9.69	3	
Patient (h) + albacore	4.31	3	55.50%
Patient (h) + yellowfin	1.65	2	82.90%

* Patient (c) and patient (h) were inhibited with tuna extracts and tested using ImmunoCAP test for yellowfin tuna specific allergy.

provided no evidence of an Albacore specific protein. Although some IgE binding proteins were detected in cooked and canned tuna, the histamine release data showed no biologic function. This might indicate that patients can tolerate processed fish due to the denaturing and digestion of native allergenic proteins reducing their overall allergenicity. By analyzing SDS-PAGE and immunoblot protein profiles, a Yellowfin specific protein was observed, showing the importance in diagnosing allergy using different species.

6.4   Future trends

Previous research on seafood allergies has concentrated on shellfish and some finfish such as cod. Gad c 1, the major allergen from cod, has been extensively characterized by many researchers. However, the protein causing the risk of eliciting a response may not be the same as the sensitizing protein, i.e., inducing the production of specific IgE antibodies. This complicates the diagnosis and the determination of which IgE binding antibodies detected in the immunochemical assays are clinically relevant. Due to the high level of cross-reactivity of fish allergens, characterization of different species of fish has been limited. To obtain a better understanding of fish allergy, it is important to characterize the IgE binding proteins and determine what makes them clinically relevant in order to provide more effective diagnosis and treatment. This will also enable us better to predict the risk of new allergens from genetically modified food.

Since the trend to lead healthier lives through diet is growing, the understanding of a potential sensitizing capacity of food allergenic proteins is critical. Recently, the United States Food and Drug Administration (FDA) addressed the concern about food allergies by stating that labeling on food will make it easier for the consumer to decide if the food is appropriate to eat. The FDA believes that the disclosure of all ingredients on the food label is essential for the reduction of adverse allergy reactions. The FDA along with the Food Allergy Issues Alliance established eight major food allergens in which fish is included. The combined effort of manufacturers and FDA regulations will allow the consumer to identify possible sources of allergens in foods.

6.5   Sources of further information and advice

The availability of different fish species for commercial consumption is possible only through understanding and research of conservation efforts. There are several conservation programs specifically for tuna, which are aimed at providing education through research. The Tuna Research and Conservation Center (TRCC) (www.tunaresearch.org) is a program that allows research on tuna in captivity. Although this program is based off the coast of California, there are many international conservation efforts for other locations of tuna. Kewalo Research Facility in Hawaii (www.nmfs.hawaii.edu) is an important organization for the understanding of the behavior and physiology of captive tuna due to the large commercial fisheries in Hawaii. The International Commission for the Conservation of Atlantic Tuna (www.iccat.org) and the Commission for the Conservation of Southern Bluefin Tuna (www.home. aone.net.au/ccsbt) are other organizations dedicated to the understanding and conservation of this fish. The understanding of the physiology of tuna is an important step in characterizing its proteins. Allergies to tuna can be controlled only if the source of the tuna is understood. This is true for all food allergies.

6.6   References

1.   WEINBERG JM, HAIMOWITZ JE, SPIERS EM, MOWAD CM, Fish skin-induced dermatitis, *J Eur Acad Dermatol Venereol*, 2000 **14** 222–3.
2.   PATIENT/PUBLIC RESOURCE CENTER, Tips to remember: food allergy, AAAAI, 1996 1–5.
3.   KROMHOUT D, BOSSCHIETER EB, COULANDER C, The inverse relation between fish consumption and 20-year mortality from coronary heart disease, *N Engl J Med*, 1985 **312** 1205–9.
4.   BURR ML, FEHILY AM, GILBERT JF, ROGERS S, HOLLIDAY RM, SWEETNAM PM, ELWOOD PC, DEADMAN NM, Effects of changes in fat, fish, and fibre intakes on death and myocardial reinfarction: diet and reinfarction trial (DART), *Lancet* 1989 **2** 757–61.
5.   AAS K, Fish allergy and the codfish allergen model, In: Brostoff J, Challacombe ST, eds, *Food allergy and intolerance*, London: Balliere Tindall, 1987 356–60.
6.   CRESPO JF, BLANCO C, CONTRERAS J, PASCUAL C, MARTIN ESTEBAN M, Food allergy: a clinical and epidemiological study (Abstract), *J Allergy Clin Immunol*, 1992 **89** 192.
7.   JEEBHAY MF, ROBINS TG, LEHRER SB, LOPATA AL, Occupational seafood allergy: a review, *Occup Environ Med*, 2001 **58** 553–62.
8.   GOETZ DW, WHISMAN BA, Occupational asthma in a seafood restaurant worker: cross-reactivity of shrimp and scallops, *Ann Allergy Asthma Immunol*, 2000 **85** 461–6.
9.   BUGASJSKA-SCHRETTER A, ELFMAN L, FUCHS T, KAPIOTIS S, RUMPOLD H, VALENTA R, SPITZAUER S, Parvalbumin, a cross-reactive fish allergen, contains IgE-binding epitopes sensitive to periodate treatment and Ca^{2+} depletion, *J Allergy Clin Immunol*, 1998 **101** 67–74.
10.   DE MARTINO M, NOVEMBRE E, GALLI L, DE MARCO A, BOTARELLI P, MARANO E, VIERUCCI A, Allergy to different fish species in cod-allergic children: in-vivo and in-vitro studies, *J Allergy Clin Immunol*, 1990 **86** 909–14.
11.   PASCUAL C, ESTEBAN MM, CRESPO JF, Fish allergy: evaluation of the importance of cross-reactivity, *J Pediatr*, 1992 **121** S29–35.
12.   MCCANTS ML, HELBLING A, SCHWARTZ HJ, LOPEZ M, LEHRER SE, Skin test and RAST reactivity to seafood, *J Allergy Clin Immunol*, 1992 **89** 194.

13.   YAMADA S, NOLTE H, ZYCHLINSKY E, Identification and characterization of allergens in two species of tuna fish, *Annals of Allergy, Asthma, & Immunol*, 1999 **82** 395–400.
14.   INVITROGEN, NuPAGE Bis-Tris Gel Instruction Booklet, 1996–2000.
15.   BERNHISEL-BROADBENT J, STRAUSE D, SAMPSON H, Fish hypersensitivity II: Clinical relevance of altered fish allergenicity caused by various preparation methods, *J Allergy Clin Immunol*, 1992 **90** 622–9.

7

Identifying heavy metals in fish

J. Oehlenschläger, Institute for Fishery Technology and Fish Quality, Hamburg

7.1 Introduction

Many elements, which are present in seafood (Oehlenschläger J, 1997) are essential for human life at low concentration (Fraustro da Silva JJR and Williams RJP, 1993), however, they can be toxic at high concentrations. Other elements like mercury, cadmium and lead show no known essential function in life and are toxic even at low concentration when ingested over a long period, therefore many consumers regard any presence of these elements in fish as a hazard to health. These elements were present in the aquatic environment long before human beings existed. This is in contrast to the organic residues, which are all anthropogenic xenobiotics and have been brought into the environment by man with the exception of dioxines and PAHs (Polycyclic Aromatic Hydrocarbons), which have been formed by combustion and natural processes, respectively.

The presence and concentration of heavy metals in the environment and more specifically in the aquatic environment and in its biota, namely in animals and plants which are used for human food is based both on natural and anthropogenic sources. While the natural concentration of these elements has been present in the world oceans and freshwater reservoirs due to marine vulcanism, geological anomalies (Gonzalez F et al., 1998; Falconer CR, Davies IM and Topping G, 1986) and geothermal events, anthropogenic pollution started with the beginning of intensive metallurgy in the industrial revolution. Later, acidic rain as a result of industrial pollution has mobilized heavy metals from minerals and contributed more to the overall concentration.

Fish and other seafood have always contained certain amounts of heavy metals as a consequence of living in water. The proportion between the natural background concentration of heavy metals and the anthropogenic heavy metals in fish varies from element to element. It can be stated that in the open seas, which are still almost unaffected by pollution, fish mostly carry just the natural burden of heavy metal concentration. In heavily polluted areas, in waters which have no sufficient exchange with the world oceans (e.g. Baltic Sea, Mediterranean Sea), in estuaries, in rivers and especially in places which are close to sites of industrial activities, the heavy metal concentrations actually found are exceeding the natural load (Kalay M *et al.*, 1999; Dobson J, 2000; Claisse *et al.*, 2001; Prudente M *et al.*, 1997).

There is a vast literature on the content of toxic heavy metal in fish, crustaceans and molluscs. The overwhelming majority of these papers deal with reports about concentrations of heavy metals which are unusually high due to anthropogenic activities and which are found in areas where an accumulation is favored by the natural conditions (insufficient water exchange, shallow waters, estuaries, rivers, inshore waters etc.). Organs and tissues, which accumulate and store heavy metals are the parts mostly investigated, while the muscle tissue (the fillet), which is mostly the only part of the fish which is actually eaten, has been of less interest because of its low burden. Only a small amount of information is available about the heavy metal content in the edible part consumed by humans of the food fish which are caught in the open ocean and which contain only the natural background concentration in their muscles.

A major problem in evaluating published data on heavy metal concentrations in fish is the inaccuracy and the poor precision of old data. Analysis of heavy metals is difficult, especially at the low levels that are found in marine animals. Data from the past are mostly affected by the fact that researchers have not worked under clean lab conditions, that the methodology was not fully developed, that the environmental conditions influenced the results, that a critical evaluation of data was not standard procedure and that the limits of detection were much higher than today. Many results reported in the literature are therefore much too high and the reliability of old data is very questionable. There are thousands of publications where a detection limit is given when the actual content was too low to be determined (e.g. cadmium <0.2 mg/kg). These values have been taken as they are in secondary literature and reported as 0.2 mg/kg although the actual and true content might have been 0.001 mg/kg. This is the mechanism by which so much extremely erroneous data has entered books, literature and unfortunately also newspapers which as a consequence has influenced public opinion about heavy metal in food fish. Modern analysis that is based on effective digestion procedures (Lamble KJ and Hill SJ, 1998; Stoeppler, 1994), sophisticated instrumental analytical methods, newly developed analytical techniques, speciation of bio-inorganic compounds (Szpunar J, 2000; Ebdon *et al.*, 2001), on clean room technology, good laboratory praxis, standard operation procedures, continuous evaluation of data, inter-laboratory comparison analyses, ring tests, use of certified standard reference materials (Jenks PJ, Rucinski RD and Jerzak H, 2001), to prove accuracy etc., nowadays provides highly reliable data about heavy metal content in fish (Neidhart B and Wegscheider W, 2002). Old data should be considered carefully and in case of doubt not be cited or used.

Fish and other aquatic animals take up heavy metals (Phillips DJH, 1995) from their food and also from the water they pass through their gills. The uptake of metals is often dependent on the amount of food ingested and on the heavy metal content of the food or prey. It was shown that in areas of the sea where there is a high concentration of phytoplankton the accumulation by fish of heavy metals was higher than in areas where a low concentration of phytoplankton was present. Accumulation takes a long time and results in high concentrations in aged and thus big fish. Some species, mainly predatory fish, which are relatively long lived, are known for storing higher amounts of heavy metals in different organs. This age accumulation may lead to a high burden of heavy metals in large specimens of redfish or ocean perch (*Sebastes* species), Atlantic and Pacific halibut, tuna, sharks, marlins, swordfish and other predatory species that can reach 25 years of age and more. High concentrations of heavy metals are only rarely found in fish muscle. When this situation appeared it was always detected later that it was based on an extremely high degree of pollution (e.g. cadmium: Itai Itai, mercury: Minamata). The main organs, which are used in fish for storage and detoxification of heavy metals, are the liver, the kidney and the bones. Both organs are normally not used for human consumption in Europe (with the exception of canned cod liver) and America because it is mostly the fillet (the muscle tissue) which is consumed . In Asia, however, many organs are eaten, either as part of the whole fish or in special preparations. Fortunately it is mostly younger, smaller fish that are consumed whole but roes and other gut contents often are used in sauces or fermented, or salted, dishes. In cases where intestines, for example, cod liver are consumed, special emphasis must be placed on checking the burden prior to processing into food.

In this chapter the heavy metal concentration of mercury, cadmium, lead, zinc, copper, tin and aluminium in fish and – where appropriate – in other aquatic animals will be described, methods of analysis will be shown and concentrations typically found in edible parts of commercially used food fish will be described.

7.2   Mercury

Man released mercury into the environment by the actions of the agriculture industry (fungicides, seed preservatives), by pharmaceuticals, as pulp and paper preservatives, catalysts in organic syntheses, in thermometers and batteries, in amalgams and in chlorine and caustic soda production. There is an annual release of 40,000 to 50,000 t into the atmosphere and of approximately 4,000 t into the sea. The mean concentration in seawater is 0.1–2 ng Hg/L, in coastal waters 0.5–10 ng Hg/L and in rivers up to 70 ng Hg/L.

The first well-documented intoxication (Clarkson TW, 1998) with mercury-contaminated fish was the Minamata-disease. This intoxication was identified after thorough investigations of intoxication by methyl mercury, which had accumulated in fish after being released from industrial vinyl chloride and acetaldehyde production with industrial wastewater into the coastal waters. Contaminated fish contained mercury at a level of 50–250 mg/kg wet weight and molluscs at 50–200 mg/kg. The concentrations of mercury in fish caught in the open sea are much lower and although there have been some serious local problems with contamination there are no signs of worldwide mercury contamination. Analyses on specimens of tuna from museums, which were more than 90 years old, revealed almost the same mercury contents as reported in recently caught tuna. Comparison of early analytical results obtained by Stock (Stock A and Cucuel F, 1934) in the 1930s with recent results shows that fish caught at the end of the century contain almost the same amounts of mercury.

The toxicity of mercury depends on its chemical form (ionic < metallic < organic). Mercury is present in fish predominantly in its organic form as dimethylmercury. This is a lipophilic compound, which has a tendency to accumulate in fatty tissue. Therefore elevated mercury concentrations are mainly found in liver of lean species and in fatty fish species. Methyl mercury has a tendency to accumulate with fish age. This leads to higher mercury concentrations in old fatty predatory species like tuna, halibut, redfish, shark, and swordfish. Processing of these species into food products like canned fish, led in the 1960s to problems with mercury levels above legal limits that today are 1 mg/kg methyl mercury in USA, 0.5 mg/kg in Canada and 0.5 mg/kg for most species except 1 mg/kg for 22 predatory species in the European Union. These limits were established by a number of countries to protect people against health hazards from long-time ingestion of mercury contaminated seafood. Today the world's fish-processing industry no longer uses big specimens, but small to medium sized fish that have much lower mercury concentrations. Recent investigations (Krüger K-E and Kruse R, 1984) have shown that the mercury concentration in canned tuna is nowadays at a very low level (0,21 mg/kg arithmetic mean from 76 market samples), in tuna from all parts of the world. USFDA (United States Food and Drug Administration) still recommends that pregnant women should not consume shark, swordfish, king mackerel or tilefish, and the Canadian government recommends only one portion per month.

As mentioned, mercury can accumulate in fish and invertebrates and the concentration tends to increase not only with age but also with increasing trophic level. While inorganic mercury is the dominant form of mercury in the environment and is easily taken up, it is also depurated rapidly, while methylmercury accumulates quickly too but depurates very slowly. Due to this most mercury in fish muscle is present as methylmercury (> 95%) (Bloom NS, 1992; Joins CR et al., 1997; Joins CR et al., 1995). Selenium was found to be a potent agent against mercury intoxication in fish. In many species a high selenium concentration was found and in most marine fish the amount of selenium in muscle tissue is in excess of that of mercury (Barghiani et al., 1991; Cappon CJ and Smith JC, 1981; Oehlenschläger J and Priebe K, 1989).

The contents of mercury in marine fish have been interpreted as a sign of worldwide anthropogenic environmental pollution, which, however, is not correct. The methyl mercury contents of marine fish from the open sea are mostly not based on environmental pollution, however, in contaminated waters accumulation of mercury in biota was reported (Joiris CR et al., 1999; Joiris CR et al., 2000; Kawaguchi T et al., 1999; Abreu SN et al., 2000; de Clerck R et al., 1995). In marine fish a correlation of mercury accumulation with the biomass content of the seawater was found. Mercury compounds are bound to plankton and fish eat plankton. If there is a lot of plankton present a dilution effect takes place and fish ingest less mercury. This holds primarily for herbivorous fish but also for carnivorous fish, which feed on herbivorous fish. Recent investigations of different fish species (cod, saithe, haddock, herring, halibut) from the Barents Sea and from the Greenland Sea exhibited mercury concentrations between 0.01 mg/kg and 0.06 mg/kg and even the average mercury content of redfish amounted only to 0.13 mg/kg (Solberg, 1997).

Mercury levels in molluscs are usually in the range between 0.02–0.05 mg/kg wet weight. Invertebrates generally carry a much lower proportion of methylmercury compared to fish. All freshwater species also accumulate mercury (Mason RP, Laporte J-M and Andres S, 2000).

7.2.1 Analysis

For later analysis of mercury (Shrivastran A K and Tandon S K, 1982; Bortoli A et al., 1995) it is necessary to homogenize the sample to get a more representative sub-sample from the homogenized tissue. Special care must be given to the effects of storage conditions on the contents of mercury (de Boer J and Smedes F, 1997). Homogenization can be done by aqueous homogenization using a blender or a homogenizer or by lyophilization, which gives a dry powder, enriched in mercury content. For the analysis of total mercury a wet ashing step is needed to destroy the organic matrix. Usually a combination of mineral acids like nitric acid and sulfuric acid (e.g. 20% nitric acid/80% sulfuric acid at 60°C) is used. For the analysis of methyl mercury or any other speciation of organic mercury there are numerous mercury extraction procedures proposed in literature. This is normally a three-step procedure: the first step is treatment with a matrix liberation agent (e.g. 1% sodium chloride, 1 N hydrochloric acid), followed by an extraction with an organic solvent (e.g. benzene or toluene) and finally a cleanup step with a combination of organic solvent and sodium thiosulphate. The recovery rate is often between 70–100%.

Recently some new sample preparation techniques have been proposed (Cappon CJ, 1994). The following analytical methodologies have been used for the determination of total mercury: spectrophotometry, polarography, x-ray fluorescence, neutron activation analysis, atomic absorption spectroscopy (flame, furnace and cold vapour), atomic emission spectroscopy (ICP and fluorescence). Most of the methods, with the exception of some atomic spectrometric techniques, cannot detect or measure mercury at low levels.

Furthermore, these methods cannot be used for mercury speciation, are time consuming and expensive and are subject to numerous interferences from the sample matrix. However, some classical AAS methods can be used for indirect methyl mercury analysis, since in fish most mercury (> 95%) is present in the form of methyl mercury.

Modern speciation techniques used for the determination of methyl mercury are based on either gas chromatography (Welz B and Sperling M, 1997) and more recently on high-performance liquid chromatography. In modern gas chromatography fused silica capillary columns are used and selective and sensitive detectors like AAS, DCP or ICP achieve the detection. When using fused silica capillary columns special care must be taken for the optimization of sample injection and the injector port conditions. Today HPLC is used for mercury speciation mostly in the reversed-phase mode under isocratic conditions because it offers more flexibility in the choice of the mobile phase modifiers and better detector compatibility as organic modifiers methanol and acetonitrile are used. Different agents for better resolution have been introduced such as 2-mercaptoethanol, EDTA, diethyldithiocarbonate, pyrrolidine dithio-carbonate and 6-mercaptopurine. The detector used is the electrochemical detector (EC) (Evans O and McKee GD, 1987).

For analytical quality control a number of certified standard reference materials (Quevauviller Ph, 1995) for mercury assay are available from BCR, Community Bureau of Reference (European Union), National Institute of Standards and Technology (USA), National Research Council of Canada (Canada), International Atomic Energy Agency, Monaco: oyster tissue (0.057 mg/kg) albacore tuna (0.95 mg/kg), dogfish muscle (0.789 mg/kg), dogfish liver (0.225 mg/kg), lobster hepatopancreas (0.13 mg/kg) and lyophilized fish tissue (0.52 mg/kg).

7.3   Lead

The situation with lead is completely different from that with mercury. Lead as a toxicologically relevant element has been brought into the environment by man in extreme amounts, despite its low geochemical mobility and has been distributed worldwide (Branica M and Konrad Z, 1980). The use of lead by man goes back approximately 9,000 years, but the maximum increase was caused by the start of industrial lead mining and metallurgy around 1750 and the increase of individual traffic around 1940 with the use of lead tetra alkyl compounds as a gasoline additive; 95%–98% of all lead present today in the environment can be traced back to anthropogenic activities.

While the lead content in deep sea water of the South pacific amounts to 1–2 ng Pb/L and in Antarctic deep sea water to 0.4 ng/L (Betti M and Papoff P, 1988), which seems to be the natural background level, the lead content of surface water of the central North Atlantic and North Pacific is around 5–50 ng/L. However, the uptake of lead through the food chain is of little importance since the concentration of lead in fish does not increase with trophic level and age but rises with increasing concentration in the water. Many of the data published for the lead content of the edible parts of fish have been much too high and are not reliable because control of contamination and quality control of analytical procedures and the results was insufficient in the past. Most values reported reflect more the general lead pollution in the environment than that in the tissue analysed. This was because lead was and is ubiquitously distributed on land, in laboratories, equipment and utensils. Until the 1970s many older laboratories had some lead-topped benches and lead-lined acid traps. Clean room technology and closed analytical systems later allowed an accurate analysis of lead content in fish muscle. These correct analyses demonstrated that the lead content in fish muscle from fish that is caught in the open sea is still very low amounting to 2–10 μg/kg. Fish from the North Sea or the Baltic Sea exhibit somewhat higher contents (20–50 μg/kg) but even these results are much lower than those reported earlier.

Fish deposit lead mainly in bones, while soft tissues like heart, gonads and gastrointestinal organs do not show elevated quantities of lead. Bones are compartments that are not usually consumed or processed into food for humans. The environmental pollution by lead in some parts of the world can therefore be shown only by analysing lead content in bones and not in muscle tissue. A recent overview about typical concentrations of lead in fish and fish organs at different locations is listed by Solberg T et al. (1997). Elevated lead content in muscle tissue is reported only from areas with intensive industrial and agriculture activities and input of untreated municipal and industrial waste waters (Wong CK, Wong PPK and Chu LM, 2001). The lead content in invertebrates like molluscs and crustaceans is higher with an average lead content around 1 mg/kg wet weight with a large variation around this average. This is caused by an active accumulation of lead in the digestive tract, hepatopancreas, of both molluscan and crustacean shellfish. In shellfish, the digestive tract has to be removed prior to consumption, preferably immediately after catch to prevent a migration of lead from intestine into muscle tissue which has been observed during prolonged storage. However, this can be difficult in small sized species, where only a tiny digestive tract is present. Lead is present in the environment in both inorganic and organic form. The organic species is mostly tetralkyl-lead, which was found in a number of freshwater and marine fish species at concentrations of 5–100 μg/kg.

The legal limits (EU-Directive 466/2001 from March 3, 2001, in force since April 5, 2001) for lead in the European Union are: fish muscle 0.2 mg/kg; muscle from eel, horse mackerel, sardine and a few other species 0.4 mg/kg; crustaceans 0.5 mg/kg; mussels 1 mg/kg; cephalopods without intestines 1 mg/kg (all on a wet weight basis).

7.3.1 Analysis

Lead analysis is extremely difficult since the ubiquitous occurrence of lead causes contamination whenever possible. In the sampling step it is necessary to use glass or titanium knives for sample dissection and to use conditioned Teflon or Polypropylene bottles. A pre-concentration step (e.g. removal of water by lyophilization) is recommended when the concentration of the analyte is expected to be very low. The digestion is mostly done by wet oxidation. Dry ashing, which was commonly used in the past, is subject to uncontrolled contamination sources. However, modern dry ashing devices like ashing in a microwave activated oxygen plasma at reduced pressure is an excellent alternative to wet digestion, but rather expensive. Fish tissues can be wet digested by concentrated nitric acid followed by an oxidation step with hydrogen peroxide. Many different analytical methods for the determination of lead in fish and other seafood have been used (Lobinski R and Adams FC, 1994).

In many countries the method of analysis to be used is often legally prescribed and as a consequence inaccurate methods which are unsatisfactory for accurate lead analysis are still in use. For lead analysis spectrophotometry, flame atomic absorption spectrometry, emission spectrometry, x-ray fluorescence, neutron activation analysis, and electrochemical techniques have all been employed. Today the most promising techniques are electrothermal atomic absorption spectrometry, inductively coupled plasma-atomic emission spectrometry (ICP-AES), inductively coupled plasma mass spectrometry (ICP-MS), and differential pulse anodic scanning voltammetry (DPSAV). Details about the most widely used method, AAS, can be found in monographs about this method (Welz B and Sperling M, 1997). For the speciation of organolead compounds gas chromatography – atomic absorption analysis (GC-AAS) and gas chromatography – microwave induced plasma emission spectrometry (GC MIP-AES) has been developed.

As in other analytical cases the analytical problems associated with analysis of lead are also often underestimated. This becomes evident in the results of inter and intra-laboratory test programs where results can disagree widely. Care must be taken with the sample collection, sample choice, sample storage, and sample preparation and pre-treatment, which are integral parts of the complex analysis. Quality control of results and critical evaluation of data prior to publication and – more importantly – a reluctance in publishing suspect data could improve the actual situation in which inaccurate data on lead content still dominate.

7.4   Cadmium

Cadmium is one of the most toxic heavy metals for human beings. It is present in the earth's crust together with zinc and came into the environment through the centuries due to zinc metallurgical mining and practices. Toxicity of cadmium became evident when the Itai-Itai disease in Japan was detected. This was caused by cadmium-containing rice, which was highly contaminated by water that had been used for mining. There is a group of proteins called metallothioneins, which can protect against cadmium toxicity (Klaasen CD, Liu J and Choudhuri S, 1999).

Cadmium is also widely distributed in the aquatic environment, and bioaccumulation of cadmium by aquatic organisms is widely recognized. Cadmium content in the edible part of fish (muscle tissue) is generally very low, while fish deposit cadmium in organs like kidney and liver. These organs can be heavily contaminated and should preferably not be consumed. The cadmium content of the edible part of marine fish from the central North Atlantic, from the Barents Sea and from waters around Greenland amounts to 0.5–5 μg/kg wet weight. In coastal areas and in the Baltic Sea cadmium contents can be higher, reaching 10–20 μg/kg. All these values are much lower than the legal limits or recommended limits set by governments. The concentrations found are all in the range of 10–20% of legal limits.

The situation with marine invertebrates (Rainbow PS, 1985; Rainbow PS, 1997; Ruangwises N and Ruangwises S, 1998; Ismail A, Jusoh NR and Ghani IA, 1995) like molluscs and crustaceans is different. Molluscs, especially cephalopods, are active cadmium accumulators (Francesconi KA, Moore EJ and Joll LM, 1993; Bargagli et al., 1996; de Gregori I et al., 1996; Aboul Naga WM, 1996; Odzak N et al., 1994). This leads to the phenomenon that cephalopods (octopus, squid and cuttlefish) can store huge amounts of cadmium in their intestines (up to 30 mg/kg) while their muscle tissue (mantle and tentacles) contains the same low amount of cadmium found in fish muscle (Oehlenschläger J, 1991; Schulz-Schroeder G and Schering B, 1995). It is therefore of utmost importance to remove all intestines from cephalopods immediately after catch. If this is not done, cadmium migrates from the intestines into the muscle tissue and contaminates this tissue to a degree that the muscle later cannot pass legal limits. Mussels show the same effect at a lower level; they must be checked regularly for their cadmium content. The higher cadmium content in molluscs is the justification for higher legal limits of cadmium for this group of seafood. Oysters too can accumulate cadmium from industrial contamination and the levels must be regularly monitored.

The legal limits (EU-Directive 466/2001 from March 3, 2001, in force since April 5, 2001) for cadmium in the European Union are: fish muscle 0.05 mg/kg; muscle from eel, sardine, horse mackerel and a few other species 0.1 mg/kg; crustaceans 0.5 mg/kg; mussels 1 mg/kg and cephalopods without intestines 1 mg/kg (all on a wet weight basis).

7.4.1 Analysis

Sampling and sample preparation for later cadmium analysis has to be as precise as with other heavy metals, however, the risk of cadmium contamination is less than in the case of lead. Special caution is necessary when plastic articles and utensils are used because cadmium is used as a stabilizer in plastic (PVC). For digestion dry ashing in open devices should not be used since it leads to uncontrolled losses and/or contamination during the digestion procedure. Today wet ashing in closed vessels that are temperature, and/or pressure controlled is the method of choice. Special emphasis must be placed on the purity of the digestion aids used (nitric acid, perchloric acid, sulfuric acid, hydrochloric acid and hydrogen peroxide).

New digestion devices have recently become available that make use of microwave energy to digest samples in pressurized Teflon vessels. The advantage of these devices is that the sample heats up rapidly and the whole digestion takes only minutes. Analytical methods used in the determination of cadmium are (Ray S, 1994): spectrophotometry, atomic absorption spectrometry, atomic emission spectrometry, ICP-AES, electroanalytical techniques such as DPSAV (Zuman P, 2000; Bard AJ and Zoski CG, 2000), neutron activation analysis and ICP-MS. AAS is the most commonly used technique for the determination of cadmium. While flame AAS is not sensitive enough to determine trace amounts of cadmium and needs a preconcentration step, ETAAS (electrothermal AAS, earlier called GF graphite furnace AAS) has overcome the fundamental limitations of conventional flame techniques to provide a much better sensitivity. Electrothermal AAS has increased the detection limit for cadmium to 0.001 mg/L from 2 mg/L with conventional flame AAS. Use of matrix modifiers (Carnrick GR, Schlemmer G and Slavin W, 1991) by which some interference from the matrix can be overcome, pyrolytic coating graphite and use of the L'vov platform together with effective background correction (e.g. Zeeman correction) have further improved ETAAS. DPSAV is a method that was neglected for many years but which gives excellent results when the matrix is completely destroyed. This method has another advantage as it is an oligoelement method, which allows the simultaneous determination of up to four heavy metals (lead, cadmium, copper and zinc), in a single run (Oehlenschläger J, 1994).

7.5   Copper

The mobility of copper in the environment is much less compared with cadmium and zinc. Copper is not toxic for humans in low concentrations and is an essential element for the human being (Linder MC and Hazegh-Azam M, 1996). A continuous supply of copper in the diet is therefore necessary. An excellent source of copper is oysters, which accumulate copper in digestive glands and kidneys. Elevated concentrations of copper in aquatic food have been reported only when waste water from human mining activities has contaminated areas from which aquatic animals were harvested. High amounts of copper are present in crustaceans, decapods, gastropods and cephalopods that use copper (White SL and Rainbow PS, 1985) in their haemocyanins to carry oxygen to their tissues. Despite the high copper concentrations in some molluscan and crustacean shellfish, copper concentrations in aquatic food present no problem for human health. Fish in polluted areas contains higher amounts of copper than fish from unpolluted areas, a principal accumulation of this element in the trophic chain could, however, not be demonstrated. The concentration of copper in fish muscle is at an average 0.2–0.5 mg/kg wet weight while organs contain more copper (in decreasing order: liver, scales, spleen, kidney, gills). Excessive copper is stored mainly in the liver.

7.5.1 Analysis

For the determination of copper in fish the same principles used in trace analysis have to be followed. However, the risk of contamination during sampling, handling of samples and sample pre-treatment is much lower in copper analysis than, for example, in cadmium analysis due to the relatively high concentrations found in aquatic foods (cadmium:copper = 1:1000) compared to environmental level. The analytical methods used nowadays are mostly ETAAS and DPSAV and sometimes ICP-AES (Bassari A, 1994). Because of the high copper concentrations in fish, flame AAS can also be used. Spectral interferences in AAS are low.

7.6 Zinc

Zinc, like copper, is an essential element for humans and forms an integral part of many enzymes. It is present in fish and other seafood in mg/kg amounts and there have been no reports of concentrations in the edible parts of food fish that form a hazard to health. With an average zinc content of 3–5 mg/kg, wet fish zinc is a good source for this essential element. An exceptionally good source is the wolffish (*Anarhichas* spec.) which contains up to 9 mg/kg and molluscs (Romeo M *et al.*, 2000) like the oyster, which often may have tenfold the level of wolffish, since zinc forms part of the enzyme carbonic anhydrase responsible for laying down the oyster shell (Ratkowski DA *et al.*, 1974).

7.6.1 Analysis

Because of its high concentration in fish and other seafood no difficulties in the determination of zinc in this matrix have been reported.

7.7 Tin

There is evidence that tin is essential for mammalian growth. Tin has been used for the production of containers for the canning industry. Tin may migrate into the can's content and contaminate this when the enamel layer is missing or is insufficient, if the content had a too high pH value and if the can was stored open to the air for a time. The use of enameled aluminium cans, glass containers and the development of better enamel layers in many countries had the effect that tin intoxication by ingestion of contaminated content of cans is today extremely rare. Organotin compounds (Fent K, 1996), mainly tributyltin (TBT) have been used as stabilizers for PVC, as molluscicides, fungicides and insecticides and as a component of antifouling paints for ships and have found their way into the aquatic environment. TBT has been found in edible parts of fish (Harino H, Fukushima M and Kawai S, 2000; Takahashi S, Tanabe, S and Kubodera T, 1997; Kannan *et al.*, 1995) and in mussels (Binato G *et al.*, 1998), however, there is an ongoing controversial discussion (Belfroid A C *et al.*, 2000; Evans SM, 1999; Goldberg ED, 1986) about the health risk of low concentrations of tin and tin compounds found in food fish. The use of TBT in paints has now been banned in many countries. Average total tin concentrations found in fish range between 0.4–8 mg/kg. TBT concentration in fish species and molluscs from the North Sea was variable depending on species from < 1 μg/kg (plaice) to > 100 μg/kg (mackerel, herring) wet weight (Kruse R, 2001).

7.7.1   Analysis

Tin can be analysed by a number of analytical methods (Sturgeon RE and Siu KWM, 1994; Caricchia AM *et al.*, 1993) such as spectrophotometry, fluorimetry, DPSAV, x-ray fluorescence, ICP-MS, FAAS, ETAAS, HG (hydride generation) AAS, for organic tin compounds gas chromatography (Ishizaka T *et al.*, 1989), and HPLC. A main problem encountered in the trace analysis of tin is the presence of considerable amounts of tin in reagents used for analysis (e.g. perchloric acid, sulfuric acid, and water). For the quality evaluation of tin analysis only a very low number of certified standard reference materials are available.

7.8   Aluminium

Aluminium is of interest since it was argued that high aluminium intake may be related to Alzheimer's dizease. This is still questionable, but the fact that aluminium is one of the most abundant metals on earth and that it has a lot of contact with fish (aluminium cans, processing machinery, cooking utensils, aluminium foil, etc.) has caused many researchers to investigate aluminium content in fish. The aluminium content of fish muscle from the open sea is close to 0.1 mg/kg wet weight. Fish muscle from coastal waters in the vicinity of an aluminium smelting plant showed elevated levels in muscle up to 1 mg/kg fresh weight (Ranau R, Oehlenschläger J and Steinhart H, 2001a). Higher concentrations were found in organs like gills. Thorough research in canned products showed that the aluminium content in spices, vegetables and sauces used as ingredients in canned fish products was always higher than the aluminium content in fish. Only after prolonged storage (> 4 years) was it shown that the aluminium content in fish was higher than the aluminium content in the other components. Studies on historical samples of canned products, which were up to 30 years old showed that there was a continuous increase in aluminium concentration in the fish part of canned fish products up to 30 mg/kg fish after 30 years (Ranau R, Oehlenschläger J and Steinhart H, 2000). This shows that aluminium migrates from the walls into the content and that the normal best-before date for canned fish (4 years) is sufficient. Another recent investigation revealed that fish wrapped in aluminium foil takes up a considerable amount from the foil when being grilled (Ranau R, Oehlenschläger J and Steinhart H, 2001b).

7.8.1   Analysis
Aluminium is preferably determined by ETAAS after ashing in a closed system (Ranau R, Oehlenschläger J and Steinhart H, 1999). Care must be taken when crustaceans and molluscs are analysed that the digestive tract, which contains small amounts of sand, is completely removed.

7.9   Future trends

It seems that the burden of heavy metal concentration in fish and other seafood is becoming lower (Guns et al., 1999). The ban of tetraalkyl lead in gasoline, successful efforts to avoid high burdens of mercury in rivers, substitution of mercury and lead in batteries, ban of the use of tributyltin and new metallurgic techniques that are less polluting have all reduced heavy metal release into the environment considerably. However, new heavy metals have appeared on the scene: these are noble metals (platinum group) used as catalysts in cars (Hoppstock K, 2001; Balcerak M, 1997) and manganese compounds as additives in gasoline.

The analytical techniques are developing rapidly and very sophisticated instruments are on the market that allow highly reliable and accurate results. However, this equipment is not generally used because of its high price. We will therefore continue to struggle with inaccurate data produced by old instrumentation and, unfortunately, insufficiently educated and trained personnel.

Legislation, which has concentrated for decades only on legal limits of mercury or methyl mercury, is now being worked on to set limits also for lead and cadmium. Hopefully, the legal limits reflect the low concentrations that are encountered in the overwhelming proportion of fish caught in the open sea and do not take as a basis for decision-making the few too high concentrations found occasionally in other regions being influenced by anthropogenic activities.

7.10   Sources of further information and advice

Further information about the content of heavy metals in fish and other seafood and its determination can be found in a few of monographs and chapters in books on the subject (Eisler R, 1981; Kiceniuk JW and Ray S, 1994; Alfassi ZB, 1994). Some scientific journals, such as Marine Pollution Bulletin (Elsevier

Science Ltd.), *Bulletin of Environmental Contamination and Toxicology* (Springer-Verlag), *Archive of Environmental Contamination and Toxicology* (Springer Verlag), *Chemosphere* (Elsevier), *Science of the Total Environment* (Elsevier), *Environmental Pollution* (Kluwer) specialize in marine pollution. There is a vast amount of literature on the topic, however, it is almost impossible for a non-expert in analytical science to judge if data presented are reliable and accurate. It is recommended that advice be sought from critical reviews about this topic, which take into consideration only those sources that are analytically sound.

A lot of information can be found on the Internet but the majority of it is secondary or tertiary literature. This information is primarily available for educational purposes, or for press release and whilst it may be suitable for these purposes, it often lacks scientific accuracy and should be treated with this in mind. A lot of information is hidden in the offices of the seafood industry that regularly analyses the heavy metal content of their raw material. Unfortunately this source is mostly inaccessible. Some of the best information sources are proceedings and reports issued after conferences and meetings that specialize in this subject. The distribution of those reports is often restricted but they are available through libraries. As guidance for the heavy metal content in the edible parts of fish, it can be said that the lower the values reported the higher the probability that they are accurate. Finally, consumers of fish should always be aware of the fact that fish is an important source of nutrients and that the proven benefits from fish consumption outweigh the potential risks from chemicals.

7.11   References

ABOUL NAGA WM (1996), Comparative study of trace metals accumulated in the muscle tissues of the most common and marketable sea food in Alexandria waters, *Int J Environ Health Res*, 6, 289–300.

ABREU SN, PEREIREA E, VALE C and DUARTE ACr (2000), Accumulation of mercury in sea bass from a contaminated lagoon (Ria de Aveiro, Portugal), *Mar Poll Bull*, 40 (4), 293–7.

ALFASSI ZB (1994), *Determination of trace elements*, Weinheim, VCH.

BALCERAK M (1997), Analytical methods for the determination of platinum in biological and environmental materials, *Analyst*, 122, 67R–74R.

BARD AJ and ZOSKI CG (2000), Voltammetry retrospective, *Anal Chem* 62, 346A–352A.

BARGAGLI R, NELLI L, ANCORA S and FOCARDI S (1996), Elevated cadmium accumulation in marine organisms from Terra Nova Bay (Antarctica), *Polar Biol*, 16, 513–20.

BARGHIARNI G, PELLEGINI D, D'ULIVO A and DERANIERI S (1991), Mercury assessment and its relation to selenium levels in edible species of the Northern Tyrrhenian Sea, *Mar Poll Bull*, 22, 406–9.

BASSARI A (1994), A study on the trace element concentrations of Thunnus thynnus, Thunnus obesus and Katsuwonus pelamis by means of ICP-AES, *Toxicol Environ Chem*, 44, 123–7.

BELFROID AC, PURPERHART M and ARIESE F (2000), Organotin levels in seafood, *Mar Poll Bull*, 40 (3), 226–32.

BETTY M and PAPOFF P (1988), Trace elements: Data and information in the characterisation of an aqueous ecosystem, *CRC Crit Rev Anal Chem* 19, 271–322.

BINATO G, BIANCOTTO G, PIRO R and ANGELETTI R (1998), Atomic absorption spectrometric screening and gas chromatographic-mass spectrometric determination of organotin compounds in marine mussels: an application in samples from the Venetian Lagoon, *Fresenius J Anal Chem, 361, 333-337.*

BLOOM NS (1992), On the chemical form of mercury in edible fish and marine invertebrate tissue, *Can J Fish Aquat Sci* 49, 1010–17.

BORTOLI A, GEROTTO M, MARCHIORI M, MUNTAU H and REHNERT A (1995), Critical comparison of methods for mercury determination in fish, *Mikrochim Acta*, 119, 305–310.

BRANICA M and KONRAD Z (1980), *Lead in the marine environment*, Oxford, Pergamon Press.

CAPPON CJ and SMITH JC (1981), Mercury and selenium content and chemical form in fish muscle, *Arch Environ Contam Toxicol* 10, 305–19.

CAPPON CJ (1994), Mercury and organomercurials, in: Kiceniuk JW and Ray S, *Analysis of contaminants in edible aquatic resources*, Weinheim, VCH, 175-205.

CARICCHIA AM, CHIAVARINI S, CREMISINI C, MORABITO R and UBALDI C (1993), Analytical methods for the determination of organotins in the marine environment, *Intern J Environ Anal Chem* 53, 37-52.

CARNRICK GR, SCHLEMMER G and SLAVIN W (1991), Matrix modifiers: their role and history for furnace AAS, *American Lab* 1991, 120–31.

CLAISSE D, COSSA D, BRETAUDEAU-SANJUAN J, TOUCHARD G and BOMBLED B (2001), Methylmercury in molluscs along the French coast, *Mar Poll Bull*, 42 (4), 329-332.

CLARKSON TW (1998), Human toxicology of mercury, *J Trace Elements Exp Med* 11, 303–317.

DE BOER J and SMEDES F (1997), Effect of storage conditions of biological materials on the contents of organochlorine compounds and mercury, *Mar Poll Bull*, 35 (1–6), 93–108.

DE CLERCK R, VYNCKE W, GUNS M and VAN HOEYWEGHEN P (1995), Concentrations of mercury, cadmium, copper, zinc and lead in sole from Belgian catches (1973–1991), *Med Fac Landbouww Univ Gent*, 60(1), 1–6.

DE GREGORI I, PINOCHET H, GRAS N and MUNOZ L (1996), Variability of cadmium, copper and zinc levels in molluscs and associated sediments from Chile, *Environ Poll*, 92(3), 359–68.

DOBSON J (2000), Long term trends in trace metals in biota in the Forth estuary,

Scotland, 1981–1999, *Mar Poll Bull* 40 (12), 1214–1220.

EBDON L, PITTS L, CORNELIS R, CREWS H, DONARD OFX and QUEVAUVILLER P (2001), Trace element speciation for environment, food and health, Royal Society of Chemistry, Cambridge, 391 pages.

EISLER R (1981), *Trace metal concentrations in marine organisms*, New York, Pergamon Press.

EVANS O and MCKEE GD (1987), Optimization of high-performance liquid chromatographic separations with reductive amperometric electro-chemical detection: Speciation of inorganic and organic mercury, *Analyst* 112, 983–8.

EVANS SM (1999), Tributyltin pollution: the catastrophe that never happened, *Mar Poll Bull* 38 (8), 629–36.

FALCONER CR, DAVIES IM and TOPPING G (1986), Cadmium in edible crabs (*Cancer pagurus* L.) from Scottish coastal waters, *Sci Total Environ* 54, 73–183.

FENT K (1996), Ecotoxicology of organotin compounds, *Crit Rev Toxicol* 26 (1), 1–117.

FRANCESCONI KA, MOORE EJ and JOLL LM (1993), *Aust J Mar Freshwater Res* 44, 787–97.

FRAUSTRO DA SILVA JJR and WILLIAMS RJP (1993), *The biological chemistry of the elements The inorganic chemistry of life*, Oxford, Clarendon Press.

GOLDBERG ED (1986), TBT an environmental dilemma, *Environment*, 28 (8), 17–24.

GONZALEZ F, SILVA M, SCHALSCHA E and ALAY F (1998), Cadmium and lead in a trophic marine chain, *Bull Environ Contam Toxicol*, 60, 112–8.

GUNS M, VAN HOEYWEGHEN P, VYNCKE W and HILLEWAERT H (1999), Trace metals in selected bentic invertebrates from Belgian coastal waters (1981–1996), *Mar Poll Bull*, 38 (12), 1184–93.

HARINO H, FUKUSHIMA M and KAWAI S (2000), Accumulation of butyltin and phenyltin compounds in various fish species, *Arch Environ Contam Toxicol*, 39, 13–19.

HOPPSTOCK K (2001), Platinum group elements in the environment (in German), *Nachrichten Chemie*, 49, 1305–9.

ISHIZAKA T, NEMOTO S, SASAKI K, SUZUKI T and SAITO T (1989), Simultaneous determination of tri-n-butyltin, di-n-butyltin, and triphenyltin compounds in marine products, *J Agric Food Chem* 37, 1523–7.

ISMAIL A, JUSOH NR and GHANI IA (1995), Trace metal concentrations in marine prawns off the Malaysian Coast, *Mar Poll Bull*, 31(1–3), 108–10.

JENKS PJ, RUCINSKI RD and JERZAK H (2001), Commercial approaches to the certification of reference materials for environmental analysis, *Spectroscopy Europe*, 13 (6), 10–20.

JOINS CR, ALI IB, HOLSBEEK L, BOSSICART M and TAPIA G (1995), Total and organic mercury in Barents Sea pelagic fish, *Bull Environ Contam Toxicol*, 55, 674–81.

JOINS CR, ALI IB, HOLSBEEK L, KANUYA-KINOTI M and TEKELE-MICHAEL Y (1997),

Total and organic mercury in Greenland and Barents Seas demersal fish, *Bull Environ Contam Toxicol*, 58, 101–7.

JOIRIS CR, HOLSBEEK L and KAROUSSI MOATEMRI L (1999), Total and methylmercury in sardines *Sardinella aurita* and *Sardina pilchardus* from Tunisia, *Mar Poll Bull*, 38 (3), 188–92.

JOIRIS CR, DAS HK and HOLSBEEK L (2000), Mercury accumulation and speciation in marine fish from Bangladesh, *Mar Poll Bull*, 40 (5), 454–7.

KALAY M, AY Ö and CAHLI M (1999), Heavy metal concentrations in fish tissues from the northeast Mediterranean sea, *Bull Environ Contam Toxicol* 63, 673–81.

KANNAN KS, TANABE S, TATSUKAWA R and WILLIAMS RJ (1995), Butyltin residues in fish from Australia, Papua New Guinea and the Solomon Islands, *Int J Environ Anal Chem*, 61, 263–73.

KAWAGUCHI T, PORTER D, BUSHEK D and JONES B (1999), Mercury in the American oyster *Crassostrea virginica* in South Carolina, USA, and public health concerns, *Mar Poll Bull*, 38 (4), 324–7.

KICENIUK JW and RAY S (1994), *Analysis of contaminants in edible aquatic resources*, Weinheim, VCH.

KLAASEN CD, LIU J and CHOUDHURI S (1999), Metallothionein: An intracellular protein to protect against cadmium toxicity, *A. Rev Toxicol* 39, 267–94.

KRÜGER KE and KRUSE R (1984), Der Quecksilbergehalt in Thunfischkonserven aus Südeuropa, Afrika und Asien, *Arch Lebensm Hyg* 35, 55–8.

KRUSE R (2001), unpublished results.

LAMBLE KJ and HILL SJ (1998), Microwave digestion procedures for environmental matrices, *Analyst* 123, 103R–133R.

LINDER MC and HAZEGH-AZAM M (1996), Copper biochemistry and molecular biology, *Am J Clin Nutr*, 63, 797S–811S.

LOBINSKI R and ADAMS FC (1994), Lead and organolead compounds, in: Kiceniuk JW and Ray S, *Analysis of contaminants in edible aquatic resources*, Weinheim, VCH, 115–56.

MASON RP, LAPORTE J-M and ANDRES S (2000), Factors controlling the bioaccumulation of mercury, methylmercury, arsenic, selenium, and cadmium by freshwater invertebrates and fish, *Arch Environ Contam Toxicol* 38, 283–97.

NEIDHART B and WEGSCHNEIDER W (2002), *Quality in chemical measurements – Training concepts and teaching materials*, Ijmuiden, Springer for Science.

ODZAK N, MARTINCIC D, ZVONARIC T and BRANICA M (1994), Bioaccumulation rate of Cd and Pb in *Mytilus galloprovincialis* foot and gills, *Mar Chem*, 46, 119–31.

OEHLENSCHLÄGER J (1991), Schwermetallgehalte in den Tuben (verzehrbarer Anteil) und Eingeweiden von Tintenfischen (Cephalopoden), *FIMA SchrReihe* Band 21, 105–15.

OEHLENSCHLÄGER J (1994), Experience with differential pulse scanning anodic voltammetry (DPSAV) in trace metal analysis in biological matrices, *Baltic Sea Environ Proc.* 58, 92–6.

OEHLENSCHLÄGER J (1997), Marine fish – a source for essential elements? In: Luten JB, Börresen, T and Oehlenschläger J (eds.), *Seafood from producer to consumer, Integrated approach to quality*, Amsterdam, Elsevier Science BV, 641–52.

OEHLENSCHLÄGER J and PRIEBE K (1989), Quecksilber- und Selengehalte in Weißem Heilbutt *(Hippoglossus hippoglossus)* aus verschiedenen Fanggebieten des Nordatlantiks, *Lebensmittelchem Gerichtl Chem*, 43, 13–15.

PHILLIPS DJH (1995), The chemistries and environmental fates of trace metals and organochlorines in aquatic ecosystems, *Mar Poll Bull* 31 (4–12), 193–200.

PRUDENTE M, KIM E-Y, TANABE S and TATSUKAWA R (1997), Metal levels in some commercial fish species from Manila Bay, the Philippines, *Mar Poll Bull*, 34 (8), 671–4.

QUEVAUVILLER P (1995), Certified reference materials for specific chemical forms of elements, *Analyst*, 120, 597–602.

RAINBOW PS (1985), Accumulation of Zn, Cu and Cd by crabs and barnacles, *Estuarine Coastal Shelf Sci*, 21, 669–86.

RAINBOW PS (1997), Trace metal accumulation in marine invertebrates: marine biology or marine chemistry?, *J Mar Biol Ass U.K.* 77, 195–210.

RANAU R, OEHLENSCHLÄGER J and STEINHART H (1999), Determination of aluminium in the edible part of fish by GFAAS after sample pretreatment with microwave activated oxygen plasma, *Fresenius J Anal Chem* 364, 599–604.

RANAU R, OEHLENSCHLÄGER J and STEINHART H (2000), Aluminiumgehalte im verzehrbaren Anteil von Fischdauerkonserven während Langzeitlagerung bei Raumtemperatur, *Arch Lebenm Hyg* 51, 142–50.

RANAU R, OEHLENSCHLÄGER J and STEINHART H (2001a), Aluminium content in edible part of seafood, *Eur Food Res Technol* 212, 431–8.

RANAU R, OEHLENSCHLÄGER J and STEINHART H (2001b), Aluminium levels of fish fillets baked and grilled in aluminium foil, *Food Chem* 73, 1–6.

RATKOWSKI DA, THROWER SJ, EUSTACE IJ and OLLEY J (1974), A numerical study of the concentration of some heavy metals in Tasmanian oysters, *J Fish Res Bd Can* 31 (7), 1165–71.

RAY S (1994), Cadmium, in: Kiceniuk JW and Ray S, *Analysis of contaminants in edible aquatic resources*, Weinheim, VCH, 91–113.

ROMEO M, SIDOUMOU Z and GNASSIA-BARELLI M (2000), Heavy metals in various molluscs from the Mauritanian coast, *Bull Environ Contamin Toxicol*, 65, 269–76.

RUANGWISES N and RUANGWISES S (1998), Heavy metals in green mussels *(Perna viridis)* from the Gulf of Thailand, *J Food Prot*, 61, 94–7.

SCHULZ-SCHROEDER G and SCHERING B (1995), Cadmiumbelastung von Tintenfischerzeugnissen, *Arch Lebensm Hyg*, 46, 40–3.

SHRIVASTRAN AK and TANDON S G (1982), The determination of mercury: A mini-review, *Toxicol Environm Chem*, 5, 311–29.

SOLBERG T, BECHER G, BERG V and ERIKSEN GS (1997), *Kartlegging av miljoegifter i fisk og skalldyr fra nord-omraadene*, SNT-rapport 4, Oslo, Statens naeringsmiddeltilsyn.

STOCK A and CUCUEL F (1934), Die Verbreitung des Quecksilbers, *Naturwissenschaften* 22, 390–3.

STOEPPLER M (1994), *Probenahme und Aufschluss – Basis der Spurenanalytik*, Berlin, Springer-Verlag.

STURGEON RE and SIU KWM (1994), Tin and organotin, in: Kiceniuk JW and Ray S, *Analysis of contaminants in edible aquatic resources*, Weinheim, VCH, 225–55.

SZPUNAR J (2000), Bio-inorganic speciation by hyphenated techniques, *Analyst* 125, 963–88.

TAKAHASHI S, TANABE, S and KUBODERA T (1997), Butyltin residues in deep-sea organisms collected from Saruga Bay, Japan, *Environ Sci Technol* 31, 3103–9.

WELZ B and SPERLING M (1997), *Atomabsorptionsspektrometrie*, Weinheim, Wiley-VCH.

WHITE Sl and RAINBOW PS (1985), On the metabolic requirements for copper and zinc in molluscs and crustaceans, *Mar Environ Res*, 16, 215–29.

WONG CK, WONG PPK and CHU LM (2001), Heavy metal concentrations in marine fishes collected from fish culture sites in Hong Kong, *Arch Environ Contam Toxicol* 40, 60–9.

ZUMAN P (2000), Current status of polarography and voltammetry in analytical chemistry, *Anal Lett* 33 (2), 163–74.

8

Fishborne zoonotic parasites: epidemiology, detection and elimination

K.D. Murrell, Danish Centre for Experimental Parasitology, Frederiksberg

8.1 Introduction

The awareness, knowledge and literature pertaining to fishborne zoonotic parasites are growing and are now quite substantial. This is due in part to the extensive range of parasites (primarily helminths or 'worms') of fish that can be transmitted to humans and to the world-wide concern over food safety and quality. Over 50 species of helminth parasites have been implicated as zoonotic resulting from eating undercooked fish, crabs, snails and bivalves (Deardorff 1991). These helminths are represented by nematodes, trematodes, cestodes and acanthocepalans. It has been estimated by WHO (1995) that there are 750 million people at risk to fish and invertebrate-borne trematode ('flukes') parasites alone. The anisakid nematodes, for example, are reported to cause thousands of cases of human illness each year in Japan and the United States (Todd 1989; Ishikura and Kikuchi 1990; Williams and Jones 1994). In a study in Japan, Ishikura *et al.* (1992) reported a total accumulation of 11,629 cases of gastric, 567 cases of intestinal and 45 cases of extra-gastrointestinal anisakiasis, as well as 355 cases of gastric pseudo-terranoviasis. More recently, Ishikura *et al.* (1998) reported that worldwide the total number of cases in the literature is 33,747. Over 500 cases of anisakid ileitis have been reported from other countries.

The focus of this chapter is on the detection and mitigation of the major fishborne zoonoses. Although the number of potentially zoonotic species is large, most are relatively infrequent, limited in distribution and, importantly, susceptible to the same measures for decontamination of fish and fish products employed for other pathogens, therefore only the most important species will be addressed. Likewise, because the emphasis is on measures that can be taken during harvesting, processing or post processing (e.g. by the consumer) to reduce the risk of infection, descriptions of the life cycles, epidemiology and distributions of these major parasites will be brief. The reader is referred, instead, to several excellent treatments of these aspects (Ko 1995; Williams and Jones 1994; Cross 2001; Kumar 1999; and Anderson 2000).

This chapter details current programs (8.3) to reduce risk of infection, including good manufacturing practices (GMPs) and hazard analysis and critical control point (HACCP) systems. These approaches embrace both detection methods and processing procedures to achieve decontamination. Finally, a discussion on future trends is presented which attempts to identify forces that may exacerbate the risk of fish and human infection.

8.2   Parasites of marine fish

8.2.1   Nematodes (roundworms)

These are long, cylindrical, non-segmented worms enclosed in a cuticle (see Fig. 8.1 for an example). They are sexually dimorphic and typically have multiple host life cycles, each stage of which is preceded by a molt (4 in total). The group is large, widespread and common in fish.

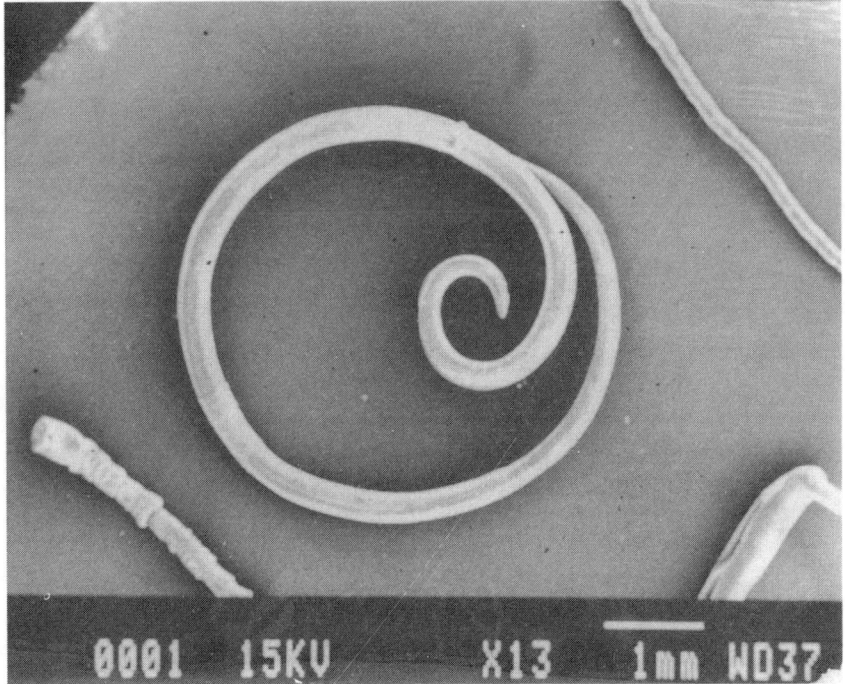

Fig. 8.1   The third stage larvae of *Anisakis* spp. from fish muscle. (Reprinted with permission from Buchmann K, Bresciani J and Beyerholm B, 2001, *Parasitic Diseases of Freshwater Trout*, DSR Publishers, Copenhagen.)

The most important of the nematode zoonotics of marine fish are members of the family Anisakidae (the anisakids), *Anisakis, Contraceacum* and *Pseudoterranova*, of which *A. simplex* (Fig. 8.1) is considered the most pathogenic, although *P. decipiens* is frequently reported to cause symptoms. More than 95% of cases manifest as acute gastric anisakidosis, in which severe epigastric pain is experienced shortly after ingestion of fish carrying parasite larvae incorporated in the fish muscle (McCarthy and Moore 2000).

Life cycle
The life cycle of *A. simplex* is complex (Fig. 8.2). The adult worms reside in the stomachs of marine mammalian hosts, mainly whales, dolphins and porpoises. Eggs produced by female worms pass in the feces and embryonate in the ocean waters. After hatching from the egg, the larvae enter small marine micro-

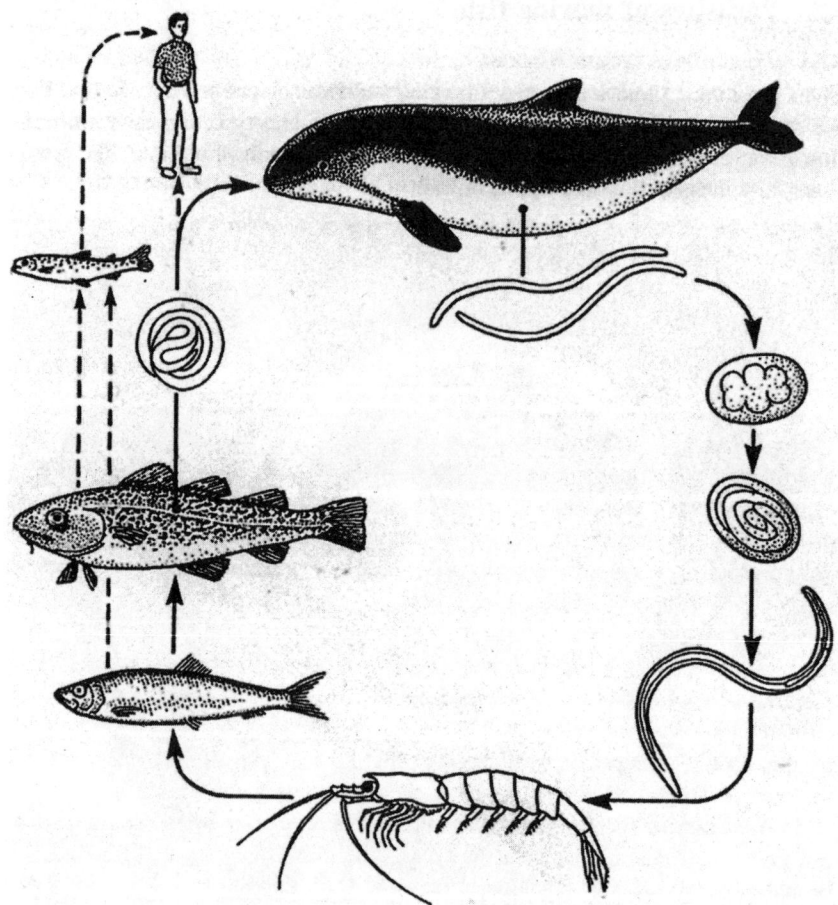

Fig. 8.2   Life cycle of *Anisakis* spp. (Reprinted with permission from Buchmann K, Bresciani J and Beyerholm B, 2001, *Parasitic Diseases of Freshwater Trout*, DSR Publishers, Copenhagen.)

invertebrates (such as crustaceans of the family Euphausidaé) and develop into third-stage larvae. When the crustacean is eaten by a fish or squid (paratenic host), the larvae are released and migrate out of the gastrointestinal tract, and enter the mesenteries, viscera or muscle. If the infected fish or squid is then eaten by a marine mammal, the larvae are released and become established in the host's stomach, where they bury the anterior portion of their bodies into the gastric mucosa.

P. decipiens, the so-called cod or sealworm, has also been the subject of considerable research because its fish intermediate host includes commercially important species such as Atlantic cod (Anderson 2000). Although, its life cycle is similar to that of *A. simplex*, its definitive hosts are more commonly the Pinnipedia (seals, sea-lions, walruses), and the crustacean first intermediate host is the copepod.

Epidemiology
The species of fish which may harbor the larval stage (3rd stage larvae) in their muscles are numerous (Williams and Jones 1994). Major fish hosts for *A. simplex* are herring (*Clupea harengus*), cod (*Gadus* spp.), mackerel (*Scomber* spp.), salmon (*Oncorhynchus* spp.) or squid (*Todarodes* spp.), and for *P. decipiens*, cod, halibut (*Hippoglossus hippoglossus*), flatfish (family Pleuronectidae) or Pacific red snapper (*Sebastes* spp.). See Ko (1995) for fuller description of potential fish host species. Humans acquire larval *A. simplex* and *P. decipiens* by eating raw or inadequately cooked, salted, pickled or smoked fish. Major sources of infection are traditional preparations such as raw herring, lomi lomi salmon (a preparation of raw, marinated salmon), ceviche, sushi, sunomono, isushi and sashimi (Adams *et al.* 1997).

Anisakiasis was first recognised in Holland in the 1960s (Van Thiel *et al.* 1960), where people frequently eat lightly pickled or raw herring. The disease was initially associated with the use of refrigerated or iced fish, which had been caught in the ocean and subsequently kept cold on the fishing boats for several days prior to reaching port. Before the use of refrigeration or ice, the boats did not stay out for such long periods and the fish's viscera was removed shortly after being caught, either on the boat or on shore. The introduction of refrigeration and the ensuing longer chilling period meant that after the fish had died in the cold, anisakid larvae could migrate out of the viscera (liver, mesenteries) and into the fish's muscles. On reaching port and the market, although the fish were eviscerated, the larvae had already moved into the muscle by this time. Furthermore, smoking or pickling the fish was inadequate to kill the worms which had already reached the muscle (Adams *et al.* 1997).

In Asia, marine fish are usually eviscerated shortly after being caught, and are not iced or refrigerated, because they reach the market in a matter of hours. This probably accounts for the lower incidence of human infections in Asian countries other than Japan. Furthermore, few other populations in Asia relish raw marine fish as the Japanese do. However, a total of 107 cases of anisakiasis were reported in Korea between 1989 and 1992 (Im *et al.* 1995). New cases still occur in Europe, especially in areas where it has not been reported previously, for example, in Southern Italy (Maggi *et al.* 2000).

Pathology
In humans, *A. simplex* larvae enter the gastric or intestinal mucosa and cause an abscess or eosinophilic granuloma (Ishikura *et al.* 1992). The worms may also enter the peritoneal cavity and other organs. Some worms may not invade tissue but instead may pass out with feces, vomit or pass up the oesophagus. *P. decipiens* larvae may also invade tissue, but rarely attempts to lodge in the oesophagus; the larvae may, however, cause 'tickling throat syndrome' in which a tickling sensation occurs and the patient may then cough up the larvae.

The symptoms of anisakiasis resemble an acute abdomen, gastric ulcer or neoplasm. Parasitological diagnosis is made by finding the worms or demonstrating the parasite in sectioned biopsied tissue. Serological tests are not conclusive. The treatment for most infections is removal of the parasite surgically or by the use of forceps through fibre-optic endoscopy. The prognosis is good once the parasite has been removed.

8.3   Parasites of freshwater fish: nematodes

8.3.1   Gnathostomiasis
Zoonotic infections by nematodes of the genus *Gnathostoma* are increasingly being reported. Several species of this parasite are transmitted by larvae encysted in the muscles of a variety of freshwater fish; the most important species are *G. spinigerum*, *G. binucleatum*, *G. hispidum* and *G. doloresi*. The final or definitive hosts for these parasites are usually carnivorous mammals, including cat, dog and pig (Miyazaki 1966). The adult worms are short, stout with a subglobose head armed with 7–9 transverse rows of hooklets. Spines extend down the body. This armature makes tissue migration by the third-stage larvae quite damaging to the host's organs and tissues.

Life cycle
The life cycle is complex and may involve a wide range of intermediate hosts (Cross 2001). Adult worms are located in the stomach wall of the definitive host (cats, dogs, etc.). Eggs are produced which pass out of the intestine in the feces. The first-stage larva hatches from the egg and is eaten by a freshwater copepod of the genus *Cyclops*, within which it develops into a second-stage larva. If the copepod is eaten by a second intermediate host (fish, birds, amphibians, reptiles, and mammals) the parasite enters the tissue and molts to a third-stage larva. When the second intermediate host is eaten by a definitive host (in which sexual reproduction occurs) it begins a complex migration. The parasite is digested free from the host tissue, penetrates the stomach wall, and migrates to the liver, and then to other organs, eventually returning to the peritoneal cavity. It then penetrates the stomach to form a tumor-like mass. The worms reach maturity

and produce eggs in approximately six months. Larvae in humans (a non-definitive host species) do not reach maturity.

Epidemiology

Although statistics on prevalence are not extensive, this zoonosis is present throughout much of Southeast Asia, particularly where the eating of intermediate hosts (fish, frogs, snakes, poultry, etc.) raw or undercooked is a common practice (Japan, Thailand, China) (Nawa 1991). Infections occur in all age groups and sexes. In Japan, outbreaks have increased in frequency, and in one outbreak, it was linked to loaches imported from China (Akahane *et al.* 1982).

Perhaps the most notable region of increase for this zoonosis is among the Latin American countries of Argentina, Peru, Ecuador and Mexico. In the later country, gnathostomiasis is now recognized as a significant public health problem (Akahane *et al.* 2000). As chronicled by McCarthy and Moore (2000), the dramatic increase in Mexican towns along the Papaloan river basin in the Gulf of Mexico is associated with the construction of the Presidente Miguel Aleman dam and the stocking of the reservoir in 1964 with imported tilapia fish (apparently infected). These events coincided with the rise of the popularity of raw fish in the form of ceviche or callos. From the first case in 1970, many cases have now been reported in the six coastal states bordering either the Pacific or Gulf states. In the city of Sinaloa (in the North-Pacific coast of Mexico), over 700 cases of gnathostomiasis have been reported since 1989 (Diaz-Camacho 2000). Most cases in Mexico are considered to be due to *G. binucleatum*. More recently, several cases of human gnathostomiasis were reported from travellers returning from Tanzania (McCarthy and Moore 2000).

Pathology

In humans, after ingestion of the second intermediate host, the larva migrates through the tissues and into any organ. It may enter subcutaneous tissue and become a 'larval migrans'; lesions develop along the worm's migratory track, causing inflammation, necrosis, and hemorrhage. Toxic products, immunological responses, and mechanical injury cause inflammatory reactions, eosinophilia, swelling, pain, and edema. A reaction is transient and disappears in a few days but may reappear some time later at a new location. The parasite may enter an eye, causing subconjunctival edema, hemorrhage, and retinal damage. Invasion into the CNS leads to encephalitis, myelitis, radiculitis, and subarachnoid hemorrhage (Chitanandh and Rosen 1967). Intracerebral hematoma and transitory obstructive hydrocephalus also have been described.

8.4   Parasites of freshwater fish: cestodes

These segmented flatworms ('tapeworms') are hermaphroditic and are characterized by a scolex ('head') that attaches to the host's intestinal epithelial lining, and a long tape-like body (strobila) made up of segments containing the reproductive organs and eggs. They have, generally, multiple host life cycles (see Fig. 8.3).

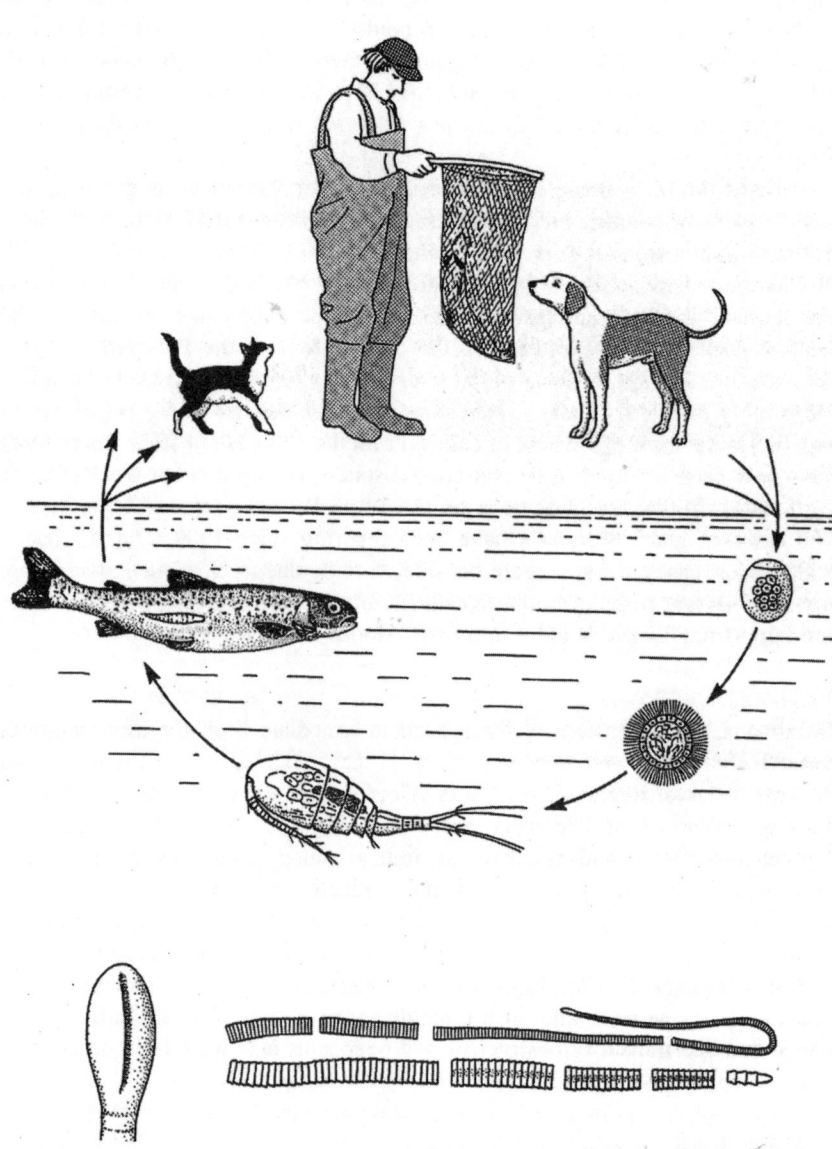

Fig. 8.3   Life cycle of *Diphyllobothrium latum*. (Reprinted with permission from Buchmann K, Bresciani J and Beyerholm B, 2001, *Parasitic Diseases of Freshwater Trout*, DSR Publishers, Copenhagen.)

8.4.1 Diphyllobothrium

Of the cestode parasites of fish, the only significant zoonotic species belongs to the genus *Diphyllobothrium*, which is most often transmitted from freshwater fish, although marine fish may be involved. This zoonotic parasite has a wide distribution in the temperate and sub-Arctic regions of the Northern Hemisphere, although certain species occur in South America, Asia and Africa. World-wide there are at least 13 species of *Diphyllobothrium* that have the potential to infect humans; the most important of these is *D. latum* (Northern Hemisphere), followed by *D. dendriticum* (Subarctic), *D. nihonkaiense* (Japan) and *D. pacificum* (Chile and Peru).

Life cycle

These 'fish tapeworms' are among the largest helminth parasites of humans. They reside in the small intestine and range from 2 to 15 meters in length, with a normal width of 20 mm. The scolex (or 'head') is 2 mm in length and has a dorsal-ventral sucking groove (or bothrium) that allows it to hold to the intestinal wall. The rest of the body is segmented, beginning with a neck, followed by immature, mature (sexually) and gravid (full of eggs) proglottids or segments; there may be up to 3,000 proglottids.

The life cycle (Fig. 8.3) of *Diphyllobothrium* is also complex (Rausch *et al.* 1998), and requires three hosts for completion. However, additional intermediate or paratenic hosts may be involved. After hatching of the egg in water, the motile embryo (coracidium) is ingested by a minute crustacean (copepod), in which the first-stage larva (procercoid) develops. When the copepod containing the procercoid is ingested by the second intermediate host, a fish, further development leads to the plerocercoid, which is infective for the final host. The localization site of the plerocercoid in the second intermediate host differs with species of *Diphyllobothrium* and, to some extent, species of fish.

After the final host (mammal) ingests the plerocercoid, the adult worm develops in the intestine quite rapidly. Host specificity is broad in these cestodes, in that a number of species of fish-eating mammals can serve as the final or definitive host for these parasites, although humans appear to be the primary host for *D. latum*. Other, important definitive hosts for this genus are dogs, minks, cats, pigs, bears, seals, and sea lions.

Epidemiology

Reliable figures on the prevalence and incidence of diphyllobothriasis are difficult to obtain. In most regions, the incidence of this disease has decreased in the last few decades, however, in the Volga basin of Russia the incidence remains high (Cross 2001). Between 1977 and 1981, the US Centers for Disease Control reported between 125–200 cases per year in the United States although the actual number of infections is thought to be much greater (Deardroft and Kent 1989). In the Pacific Coast states of the USA, of the 52 cases reported in one year, salmon (*Oncorhynchus*) was implicated in 82% of cases (Williams and Jones 1994). In Canada approximately 25 cases per year are recorded, some of which are from marine fish. Plerocercoids are commonly found in pike, salmon, trout, sugar, ruff, and whitefish (Rausch *et al.* 1998); up to 70% of northern and wall-eye pike may be infected (Bogitsch and Cheng 1990).

In Japan, the incidence seems to be increasing since the 1970s, with about 100 cases now recorded annually (Oshima 1984). In the former USSR, 12.4% of the population of the Khatangskii region were reported to be infected with *D. latum* and *D. dendriticum* (Klebanovskii *et al.* 1977). Romanenko (1986) reported a prevalence in humans of 6.3% near the Krasnoyarsk reservoir. According to Williams and Jones (1994), the most important fish species in transmission to man are:

Eurasia	Pike *(Esox lucius)*; burbot *(Lota lota)*; perch *(Perca fluviatalis)*; and ruff *(Acerina cernua)*.
North America	Northern pike *(E. lucius)*; wall-eyed pike *(Stizostedion vitreum)*; sand pike *(S. canadense griseum)*; burbot *(L. maculosa)*; and yellow perch *(P. flavescens)*. Also, various species of salmonids *(Oncorhynchus)* along the Pacific coast.
Japan	*Oncorhynchus masu*; *O. gorbuscha*; *O. keta*; and *O. nerka*.
Chile and Peru	*Sciaena callaensis*; *Cynoscion analis*; *Merluccius gayi peranus*; and *Genypterus maculatus*.

The important risk factor for this zoonosis is the consumption of raw or undercooked fish, including inadequate smoking or pickling. Related factors in the epidemiology of this zoonosis are the intrusion of man into the wild habitat of the host fish and contamination of water with infected human waste. Although (non-human) fish-eating mammals can serve as reservoir hosts, it is humans that are the primary definitive host and that are mainly responsible for establishing and maintaining endemicity in human populations, through the contamination of water with feces. It is vital that sewage from lake-side dwellings, hotels and ships be properly treated before entering the water (Murrell 1995). Deworming of domestic pets is also beneficial. Greater control of the latter, however, will not eliminate the risk of diphyllobothriasis in marine systems because of the existence of alternative mammalian hosts for species of *Diphyllobothrium*. Therefore, post-harvest controls such as proper fish processing and preparation of food are the most effective in preventing diphyllobothriasis (see Section 8.3).

Pathology
Although the worm is quite large, it causes little pathology. However, *D. latum* can compete with the host for vitamin B_{12}, which may lead, after several years, to megablastic anemia. This is seen primarily in Scandinavian countries (especially Finland). Other clinical problems occasionally occur (Cross 2001). Toxic worm secretions may affect the central nervous system causing peripheral and spinal nerve degeneration (Rausch *et al.* 1998). Other species of *Diphyllobothrium* are not known to be associated with anemia. Diagnosis of

infection by *D. latum* and other diphyllobothriid cestodes usually depend on finding the eggs in the feces of the host.

8.5   Parasites of freshwater fish: trematodes

These endoparasites are non-segmented flatworms or 'flukes' (platyhelminths), which are characterized by possession of oral, and usually, ventral suckers and a life cycle requiring one or more intermediate hosts. They are, like the cestodes, mostly hermaphroditic. They are widespread throughout the vertebrate and invertebrate classes (see Kumar 1999). Among the fish-borne zoonotic trematodes, only five genera are of major importance: *Clonorchis* and *Opisthorchis*, which infect liver and bile ducts; and *Heterophyes*, *Haplorchris* and *Metagonimus* which are intestinal parasites.

8.5.1   Clonorchiasis

The causative agent of this zoonosis, *Clonorchis sinensis* is widespread throughout Asia and the former Soviet Union, with an estimated 290 million people at risk and seven million infected (WHO 1995). Although over 100 species of fish can harbor the infective (metacercaria) stage, most human infections are derived from cyprinids reared in aquaculture enterprises. The increase in fish farming in Laos, Thailand, Korea, Vietnam, Japan, and China appears to be an important factor in the increasing prevalence of this zoonosis.

Life cycle
The definitive or final hosts include man, dogs, cats, pigs, and rats (Fig. 8.4). When eggs pass out of the feces and reach water, they are ingested by snails and hatch releasing a ciliated stage (miracidia). This stage penetrates into the tissues of certain organs, where they undergo asexual multiplication, producing a stage termed the cercaria (possessing a tail). The cercariae emerge from the snail and swim about until locating a suitable fish intermediate host. They penetrate beneath the scales of the fish and encyst (metacercaria) in the muscles and subcutaneous tissues. Humans and other suitable hosts become infected by ingesting infected raw fish. The metacercaria, digested free from the fish tissue in the host's stomach, move into the small intestine and then enter the common bile duct (usually 4–7 hours after ingestion) and mature in about four weeks.

Epidemiology
The occurrence of human clonorchiasis is closely associated with ethnic customs which favor consumption of raw or inadequately processed or cooked fish (Kumar 1999). A second major factor in the epidemiology of this zoonosis is the contamination of fish ponds and natural water bodies with infected human and animal waste. Also of importance is the preference for cultivation of cyprinids

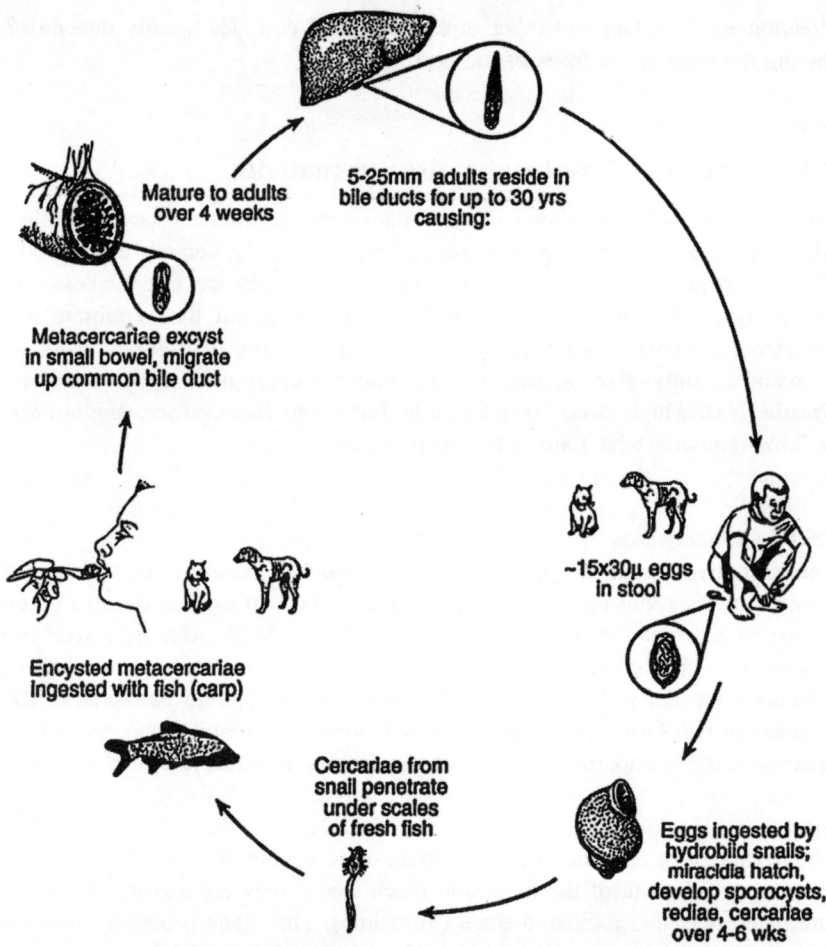

Fig. 8.4 Generalized life cycle for fish-borne liver flukes (*Clonorchis sinensis*, Opisthorchis viverrini and *O. felineus*). Adapted from MacLean, JD, Cross JH and Mahanty S, 1999, in *Tropical Infection Diseases*, Guerrant RL, Walker DH and Weller PF, eds. Churchill Livingston, Philadelphia, p. 1039.

(carp), an excellent intermediate host for *C. sinensis*. *Clonorchis sinensis* is common in Asia. According to Li (1991), in China more than 4 million people are infected in Guangdong and Guangxi provinces. Pigs, which may serve as reservoir hosts, also have high prevalences of infection judging by the results of various surveys in China (e.g. 11–35%) (Li 1991). Soh (1984) considers *C. sinensis* to be important in Korea where the prevalence may exceed 10% in many rural areas. In Hong Kong, the prevalence reported for some villages is 13% (Ko 1995). The Red River delta of Vietnam is also recognized for its high prevalence rates (28%) (Kumar 1999); about a million Vietnamese are considered infected.

Pathology
Infected people complain of indigestion, epigastric discomfort, and diarrhea. If the adult worm invades the pancreatic duct, acute pancreatitis may result (Mas-Coma *et al.* 2000). Chronic infection may lead to cholangiocarcinoma (Cross 2001).

8.5.2   Opisthorchiasis
A related human fish-borne liverfluke disease is opisthorchiasis, caused by *Opisthorchis viverrini* and *O. felineus* (the cat liver fluke). These parasites are found over a wide geographic range; *O. viverrini* occurs chiefly in Southeast Asia and *O. filineus* is generally found in Eastern Europe, Poland, Germany, and Siberia. In Southeast Asia, *O. viverrini* is not only a medical problem, but also an important public health-related economic cost; in Thailand, the wages lost due to opistorchiasis (*O. viverrini*) in young men are estimated at USD 65–85 million per annum (WHO 1995; Ko 1995; Loaharanu and Santasiri 1991). Opistorchiasis caused by *O. felineus* is the most prevalent and widely distributed human trematode infection in the Russian Federation and is also a public health problem in Kazakhstan and Ukraine (WHO 1995).

Life cycle
The epidemiology and life cycle of the opistorchid flukes are similar to that of *C. sinensis* (Fig. 8.4). The infection is acquired by eating improperly prepared fish harboring the infective muscle stage (metacercaria). Cyprinid fish species are also major hosts; in endemic regions, infection rates may be as high as 50–90% with fish harboring 20–50 metacercariae each (Ko 1995). Interestingly, *O. felineus* infections are occasionally reported to occur when fish are eaten on the first day of fish salting, underscoring that salting must be carried out carefully to overcome the resistance of the metacercarial stage. Cats and dogs (reservoir hosts) are commonly infected in endemic areas, complicating efforts to protect snail-inhabiting water sources.

Epidemiology
The primary causes of human infection is the consumption of raw fish. *Cyclocheilicthys siaja*, *Hampala despar*, and *Puntius orphoides* are important fish hosts in Southeast Asia, while in Europe and Siberia the important species are *Leuciscus rutilus*, *Blicca bjoerkna*, *Cyprinus carpio*, *Tinca tinca* and *Barbus barbus* (Ko 1995; Cross 2001). In Thailand, fish raised in ponds fertilized with human and animal feces are at high risk. An important human risk factor is the feeding of raw or partly cooked fish to infants by mothers; although only a few worms are typically found in young children, the continued consumption of raw fish may lead to the accumulation of large worm burdens, which appear to stabilize around 30–40 years of age (Cross 2001). The worms may live for 20 years.

Pathology
Opisthorchis viverrini infections cause clinical symptoms similar to *Clonorchis sinensis* (Mas-Coma *et al.* 2000). However, opisthorchiasis may have greater public health significance because it is more often associated with cholangiocarcinoma than clonorchiasis (Haswell-Elkins *et al.* 1992), and is a major cause of death in rural Northeast Thailand. The International Agency for Research on Cancer has declared this parasite a Group 1 carcinogen.

8.5.3   Intestinal trematodes
There are a large number of small (usually less than 2.5 mm long) trematode species that can be transmitted from fish to humans and localize in the intestine. Twenty-eight Asian countries report the presence of over 70 species of intestinal trematodes infecting humans. In the trematode family Heterophyidae, over 10 species infect man, the most important of which are *Heterophyes heterophyes*, *Haplorchris* spp., and *Metagonimus yokogawi*. These species, plus a number of closely related species, are highly prevalent in the Middle East (especially the Nile Delta) and Southeast Asia (Mao 1988, Khamboonraung 1991, Edoardo 1991, Giboda *et al.* 1991, Chai and Lee 1991). Village surveys in Laos have found heterophyid infection rates as high as 86% (Giboda *et al.* 1991).

Life cycle
These parasites are typical snail-borne trematodes, utilizing a variety of freshwater fish as second intermediate hosts. As with *Clonorchis* and *Opisthorchis*, the metacercarial infective stage is encysted in the fish host's muscle, and when consumed raw or improperly cooked, the fluke completes its development in the host's intestine (Ko 1995). The life cycle is completed when eggs of the intestinal worms are shed in the host's feces. If they reach water, they may be ingested by snails in which the asexual stages occur. The cercariae, which eventually emerge, seek a suitable fish host as intermediate hosts. A very large number of bird and mammal species may serve as final hosts for this parasite species assemblage. At least 45 genera of fish may serve as intermediate hosts, but the major hosts for *Heterophyes* spp. include mullet species *Mugil cephalus*, *M. capito*, *M. japonicus*, *Talapia milotica*, *Aphanius fasciatus*, *Acanthogobius*, *Sciaena aquilla*, *Solea vulgaris*, and *Arius manilensis* (Ko 1995). For *Metagonimus* spp., fish hosts commonly include *Leuciscus nakuensis*, *Odontobutis obscurus*, *Plecoglossus ativelis*, *P. parva*, *Salmo perryi*, *Carassius carassius*, *Tribolodon* spp., and *Lateolabrax japonicus* (Ko 1995; Kumar 1999).

Epidemiology
Thirty countries and areas have reported infections with these intestinal trematodes (WHO 1995). The restricted, usually local nature of infection foci, makes it difficult to determine the overall prevalence or population at risk, but WHO estimates a minimum of 1 million people are infected (WHO 1995). This

breaks down to the following estimates: Egypt (heterophyiasis) 10,000; China (heterophyiasis) 230,000; Japan (metagonimiasis) 150,000; Korea (metagonimiasis) 500,000; Russian Federation (metagonimiasis) 12,500. Various prevalence surveys indicate these parasites are not infrequent in Vietnam, Laos, Taiwan, Philipinnes, Indonesia, Israel, the Balkans, Spain, Tunesia, Sudan, Turkey, Iran, and India (Kumar 1999; Cross 2001).

Infection is acquired by eating raw or partially cooked fish. Infections may also be acquired by the ingestion of freshly salted fish, a common food practice; the metacercaria can remain infective in freshly salted fish for a week (Kumar 1999). A variety of fish-eating birds and mammals can serve as hosts, acting as reservoirs of infection; dogs are a common reservoir in many areas.

Pathology
Heterophyes heterophyes   The disease has been described in detail by MacLean *et al.* (1999): An average of nine days following ingestion of the metacercaria, dyspepsia, colicky abdominal pain, diarrhea, and eosinophilia may occur. A mild focal inflammatory reaction and superficial erosions are produced at the site of attachment. The flukes appear to live for less than a year. The fluke may penetrate through the mucosa and eggs may embolize from these intramucosal sites via lymphatics to the systemic vascular system. Eggs of three different heterophyid species have been recovered from capillaries of brain, heart, lungs, spleen, and liver. Space-occupying lesions in the brain and spinal cord produce pathologic changes. Myocarditis can follow the occlusion of myocardial vessels by eggs and the resultant granulomatous and fibrotic host reaction. Thickened mitral valves containing ova have been reported.

Metagonimus yokogawi   This species exhibits a similar pathological pattern in humans. Worms can be detected in the crypts of Lieberkühn in the middle section of the small intestine (early stages) and between the villi in older infections. These worms have not been seen in the submucosa in immunocompetent hosts. Usually hosts are asymptomatic, but in heavy infections, abdominal symptoms may occur (diarrhea, weight loss) (Ichiki *et al.* 1990).

Other fish-borne parasites
There exist a number of other trematode and nematode zoonoses that may occasionally be encountered, particularly in wild fish. Examples are the echinostomes (a group of about 16 *Echinostoma* species with diverse life cycles) (Graczyk and Fried 1998), *Nanophytes salmincola* (a cause of so-called 'salmon poisoning' in dogs of the Northwest USA) (Cross 2001) and *Metorchis conjunctus* (transmitted by raw fish such as *Catostomus commersoni* in Canada) (MacLean *et al.* 1996). The literature should be consulted for further details.

Since the 1960s, there have been a number of cases of human infection in Southeast Asia, Iran and Egypt by the nematode *Capillaria philippinensis*. Although the wild definitive host is probably a fish-eating bird, monkeys and gerbils have been experimentally infected (Cross 2001). Freshwater fish have been shown to transmit the larval stage to people in the Philippines and Thailand; the fish host range for this parasite is probably broad.

8.6   Prevention and decontamination: marine fish

Prevention of infection for fish-borne parasites can easily be achieved by avoidance of raw, insufficiently cooked or inadequately processed fish. However, the continuing demand for raw or lightly processed food, often related to strong cultural and social traditions, often renders this approach unrealistic, and other strategies are called for. Perhaps the most well-developed prevention and control methods are those for marine fishes, mainly for the anisakid nematodes.

Anisakid nematodes, being part of the natural environment of their hosts, wild marine fish such as herring, cod, salmon and tuna are difficult to prevent or eliminate completely from fish. However, there are effective measures that can be taken to minimize or mitigate the risks of infection; these involve either physically removing the parasite from fish or inactivating them *in situ*. These measures may be applied either during harvesting, processing or post-processing treatment by the customers. These measures have been described below in detail for marine fish by Adams *et al.* (1997) and Adams and De Vlieger (2001), and are presented below:

Harvesting
Actions to reduce the likelihood or abundance of parasite contamination of a seafood product can be taken prior to harvesting. The type, size and biology of fish (e.g., groundfish, pelagic or anadromous fish) intended for capture should be considered. For example, the ecology, particularly feeding habits, of groundfish, such as arrowtooth flounder (*Atheresthes stomias*) and many species of sole (family Pleuronectidae) tend to result in the acquisition of large numbers of larval anisakid nematodes. Additionally, zoonotic parasites (e.g., anisakids and diphyllobothriids) accumulate within the host over the lifetime of the fish. Therefore, the selective harvesting of younger fish of such species will reduce the likelihood of large numbers of parasites being present. Young (1972) found that at one study site, 83% of the cod were harvested before they attained 60 cm in length, an age when the fish were too young to have acquired infections of anisakids. Certain fish stocks or geographic locations develop reputations for having excessive numbers of parasites, and are either avoided by fishing vessels or the harvested fish are heavily processed into minced products (or analogue as used for surimi). To reduce the need for extensive processing, some firms specifically purchase only smaller fish of a species known to have such parasite problems.

Fish occupying inland waters which are also frequented by marine mammals (including rookeries and haul-out areas) often exhibit appreciably larger numbers of parasites in the edible flesh. For example, populations of the grey seal (*Halichoerus grypus*) have been markedly increasing during the past 35 years in the North Atlantic and this expansion has resulted in an apparent greater incidence of *P. decipiens* in cod and other bottom-dwelling fish from these locations. Harvesting of these types of fish should be avoided, or the fish should be thoroughly processed. To avoid the additional cost incurred by extensive processing to ensure that the anisakids are removed, a processor may prefer to pay a premium for fish caught outside marine mammal areas.

During harvesting, the method of capture may contribute indirectly to the necessity to remove parasites during processing. Fish which are caught with long-lines rather than with nets tend to be fresher because they are bled immediately after death and are then chilled or frozen. The resulting product has a whiter flesh and is easier to candle (view over a strong, bright light) for physical identification and removal of parasites.

After capture, the fish may be stored or chilled, or may receive some preliminary processing (e.g., heading and gutting). The larval helminths of concern accumulate primarily in the viscera and secondarily in the edible flesh. If the worms are prevented from migrating into the flesh after the death of the fish host by rapid chilling or gutting, the number of parasites moving into the flesh will be reduced. Some fish, such as salmon, herring and arrowtooth flounder, appear to be more susceptible to post-mortem migration than other species therefore particular care should be taken with these species. Gutting allows the physical removal of the viscera, along with any parasites present; chilling or freezing, however, may be carried out with the viscera intact. In addition to preventing the migration of parasites, rapid chilling or freezing also extends the shelf-life and freshness of the seafood. Freezing the fish at appropriate temperatues and durations will kill the parasites and eliminate the possibility of infection. None of these methods will eliminate the presence of parasites which migrated into the flesh prior to harvesting, however.

For those fish known to be potential hosts for parasites, processors may place further restrictions on the harvesters, especially for those vessels that remain at sea for extended periods of time, capturing several different species of fish. For fish such as arrowtooth flounder, the processor may request that the time at sea be reduced or that the fish be caught and held only during the last few days of fishing, to decrease the period of post-mortem anisakid migration. The processor may also 'test' a sample of the catch to determine the extent of parasite migration. A failing grade occurs when nematodes are found above the lateral line of the fillets; the catch may then be refused by the processor.

For a few commercially valuable species, particularly the salmonids, aquaculture can produce stock in which the presence of those parasites of public health concern is decreased or eliminated. In aquaculture, the fish can be trained to accept only pelleted feed. Such fish do not appear to recognize crustaceans and smaller fish as possible prey. Under these conditions, the life cycles of anisakid nematodes and tapeworms are interrupted and fish remain free of these parasites. However, if the fish are acquired from a hatchery which is inhabited by *Juga plicifera*, the snail intermediate host needed for the *Nanophyetus* life cycle, the salmon may arrive at the net pens already infected with the trematode. Hatcheries which use only well-water and which are free of the snails, however, can produce salmon free of this trematode. In a study involving two net-pen sites and 237 Atlantic salmon (*Salmo salar*), the viscera and edible flesh of all fish were free of anisakids, diphyllobothriids and *N. salmincola* (Deardorff and Kent 1989). For those culinary dishes which incorporate raw or undercooked seafood, the use of salmon raised in aquaculture conditions is a reasonable alternative to wild-caught salmon with regard to parasites of public health concern.

8.6.2   Processing

Depending on the final product and market, a processor may perform several steps in the handling of a catch, including heading and gutting, filleting, skinning, candling and trimming. Candling, a process in which a fish fillet is placed on a light table to enable the detection of parasites, is not completely effective in revealing the presence of all parasites. (See Freshwater Fish section below for details on the candling method approved by the Association of Official Analytical Chemists.) Factors which may interfere with parasite detection are as follows:

- thickness of the fillet
- presence of skin on the fillet
- oil content
- pigmentation
- the level of experience of the operator.

In a comparison of four types of white-fleshed fish – rockfish (*Sebastes* spp.), arrowtooth flounder (*Atheresthas stomias*), sole (family Pleuronectidae) and true cod (*Gadus macrocephalus*) – candling detected 53–79% of the infected fillets and from 43–76% of the anisakids present (Adams *et al.* 1997). Parasites observed by candling may be removed either with a probe or forceps or by trimming the portion of fillet affected. The area of the fillet which generally has the most parasites is the belly flap region surrounding the viscera (the first area into which the parasite can migrate). Some processors automatically cut away the belly flap without candling, while others may candle the fillets and trim only those containing visible parasites.

Irradiation has been proposed as an additional step prior to marketing to increase food safety for many food products, including fish (Loaharau and Murrell 1994). However, several issues need to be considered before this technology is adopted. Many consumers do not understand how irradiation works (e.g., expecting food to become radioactive), and are reluctant to purchase food which has been treated in this manner. Of primary importance is the fact that the amount of irradiation necessary to eliminate a hazard varies according to the organism. Irradiation of pork to eliminate the possibility of trichinellosis is highly effective at 0.15–0.30 kGy, in that the larvae do not need to be killed, but simply rendered sexually sterile (Brake *et al.* 1985). Unfortunately, irradiation may be inappropriate for the anisakid nematodes. It is the larvae which are present within the fishery products that cause anisakiasis, thus the level of irradiation may have to be high enough actually to kill the nematodes (0.5 MegaRads or 10 kGy) to prevent migration. At very high levels of irradiation, fish acquire an unpalatable texture and taste.

In addition to fresh and frozen seafood on the market, products which have been prepared in brine, salt, pickle or marinade are available, (consumers may also utilize these steps at home to prepare a seafood dish). In general, two components are present in pickling solutions and marinades: salt and acid. The acid level present in brining solutions has no appreciable effect on anisakid nematodes, which is not surprising since anisakids live primarily in the stomach of marine mammals. Therefore, attention should be given to the salt level of any brining or pickling solution. Of the helminths described above, the nematodes are the most resistant to brining solutions. Dry salt is lethal to anisakids within ten minutes of direct contact. In a 22% salt solution (saturated), the worms will die in ten days. As the concentration of salt drops, the time required to kill the nematodes increases; in a 15% salt and 7% acid solution, 97% of the worms are killed after 30 days, while at a 6% salt and 4% acid solution (similar to most pickling solutions) more than 70 days are required. For many pickled fish products (e.g., pickled herring, ceviche), freezing the fish prior to pickling or marinating is recommended.

Within the framework of seafood Hazard Analysis Critical Control Point (HACCP), freezing provides the most practicable control measure for parasites. In the Final Rule for seafood HACCPs, processors are required to institute a control plan for parasites on the assumption that the product will be consumed without adequate removal of the hazard, or for products marketed for such consumption. Therefore, a control step is required for fishery products which are known to have a parasite problem and which are generally sold to be consumed raw or undercooked. Freezing, carried out at only one of several production steps, will fulfil the HACCP requirement. Fish frozen at the harvesting stage, for example, need not be refrozen by the processor as a parasite control measure. During the production of Nova Scotia-type smoked salmon, the salmon is frozen after smoking to enable the thin-slicing of the fish which is characteristic of the final product.

Freezing as a means of killing parasites is time/temperature dependent. Generally, for parasitic worms, 15 hours in a blast freezer at −35°C (−31°F) or seven days at −20°C (−4°F) will be adequate. Home freezers should be checked for temperature and sufficient air-flow to ensure effective freezing. Heat inactivation of parasites is, however, the single most effective method for eliminating the risk of parasitic infection and it can be implemented during processing or at home by the consumer. However, macroscopic parasites, such as worms, will still be present within the product and may be visible to the consumer, although this is an aesthetic issue and not a health problem. Heat inactivation of parasites is also time/temperature dependent (above a minimum temperature). For conventional cooking (including baking), the internal temperature of the thickest part of the product should reach a minimum of 63°C (145°F) for 15 seconds or longer. Cooking with a microwave oven requires a higher temperature to kill all the parasites due to the uneven heating which occurs: a temperature of 77°C (170°F) in the thickest part of the product is recommended. Turning the food during cooking, or covering the food and adding liquid while cooking, helps to stabilize the uneven heating of the fish.

Control of parasites in smoked products is dependent on the type of smoking performed. Hot smoking of products, particularly fishery products, can effectively inactivate parasites because the product is cooked during the smoking process. However, care should be taken to ensure that the product reaches an appropriate temperature for a sufficient period. Unlike bacteria, helminth parasites are not affected by the seasoning and flavorings (liquid smoke) which are used on a raw product to impart a smoked flavor. In cold-smoking operations which use real smoke, the product contained within a chamber, generally the temperature achieved is not sufficient to kill parasites. Brining or salting prior to cold-smoking may also have a limited effect. Many operations use fish which has been frozen previously, or the product is frozen after smoking for either storage and transportation, or to enable thin slices of product to be cut, and in these situations the freezing of the product could meet the control requirements if proper temperatures and time durations are ensured. It must be noted, however, that the presence of killed anisakid larvae in the fish muscle may still represent a health hazard (allergic reactions) (Audicana et al., 2002).

8.7   Prevention and decontamination: freshwater fish

Parasitic zoonoses involving freshwater fish include both wild fish and those reared in aquaculture systems. For this reason, comprehensive control is multidimensional and complex. This also accounts for the fact that control of these zoonoses lags behind that for marine fish parasites. Such control requires a national strategy that links together various agencies, or sectors and activities (WHO 1995). The greatest problem encountered in controlling infections in farmed fresh water fish is the prevention of pond contamination with human and animal feces.

The prevention of infection of fish depends on environmental control of surface waters where fish are caught, hygienic aquaculture, and the control or elimination of the first intermediate hosts (snails). In addition, the treatment of the definitive host (animal and human reservoir) is also an important control measure. There are many potential difficulties in the implementation of these measures. For example, monitoring and control of large bodies of surface water (rivers, lakes) may be impracticable in developing countries. However, adoption of certain aquaculture practices to exclude foodborne trematodes may be possible (FAO 1974).

Priority must be given to the prevention of contamination of water with domestic animal and human excreta. The increasing use of wastewater in agriculture demands a high degree of management. Proper disposal of excreta is fundamental to preventing contamination of water containing susceptible parasitic intermediate hosts. This does not mean that the hygenic disposal of treated waste, sewage, and sludge through use in well-managed fish ponds is unacceptable. Although water quality guidelines specify the complete absence of helminth eggs, this is rarely practicable for small subsistence ponds in Asia. However, this should be the goal.

Conventional wastewater treatment systems are not generally effective in removing helminth eggs in wastewater. Rapid sand filtration has been shown to be effective against some eggs, although its effectiveness at removing all trematode eggs is not known. The general guidelines for treating of sludge from wastewater treatment specify heating at more than 55°C; however, the effectiveness of such heating has not been specifically tested against trematode eggs (WHO 1995).

WHO and FAO have promulgated recommendations for the application of HACCP systems for application in fish farming to control parasites (Ahmed 1992; WHO 1995):

Application of the HACCP approach to the control of fishborne parasites should start with assessment of the risk that certain species of fish (e.g. carp) caught or harvested in endemic areas are infected. The risk of disease should also be determined; this will depend on the specific parasite and whether the possibility of acquiring the infection through eating fish is eliminated by cooking or another preparation process before consumption. A risk will exist if fish from endemic areas are consumed raw or inadequately processed. The HACCP approach is unique for every product and for every production unit: in each case a detailed study of the food producing/processing/marketing/preparation flow is necessary to identify hazards and the critical control points.

When the HACCP analysis is complete and the program is ready for implementation, training must take place. All the people involved in the programme, from fish farmers, traders, and processors to food-sellers, restaurant managers, etc., must understand the principles, and have a clear idea of their own role in the system. For the HACCP approach to be fully operational and generally applied, increased communication to promote understanding and collaboration between the scientific community, the general public, and the food regulatory agencies will be required.

A recent effort to implement a HACCP program for control for *Opistorchis viverrini* in cultured carp (*Puntius gonionotus*) has been reported from Thailand (Khamboonraung *et al.* 1997). Using a multidisciplinary team approach (public health, parasitologists, fish aquaculturists and food hygienists), it was shown that the HACCP strategy can prevent and control *O. viverrini* in cultured carp, especially with an emphasis on eliminating contamination of pondwater with helminth eggs.

Specific procedures for detection and decontamination of freshwater fish are not as well developed as that for marine fish (see above). However, procedures for examination of fish for certain fish-borne trematodes have been recently published (see recommendations of the WHO Study Group (WHO 1995, Annex b), and are summarized below:

8.7.1 Inspection techniques

1. Compression method
   (a) Dissect the fish into four parts: fins, scales, subcutaneous tissue, and flesh. Each part is examined by compression and microscopy to determine the distribution of encysted larvae. The addition of a few drops of tap water or saline may be necessary, in particular for fin and scale specimens. For *Clonorchis sinensis* and *Opisthorchis* spp., the subcutaneous tissue and flesh should be examined. For *Metagonimus yokogawi*, the fins and scales should be examined first. Examine them at low magnification (30× or 50×) using a binocular dissecting stereomicroscope with light transmitted from beneath the stage. For flesh, compress a piece of flesh (0.5–1.0 g) between two large glass plates and examine as described above.
   (b) Differentiate the species of encysted larvae (metacercariae) morphologically by examining the size and shape of the cysts and the characteristic features of the internal organs. If only representative metacercariae are needed (e.g., for identification), they may be dissected from the specimens using forceps and dissecting needles. Diagnostic reference guides are available.
   (c) Estimate the total number of encysted larvae, if required, in the fish by multiplying the weight of the fish in grams by the number of encysted larvae observed per gram of flesh.
2. Digestion method (for isolation of parasitic larvae)
   (a) Divide the fish into five parts: head, anterior trunk, posterior trunk, tail, and subcutaneous tissue. Each part is digested separately with artificial gastric juice and the isolated larvae are counted to determine their distribution. The following steps should be applied to each part in turn.
   (b) Grind a large piece of the fish (10–20 g) using a meat grinder, and mix it with 200–300 ml of artificial gastric juice (0.6% HCl and 1% pepsin in distilled water).
   (c) Incubate the mixture in a flask with glass beads at 37°C for 3–4 hours with occasional shaking, or use a stirring plate with a magnetic stirring rod placed in the flask. Note that if a stirring plate is used or if the flesh of the fish is soft, digestion may take less than three hours.
   (d) Isolate the freed larvae from the debris by repeated sedimentation in tap water until the supernatant becomes clear (the liquid will usually need

to be decanted at least three times). Detection of parasites may be facilitated by pouring the sediment into a dark colored pan or by placing the glass container over a dark piece of paper. Collect the isolated larvae on a watch glass and count them under a binocular dissecting stereomicroscope.

(e) When necessary, metacercariae may be kept in saline at 4–5°C for up to 30 days.

3. Candling method for fish flesh
   The following description is derived from method 985.12 in Helrich K, ed. *Official methods of analysis of the Association of Official Analytical Chemists*, 15th edn (Arlington, VA, Association of Official Analytical Chemists, Inc., 1990). The method is similar to that for marine fish fillets to detect *Anisakis* spp. It is not suitable, however, for detection of metacercariae of *Opisthorchis* spp. or *C. sinensis* because of their small size. The accuracy of the results can be limited by thickness, pigmentation, and high oil content of the flesh, and by the presence of skin on the fillet. Some larvae will remain undetected by this procedure, but its advantages are that it is quick and inexpensive, it does not destroy the product, and little training is required. The procedure requires a candling table, i.e. a 'cool white' diffused light source within a box-like structure, topped by a translucent working surface (e.g., frosted glass plate or acrylic plastic).

   (a) Place a skinned fillet (or fish steak) on the lighted working surface. Thick fillets can be cut lengthwise to facilitate the transmission of light through the flesh.
   (b) Examine the fillet for parasites. Larvae close to, or on, the surface should be readily visible. Embedded larvae may appear as shadowed spots in the flesh. Removal of these 'spots' can verify the presence of the parasites.
   (c) An accurate count of the number of metacercariae present can be made following digestion of the infected fillet.

Aside from the organoleptic methods described above, there are few other methods for detecting fishborne parasites, at least, none are being used by fish inspection services. An ELISA test for screening fish for anisakid larvae (*A. simplex*) in saithe fish has been reported (Huber *et al.* 1989). However, further development work on this test, particularly optimizing its specificity, does not appear to have been made (Ko 1995). Although the practicality and flexibility of molecular based diagnostic methods is ever increasing, work in this area for detection of fish parasites appears to be of little interest (see WHO 1995, Annex 7).

8.7.2 Processing

The appropriate technologies for inactivating parasites of freshwater fish are, in general, not well developed except for certain trematode metaceriae (Farkas 1987). Heat treatment is considered effective for fish if the flesh is heated until it

Table 8.1 Processing conditions under which inhibition of infectivity of foodborne trematode metacercariae has been observed (Adapted from WHO 1995)

Process or variable	Parasite	Processing parameters	
		Temperature	*Time*
Heating	*O. viverrini* free metacercariae	50°C	5 hours
		70°C	30 min.
		80°C	5 min.
Acidity[a]		*Concentration*[c]	*Time*
commercial vinegar	*O. viverrini* free metacercariae	4%	1 hour
acetic acid	*O. viverrini* free metacercariae	4%	1.5 hours
lactic acid	*O. viverrini* free metacercariae	4%	1.5 hours
citric acid	*O. viverrini* free metacercariae	4%	1 hour
		Concentration[c]	*Time*
Salting (NaCl)[b]	*O. viverrini* free metacercariae	0.9%	10 days
		10%	3.6 hours
		20%	12 hours
		30%	1 hour
	Opisthorchis metacercariae in fish	13.6%	24 hours
		Temperature	*Time*
Freezing	*Clonorchis* and *Opistorchis* metacercariae in fish	−10°C	5 days
	O. felineus metacercariae in fish	−28°C	32 days
		−35°C	14 hours
		−40°C	7 hours

[a] pH not measured.
[b] a_w (water activity) not measured.
[c] Percentage salinity in fermented fish.

is firm in texture and white or pale in color throughout (Table 8.1). However, *Opisthorchis* spp. may require at least 30 minutes at 70°.

Freezing can also be effective for metacercariae (Table 8.1), as is canning, salting, drying, smoking, fermenting, etc., according to appropriate codex or ministry standards, although this is not always feasible in countries where these parasites are endemic. Irradiation of fish can also be effective for fishborne trematode and cestode parasites (Farkas 1987; Loaharanu and Murrell 1994), even with low doses (less than 1 Krad). This technology is applicable to fish handed in bulk. The adoption of irradiation has been slow, primarily because of consumer attitudes, regulatory actions, economics and logistical issues. In most situations, it will not be cost effective to irradiate for a specific parasitic zoonosis, but is more likely to be employed to ensure safety from a spectrum of pathogens, especially bacterial pathogens.

8.8   Future trends

Predicting the future trends for zoonotic parasites has proved risky, because of uncertain global influences such as demographics, environmental changes, and industrial activities. Zoonotic parasites exist within a continuum among wild and domestic animals and human populations and pertubations to their equilibrium may come from agricultural and industry intensification, environmental alterations, translocations of humans and animals, human travel and the export and distribution of food (Daszak *et al.* 2000). Fishborne parasites are influenced by such causal factors and changes in the seafood and freshwater fish farming practices are of particular importance. There are expectations that world-wide demand for fish and fish products will increase (Deardorff 1991; FAO 1992; WHO 1995), particularly for freshwater fish as Asian producers seek more lucrative markets in the developed world. This may have the effect of raising the risk in urban areas because of shipping unfrozen fish by air by these producers in order to gain a competitive edge (Nawa *et al.* 2001; Käferstein 1994). There is also the risk that zoonotic parasites introduced into new areas may gain a foothold and spread, a risk exacerbated by increased global travel, especially from tourism (Ko 1995).

Increased demand for fish from endemic areas is likely also to be stimulated by changing consumer tastes, particularly for raw or lightly cooked or processed fish (e.g., sushi, ceviche, etc.) (Deardorff 1991; Ko 1995). A well-known example of a fish-transmitted parasite that became established in the United States because of food preferences by immigrants is *Diphyllobothrium*; this transplant from Scandinavia persists and continues to infect North Americans.

Seemingly unrelated environmental events may also exacerbate these zoonosis problems. Ishikura *et al.* (1998) report that global climate changes have had a profound effect on the incidence of anisakidosis in the Western Pacific. The El Niño events especially have caused decreases in the abundance and distribution of the intermediate and mammal hosts (e.g., sea lions), thereby affecting the parasite's life cycle and risk of transmission. In contrast, the passage in 1972 of the Marine Mammal Protection act has resulted in a rapid growth of some marine mammals, including sea lions, dolphins and seals, which has increased the risk for infection of fishes (Deardorff 1991; Ishikura *et al.* 1998).

These examples of alterations in nodes of the 'food web' complex underscore the need for comprehensive approaches in assessing risk attendant to anticipated changes in fish production and marketing, and also in activities that may affect the aquatic ecosystems. The growing public awareness of food safety risks will demand a more effective risk assessment process. The fish production processing and marketing industry will need to improve monitoring and inspection proficiency to meet consumer expectations. An important step towards ensuring safety and quality would be a uniform, global marine and freshwater fish inspection system (Deardorff 1991).

8.9   References

ADAMS AM and DEVLIEGER DD (2001). Seafood Parasites: Prevention, inspection and HACCP, in Hui YH, Sattar SA, Murrell KD, Nip W-K, Stanfield PS. *Foodborne Disease Handbook*, 2nd edn, New York, Marcel Dekker, 407–23.

ADAMS AM, MURRELL KD and CROSS JH (1997). Parasites of fish and risks to public health. *Rev.Sci.Tech.Off.Int.Epiz*, 16, 652–60.

AHMED FE (1992). Review: Assessing and managing risk due to consumption of food contaminated with microorgnisms, parasites, and natural toxins in the US. In *J.Food.Sci.Technol.*, 27, 243–60.

AKAHANE H, IWATA K and MIYAZAKI I (1982). Studies on *Gnathostoma hispidum*. Fedchenko, 1872 parasitites in loaches imported from China. *Jap. J.Parasitol.*, 31, 507–16 (in Japanese).

AKAHANE H, KOGA M, ARGUMEDO RL, SARABIAN DO, PRIETO LG, DIAZ-CAMACHO SP, NUMTANONG S, WAIKAGULI J and KURAMOCHI T (2000). On *Gnathostoma binucleatum*: A causative agent of the Mexican gnathostomiasis. *Proceedings 3rd Seminar on Foodborne Parasitic Zoonoses*, Bangkok, Mahidol University, p. 89.

ANDERSON RC (2000). *Nematode Parasites of Vertebrates*, 2nd edn, CABI Publishing, New York.

AUDICANA MT, ANSOTEGUI IJ, FERNANDEZ DE CORRES L and KENNEDY MW (2002). *Anisakis simplex*: dangerous – dead and alive? Trends in Parasitol, 18, 20–25.

BOGITSCH BJ and CHENG TC (1990). *Human Parasitology*, Philadelphia; Sauders College Publishing.

BRAKE RJ, MURRELL KD, RAY EE, THOMAS JD, MUGGENBERG BA and SIVINSKI JS (1985). Destruction of *Trichinella spiralis* by low-dose irradiation of infected pork. *J. Food Safety*, 127–43.

CHAI JY and LEE SH (1991). Intestinal trematodes infecting humans in Korea. *Southeast Asian J.Trop.Med.Hyg.*, 22, 163–70.

CHITONANDH H and ROSEN L (1967). Fatal encephalomyelitis caused by the nematode *Gnathostoma spinigerum*. Am. J. Trop. Med. Hyg., 16, 638–45.

CROSS JH (2001). Fish and invertebrate-borne helminths, in Hui YH, Sattar SA, Murrell KD, Nip W-K and Stanfield PS. *Foodborne Disease Handbook*, 2nd edn, Marcel Dekker Inc., New York, pp. 249–88.

DASZAK P, CUNNINGHAM AA and HYATT AD (2000). Emerging infectious diseases of wildlife – threats to biodiversity and human health. *Science*, 287, 631–4.

DEARDORFF TL (1991). Epidemiology of marine fishborne parasitic zoonoses. *Southeast Asian J.Trop.Med.Publ.Hlth*, 22, 146–9.

DEARDORFF TL and KENT ML (1989). Prevalence of larval *Anisakis simplex* in pen-reared and wild-caught salmon (salmonidae) from Puget Sound, Washington. *J.Wildlife Dis*, 25, 416–9.

DIAZ-CAMACHO SP, DE LA CRUZ-OTERO MC, ZANUETA-RAMOS ML, OSUNA-RAMIREZ I, BAYLISS-GAXIOLA S, NAWA Y and WILLMS K (2000). Epidemiological survey of gnathostomosis in Sinaloa, Mexico (2000). *Proceedings, 3rd Seminar on Foodborne Parasitic Zoonoses*, Bangkok, Mahidol University, p. 88.

EDOARDO SL (1991). Foodborne parasitic zoonoses in the Philippines. *Southeast Asian J. Trop. Med. Publ. Hlth*, 22, 16–22.

FAO (1974). Fish and shellfish hygiene. Report of a WHO Expert Committee convened in cooperation with FAO. WHO Technical Report Ser., No. 550.

FAO (1992). The state of food and agriculture. FAO Agriculture series No. 25. Food and Agricultural Organizations of the United Nations, Rome.

FARKAS J (1987). Decontamination, including parasite control, of dried, chilled and frozen foods by irradiation. *Acta alimentaria*, 16, 351–84.

GIBODA M, DITRICH O, SCHOLZ T, VIENGSAY T and BOVAPHAN (1991). Current status of foodborne parasitic zoonoses in Laos. *Southeast Asian J.Trop.Med.Publ.Hlth*, 22, 56–61.

GRACZYK TK and FRIED B (1998). Echinostomiasis: a common but forgotten food-borne disease. *Am.J.Trop.Med.Hyg.* 58, 501–4.

HASWELL-ELKINS MR, SITHITHAWORN P and ELKINS D (1992). *Opisthorchis viverrini* and cholangiocarcinoma in Northeast Thailand. *Parasitol. Today*, 8, 86–9.

HUBER C, MARTLBAUER E, PRIEBE K and TERPLAN G (1989). Entwicklung und Anwendung eines ELISA zum Nachweis von Antikorpen gegen *Anisakis simplex* (Nematoda) beim Seelachs *Pollachius virens*. *Deut. Vet. Gesellschaft*, 28, 272–5 (In German).

ICHIKI Y, TANAKA T, HARAGUCHI Y, TANAKA K and NAWA Y (1990). A case of severe metagonimiasis with abdominal symptoms. *Jap. J. Parasitol.*, 39, 72–74.

IM KI, SHEN H, KIM B and MOON S (1995). Gastric anisakiasis in Cheju-do, Korea. *Kor.J.Parasitol.*, 33, 179–86.

ISHIKURA H and KIKUCHI K (1990). *Intestinal Anasakinsis in Japan*. Springer-Verlag, Tokyo.

ISHIKURA H, KIKUCHI K, NAGASAWA K, OOIWA T, TAKAMIYA H, SATO N and SUGANE K (1992). Anisakidae and anisakidosis, in Sun T, *Progress in clinical parasitology*, Vol. III, Springer Verlag, New York, 43–102.

ISHIKURA H, TAKAHASHI S, YAGU K, NAKAMURA K, KON S, MATSURA A, SATO N and KIKUCHI K (1998). Epidemiology: global aspects of anisakidosis. In Tada I, Kojima S and Tsuji M. *Proceedings, ICOPA IX*, Monduzi Editore Sp.A., 379–82.

KÄFERSTEIN F (1994). Comments, In Maurice J, *Is something lurking in your liver? New Scientist*, 19 March, 26–31.

KHAMBOONRAUNG C (1991). On emerging problems in food-borne parasitic zoonoses: impact on agriculture and public health. *Southeast Asian J.Trop.Med.Publ.Hlth.*, 22, 1–7.

KHAMBOONRAUNG C, KEAWVICHIT R, WONGWARAPAT K, SUWANRANGSI S, HONGPROMIGART M, SUKHAWAT K, TONGUTHAI K and LIMA DOS SANTOS CA (1997). Application of hazard analysis critical control point (HACCP) as a possible control measure for *Opisthorchis viverrini* in cultured carp (*Puntius gonionotus*). *Southeast Asian J. Trop. Med. Publ. Hlth.*, 28 Suppl. 1, 65–72.

KLEBANOVSKII VA, SMIRNOV PL, LABANOVSKAYA IA and OBGOL'TS AA (1977). Human helminthiasis in eastern Taimyr (Katangskii region), In *Probl.epidem.profilak.pirod.bolez.*, Zapolyar'e, (Sb. Nauch.rabot), Omsk, USSR: Omskii Meditsinskii Institut, 144–64 (In Russian).

KO R (1995). Fish-borne parasitic zoonoses, In Woo PTK, *Fish Diseases and Disorders*, Vol. 1, CAB International, Cambridge, UK, 631–71.

KUMAR V (1999). *Trematode infections and diseases of man and animals.* Kluwer Academic Publishers, Dordrecht, and ITG Press, Antwerp.

LI X (1991). Foodborne parasitic zoonoses in the People's Republic of China. *Southeast Asian J. Trop. Med. Publ. Hlth.*, 22 Suppl., 31–4.

LOAHARANU P and MURRELL D (1994). A role for irradiation in the control of foodborne parasites. *Tr. Food. Sci. Technol.*, 5, 190–5.

LOAHARANU P and SANTASIRI S (1991). Preliminary estimates of economic impact of liver fluke infection in Thailand and the feasibility of irradiation as a control measure. *Southeast Asian J. Trop. Med. Publ. Hlth.*, 22, 384–90.

MCCARTHY J and MOORE TA (2000). Emerging helminth zoonoses. *Int.J.Parasitol.*, 30, 1351–60.

MACLEAN JD, ARTHUR JR, WARD BJ, GYORKOS TW, CURTIS MA and KOKOSIN E (1996). Common-source outbreak of acute infection due to the North American liver fluke *Metorchis conjunctus*, *Lancet* 347, 154–8.

MACLEAN JD, CROSS J and MAHANTY S (1999). Liver, lung and intestinal fluke infections, in Guerrant RL, Walker DH and Weller PF, *Tropical Infectious Diseases*, Vol. II, Churchill Livingstone, Philadelphia, pp. 1039–57.

MAGGI P, CAPUTI-IAMBRENGHI O, SCARDIGNO A, SCOPPETTA L, SARACINO A, VALENTE M, PASTORE G, ANGARANO G (2000). Gastrointestinal infection due to *Anisakis simplex* in southern Italy. *Eur. J. Epidemiol.*, 16, 75–8.

MAO S (1988). Control of parasitic diseases in China. *Chinese Med. J.*, 100, 445–53.

MAS-COMA S, CASTELLO MDB, MARTY AM and NEAFIE RC (2000). Hepatic trematodiasis. In Meyers WM, Neafie RC, Neafie AM and Wear DJ. *Pathology of Infectious Diseases*, Vol. 1, Helminthiases, Armed Forces Institute of Pathology, Washington DC, 69-92.

MIYAZAKI I (1966). *Gnathostoma* and gnathostomiasis in Japan. In Morishita K, Komiya Y, and Matsubayashi H. *Progress of Medical Parasitology in Japan*, Vol. III, Meguro Parasitol. Mus, Tokyo, pp. 530–86.

MURRELL KD (1995). Foodborne parasites. *Int. J. Environ. Hlth. Res.*, 5, 63–85.

NAWA Y (1991). Historical review and current status of gnathostomiasis in Asia. *Southeast Asian J. Trop. Med. Publ. Hlth.*, 22, 217–19.

NAWA Y, NODA S, UCHIYAMA-NAKAMURA F and ISHIWATA K (2001). Current Status of foodborne parasitic zoonoses in Japan. *Southeast Asian J. Trop. Med. Hyg.* (Suppl.), in press.

OSHIMA T (1984). Anisakiasis, diphyllobothriasis and creeping disease –

changing pattern of disease in Japan, In Ko RC, *Current perspectives in parasitic diseases*, University of Hong Kong, 93–102.

RAUSCH RL, CROSS JH and CROSS AW (1998). Other cestodiasis and diphyllobthriasis in Wallace RD, Doebbeling BN and Last JM, *Public Health and Preventive Medicine*, Appelton and Lange, Stamford, Conn, pp. 377–8.

ROMANENKO NA (1986). Emergence and current status of a diphyllobothriasis focus in the Krasnoyarsk water reservoir. *Med. Parazit* No.1, 69–73 (In Russian).

SOH CT (1984). The current status of human parasitic infections in Korea. In Ko RC, *Current Perspectives in Parasitic Diseases*, University of Hong Kong, 83–92.

TODD ECD (1989). Preliminary estimates of costs of foodborne disease in the United States. *J. Food. Protect.*, 52, 595–601.

VAN THIEL PHF, KUIPERS FC and ROSKAM R (1960). A nematode parasite of herring, causing acute abdominal syndromes in man. *Trop. Geograph. Med.*, 2, 97–113.

WHO (1995). *Control of foodborne trematode infections*. WHO Tech. Rept. Ser. 849.

WILLIAMS H and JONES A (1994). *Parasitic worms of fish*. Taylor and Francis, London.

YOUNG PC (1972). The relationship between the presence of larval anisakine nematodes in cod and marine mammals in British home waters. *J. Appl. Ecol.*, 9, 459–485.

9

Rapid detection of seafood toxins

G. Palleschi, D. Moscone, L. Micheli and D. Botta, University of Rome

9.1 Introduction

Filter-feeding molluscs such as clams, oysters and mussels can become toxic to humans during the so called 'red tides'. The phenomenon of 'red tide' is caused by the fast growth of a kind of microscopic and single-celled algae, which are usually not harmful. Unfortunately, a small number of species (HAB = Harmful Algae) produce potent toxins that can be transferred throughout the food chain, affecting and even killing zooplankton, shellfish, and eventually humans that feed on them either directly or indirectly.

The growing threat of seafood intoxication has become evident in recent times. On-site refrigeration and transportation has removed the problem far from the incidence point. In Europe the economy of many coastal cities is linked to seafood production, hence toxin detection is of extreme importance. There are four human illnesses associated with shellfish and toxic blooms: paralytic (PSP), neurotoxic (NSP), amnesic (ASP) and diarrhoeic (DSP) shellfish poisonings. Their occurrence is extremely rare, however, regulations are in place effectively to protect the consumers from shellfish toxins.

The methods most widely used for the assay of naturally occurring toxins are high performance liquid chromatography (HPLC) and mouse time to death bioassay (MBA), but these methods are slow, expensive and not sufficiently rugged for routine use or for analysis in the field. The need for rapid, on-site determination of seafood poisons favours the development of biosensors in this area. The toxins of major interest are non-protein molecules such as saxitoxin (STX), okadaic acid (OA) and domoic acid (DA). These molecules cannot be destroyed by normal cooking, freezing or smoking. In this context a European research project was approved, in which the University of Rome 'Tor Vergata' as co-ordinator (Prof. G. Palleschi, Italy), the University of Cork (Prof. G.G. Guilbault, Ireland) and the University of Lyon (Prof. P. Coulet, France) as scientific partners took part. The industrial partner (Domotek, Italy) was a company whose expertise was to construct electrochemical portable instrumentation useful for the developed immunosensors. The screen-printed electrodes have been developed by Biosensors Laboratory of the University of Florence (Prof. M. Mascini, Italy). The developments, analytical evaluations and applications, reported in this chapter, are the work of all the mentioned research groups involved.[1]

In this chapter, new analytical procedures based on the use of electrochemical disposable immunosensors for the detection of seafood toxins are shown. Preliminary, spectrophotometric and electrochemical Enzyme-Linked ImmunoSorbent Assays (ELISA) have been optimised for the determination of toxins, then disposable immunosensors for the measurement of these toxins have been assembled using specific antibodies.

A simple pocket instrumentation has been constructed and used for the *in-situ* determination of seafood toxins. Results, compared with those obtained using conventional instrumentation, showed the possibility to measure toxins in seafood collected in contaminated seawater, directly in the field. The development of electrochemical disposable immunosensors showed advantages in terms of sensitivity, rapidity and cost-effectiveness compared with previous analysis methods, such as MBA and HPLC, and were particularly useful for rapid screening tests.

9.2 Immunosensors

Immunosensors are analytical devices, which selectively detect analytes and provide a concentration-dependent signal. Electrochemical immunosensors employ either antibodies or their complementary binding partners (antigens) as biorecognition elements in combination with electrochemical transducers.[2]

Immunoassay techniques are based on the ability of the antibodies to form complexes with corresponding antigens. The property of highly specific molecular recognition of antigens by antibodies leads to highly selective assays based on immune principles. The extreme affinity of antigen-antibody interactions results also in a great sensivity of immunoassay methods. The most common type of immunoassay is known as enzyme linked immunosorbent assay or ELISA. This method is a conventional solid phase immunoassay technique, where the antigen-antibody interaction occurs at a solid phase surface. There are different schemes of ELISA and the most popular is the competitive (direct and indirect) binding immunoassay method.[3] Direct competitive assays are based on a competition between antigens labelled (added) with enzyme and unlabelled (sample) for binding sites of antibodies, immobilised on the support, while in an indirect format the competition occurs between free and immobilised antigen *versus* the binding sites on the antibody labelled with enzyme. The number of labels associated with the solid phase is inversely related to the concentration of antigen.

The spectrophotometric ELISA is performed with a 96-well microtiter plate, in which samples can be processed simultaneously. All plates and wells are made from polyvinyl chloride (PVC). The method of detection for ELISA systems involves the addition of a chromogen to a completed assay and a coloured product formation is indicative of the amount of analyte present.

In the electrochemical ELISA, the selectivity of immunological analysis is combined with the sensivity of the electrochemical detection. Electrochemical sensors can make use of a number of different measurement techniques and different immunosensors are developed using potentiometry, amperometry or voltammetry.[4,5,6] The most widely used systems have their immunoreactives located adjacent to the electrode and bound to a membrane or directly immobilised on their surface.

Label enzymes used on such electrochemical immunoassays are usually oxidoreductases such as horseradish peroxidase (HRP) or hydrolytic enzymes, such as alkaline phosphatase (AP) that yield electroactive species as products of the enzymatic reaction.

Electroanalytical immunosensors provide an exciting and uncomplicated opportunity to perform food analyses away from a centralised laboratory. The most common disposable biosensors are those produced by thick-film technology. Particular attention was directed toward the screen printed electrodes (SPE), because they can combine the easiness of use and portability with simple and inexpensive fabrication techniques. The modest cost of SPEs has further enhanced their desirability because it allows the device to become disposable. The use of disposable electrochemical immunosensors can be coupled with a portable instrument for the detection of seafood toxins. A prototype of this instrument was constructed and evaluated for test measurement using Differential Pulse Voltammetry (DVP) and potentiostatic methods. The prototype was also compared with the bench laboratory instrumentation.

9.3   Domoic acid detection

Domoic acid (DA) is a marine toxin (produced by the phytoplankton species, *Nitzschia pungens*) and the main toxic agent associated with incidents of amnesic shellfish poisoning (ASP) on the east and west coasts of North America. This rare, naturally occurring aminoacid is a member of a group of potent neurotoxic aminoacids that act as agonists to glutamate, a neurotransmitter in the central nervous system. *Pseudonitzschia* phytoplankton species have been identified as a source of DA in toxic seafood incidents.[7] Isomerisation of DA can occur photochemically and thermally, the latter being significant for cooked seafood. These isomers show a varying degree of toxicity.[8] Reliable methods for the analysis of DA and its isomers in seafood products are vitally important for the protection of the public.

9.3.1   Analytical methods

Mouse bioassay is the commonly used method for the detection of other seafood toxins and is not sufficiently sensitive to detect the regulatory level of $20\mu g$ DA/g of edible tissue. Methods most commonly employed for determining the DA in contaminated samples involve high-performance liquid chromatography (HPLC) with a variety of sample extraction techniques.[9,10] Liquid chromatography with ultraviolet absorbance detection (LC-UVD) is currently the preferred technique for the determination of DA in shellfish.[11,12] DA may be extracted from shellfish tissues by the AOAC hot acid method[13] or by blending with aqueous methanol,[14,15] the latter being most commonly used because it is better suited to trace analysis and combines well with a highly selective clean-up based on strong anion exchange. The detection limit of the toxin in an extract solution is 10–80ng/mL and depends on the sensitivity of the UV detector. The detection limit in the original tissue is dependent on the method of extraction clean-up.

A new rapid, sensitive and disposable amperometric immunoassay (ELISA) method has been employed for determination of DA in various marine samples, without the clean-up step, fundamentally in chromatographic assay.

Spectrophotometric study

The development of spectrophotometric ELISA before the electrochemical study allowed the definition of the working ranges, limit of detection and cross-reactivity evolution prior to the additional testing with the electrode. The test was performed in a 96-well microplate according to the details in Garthwaite.[16] This immunoassay was performed in an indirect competitive assay involving antibodies against DA and DA conjugated to bovine albumin serum (BSA-DA) for coating, provided by Toxicology and Food Safety AgResearch (NZ), and horse anti-goat IgG Alkaline Phosphatase conjugate (IgG-AP) for the detection. The enzyme substrate was 4-nitrophenyl phosphate (4-NPP).

Figure 9.1 shows the results obtained using the applied indirect spectrophotometric ELISA test with all the analytical parameters optimised. The operative range was calculated with a 4-parameter logistic model given by equation (9.1):

$$Y = d + [(a - d)/(1 + (X/c)^b]$$ (9.1)

where d and a are the asinptotic values for maximum and minimum (higher concentration and zero concentration), c corresponds to the analyte concentration (x), which gives $Y = (a + d)/2$, and determines the centre of the curve (IC$_{50}$), b gives the slope of the curve.[17] The errors involved were all below 10% $(n = 3)$ and the linear range for such an assay was deemed to be between 1 and 185ng/mL of DA.

Electrochemical study

A disposable amperometric biosensor, based on SPE coated with BSA-DA, has been prepared for measuring DA in mussel tissue according to Kreuzer.[1]

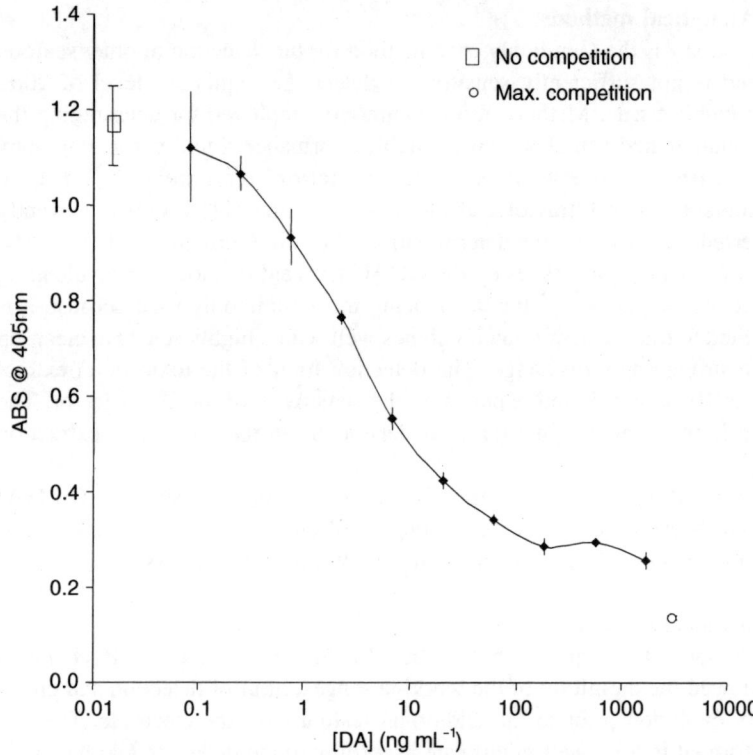

Fig. 9.1 Competitive curve for domoic acid with indirect test using spectrophotometric detection; AP as label.

Chessboard titrations were performed on electrodes to assess the optimum conditions yielding sufficient current ($\sim 1\mu A$). Dilutions of both BSA-DA and DA sheep serum were prepared, followed by excess α-sheep IgG-AP and SPE measurement at + 300mV *vs*. Ag/AgCl. Once these conditions were optimised, indirect competitive analysis on SPEs was developed. A typical assay can be seen in Fig. 9.2, with an approach to the maximum signal seen at low DA concentrations. The inset of Fig. 9.2 also shows a linear range analysis of another assay between 10 and 160ng ml^{-1} DA with an accompanying regression coefficient of 0.997. Errors associated with each standard were generally all below 6% ($n = 2$).

The recovery of spiked mussel samples had to be determined with ELISA methods. Both spectrophotometric and sensor assay were used in an attempt to discover the accuracy of the DA system. DA was extracted from homogenised tissue according to Garthwaite.[16] As the calibration curves have been accurately determined, dilutions of extracted tissue containing the toxin were prepared in a way to ensure they would fall within the range of the calibration curve.

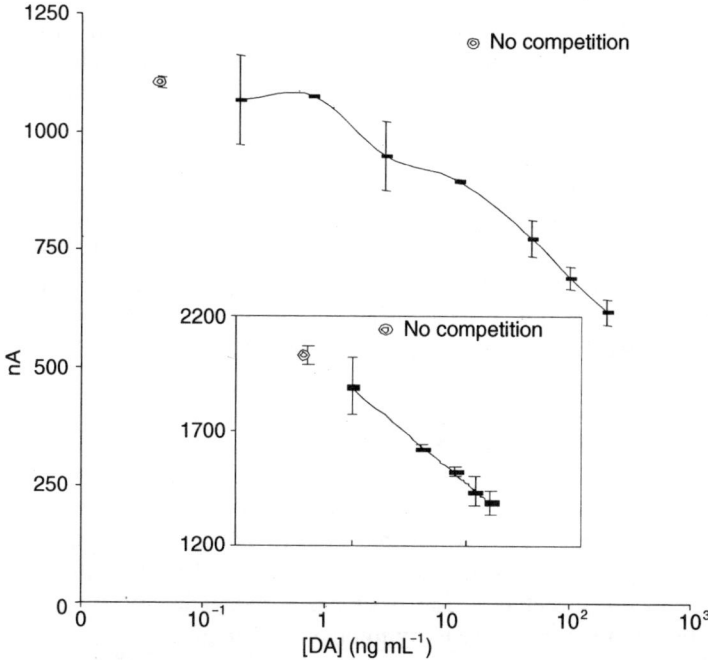

Fig. 9.2   Indirect competitive immunosensor for domoic acid with approach toward maximum control (zero analyte). Inset shows linear range of another assay between 10 and 160ng ml^{-1} domoic acid ($r^2 = 0.997$).

Percentage recovery tests were performed on established calibration curves to assess the accuracy of unknown DA samples. The values returned were obtained by equation (9.2).

$$\%\text{Recovery} = \left(\frac{[DA] \text{ in sample 1,2} - [DA] \text{in blanks 1,2}}{[DA] \text{ in original sample}} \right) \times 100 \quad (9.2)$$

The spectrophotometric procedure exhibited good recoveries ranging from 99–112% of the true [DA] concentration. The sensor however lacked accuracy but this was attributed to the number, or lack of electrodes per standard, clearly reflected in the spectrophotometric results. Spectrophotometric ELISA has a higher throughput due to the convenience of standard preparation and measurement, whereas working with SPEs is more difficult in this matter. Nevertheless, the values were acceptable ranging from 91–125% of the true [DA] concentration.

9.4   Okadaic acid detection

Okadaic acid (OA) and its structural homologous are the toxins responsible for most human diarrhetic shellfish poisoning (DSP)-related illnesses. The acid was first isolated from two sponges, namely *Halichondria okadai* and *H. melanodocia* and subsequently found in the dinoflagellates *Prorocentrum lima*[18] and *Dinophysis spp.*[19] OA is a cyclic fatty acid (C_{38}) whose structure was first discovered by Tachibana.[20]

In 1982, Shibata and co-workers discovered that OA caused long-lasting contraction of smooth muscle from human arteries.[21] OA also causes diarrhoea by stimulating the phosphorylation of proteins controlling sodium secretion by intestinal cells.[22] Prolonged exposure and continuous uptake of sub-acute levels of okadaic acid and dinophysistoxins should be avoided due to their potent tumour promoting activity.[23] Acute toxicity of various toxins from the OA group after intraperitoneal injection in mice[24] was $200 \mu g/kg$ of mouse.

9.4.1   Analytical determination

The most common DSP assays are biological-based techniques. Non-specific toxin detection and the risk of false positives caused by fatty acids have led to the search for more simple and reliable assays[25] not to mention the inherent dislike of using live animals. Immunoassays were also developed including the RIA[26] and ELISA[27] methods. The latter has a shorter assay time with good quantitation and little cross-reactivity to other DSP toxins.

HPLC techniques, based on the methods of Lee[28] and Aase,[29] will detect OA and DTXs by the spectrophotometric detection of 9-antharyldiazomethane derivatives. The level of sensitivity lies at approximately $15 \mu g$ OA/100g shellfish tissue (37.5ng ml^{-1} or 4.6×10^{-8} M OA). Methods combining HPLC and mass spectroscopy have proved to be sensitive (2ng OA detectable).[30] Nevertheless, it has been shown that up to 50% of the toxin can be lost in the derivatisation and extraction procedure.

More recent advances have led to a relatively rapid radioactive protein phosphatase (PP) assay, developed by Honkanen.[31] It was used to detect OA in oyster extracts, and samples containing OA $\geq$ 0.2ng/g were considered positive. Results correlated well with L.C. determination. Tubaro[32] developed a colorimetric assay that could detect OA as low as 2ng/g of digestive glands using PP2A. In 1997, a patent was granted for a fluorimetric PP assay that utilises 4-methylumbelliferyl phosphate as substrate with a 20-fold improved sensitivity.[33]

For economical and human health reasons, the presence of OA should be detected immediately at the production site and as quickly as possible, without pre-treatment and with an appreciable sensitivity regarding the EU critical limit, 40–60ng OA/g of mussel tissue.[13] In light of this, we report here the results of two different studies for a rapid and sensitive determination of OA in mussel tissues: a chemiluminescent immunosensor integrated in a flow injection analysis system and an immunosensor based on the use of SPEs.

Chemiluminescent immunosensor

Chemiluminescence is at present widely used as the signalling system in ELISA detection because of the high sensitivity of light measurements. The peroxidase from horseradish (HRP), which catalyses the luminescent luminol oxidation in the presence of hydrogen peroxide is the most used chemiluminescent enzyme for antibody/antigen labelling. The chemiluminescence immunosensor, integrated in a flow system, was developed for the detection of OA, as reported by Marquette.[34] The immunosensor equipped with BSA-OA was immobilised on commercially available polyethersulfone membranes (UltraBind™ membranes, Pall German Science), prepared by the soaking procedure. Anti-OA monoclonal antibodies were labelled with horseradish peroxidase (HRP) for their use in a competitive assay, in which the free antigen of the sample competes with immobilised OA. The bioanalytical system exhibits low non-specific binding of antibodies in the presence of mussel homogenate.

The immunosensor was used in a semi-automated analysis procedure in which the sample containing the free OA was injected in the flow system concomitantly with the labelled antibodies. With an overall measurement time of 20 minutes, the immunosensor has a detection limit of $0.2\mu g$ OA/100g mussel homogenate ($0.1\mu g$ OA/L in the assay solution), and the measurements could be performed over three decades of concentration, from 0.2–$200\mu g$ OA/100g of mussel homogenate (Fig. 9.3). The detection range was determined from the linearised calibration curve, obtained with the log it function (9.3):

$$\text{logit} = \ln[B/(B_0 - B)] \qquad\qquad 9.3$$

where B_0 is the maximum signal in the absence of free antigen, and B is the signal measured in the presence of a fixed antigen concentration.

The performances of this chemiluminescent immunosensor appeared then to be sufficient for the detection of OA since the critical content in mussel tissue is $40\mu g$ OA/100g (EU critical content). The operational stability of the sensor for the detection of critically contaminated mussels ($40\mu g$/100g of mussel homogenate) has been investigated over 38 OA successive determination cycles. A stable response was obtained under the first 38 measurements with a CV of 11.7%. Moreover, no significant change in the immunosensor performance was observed during short- and long-term storage of the membranes (1 month). The performances of five immunosensors (i.e. equipped with five different membranes) showed good repeatability with a variation of the blank signal and of the critically contaminated signal equal to 7% and 12.6%, respectively.

SPE immunosensor

The electrochemical enzyme immunoassay for OA has been performed using as signal transducer the carbon working electrode of the disposable sensors, which was also used as solid phase for reagent immobilisation. It was important to determine the working range for the OA immunoassay before the assembling of SPE, which was done using the spectrophotometric ELISA, developed by Kreuzer.[35] Then the results were used for the electrochemical studies.

The SPE were prepared and stored desiccated at room temperature. Then the reagents were dropped on the working electrode for the immobilisation. After all

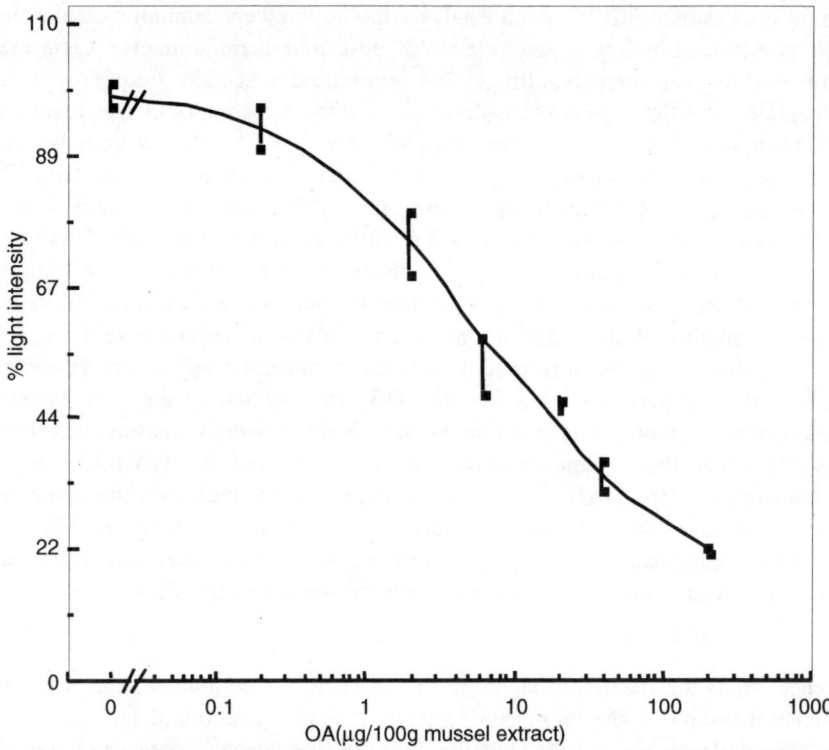

Fig. 9.3 Free okadaic acid calibration curve in mussel homogenate. Each concentration has been assayed in duplicate.

the steps of the immunoassay were performed, $100\mu L$ of the enzyme buffer solution containing the electrochemical substrate p-aminophenyl phosphate (p-APP) were added to the electrode. Signal detection was performed by placing the SPE into a stirred electrochemical batch cell containing the substrate buffer solution and injecting p-APP to a final concentration of 1mM. The current was monitored at + 300mV (applied potential) vs. Ag/AgCl reference electrode. The indirect competitive ELISA format[35] on SPEs yielded the best results in terms of linear range and limit of detection. The optimum conditions were used to obtain large signals and the final step of the immunosensor involved the competition of labelled antibody (Ab*) with a dilution range of the free analyte (OA).

Good accuracy, in terms of r^2 (> 0.991), was obtained for all the assays performed in the linear range between 4 and 125ng/mL of OA. Figure 9.4 reports results typical of such assays. The repeated measurement of each OA standard was somewhat limited by the time necessary to measure each electrode. This could be overcome by using a multi-electrode potentiostat, which allows the limiting of errors involved with each point. Another important aspect of these OA-SPEs was the detection limit. This parameter was again determined by several measurements and yielded a value between 1 and 4ng/ml OA. This was

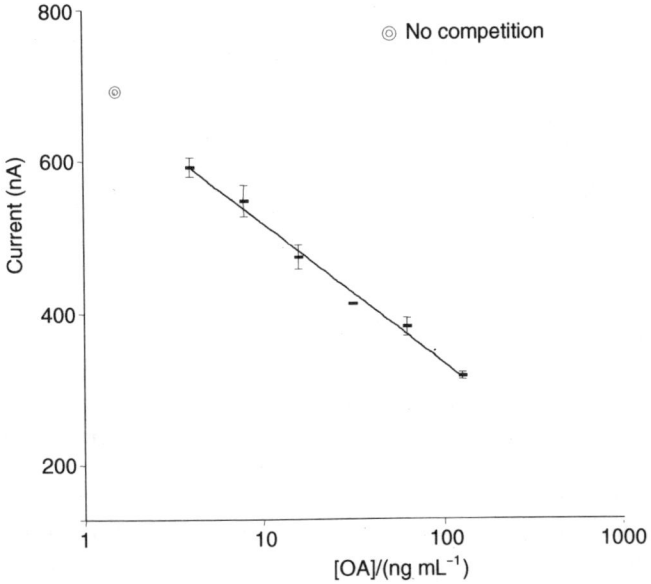

Fig. 9.4   Typical competitive immunoassay for okadaic acid using SPE with amperometric detection at + 300 mV *vs.* Ag/AgCl.

in the range of nanomolar quantities of OA (1.24×10^{-9} mol/L) and when compared to fluorimetric assays showed more sensitive results with no need of prior derivatisation.[36]

The recoveries of OA from spiked mussel samples were studied using the OA immunosensor. As the calibration curves had been determined, dilutions of organic solvent from the extraction procedure containing the toxins were prepared in such a manner that they would fall within the range of the calibration curve. Blank extracts were also prepared where zero toxin concentration should be present. These blanks would be indicative of the matrix effect resulting from the mussel tissue and fats also in the organic solvent used for the extraction. The recovery percentage using screen-printed electrodes for real samples yielded values generally ±10% of the true value. Assays were generally completed in 60 minutes and measurements within five minutes. The incubation period could also be shortened to approximately 30 minutes, thus making the overall procedure no longer than 35 minutes.

9.5   Saxitoxin detection

Saxitoxin is one of the most lethal non-protein toxins known (LD_{50} $9 \mu g/Kg$[37]) and is one of the 'Paralytic Shellfish Poisons' (PSP), produced by several marine dinoflagellates and freshwater algae. Contamination of shellfish with saxitoxin has been associated with harmful algal blooms throughout the world. In humans, paralytic shellfish poisoning causes dose-dependent perioral numbness or tingling sensations and progressive muscular paralysis, which may result in death through respiratory arrest.[38] According to the Food and Drug Administration, the maximum acceptable level for paralytic poison in fresh, frozen or tinned shellfish is up to 400 Mouse Units (MU) or about 40–80μg/100g edible portion.[13] This value is equivalent to twice the minimum detection level of the mouse bioassay, the first and still most common PSP toxin testing method is also the official AOAC method.[39]

9.5.1 Analytical methods

The Mouse Bioassay (MBA) is the official method for the determination of the PSP in seafood but this is neither specific nor sensitive; it requires a continuous supply of mice and results are affected by test conditions such as animal strain and sample extract preparation. Other methods include fluorimetric assay[40] and liquid chromatography.[41,42] The latter requires expensive equipment for pre-[41] or post-column[42] analyte oxidation. Additionally, samples must be analysed one at a time, and so the method is unsuitable for routine on-site testing.

Immunochemical methods have advantages in terms of both sensitivity and speed, and are therefore of increasing importance in food control as rapid screening tests. Because of the highly specific antigen-antibody interaction, several laboratories have attempted to develop an immunoassay for PSP.[43,44] In the following paragraphs the optimisation and comparison of spectro-photometric and electrochemical competitive ELISA formats (direct and indirect) for the detection of saxitoxin (STX) are reported.

Spectrophotometric study

The tests were performed in a 96-well microplate using toxin-specific polyclonal antibodies produced in our laboratory.[45] The antibodies were obtained from rabbits immunised with saxitoxin-keyhole limpet hemocyanin (STX-KLH). In indirect ELISA format, saxitoxin, conjugated to bovine serum albumin (BSA-STX), was coated onto the microtitre plate and incubated with standard toxin and anti-STX antibody. A goat anti-rabbit IgG Peroxidase conjugate (IgG-HRP) was used to enable the detection. In the direct ELISA format, STX standard, STX conjugate to Horseradish Peroxidase (STX-HRP) and the enzyme substrate/chromogen solutions were sequentially added to the microplate after antibody coating.

The operative range was calculated with 4-parameter logistic model by the equation (9.1). The detection limit, defined as the concentration of toxin standard equivalent to three standard deviations at A_0 (no competition), was 3×10^{-3} and 1×10^{-2} ng/mL for direct and indirect ELISA formats, respectively (Figs 9.5 and 9.6). In both tests, the linear regression showed ranges between 5×10^{-3} and 4×10^{-1} ng/mL (top right insert Figs 9.5 and 9.6).

The stability of the coating reagents was evaluated using microplates coated with conjugated antigen or antibody, blocked and then stored at 4°C. Assays

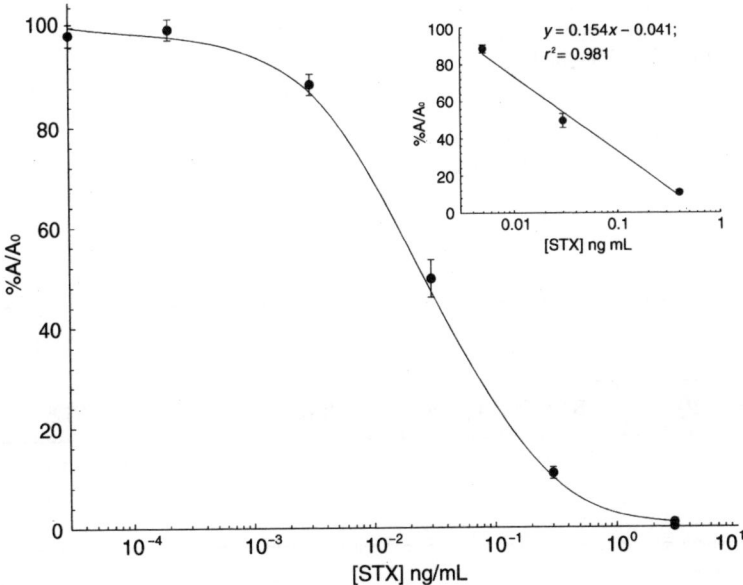

Fig. 9.5   Direct competitive ELISA for saxitoxin. Antibody against saxitoxin (10μg/mL) was coated on the ELISA plate and STX-HRP (1:30) was used as competitor.

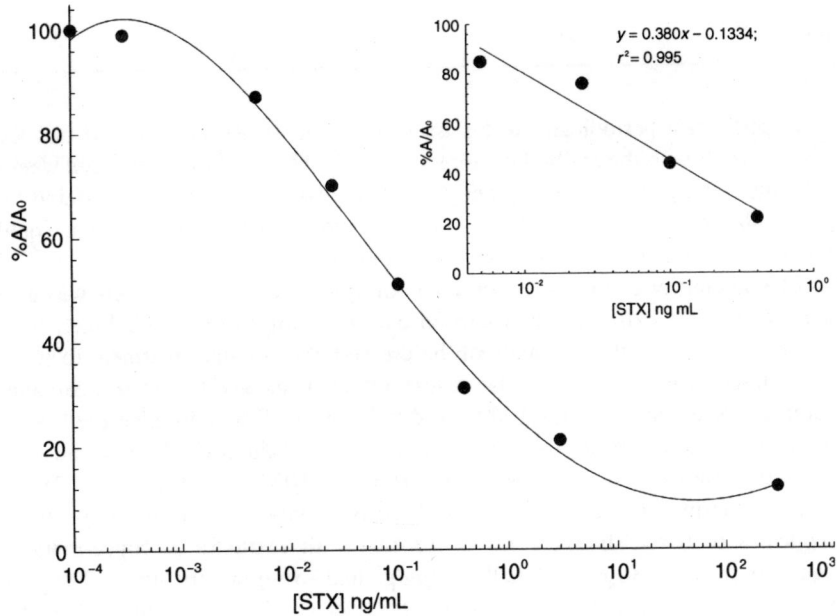

Fig. 9.6   Indirect competitive ELISA for saxitoxin. BSA-STX (3μg/mL) was coated on the ELISA plate.

Table 9.1   Precision (RSD) and accuracy (RE%) for saxitoxin in mussel, determined by ELISA and LC: (a) results with ELISA method; (b) results with LC method; (c) comparison ELISA/LC methods.

(a)

STX added µg/g	STX found µg/g	RSD	RE%
0.2	0.19	4	3
0.4	0.43	1	7
0.8	0.83	3	4

(b)

STX added µg/g	STX found µg/g	RSD	RE%
0.2	0.18	4	−10
0.4	0.42	5	5
0.8	0.88	5	8

(c)

STX added µg/g	Direct ELISA/LC RE (%)
0.2	6
0.4	2
0.8	−6

were performed periodically using assessed protocols. Results showed that the plates coated with the antibodies (direct test) could be used for up to three weeks after the coating step, while the antigen immobilised on the wells was stable for only 24h (indirect test). Better results obtained from the direct format could therefore be due to this higher antibody stability.

The suitability of the assay for saxitoxin quantification in mussels was also studied. Sample extraction was carried out according to the AOAC method.[13] Samples were spiked with saxitoxin before and after sample treatment to study the extraction efficiency and the matrix effect, respectively. After treatment, samples were analysed at 1:1000 v/v dilution in PBS to minimise the matrix effect and to detect the established limit of 40µg of saxitoxin in 100g of mussels. The saxitoxin extraction efficiency was from 72–102% (see equation (9.2)).

Repeatability and accuracy of ELISA assays were evaluated by means of six replicates of tissue. Blank controls fortified with saxitoxin at concentrations equal to twice (0.8µg/g), to half (0.2µg/g) and to regulatory limit (0.4µg/g), were prepared and extracted on three days for each concentration ($n = 18$). Precision was calculated by the relative standard deviation (RSD%) for the replicate measurements and the accuracy (relative error, RE%) was calculated by assessing the agreement between measured and nominal concentrations of the fortified samples. Results were confirmed by the analysis of the same extracts using a previously validated LC method.[42] Values reported in Table 9.1 (a, b, c), together with the accuracy of the spectrophotometric ELISA *versus* LC, showed good agreement.

In conclusion, ELISA assays were shown to be suitable screening tools for routine analysis of saxitoxin in mussels. In fact, compared to the LC method, spectrophotometric direct ELISA showed similar precision but better accuracy and speed, and lower cost. Additionally, this method does not require sample purification.

SPE immunoassay

The electrochemical enzyme immunoassay for STX has been performed using the carbon working electrode of the disposable sensors as solid phase for reagent immobilisation and as signal transducer. In a first phase, the spectrophotometric ELISA protocols were applied, but the results were unsatisfactory. In order to obtain the best signal/noise ratio and the highest sensitivity, several trials were performed for each test to optimise the analytical parameters – all tests were repeated several times in order to confirm the obtained data. This immunosensor was employed in a direct competitive assay involving STX labelled with antibody. The enzyme substrate used was $3,3',5,5'$-tetramethylbenzidine (TMB) plus hydrogen peroxide and the product of the reaction was detected by chronoamperometry at -100 mV for 60s.

The calibration curve for STX was carried out in the concentration range $0 \div 10^3$ng/mL and the results showed a sensivity in the range $1 \div 10^3$ ng/mL of the toxin (Fig. 9.7). STX levels determined by the proposed electrochemical immunoassay compared favourably with a spectrophotometric method.

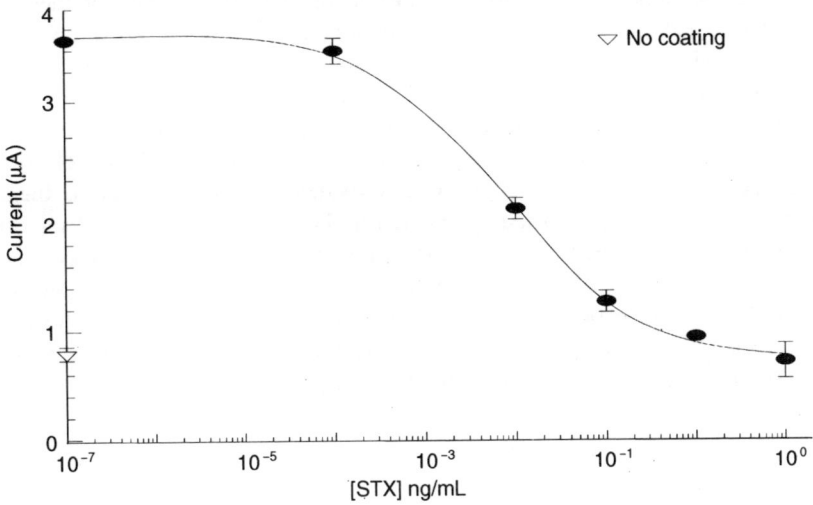

Fig. 9.7 Competitive curve with direct test for saxitoxin using SPEs.

9.6   Prototype evaluation

A small portable instrument (calculator size) microprocessor controlled with an LCD display and easy to use was constructed for toxin measurement with disposable strips. This system was provided with one site for disposable strip connection. A battery applies a selected potential at the electrodes screen printed on the strip, and the current due to the reaction occurring on the strip is recorded and displayed as the concentration of toxin measured. The instrument will be provided with self-calibration to make it easy for use by unskilled personnel.

This small instrument is able to perform DPV and chronoamperometric measurements in an interval range between $-1000 \div 1000$mV, useful for the determination of 1-naphthol, the product of the enzymatic reaction of AP with 1-naphthyl phosphate. It was tested in our laboratory and a comparison was carried out with the results obtained with an Autolab instrument (equipped with GPES software) and a polarographic analyser, model 433/W from Amel (Milan, Italy). An excellent agreement was observed among all results.

Indirect non-competitive and competitive tests have been performed using the SPE's working electrode also as a solid phase for the immobilisation of the reagents and the same experiments were carried out using the Autolab instrument (data not shown) and this portable instrument. The results obtained with the two devices were comparable.

9.7   Conclusion and future trends

This chapter reports the development of novel procedures based on cost-effective electrochemical instrumentation for the detection of selected seafood toxins by the use of monoclonal and polyclonal antibodies, electrode strips and a portable instrument. The work was carried out by the selection of the most common seafood toxins and the production of specific polyclonal and monoclonal antibodies. A parallel biosensor development has been initiated and carried on together with the set-up of ELISA procedures with spectrophotometric and then electrochemical detection.

The development of a procedure for measuring toxins with disposable strips was carried out, first using bench electrochemical instrumentation, then a portable prototype constructed by the industrial partner. Validation of disposable strips using established reference procedures and the portable electrochemical instrument was carried out. This research has successfully achieved the main objective: the detection of seafood toxins using disposable strips. However, further studies on the optimised antibody production and stability are required to make this system useful for mass production.

9.8   References

1.   PALLESCHI G, Final Report for European Union Fair Ct 95–1092, Rome Italy, 2000.

2.   GHINDILIS AL, ATANASOV P, WILKINS M and WILKINS E, 'Immunosensors: electrochemical sensing and other engineering approaches', *Biosensors & Bioelectronics*, 1998 **13**(1) 113–31.

3.   HARLOW E and LANE D, *Antibodies: A Laboratory Manual*, New York, CRC Press, Cold Spring Harbor, 1988.

4.   SANTANDREU M, SOLÈ S, FÀBREGAS E ALEGRET S, 'Development of electrochemical immunosensing system with renewable superfaces', *Biosensors & Bioelectronics*, 1998 **13**(1) 7–17.

5.   PEMBERTON PM, HART JP, STODDARD P and FOULKE S, 'A comparison of 1-naphthyl phosphate and 4-aminophenyl phosphate as enzyme substrates for use with a screen-printed amperometric immunosensor for progesterone in cow's milk', *Biosensors & Bioelectronics*, 1999 **14**(5) 495–503.

6.   DEL CARLO M, LIONTA I, TACCINI M, CAGNINI A and MASCINI M, 'Disposable screen-printed electrodes for the immunochemical detection of polychlorinated biphenyls', *Analytical Chimica Acta*, 1997 **342**(3/2) 189–97.

7.   QUILLIAM MA and WRIGHT JL, 'The amnesic shellfish poisoning mystery', *Anal Chem*, 1989 **61** 1053A–9A.

8.   WRIGHT JLC, FALKEL M, MCINNES AG and WALTER JA, 'Identification of isodomoic acid D and two new geometrical isomers of domoic acid in toxic mussels', *Can J Chem*, 1990 **68** 22–5.

9.   LAWRENCE JF, CLEROUX C and TRUELOVE JF, 'Comparison of high-performance chromatography with radioimmunoassay for the determination of domoic acid in biological samples', *J Chromatogr A*, 1994 **662**(1) 173–7.

10.   QUILLIAM MA, SIM PG, MCCULLOCH AW and MCINNES AG, 'High-performance liquid chromatography of domoic acid, a marine neurotoxin, with application to shellfish and plankton', *Int J Envir Analyt Chem*, 1989 **36** 139–54.

11.   LAWRENCE JF, CHARBONNEAU CF, MENARD C, QUILLIAM MA and SIM PG, 'Liquid chromatographic determination of domoic acid in shellfish products using the paralytic poison extraction procedure of the Association of Official Analytical Chemists', *J AOAC Int*, 1995 **78**(2) 555–70.

12.   EILERS P, CONRAD S and HALL S, 'Domoic acid analysis', *Toxicon*, 1996 **34** 338.

13.   CUNNIFF P, *Official Methods of Analysis of AOAC International* (16th edn), Gaithersburg (USA), AOAC International, 1999.

14.   QUILLIAM MA, XIE M and HARDSTAFF WR, 'Rapid extraction and cleanup for liquid chromatographic determination of domoic acid in unsalted seafood',

J AOAC Int, 1995 **78**(2) 543–54.

15.   QUILLIAM MA, XIE M and HARDSTAFF WR, *National Research Council of Canada,* Institute of Marine Biosciences, Technical Report 64, NRCC, 1991.

16.   GARTHWAITE I, ROSS KM, MILES CO, HANSEN RP, FOSTER D, WILKINS AL and TOWERS NR, 'Polyclonal antibodies to domoic acid, and their use in immunoassay for domoic acid in seawater and shellfish', *Natural Toxins*, 1998 **6** 93-104.

17.   DIAMANDIS EP and CHRISTOPOULOS TK, *Immunoassay*, Toronto (Canada), Academic Press Inc., 1996.

18.   MURAKAMI Y, OSHIMA Y and YASUMOTO T, 'Identification of okadaic acid as a toxic component of a marine dinoflagellate Prorocentrum lima', *Bull Jpn Soc Sci Fish*, 1982 **48** 69–72.

19.   YASUMOTO T, OSHIMA Y, SUGAWA W, FUKUYO Y, OGURI H, IGARASHI T and FUJITA N, 'Identification of Dynophysis fortii as a causative organism of diarrhetic shellfish poisoning', *Bull Jpn Soc Sci Fish*, 1980 **46** 1405–11.

20.   TACHIBANA K, SCHEUER PJ, TSUKITANI Y, KIKUCHI H, VAN ENGEN D, CLARDY J, GOPICHAND Y and SCHMITZ FJ, 'Okadaic acid, a cytotoxin polyether from two marine sponges of the genus Halichondria', *J Am Chem Soc,* 1981 **103** 2469–71.

21.   SHIBATA S, ISHIDA Y, KITANO H, OHIZUMI Y, HABON J, TSUKITANI Y and KIKUCHI H, 'Contractile effects of okadaic acid, a novel ionophore-like substance from black sponge, on isolated muscles under the condition of Ca deficiency', *J Pharmacol Exp Ther*, 1982 **223** 135–43.

22.   COHEN P, HOLMES CFB and TSUKITANI Y, 'Okadaic acid: a new probe for the study of cellular regulation', *TIS*, 1990 **15** 98–102.

23.   SAKAI A and FUJIKI H, 'Promotion of BALB/3T3 cell transformation by the okadaic acid class of tumor promoters, okadaic and dinophysistoxin-1', *Jpn J Cancer Res*, 1991 **82** 518–23.

24.   YASUMOTO T, MURATA M, LEE JS and TOROGOE K, *Mycotoxins and Phycotoxins '88*, Amsterdam, Elsevier, 1989.

25.   SAJIKI J and TAKAHASHI K, 'Free fatty acids inducing mouse lethal toxicity in lipid extracts of Engraulis japonica, the Japanese anchovy', *Lipids*, 1992 **27** 988–92.

26.   LEVINE L, FUJIKI H, KIYOYUKI Y, OJIKA H and VAN VAKIS, 'Production of antibodies and development of a radioimmunoassay for okadaic acid', *Toxicon*, 1988 **26** 1123–8.

27.   USAGAWA T, NISHIMURI M, ITOH Y, UDA T and YASUMOTO T, 'Preparation of monoclonal antibodies against okadaic acid prepared from the sponge Halichondria okadai', *Toxicon*, 1989 **27** 1323–30.

28.   LEE J, YANAGI T, KENMA R and YASUMOTO T, 'Fluorimetric determination of diarrhetic shellfish toxins by high-performance liquid chromatography', *Agric Biol Chem*, 1987 **51** 877–81.

29.   AASE B and ROGSTAD A, 'Optimisation of sample cleanup procedure for determination of diarrhoeic shellfish poisoning toxins by use of

experimental design', *J Chromatogr A*, 1997 **764** 223–31.

30. PLEASANCE S, QUILLIAM MA and MAR JC, 'IonSpray mass spectrometry of marine toxins. IV. Determination of diarrhoetic shellfish poisoning toxins in mussel tissue by liquid chromatography/mass spectrophotometry', *Rapid Comm Mass Spec*, 1992 **6**(2) 121–7.

31. HONKANEN RE, STAPLETON JD, BRYAN DE and ABERCROMBIE J, 'Development of a protein phosphate-based assay for the detection of phosphate inhibitors in crude whole cell and animal extracts' , *Toxicon*, 1996 **34**(11/12) 1385–92.

32. TUBARO A, FLORIO C, LUXICH E, SOSA S, DELLA LOGGIA R and YASUMOTO T, 'A protein phosphate 2A inhibition assay for a fast and sensitive assessment of okadaic acid contamination in mussels', *Toxicon*, 1996 **34**(7) 743–52.

33. VIEYTES MR, FONTAL OI, LEIRA F, BABTISTA DE SOUSA JMV and BOTANA LM, 'A fluorescent microplate assay for diarrhetic shellfish toxins', *Anal Biochem*, 1997 **248** 258–64.

34. MARQUETTE CA, COULET PR and BLUM LJ, 'Semi-automated membrane based chemiluminescent immunosensor for flow injection analysis of okadaic acid in mussel', *Analytica chimica Acta*, 1999 **19931** 1–10.

35. KREUZER M, O'SULLIVAN C and GUILBAULT GG, 'Development of an ultrasensitive immunoassay for rapid measurement of okadaic acid and its isomers', *Anal Chem*, 1997 **71**(19) 4198–202.

36. DRAISCI R, GIANNETTI L, LUCENTINI L, MACHIAFAVA C, JAMES KJ, BISHOP AG, HEALY BM and KELLY SS, 'Isolation of a new okadaic acid analogue from phytoplankton implicated in diarrhetic shellfish poisoning', *J Chromatogr A*, 1998 **798**(1/2) 137–45.

37. BROWER DJ, HART RJ, MATTHEWS PA and HOWDEN MEH, 'Non protein neurotoxins', *Clin Toxicol*, 1981 **18** 813–65.

38. USLEBER E, SHNEIDER E, TERPLAAN G and LAYCOCK MV, 'Two formats of enzyme immunoassay for the detection of saxitoxin and other paralytic shellfish poisoning toxins', *Food Addit Contam*, 1995 **12**(3) 405–13.

39. *Gazzetta Ufficiale della Repubblica Italiana* (8-9-1990), serie generale no. 218.

40. BATES HA and RAPOPORT H, 'A chemical assay for saxitoxin, the paralytical shellfish poison', *J Agr Food Chem*, 1975 **23**(2) 237–9.

41. LAWRENCE JF and MENARD C, 'Liquid chromatographic determination of paralytic shellfish poisons in shellfish after prechromatographic oxidation', *J Assoc Off Anal Chem*, 1991 **74**(6) 1006–12.

42. OSHIMA Y, 'Postcolumn derivatisation liquid chromatographic method for paralytic shellfish toxins', *J AOAC Int*, 1995 **78**(2) 528–32.

43. CHU FS and FAN TL, 'Indirect enzyme-linked immunosorbent assay for saxitoxin in shellfish', *J Assoc Off Anal Chem*, 1985 **68**(1) 13–16.

44. DAVIO SR and FONTELLO PA, 'A competitive displacement assay to detect saxitoxin and tetrodotoxin', *Anal Biochem*, 1984 **141** 199–204.

45. MICHELI L, DI STEFANO S, MOSCONE D, MARINI S, COLETTA M, DRAISCI R,

DELLI QUADRI F and PALLESCHI G, 'Production of antibodies and development of highly sensitive formats of enzyme immunoassay for saxitoxin analysis', *Frez J Anal Chem*, 2001 (submitted).

9.9   Acknowledgement

This work was supported by the E.C. project CT 96 FAIR 1092 and by the European Concerted Action QLK3 200 01311 'Evaluation/Valuation of Novel Biosensors in Real Environmental and Food Samples'.

Part II

Analysing quality

10

Understanding the concepts of quality and freshness in fish

H. Allan Bremner, Allan Bremner and Associates, Mount Coolum

10.1 Introduction

The terms quality and freshness are commonly employed throughout the literature on fish science and technology at all levels from the most complex scientific papers to newsletters, magazine and trade newspaper articles. Meaning is use, and the meaning of a linguistic expression is its canonical, proper communicative function. That is, its potential contribution to the communicative function of the utterances of which it forms part. For many circumstances the meaning of some expressions is determined in the context of their use, but for others there can be a semantic meaning independent of its communicative use (Gärdenfors, 2000).

Every supplier purports to deal only in quality products made from fresh fish but what do people mean when they use these terms? The terms themselves are subject to abuse as if designating a product to be a quality item is an answer to any question. However, such a designation merely begs the question since it provides no firm information about the material under discussion. Indeed, the term often says more about the user than about the product. Quality is part of the image a manufacturer, supplier or catcher wishes to convey. They give the impression that the less they say about the actual material the better their situation is. In publicity and advertising the image is everything.

It would be comforting to think that in technology and science, image was less important and that substance was paramount. Nevertheless, this is not the case at all. There are numerous papers and articles in which the word quality, or freshness, is used in the title, subtitle or text, but examination of the work indicates that quality, or freshness, was never described, defined or determined according to some acceptable, or at least stated, definition. At times, some *post hoc* definition is stated that arises from the results of the work itself, as the authors may have been seeking to establish what the 'quality' of some material was and a means to describe it. This approach 'puts the cart before the horse' and although exploratory and chemometric methods are invaluable in finding out and describing hidden properties of foods, and how they are related, the results are not 'quality' *per se*.

In any science it is critical to have clear statements of the operating paradigms, if these are incorrect or poorly expressed then the interpretations of results and the conclusions and inferences drawn from them are unfounded. Examples of loose usage abound and these find their way into the technical and trade literature and the effects of incorrect usage of terms in the scientific literature can have unforeseen consequences. One example is the headline 'Diet Has Little Impact on Flesh Quality' (Anon., 2000) and this is presented as a memorable one-line 'take home' message as though it were some universal truth. If it chanced that a reader were an industry representative on an advisory board reviewing grant applications, then it is likely that applications to conduct work on the effects of diet would be dismissed out of hand. In fact, the research did not measure flesh quality at all but colour, pigment, fat content and fibre size. These are important production parameters that can relate to consumer, or other assessments of quality, but they are not quality *per se*.

These arguments have previously been set out in detail (Bremner, 1997; Bremner, 2000; Bremner and Sakaguchi, 2000) and in this chapter they will be expanded and related to other approaches to quality. In addition I will relate my ideas to quality of business and technological practices and the idea of pursuit of quality as a driving force. The chapter will mainly use quality as an example, but the majority of the ideas can be applied to freshness as well.

10.2   Quality and freshness as concepts

It is paramount to this approach to understand that quality, and freshness, are concepts and that they are not entities which can be measured such as weight, acidity, breaking strength or the like. Nor is it reasonable to propose that the quality of an article, or a service can be measured or expressed without first stating that quality is being defined in terms of the measurements or estimations. This has led to the realisation that there has been a missing link in the quality hierarchy (Bremner, 2000).

10.2.1   Defined quality, the missing link
Attempts to define fish quality have been legion, ranging from those found in dictionaries to those constructed by various authors to suit their purposes, to those promulgated by standards bodies. A very acceptable general definition is included in the ISO standards (ISO, 2000), however it is not one that can be

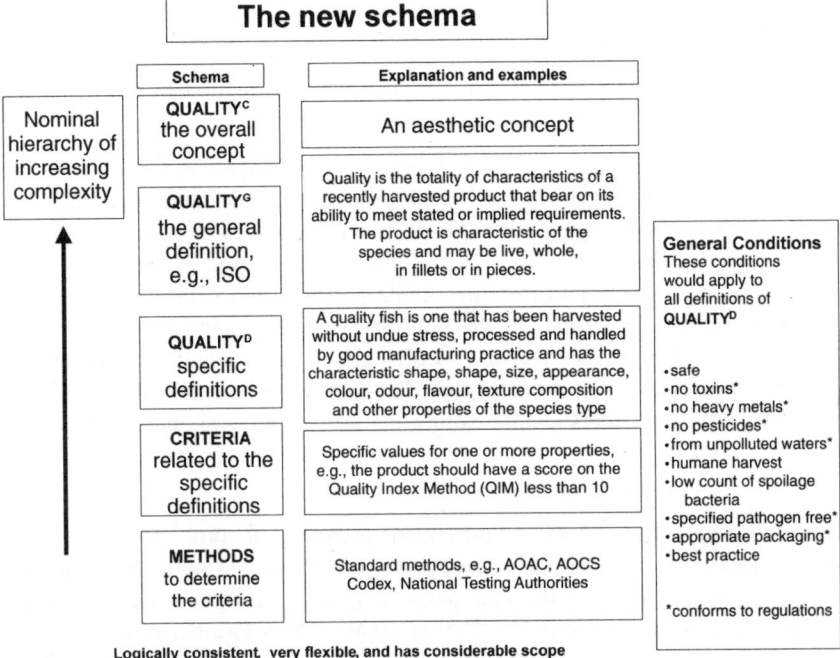

Fig. 10.1 Hierarchical organisation of quality, linking the overall concept through a general definition to a specific definition related directly to properties, attributes, methods, values and specifications.

taken and applied immediately to an individual product. What is required is a much more specific definition that can be put into direct practice and that is consistent with this ISO definition, or indeed, with other overall definitions. This has been described as a hierarchy and is set out in Fig. 10.1. Primacy in the hierarchy is given to the overall concept of quality beneath which the ISO definition fits, followed by defined quality, beneath which is methodology followed by standard values and so on.

The major value of the statements defining quality is that every stakeholder in the information knows precisely what the originators of the definition mean when they use the word quality. This is very useful in reporting research where the authors would state their opinion as to what constitutes quality. Furthermore, it does not matter if the reader necessarily agrees with the statement or not, but at least they can see the basis used for it. In many circumstances the quality statement would be worked out iteratively with the various interested parties. A general statement could take the following form.

For the following circumstances *(circumstance 1 ... circumstance x),* a quality *product name* should have properties (be) *(property 1 ... property x),* be made from *substances (1 ... x)* or grown by *system (1 ... x),* or harvested by *method (1 ... x),* be free from the following *matters (1 ... x)* , and have the desirable attributes of *(attribute 1 ... attribute x)* and so on.

The statement below provides a template for construction of more specific descriptions suited to particular products. An example for frozen cod fillets is:

For production of high quality cod fillets frozen at sea, the fish (*Gadus morhua*) shall retain the characteristic properties of the species and be caught by approved trawling methods in a preferred catch area using short towing periods and low catch volumes. The period between discharging on deck and freezing should be short, and within a few hours after catch the cod should be properly handled and filleted, packed and frozen. The frozen cod fillets must be stored at temperatures below minus 20°C for no more than 12 months before final sale. After thawing the cod fillets should exhibit the proper physical characteristics and chemical properties and desirable sensory attributes. The appropriate microbial counts should be low.

The keywords and phrases underlined provide, in turn, more detailed descriptions of circumstances, of properties, of options for methods to measure them, of values for analytes and properties that can form limits and other pertinent information or records. The information thus forms a coherent whole supporting the statement of quality in a manner consistent with the ISO (or general) definition, which is itself consistent with the overall concept of quality. This approach is currently being developed and Web and PC based software is under development to link all the information. The software will enable the construction of agreed definitions between trading partners, research organisations and supply chains. It is a tool to facilitate the mechanical provision of correct information and to provide a structured approach, but it is not a substitute for rigorous thinking. The construction of practical definitions of quality is not a simple task.

10.2.2   Defined quality and quality as a concept
It is very common to find a mixture of quality, the concept, with quality, the measurement or attribute, in the literature. Authors tend to slide back and forth between the two without actually stating the sense in which they mean the word quality to be taken. Adoption of the approach of defined quality means that quality, the concept, can be kept out of the discussion. The reason to do this is that its inclusion is the cause of much confusion and that personal and group concepts of quality are vastly different. I am not arguing that there is not a concept called quality, quite the opposite, and that it is important. I am stating that discussions about it ought to be kept separate as it belongs in the realm of philosophy. I cover this application of philosophical context later.

10.2.3   The watershed

The watershed in approach is whether quality is considered to be some essential, ineffable property of the food in question or whether it is the sum of numerous properties. I do not think that it is practical to hold both these opposing views at the one time; it is either one, or the other. If the view is held that quality is the intangible, ineffable characteristic of a fish product when it is consumed in the manner and circumstances for which it is prepared then the only way it can be evaluated is using the psychophysical approach. If the view is held that quality can be described in terms of various properties then it is these that can be measured by all the armoury of techniques available to the investigator. One view is not the same as the other and the values and results obtained by the second approach are not directly relatable to the first.

10.2.4   Circumstances and expectations

Not all agree with this approach and consider that these two views are compatible (see discussion, Meiselmann, 2001). However, any attempt to measure quality *per se* is strongly influenced by circumstances and expectations. Meiselman (2001) reported examples of how perceptions of quality changed according to the venue and circumstances in which a set meal was provided. Furthermore, it was clear that, experimentally, it would be very difficult to separate the influences of expectation from circumstances. Thus a set meal, and by analogy a given seafood product, could not have an absolute quality. There are many examples where products are rated highly if they are thought to come from an area or producer with a good reputation. It was for reasons like these that the circumstances of evaluation or production were included in the quality definition.

Something of the same considerations have been outlined in Chapter 11 dealing with shelf-life where the circumstances in which the product are to be sold dictate the appropriate techniques to be used and the usefulness of the information that is obtained.

10.3   Other approaches to concepts of quality

10.3.1   Multivariate measures of quality

In their excellent, recent textbook *Multivariate Analysis of Quality*, Martens and Martens (2001) provide dramatic illustration of the enormous complexity that exists in the real world with regard to extracting knowledge about quality from the masses of data that can be collected by measurement of food properties using sensory, instrumental and analytical means. This mass of data is just too complex for the human brain readily to make sense of, and that multivariate analysis affords a means of seeking the latent variables, hidden in the morass. For this they, and many others, use soft modelling where no hard prior assumptions are made about the relationships between the variables; the analysis itself provides the answers.

They have reached the conclusion, arising from common definitions, that there are four ways in which we can think about, or express quality and, for convenience, they are called QD1 to QD4, and I paraphrase their descriptions below.

- QD1 is qualitas, that is the essential nature, inherent characteristic or property.
- QD2 is excellence/goodness, quality as an expression for the intuitively evaluated excellence/goodness.
- QD3 is standards, that is, quality as a practical interaction between inherent characteristics of an entity, and stated and implied human needs; needs should be satisfied, requirements should be fulfilled.
- QD4 is an event, that is quality as a subjectively experienced event.

All these quality definitions, QD1 to QD4, are separate aspects of how quality can be thought of and they all bear a relation to one another (Martens and Martens, 2001). What is clear is that multivariate analytical methods are very suited for handling the data from research on quality issues, since they are capable of dealing with incomplete sets of both hard and soft data and, very importantly, with the very large sets of interrelated data, particularly from spectral investigations. The techniques are suited to process analysis in real time, or *post hoc* analysis on data from process measurements and from product samples.

10.3.2   Conceptual spaces

The ideas around which quality is based can be considered as part of cognitive science. In a recent book Gärdenfors (2000) has proposed his ideas of thought in terms of conceptual spaces. This is an attempt to integrate the two opposing current views on cognitive philosophy. These can be stated as:

1. The symbolic approach which assumes that cognitive systems can be described as Turing machines and that cognition is essentially computation involving signal manipulation.
2. The approach of associationism which says that associations between different kinds of information elements carries the main burden of representation, with connectivism being a special case of associationism that models associations using artificial neuron networks.

His view is that neither of these approaches can explain the mechanisms of concept acquisition, or concept learning which is closely tied to the notion of similarity. He has therefore suggested that a conceptual mode based on geometrical and topological representations is at least as valid as the current views. He terms this 'conceptual spaces'. A conceptual space is built on geometrical structures based on a number of quality domains. For example, spatial concepts belong to one domain, colours to another and shape to yet another and so on for temperature, brightness, weight, pitch, depth, etc. Not all domains are metric but may comprise an ordering with no distance defined. A domain is a set of integral dimensions that are separate from all other dimensions, although they may not be totally independent, for example, ripeness and colour in fruits co-vary. The fundamental role of quality dimensions is to build up the domains needed for representing concepts. This structure of quality dimensions makes it possible to talk about distances along the dimensions thus providing a relationship between similarities.

Thus a conceptual space could be built up for different categories of fish products, based on numerous dimensions and domains so that they could be related to one another. Which measurements would be used must be chosen with care, but the opportunity arises to relate the scientific with the phenomenal. As an example, the psychological colours red, green and blue could be related to measures of wavelength, cone response and tuning in the eye. Psychophysical measurements, not theoretical values, are needed to determine the structure of the phenomenal spaces.

Multidimensional scaling has been used for this and so too have neural networks, although they may require implausibly large training sets to achieve good categorisation, nor can they easily generalise what is learned from one dimension to another.

This whole idea of conceptual spaces shows promise as a means of underpinning sensory science with an intellectual framework and by providing a structure to integrate it into the broader realm of cognitive philosophy. There is a long way to go yet, but the idea is exceptionally intriguing if only because it begins to provide a means of exploring the human conceptual space concerning quality and provides a framework whereby the formerly irreconcilable variables of venue and circumstance given above (10.1.3) may begin to be reconciled.

10.3.4   Reconciling conceptual spaces with multivariate approaches

The results so far used as examples to support the idea of conceptual spaces have been derived from work on colour cognition on the colour space and from work on preferences using fuzzy algebra (amongst other techniques). Science generally has increasingly placed greater emphasis on exclusion of sense perception but for natural products such as foods this is probably the wrong direction. The enormous advances in multivariate data analysis and the sensory work with which it is combined opens up a very good opportunity to provide input into the construction of conceptual spaces.

10.4   Quality as a driving force

No matter how quality is defined its pursuit can be a major driving force in the fish industry. The drive for quality as an operating principle has been proven to provide a major boost to the operational efficiency of many industries including the mining, automotive, electronic and telecommunication sectors. Companies in these sectors look not just to the quality of their products but to all their operating systems from concept design through record keeping, stock control, book keeping, indeed, all their business systems. This has been far less evident in the fish industry itself although large multinational food corporations with interests in fish products operate quality schemes such as Total Quality Management (TQM). The Six Sigma approach has demonstrated its power in large corporations and placed emphasis on all functions in the company by demonstrating the enormous savings in efficiencies that can arise from pursuing quality and eliminating 'the hidden factory' (Harry and Schroeder, 2000).

10.4.1   Routine measures and IT

It goes without saying that the industry needs the routine tools of quality assurance (QA) linked with HACCP schemes (see Chapter 4) and quality control (QC) functions to perform the necessary measures and evaluations on its products and processes. It is important that the data arising from these are used also for diagnosis and for planning and that it not be lost in paper archives. Software packages are available for recording and analysing such data and greater use of this approach can only lead to better descriptions of quality attributes of both product and processes. The growing trend towards the electronic marketing of fish and fish products and the need for control of the quality chain will be covered in Chapter 15.

10.5   Freshness

Much of the body of this chapter has concentrated on quality and many of the same remarks relate to freshness. Freshness is another difficult concept, about which there is no set agreement. It is a term widely used in more than one context. At times, it refers to the fact that the fish has been recently caught and at times it is used in the context of unfrozen or unprocessed but sometimes 'fresh frozen' is used. Unless the context for its use is carefully defined or described it is better not to use it at all (Bremner and Sakaguchi, 2000). Much effort has been put into the development of tests for 'freshness' but most of these are fairly poor indicators of the post-catch history as they cannot integrate time-temperature effects (Bremner et al., 1987, Bremner and Sakaguchi, 2000). These efforts would have been better placed into developing systems of control and of quality chain management, including traceability (Chapter 15; Frederiksen et al., 2002).

10.6   Safety

There seems to be no reason why the ideas outlined above (10.2.1) about the use of definitions cannot also be employed to describe safety of fish products in a coherent structured manner. This approach could readily be developed using the existing information about safety of fish products (this volume for example). The collation into an interconnected, logical structure in software that can be interrogated would have enormous advantages.

10.7   Future trends

There is always a considerable degree of overlap and agreement in most discussions about quality and my efforts to set down specific definitions are a means of trying to integrate some of the various factors commonly thought of as comprising quality. It seems clear that the template for definitions can incorporate all the QD1 to QD4 definitions of Martens and Martens (2001) as some of the specific terms. Furthermore, the general multivariate analytical approach can be included as the specific means of analysing the data so that inherent concepts of scope and limitation can be included. It is also possible that more of the effects of circumstance and expectation (Meisleman, 2001) can be included too. This could lead to a broader family of definition generating templates that would be compatible with all three approaches to quality.

Whatever approaches are taken in dealing with issues of quality (and freshness) it is best to try to avoid using the word 'quality' as a catch-all term and always to be as specific as possible.

10.8   References

ANONYMOUS, 'Diet has little impact on flesh quality'. *Global Aquaculture The Advocate*, 2000, October, p. 69.

BREMNER, H A, OLLEY J and VAIL A M A, 'Estimating time-temperature effects by a rapid systematic sensory method'. In *Seafood Quality Determination*, Kramer D E and Liston J (eds) Elsevier, Amsterdam, 1987, pp. 413–35.

BREMNER, H A, 'If freshness is lost, where does it go?' In *Methods to Determine the Freshness of Fish in Research and Industry*, Olaffsdóttir G, Luten J, Dalgaard P, Careche M, Verrez-Bagris V, Martinsdóttir E and Heia K (eds). Final meeting of the Concerted Action 'Evaluation of Fish Freshness'. International Institute of Refrigeration, Paris, 1997, pp. 36–51.

BREMNER, H A, 'Towards practical definitions of quality for food science', *Critical Reviews in Food Science and Nutrition*, 2000, **40**, 83–90.

BREMNER, H A and SAKAGUCHI M, 'A critical look at whether 'freshness' can be determined'. *Journal of Aquatic Food Product Technology*, 2000, **9**, 5–25.

FREDERIKSEN M A, ØSTERBERG C, SILBERG S T, LARSEN E and BREMNER H A, 'Info-fisk. Development and validation of an Internet based traceability system in a Danish domestic fresh fish chain'. *Journal of Aquatic Food Product Technology*, 2002, 11 (2), 13–34.

GÄRDENFORS P, *Conceptual Spaces:the geometry of thought*. MIT Press, Cambridge Mass. USA, 2000.

HARRY M and SCHROEDER R, *Six Sigma*, Doubleday, NY, USA. 2000.

ISO, *Quality management systems – Fundamentals and vocabulary*. Brussels, Belgium, European Standard (EN ISO 9000:2000, Point 3.5.4), European Committee for Standardisation. 2000.

MARTENS H and MARTENS M, *Multivariate Analysis of Quality*, John Wiley & Sons, Chichester UK. 2001.

MEISELMANN H L, 'Criteria of food quality in different contexts' *Food Service Technology*. 2001, 1, 67–84.

11

The meaning of shelf-life

A. Barbosa, University of Porto, A. Bremner, Allan Bremner and Associates, Mount Coolum and P. Vaz-Pires, University of Porto

11.1 Introduction: the concept of shelf-life

11.1.1 Language considerations

Although simple definitions for shelf-life can easily be found, they are almost never based on clear definitions of what the authors mean by the expression 'shelf-life', sometimes also spelled 'shelf life' and 'shelflife', although the last one is not considered as correct in English (Fox *et al.,* 1987). When decomposing the expression 'shelf-life', the term 'shelf' suggests a waiting period in a supermarket shelf, until the sale of the food item. This means the word is restrictive, defining only part of the total. In fact, the total period is far wider, encompassing the time between the very first moment the food exists as such, for example at catch (for wild fish) or death (for farmed fish), and the stage at which it is rejected for a determined purpose. 'Life' can also be considered as controversial, as here it designates a period that usually starts with death. It suggests a period of time, and consequently clear limits for the beginning and the end of this period must be indicated. It also includes the idea of suitability for intended purpose and this too is a necessary element in any definition. In conclusion, although the expression can be restrictive, it is already widely spread both in common and scientific food literature and its use is so valuable that it must be retained. What are required are clearer descriptions of meaning and circumstance.

11.1.2 Shelf-life definition

Several definitions of shelf-life have been proposed by different authors and, broadly speaking, there is a degree of similarity between them. According to

Daun (1993), 'shelf-life is defined as the maximal period of time during which the predetermined quality attributes of food are retained'. Note the avoidance of definition of clear limits. Another definition is 'for each particular food, there is a finite length of time after production that it will retain a required level of quality organoleptically and safety wise, under stated conditions of storage' (Taoukis *et al.*, 1997). Note the word finite, clearly pointing to a precise end limit, and also that authors implicitly accept some variation, as it depends on conditions of storage. The same authors were concerned about their own definition, as they stated later in the same publication 'there is no established, uniformly applicable definition of shelf-life'. In a small dictionary on the composition and quality of fish, Waterman (1982) defined shelf-life as being the same as storage life, keeping quality, keeping time and storage time. The definition given was 'length of time that a fish or fish product of initially high quality can be kept under specified storage conditions before it becomes either significantly poorer in quality or unsuitable for sale or consumption'.

From these definitions, one could assume that shelf-life is an easy concept to define. In the definition given by Waterman (1982), however, the expression 'product of initially high quality' is also an example of a somewhat vague indication of an initial moment to start the countdown from. Closer examination reveals that the terms and expressions used to describe and define shelf-life are themselves often vague and undefined and involve value judgements, for example, quality, safety wise, significantly poorer. It is probable that, in turn, these definitions of the words would include the term shelf-life itself thus making a circular argument. This is commonly encountered in the use of a dictionary as an aid, where the definition of one term is expressed in some other term and vice versa. This occurs because the definitions tend to be developed independently rather than as parts of a whole interlocking intellectual structure. For shelf-life this structure would include the problem of establishing the limits of the definition(s) so it (they) can readily be grasped and put into practice.

This chapter is about the difficulties that can be found when the expression 'shelf-life' is used in science, namely by lack of clear definition of the first and specially the last moment of this period, and suggests some other expressions that can be used in certain specific situations and a broader approach using information technology to provide a coherent set of interrelated and consistent descriptions about shelf-life.

11.2   The beginning of shelf-life

The beginning of shelf-life is generally, but not always, consensual. Authors seem to agree it starts when fish die, which is normally assumed to be when fish are taken from water (Doyle, 1989). This is valid both for wild and farmed fish. Sometimes, however, fish die before this during entrapment in the fishing gear or after extreme struggle on the hook or in an attempt to escape from the nets. Fish can also die after being taken from the water caused by lack of oxygen when landed on the deck or when slaughtered at farms. This moment is also normally cited as the beginning of irreversible changes in the properties of the fish and it is the total sum of these degradative processes that bring about the end of shelf-life. This point at which changes begin has been referred to as the peak of 'freshness' but this term is non-specific as well and a different means of expressing this concept based on properties of the product, not a vague notion, should be used (Bremner, 2000). It is important to point out that a fish does not necessarily die when it is caught or taken from the water. It is not so simple for physiologists to define fish death since the heart may continue beating after the brain has been destroyed; so heart activity is not precise enough. Cessation of cerebral activity may be a good indicator in physiological experiments, but this is too complex for commercial practice. Muscle movements are not specific, as twitching may occur due to nerve impulses long after both heart and cerebral activities cease. For some crustaceans, activity of gills or response of an appendage to a physical stimulus may be used. For molluscs, it may be retention of ability to close or open their shell or some other type of molluscan muscular activity.

In some particular situations, for example in Eastern countries where fish is caught and sold alive, shelf-life can be considered in two periods, one from the moment when fish is put in water, in the commercial container, until point of sale (fish is still alive) when it is usually slaughtered in front of the buyer, plus a second period (fish is dead), until rejection for a particular use, for example, as sashimi, for consumption after cooking or for other purposes. Products like oysters and many other bivalve molluscs are often sold alive and kept alive until they are cooked at home or restaurant but they are also commonly eaten raw and alive, having been just taken from the shell and the term 'life' is for this particular case perfectly apt. However, spoilage organisms can proliferate on the outside of the shell in any debris present and these may cause taint to the product and shorten its shelf-life, even though it is still alive.

As some fish can be obtained only from long-distance fisheries and technologists can obtain it only after some days of prior storage, some published work on fish shelf-life has been done starting with fish already stored some days in ice (usually one to three) on the boat. It is common to find publications where the origin (zero) of the graphs about degradation processes is not a real zero in shelf-life. One to three days were spent on board, landing and transportation until the fish arrived at the laboratories, and these should be included in the time count. The term 'icedays' represents a convenient way of expressing 'equivalent days of box storage in ice' (Bremner and Sakaguchi, 2000). It is important to emphasise that precise knowledge of at least the time/temperature history of the fish must be known, otherwise the precision of the shelf-life measurements is compromised. Ideally, researchers should be on-board when the fish is caught, and treat the samples separately, in ideal conditions, to avoid the interference of several other variables that are normally impossible to control. Unfortunately, this is not always possible. Peri-mortal circumstances must be included in the description/definition of this moment, as the pre-harvest, harvest, slaughter and post-slaughter techniques affect every major property of fish flesh. Techniques like rested harvest, pre-mortal chilling and brain spiking are now commercial standard practice (Bremner and Sakaguchi, 2000), and they tend to slow the degradative changes, extending the duration of the high-quality period. Thus, in countries where this is practised, published shelf-lives are considerably longer (or closer to theoretical maximum, depending on point of view) than in countries or regions where more traditional techniques are used, like stunning in the head, electrocution or oxygen deprivation and carbon dioxide poisoning.

In conclusion, the beginning of shelf-life depends on the totality of circumstances, only in special experimental circumstances can it be described to the precise minute. In salmon aquaculture, the use of a club or of a stunning machine, coupled with gill cutting and bleeding, can narrow the time down to a particular hour. So too, can many line and gill net fisheries be this precise as the fish are taken live onboard the boat and then dealt with accordingly. For bulk catches the time of death can be taken only approximately from when the codend is discharged on deck, or from when the catch is sorted. Nevertheless, a beginning of shelf-life can generally be pinned down to within a few hours, or to a particular day. The significance of this precision depends on the fate of the catch and for which product form and market it is destined.

11.3   The end of shelf-life

Scientific writers seem to have their own personal idea about the end of shelf-life, as it is dependent on several factors such as further processing or market destination (Howgate, 1985b). Some authors are concerned with shelf-life limits: 'the definition of shelf life and the criteria for the determination of the end of shelf life are dependent on specific commodities and on the definition's intended use (i.e., for regulatory vs. marketing purposes)' (Taoukis et al., 1997). Many projects have been conducted around the concept of 'extension of shelf-life', but in many instances the shelf-life was defined by the results themselves rather than on any a priori criteria. This is legitimate where there is no previous information on the species or the circumstances. For reported results, however, to have general validity it is necessary that some standard means of expressing shelf-life is adopted.

11.3.1   Geographically influenced concepts

There are several reports in the fish area of extension of shelf-lives by several weeks induced by modified atmosphere storage when measured by taste of cooked fillets (Fey and Regenstein, 1982; Reddy et al., 1992), but considerably shorter when measured by sensorial analysis based on external characteristics of whole fish (Capell, 1999). Both conclusions are right, but the authors are talking about different ways of measuring the end of shelf-life. This example also shows the 'local value' of these scientific considerations, in this particular case, fillets are the main product in Northern European countries and in the United States of

America, while in Southern Mediterranean countries whole fish is still the main product presented to consumers. This means some scientific works are strongly influenced by geography and social culture, both in experimental design and in drawing conclusions.

11.3.2 Points of view: food industry and consumers

Fu and Labuza (1993), in a review of shelf-life prediction, stated that 'the shelf-life of a food is the time period for the product to become unacceptable from sensory, nutritional or safety perspectives'. These same authors, however, concluded that 'a universal definition of shelf-life is thus virtually impossible to establish'. For the food industry, shelf-life is based on the extent of change in properties and attributes in a food that the food company is prepared to accept for its branded product in the marketplace prior to sale. Whereas for consumers, the end of shelf-life is the time when the food no longer has an acceptable appearance or taste. One should note that these are not coincident definitions, as using the food industry viewpoint the food may still be organoleptically acceptable at the end of what they deem to be its shelf-life. The goal for the food industry is not an absolute value of the shelf-life, but rather the assurance that shelf-life is longer than the time of normal product distribution and consumption (Fu and Labuza, 1993). Food industry shelf-life is often based on the time to rejection by a pre-determined percentage of panelists, less some time period (days) within which the product will be purchased and consumed.

11.3.3 Shelf-life and food label dates

Difficult interpretations can also result when considering shelf-life and food label dates. In the legislation of the European Union, the following information is compulsory on the labelling of foodstuffs: 'the date of minimum durability or, in the case of foodstuffs which, from the microbiological point of view, are highly perishable, the "use by" date' (Anonymous, 2000). This represents the way consumers must be informed about what date food products should be consumed by, but does not mean necessarily, and in all cases, the end of shelf-life under any of the several possible definitions, although sell-by date and best-if-used-by date of foods must be based on some type of shelf-life testing (Fu and Labuza, 1993).

In fact, the date of minimum durability is calculated from the predicted end of shelf-life minus the final period in which organoleptical properties are not considered as completely satisfactory; it may correspond to the end of display or commercial life. The 'use by' date is the date when foods become, from the microbiological point of view, no longer acceptable for human consumption; it could be coincident with the end of shelf-life, but if safety is the main concern, normally the food companies use a safety margin to increase confidence. This means 'use by' dates are marked before the end of total shelf-life (total shelf-life minus the selected confidence margin). In setting a shelf-life, any company must consider what the characteristics are, such that they would no longer wish to be associated with that product. This is a commercial, marketing and business decision, that they must make for themselves, it is not a technical one.

11.4   Are there several shelf-lives?

11.4.1   Examples from literature

From a wide set of references on food (mainly seafood), some examples of expressions that can be found meaning shelf-life (Table 11.1) or similar concepts (Table 11.2) are presented. In fact there are several different interpretations and meanings associated with the expressions shown, but the most common is based on the sensorial rejection by a group of panelists. Even in these cases, sometimes the authors are talking about very different moments, as the rejection is always dependent on the product used for analysis (whole fish, fillets, other portions) and on what is asked of the panelists (to judge visually the sample, to smell, to taste). A good example of possible consequences can be taken from Fey and Regenstein (1982) who stated that 'the cooked quality of the fish was fine but the raw appearance of the whole fish was not as good. Thus, the potential end-use of the product would be of some importance: these fish might be more appropriate for filleting'.

A more extensive bibliography dealing with both fresh and frozen fish was compiled by Howgate (1985a). In that paper and others (Howgate, 1983; Howgate, 1985b) he prefers the name storage life over shelf-life and this was the phrase used predominantly up to that time. He also makes the case that storage life refers to the storage of the material, for example, whole fish, from the point of harvest to the point of final sale or of transformation into another product, for example packaged fillets. Shelf-life then refers to the packaged or processed product. This means that in the case of some trawl fish storage life, maybe ten days, would be far greater than shelf-life, maybe three or four days. Whereas, with farmed fish such as salmon the situation would be reversed.

In this chapter we have chosen to use the phrase shelf-life rather than storage life as this seems to be the term most used over the last decade, it relates more closely to usage and results appropriate for consumer information, the emphasis on research seems to have been greater on unfrozen than on frozen fish products and it can convey more the idea of a continuum from harvest to consumption. It should be pointed out that neither term is an absolute, nor is there any fundamental reason to choose one over the other. For canned products and frozen materials and products, or dried or highly salted products the term storage life may be just as appropriate since in these cases the product is deliberately stored for later use. What is clear is that, like quality (Bremner, 2000) and freshness (Bremner and Sakaguchi, 2000) the terms used, whether they be shelf-life or storage life, must be described and defined so the reader and user understands exactly what is being reported and that meaning is not left to be interpreted by the user (Howgate, 1985b).

Table 11.1 References that include the expression 'shelf-life' or 'shelf life'

Expression and context	Meaning	Measurement	Product	Ref.
Commercial shelf-life	period fish can be offered for sale	sensorial (consumer acceptability)	whole and gutted fish, cooked fillets	Gelman *et al.* (1990)
End of shelf life, microbiological shelf life, extended shelf life	microbial development	microbiological counts	general foods	Labuza *et al.* (1992)
Expected shelf life	all properties	mainly sensorial (cooked flavour)	highly perishable foods	Labuza (2000)
Extend shelf life	rejection by sensorial properties	sensorial (odour and taste)	several	Flick *et al.* (1991)
Extend shelf-life	rejection by sensorial properties	sensorial (odour)	whole and fillets of winter flounder	Field *et al.* (1986)
Extension of microbiological shelf-life	delay in the lag phase of microbial growth	microbiological counts	catfish fillets	Kim and Hearnsberger (1994)
Extension of shelf life	until unacceptability by consumers	microbial, texture and taste tests	soft dates	Nussinovich *et al.* (1989)
Extension of shelf-life	growth rate of SSOs, rate of loss of free sulphite	counts of SSOs like *Brocotrix thermosphacta*	British fresh sausage	Adams *et al.* (1987)

Table 11.1 Continued

Expression and context	Meaning	Measurement	Product	Ref.
Maximum shelf-life	up to inedibility of the fish	sensorial evaluation	whole and gutted fish, cooked fillets	Gelman *et al.* (1990)
Microbial shelf-life	rejection based on microbial development	microbial counts	cooked beef	Papadopoulos *et al.* (1991)
Predict shelf-life	based on microbial counts	*Shewanella, Pseudomonas* and *Photobacterium* counts	lightly preserved fish products	Gram and Huss (1996)
Remaining shelf life	sensorial and bacterial based rejection	cooked odour and flavour, microbial counts	vacuum-packed cod and cod fillets	Jorgensen *et al.* (1988)
Remaining shelf-life	sensorial properties	Quality Index Method	whole fish	Branch and Vail (1985)
Shelf-life	microbial-based rejection	bacterial counts	minced meat	Mattila-Sandholm and Skyttä (1991)
Shelf-life extension	rejection by taste and overall acceptability	cooked flavour, several chemical analysis	red hake and salmon MAP fillets	Fey and Regenstein (1982)
Shelf-life prediction	counts of specific spoilage organisms	*Shewanella* and *Photobacterium*	MAP cod	Dalgaard (1995)
Total shelf-life	until sensorial rejection for any food uses	sensorial evaluation	fish	Gelman *et al.* (1990)
True shelf-life	microbiological rejection	microbial count, mathematical prediction	general foods	Fu and Labuza (1993)

Table 11.2 References to other expressions meaning 'shelf-life'

Expression	Meaning	Measurement	Product	Ref.
Extension of storage life	sensorial properties	cooked flavour of fillets/ steaks	whole cod and salmon, gutted trout	Cann (1988)
Storage life	sensorial rejection	sensorial tests	albacore	Pérez-Villarreal and Pozo (1990)
Maximum storage time	sensorial (whole and cooked), chemical	freshness scores, K-value, TVB, TMA	albacore	Pérez-Villarreal and Pozo (1990)
Total storage life	scoring texture, odour and flavour	cooked fillets	gutted cod and plaice	Huss et al. (1974)
Remaining storage life on ice	rejection of cooked fish by taste	QIM	several species	Larsen et al. (1992)
Storage stability	chemical and physical characteristics	water binding capacity, texture measurements	lightly salted minced cod	Bligh and Duclos-Rendell (1986)
Keeping quality, storage life	rejection mainly by sensorial characteristics	sensorial tables, sensory panel (raw and cooked)	fish in general	Lima dos Santos et al. (1981)

11.4.2   Frozen seafood

In the particular case of frozen foods, the concepts and therefore their inter-pretation are completely different. Freezing stops the usual degradation process, and the use of spoilage effects to evaluate acceptability is no longer possible. These differences induced the creation of different terms for frozen products.

The International Institute of Refrigeration (IIR) recommendations for frozen food (IIR, 1972) introduced two new definitions: High Quality Life (HQL) which is the time from freezing of the product for a just noticeable sensory difference to develop (70–80% correct answers in a triangular sensory test), and Practical Storage Life (PSL), the period of proper (frozen) storage after processing (freezing) of an initially high-quality product during which the organoleptic ·quality remains suitable for consumption or for the process intended (IIR, 1972; Howgate, 1983; Taoukis et al., 1997). But as Howgate (1983 and 1985b) correctly points out, the test methods to describe a product of HQL are inappropriate, have poor confidence limits, and their use results in impractically short commercial shelf-lives.

A properly frozen product under proper storage conditions will never spoil due to bacterial action, but will suffer other kinds of degradation processes. Rejection must therefore be based on phenomena that occur during storage, like lipid oxidation, changes in texture induced, for example, by loss of water, or long-term effect of enzymes, as microorganisms simply cannot grow at commercial freezing temperatures (around $-18°C$).

11.4.3   Different ways of measuring shelf-life

Shelf-life has been measured in scientific investigations mainly by sensory methods, but also by microbiological, chemical and physical procedures. The sensory methods are based normally on sensory tables. The most widespread in Europe are the EC scheme at official markets (Howgate et al., 1992), the Torry scheme, when precision needs to be higher, as in scientific laboratories (Shewan et al., 1953), and recently the quality index method (QIM), where the high variability of the aquatic species is more taken into account (Branch and Vail, 1985; Bremner, 1985; Bremner et al., 1987; Luten and Martinsdóttir, 1997). When using EC and Torry schemes, end of shelf-life is determined by rejection of whole fish, based on externally visible non-destructive tests, but sometimes also including internal characteristics when the fish has been gutted. Rejection by the QIM can be estimated by comparison of the score with results obtained from evaluation of the cooked product by a panel. Independently from tables or schemes, sensory methods are considered the most appropriate, as they represent more closely the consumer's opinion and they are fast and simple, a necessary requirement as non-processed seafood is highly perishable. Other methods must be in accordance, be defined and be designed so that the results relate to those obtained by sensory methods.

Microbiological methods are based on bacterial counts. The total number of microorganisms per cm^2 or per gram does not reliably show a relationship with shelf-life. At the start of storage, or just after catch, the total bacterial count is variable between 10^1 and 10^4 cfu/g or /cm^2 but at the time of sensorial rejection, it can be between 10^5 and 10^8, generally 10^6 to 10^7 cfu/g or /cm^2 (Fu and Labuza, 1993). Since many of these organisms do not contribute to changes that we call spoilage, i.e., development of objectionable volatiles or off-flavours, they are only circumstantial indicators of shelf-life. In the majority of circumstances the major changes are due to the proliferation of specific spoilage organisms (SSO). Counts of these organisms are related to shelf-life. The most common are H$_2$S producers, *Shewanella* and *Photobacterium* and, in warmer waters, *Pseudomonas*. It is possible nowadays to calculate the remaining shelf-life using counts of SSO (see Chapter 12), but sensory analysis is still necessary to validate the system.

Chemical methods are normally destructive and are indicators of spoilage since many changes are not detectable during the first few days of storage. Tests have better discriminatory value on stored product after some days in ice and poor discriminatory capacity in the most important period of the first few days in ice. One exception to this general pattern is the K-value, which can be very useful in the early period of storage with some species since it measures breakdown products of ATP. This is greatly influenced by peri-mortal circumstances and by species and even where it is appropriate the measurements take time, are expensive and need trained laboratory workers.

Physical methods include some promising techniques as they are generally non-destructive, very fast, not labour intensive and have low test costs. However, due to the high variability among sea species, both inter- and intra-specific, conclusions obtained for single fish are extremely difficult to interpret and to extrapolate from. Some instruments such as the Torrymeter, Freshmeter and Fishtester (dielectric instruments) require tests on large numbers of fish and are not capable of integrating time and temperature effects (Bremner *et al.*, 1987). Time-temperature recording after catching and during storage provides the most useful indicator from which remaining shelf-life can be estimated. This is crucial information and various time-temperature integrators have been developed, but they are not yet commercially successful. The most appropriate steps are to have a fully functional quality chain in which good practices are used as standard procedures and estimates can be readily made from records (see Chapter 15). Few investigators have used physical methods alone to define the end of shelf-life. These methods are and should be used as complementary in relation to sensory methods.

As a conclusion, sensory methods are normally preferred, alone or combined with others. For whole fish, the EC scheme is the method most used, for fillets and other products some evaluation of the taste is common; QIM is gradually replacing the EC scheme in commercial practice in Europe (http://www.QIM-Eurofish.com). As an answer to the question: 'are there several shelf-lives?' we can say YES, there are several shelf-lives as the period considered is different under different circumstances. Furthermore, none of the methods are particularly precise and, unlike the beginning of shelf-life, the confidence limits that can be placed on an estimate of its end are very broad and generally become broader the longer is the period to be measured. Estimates of the end of shelf-life within ± 1 day are as precise as can be obtained.

11.5   Do we need the expression shelf-life?

The answer is YES, but it is only valid if clear indication of the limits (specially the end) and how they were measured and/or defined is given. The idea of shelf-life is of widespread practical use and regulations regarding 'use-by dates' and the need for consumer advice on 'best-by' dates, make this essential information. The information is critical within the industry as well, for planning and logistic purposes in production, marketing and distribution.

11.6   Future trends

Attempts have previously been made to find more terms like 'taste-life' which means shelf-life measured by taste, 'safe-life' for shelf-life measured by safety for consumers and 'commercial-life' (proposed for fish by Gelman et al., 1990 and Vaz-Pires, 1995) or 'display-life' (proposed for meat by Lavelle et al., 1995) when based on parameters that interfere with commercial acceptability.

Figure 11.1 shows some of the concepts and their corresponding proposed meanings. We suggest for future publications that these more objective terms should be used to indicate exactly shelf-life measured by taste, safety aspects or commercial purposes, but the method of determination of shelf-life should always reflect the intended product state and be clearly indicated. For example, reports of extensions of shelf-life must be looked at carefully, as what is obtained is a delay in the period taken for the product to reach the point of rejection. This is normally almost meaningless, as at the end of shelf-life the use of that seafood for normal consumption by consumers is no longer possible. Extension of the period of high quality (of high-quality life), if these terms are correctly explained, would be much more valuable both for industry and for consumers. We suggest the following checklist for what to mention when authors are measuring duration and extension of shelf-life (see Fig. 11.2).

So what is the beginning and what is the end of shelf-life? From the above considerations it is clear that for the answer to this question to have practical meaning all the relevant circumstances must be considered. Many factors all have great bearing on how to decide the answer such as (a) those that relate to the raw material and product such as catch method, perimortal circumstances, species, geography, distribution chain, and (b) to the consumer such as cultural region, product type, modes of sale, typical dishes, and so on and, (c) whether the answer is to be used for a technical, commercial, marketing, personal, nutritional, health, institutional, national or regulatory reason or for a mix of these. The methodology chosen also influences the answers obtained to all the

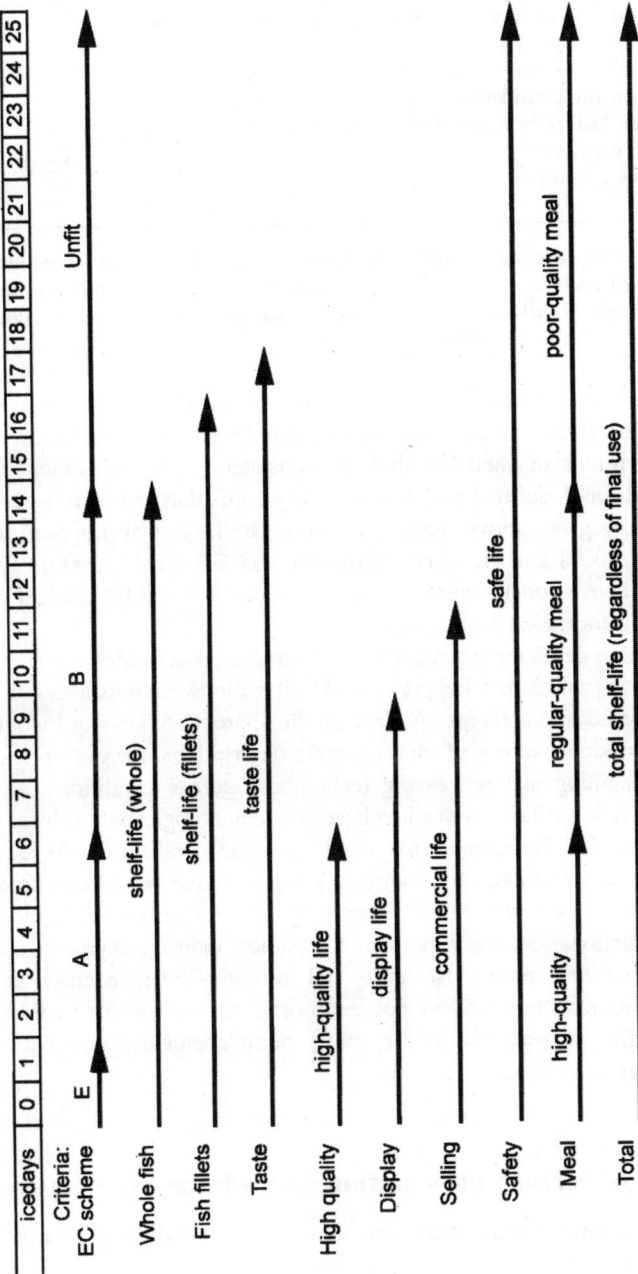

Fig. 11.1 Different concepts and assumptions of shelf-life and their corresponding limits. EC scheme presented is for iced/boxed cod (*Gadus morhua*) (Howgate *et al.*, 1992). All values are theoretical and must be understood as examples only. The exact correspondence between concepts and days in ice is then merely hypothetical.

Definition of shelf-life ☐
General purpose of the measurement ☐
Fishing method (details on stress, catch and post-catch procedures) ☐
Geographical details about fishing ☐
Moment when fish is caught ☐
Beginning of shelf-life considered ☐
Time-temperature history from catch to end of shelf-life ☐
End of shelf-life considered ☐
Destination of the product ☐

Figure 11.2   Suggested checklist for the information that should be given when shelf-life (or extension of shelf-life) is measured or quantified. For exact planning of the experiments and value of different methods for shelf-life assessments, see Lima dos Santos *et al.* (1981).

above. The answers are not likely to be the same in the various combinations of these circumstances. Therefore the approach suggested here is to use a hierarchy of structured definitions of shelf-life that are consistent with each other and which are linked through defined and described keywords that can be related to all the considerations given above. This approach is similar to that proposed for quality (Bremner, 2000) and freshness (Bremner and Sakaguchi, 2000). The complexity of the above considerations, makes it essential that it be handled by modern means of information technology.

This could involve developing a related set of structured definitions of shelf-life comprised of keywords and keyphrases to suit various circumstances that would be programmed into software. A click on the appropriate keyword would reveal a further dropdown screen of more specific descriptions, for example, of fishing method, handling and processing techniques, storage conditions, etc., that would also include further keywords related to methodologies and values. A further click on one of these keywords would provide details of technique, limits, appropriate circumstances for use, etc. This technique would overcome the problems mentioned above (11.1.2) concerning the circularity of definitions and the fact that dictionaries, organisations, companies, industry segments and regulatory bodies for that matter, use terms and meanings independently, and often in conflict with each other. Some degree of order and uniformity could be given to the situation without loss of flexibility or independence or ignoring special needs of different groups.

11.7   Sources of further information and advice

The definitions of scientific terms and expressions can be found easily in article introductions, as the authors feel they are needed as a basic starting point. From time to time and normally in other sections of the scientific magazines than the full-length articles, for example in sections like 'viewpoint sections', 'short letters', 'comments on science', etc., the scientific literature offers publications on language and precision, usually ending with suggestions to avoid confusions and misinterpretations. Examples in the fish area are Connell (1989), concerned with the use of organoleptical and sensory, which stated that 'the organoleptic properties of a food are measured by sensory assessment; the term "organoleptic assessment" should not be used'. One should notice, however, that after 12 years, incorrect use of these terms can be easily found.

A discussion on the possibility of determining 'freshness' in seafood was made recently by Bremner and Sakaguchi (2000). They recommend that the use of the term should be kept to a minimum in scientific literature but where it is used, it must be carefully defined and described.

11.8   References

ADAMS M R, BAKER T and FORREST C L (1987), 'A note on shelf-life extension of British fresh sausage by vacuum packing', *Journal of Applied Bacteriology*, 63, 227–32.

ANONYMOUS (2000), Directive 2000/13/EC of the European Parliament and of the Council of 20 March 2000, article 3, (5).

BLIGH E G and DUCLOS-RENDELL R (1986), 'Chemical and physical characteristics of lightly salted minced cod (*Gadus morhua*)', *Journal of Food Science*, 51 (1), 76–8.

BRANCH A C and VAIL A M A (1985), 'Bringing fish inspection into the computer age', *Food Technology in Australia*, 37 (8), 352–55.

BREMNER H A (1985), 'A convenient easy-to-use system for estimating the quality of chilled seafoods', *DSIR Fish Processing Bulletin*, 7, 59–70.

BREMNER H A (2000), 'Towards practical definitions of quality for food science', *Critical Reviews in Food Science and Nutrition*, 40, 83–90.

BREMNER H A and SAKAGUCHI M (2000), 'A critical look at whether 'freshness' can be determined', *Journal of Aquatic Food Product Technology*, 9, 5–25.

BREMNER H A, OLLEY J and VAIL A M A (1987), 'Estimating time-temperature effects by a rapid systematic sensory method', in Kramer D E and Liston J, *Seafood Quality Determination*, Amsterdam, Elsevier, 413–35.

CANN D C (1988), 'Modified atmosphere packaging of fishery products', *INFOFISH International*, 1, 37–39.

CAPELL C (1999), *Growth and interaction of bacteria during spoilage of fresh fish*, PhD thesis, Portuguese Catholic University, Porto, Portugal.

CONNELL J J (1989), 'Sensory assessment of fish quality', *Torry advisory note no. 91*, Aberdeen, Torry Reseach Station.

DALGAARD P (1995), 'Qualitative and quantitative characterization of spoilage bacteria from packed fish', *International Journal of Food Microbiology*, 26, 319–33.

DAUN H (1993), 'Introduction', in Charalambous G, *Shelf-life studies of foods and beverages Chemical, Biological, Physical and Nutritional Aspects*, Amsterdam, Elsevier Science Publishers B V, ix–x.

DOYLE J P (1989), 'Seafood shelflife as a function of temperature', *Alaska Sea-Gram 30*, Marine advisory program, University of Alaska.

FEY M S and REGENSTEIN J M (1982), 'Extending the shelf-life of fresh wet red hake and salmon using CO_2–O_2 modified atmosphere and potassium sorbate ice at 1°C', *Journal of Food Science*, 47, 1048–54.

FIELD C E, PIVARNIK L F, BARNETT S M and RAND A G (1986), 'Utilization of glucose oxidase for extending shelf-life of fish', *Journal of Food Science*, 51, 66–70.

FLICK G J, HONG G P and KNOBL G M (1991), 'Non-traditional method of seafood preservation', *MTS Journal*, 25 (1), 35–43.

FOX G, MOON R and STOCK P (1987), *Collins Cobuild English Language Dictionary*, London, Collins Publishers.

FU B and LABUZA T P (1993), 'Shelf-life prediction: theory and application', *Food Control*, 4 (3), 125–33.

GELMAN A, PASTEUR R and RAVE M (1990), 'Quality changes and storage life of common carp (*Cyprinus carpio*) at various storage temperatures', *Journal of the Science of Food and Agriculture*, 52, 231–47.

GRAM L and HUSS H H (1996), 'Microbial spoilage of fish and fish products', *International Journal of Food Microbiology*, 33, 121–37.

HOWGATE P (1983), 'Measuring the storage lives of chilled and frozen fish products', in *Proceedings XVI Congress of Refrigeration*, Paris, International Institute of Refrigeration, 537–45.

HOWGATE P (1985a), 'Bibliography of storage lives of wet and frozen fish', in *Storage lives of chilled and frozen fish products*, Proceedings of meetings of Commissions C2 and D3 (1–3 October 1985) Paris, International Institute of Refrigeration, 389–401.

HOWGATE P (1985b), 'Approaches to the definition and measurement of the storage life of chilled and frozen fish: a brief review', *Refrigeration Science and Technology*, 4, 45-53.

HOWGATE P, JOHNSTON A and WHITTLE K J (1992), *Multilingual Guide to EC Freshness Grades for Fishery Products*, Aberdeen, Torry Research Station.

HUSS H H, DALSGAARD D, HANSEN L, LADEFOGED H, PEDERSEN A and ZITTAN L (1974), 'The influence of hygiene in catch handling on the storage life of iced cod and plaice', *Journal of Food Technology*, 9, 213–21.

INTERNATIONAL INSTITUTE OF REFRIGERATION (1972), *Recommendations for the processing and handling of frozen foods*, 2nd edition, Paris, International Institute of Refrigeration.

JORGENSEN B R, GIBSON D M and HUSS H H (1988), 'Microbiological quality and shelf life prediction of chilled fish', *International Journal of Food Microbiology*, 6, 295–307.

KIM C R and HEARNSBERGER J O (1994), 'Gram negative bacteria inhibition by lactic acid culture and food preservatives on catfish fillets during refrigerated storage', *Journal of Food Science*, 59 (3), 513–16.

LABUZA T P (2000), 'The search for shelf life', *Food Testing & Analysis*, 6 (2), 26–36.

LABUZA T P, FU B and TAOUKIS P S (1992), 'Prediction for shelf-life and safety of minimally processed CAP/MAP chilled foods: a review', *Journal of Food Protection*, 55 (9), 741–50.

LARSEN E, HELDBO J, JESPERSEN C M and NIELSEN J (1992), 'Development of a method for quality assessment of fish for human consumption based on sensory evaluation', in Huss H H et al., *Quality Assurance in the Fish Industry*, Amsterdam, Elsevier Science Publishers B V, 351–8.

LAVELLE C L, HUNT M C and KROPF D H (1995), 'Display life and internal cooked color of ground beef from vitamin E-supplemented steers', *Journal of Food Science*, 60 (6), 1175–8, 1190.

LIMA DOS SANTOS C A M, JAMES D and TEUTSCHER F (1981), 'Guidelines for chilled fish storage experiments', *FAO Technical Paper 210*, Rome, FAO.

LUTEN J B and MARTINSDÓTTIR E (1997), 'QIM: a European tool for fish freshness evaluation in the fishery chain', in Olafsdóttir G, Luten J, Dalgaard P, Careche M, Verrez-Bagnis V, Martinsdóttir E and Heia K, *Methods to determine the freshness of fish in research and industry*, Paris, International Institute of Refrigeration, 287–96.

MATTILA-SANDHOLM T and SKYTTÄ E (1991), 'The effect of spoilage flora on the growth of food pathogens in minced meat stored at chilled temperature', *Lebensmittel-Wissenshaft und-Technology*, 24, 116–20.

NUSSINOVICH A, ROSEN B, SALIK H and KOPELMAN I J (1989), 'Effect of heating media on the microbiology and shelf-life of heat-pasteurized soft dates', *Lebensmittel-Wissenshaft und-Technology*, 22, 245–47.

PAPADOPOULOS L S, MILLER R K, ACUFF G R, VANDERZANT C and CROSS H R (1991), 'Effect of sodium lactate on microbial and chemical composition of cooked beef during storage', *Journal of Food Science*, 56 (2), 341–7.

PÉREZ-VILLARREAL B and POZO R (1990), 'Chemical composition and ice spoilage of albacore (*Thunnus alalunga*)', *Journal of Food Science*, 55 (3), 678–82.

REDDY N R, ARMSTRONG D J, RHODEHAMEL E J and KAUTTER D A (1992), 'Shelf-life extension and safety concerns about fresh fishery products packaged under modified atmospheres: a review', *Journal of Food Safety*, 12, 87–118.

SHEWAN J M, MACINTOSH R G, TUCKER C G and EHRENBERG A S C (1953), 'The development of a numerical scoring system for the sensory assessment of the spoilage of wet white fish stored in ice', *Journal of the Science of Food and Agriculture*, 4, 283–98.

TAOUKIS P, LABUZA T and SAGUY S (1997), 'Kinetics of food deterioration and shelf-life prediction', in Valentas K, Rotstein E and Singh P, *The Handbook of Food Engineering Practice*, New York, CRC Press, 361–403.

VAZ-PIRES P (1995), *Efficacy of heat treatments for reducing microbial activity*

during refrigerated storage of fresh fish, PhD thesis, Portuguese Catholic
University, Porto, Portugal.
WATERMAN J J (1982), 'Composition and quality of fish: a dictionary', *Torry
advisory note no. 87*, Aberdeen, Torry Research Station.

12

Modelling and predicting the shelf-life of seafood

P. Dalgaard, Danish Institute for Fisheries Research, Lyngby

12.1 Introduction

It has been estimated by the Food and Agricultural Organization of the United Nations (FAO) that 25% of all catches of fish are lost due to spoilage or discard and this important issue clearly deserves to be addressed. In addition to economic and management tools, various technological methods can be applied to reduce losses of fish and seafood due to spoilage. These methods include techniques to extend shelf-life, and thereby reduce spoilage problems, techniques to evaluate spoilage and assure seafood are not rejected by mistake, as well as techniques to predict shelf-life. Prediction of shelf-life as a function of product characteristics and conditions of storage and distribution can be used to evaluate the effect of normal and abusive handling of products. This shelf-life prediction may help to determine realistic distribution times or point out the steps in the distribution chain that need be improved for the product shelf-life to become sufficient. In both cases shelf-life prediction may contribute to reduce losses of seafood due to spoilage. In addition, shelf-life prediction is used in education and research where simulation of mathematical shelf-life models conveniently illustrates the effect of environmental factors, e.g., on growth of spoilage microorganisms.

Highly complex series of reactions take place during storage of seafood and the understanding of spoilage processes is in many ways incomplete. Furthermore, seafood spoilage is dynamic with changes in spoilage reactions depending on product composition and storage conditions. This dynamic nature of seafood spoilage complicates shelf-life prediction. Nevertheless, some general and simple patterns of seafood spoilage have been determined and particularly the concepts of relative rates of spoilage (RRS) and specific spoilage organisms (SSO) have allowed shelf-life models to be developed. RRS models are empirical and rely directly on the effect of temperature on shelf-life of products as determined in storage trials. In contrast, kinetic models rely on more detailed information about the reactions responsible for spoilage, e.g., the effect of environmental factors (temperature, pH, water activity, carbon dioxide, etc.) on growth of spoilage microorganisms.

In fresh and lightly preserved seafoods, enzymatic and chemical reactions are usually responsible for the initial loss of freshness attributes whereas activity of spoilage microorganisms is responsible for overt spoilage, thereby limiting shelf-life. Thus, models for growth of spoilage microorganisms can be used to predict shelf-life of these products. An SSO has been defined as the part of the total microflora responsible of spoilage of a given product. The practical importance of an SSO in relation to shelf-life prediction depends on its spoilage domain, i.e., the range of environmental conditions within which the SSO cause product rejection. Relying on the SSO-concept a simple conceptual model (Fig. 12.1) for microbial seafood spoilage has been suggested.

Figure 12.1 suggests it is possible to predict growth, production of metabolites and shelf-life of seafood based on the initial concentration of the SSO in a product. This simple conceptual model clearly includes a number of assumptions: (i) on a newly processed product, the SSO can make up only a minor part of the total microflora but the remaining microflora does not influence growth of the SSO; (ii) the SSO grow without a lag phase and produce the metabolites responsible for spoilage; (iii) the production of metabolites is directly proportional to growth; (iv) the seafood becomes sensorially spoiled when the SSO reach a minimal spoilage level (MSL); and (v) micro-organisms other than the SSO are without importance for spoilage. Figure 12.1 provides a simplified description of microbial seafood spoilage but still it can be most valuable. The conceptual model, for example, allows assumptions about the spoilage process to be tested and provided these assumptions are reasonable it can lead to a simple model for shelf-life prediction.

To predict the shelf-life of seafood some form of mathematical model is always required. This chapter includes an overview of models available for shelf-life prediction of seafood. Empirical models relying exclusively on data from product storage trials and kinetic models for microbial growth and activity are treated in separate sections. In addition, validation of models and software that facilitate the practical use of shelf-life models are presented.

12.2 Modelling of shelf-life and quality attributes determined in product storage trials

Seafood storage trials carried out under controlled conditions of temperature and atmosphere are essential in shelf-life determination. With new or modified products shelf-life can be determined only by the combined use of storage trials and sensory methods. However, sensory methods have the disadvantages of being difficult to calibrate, expensive and time consuming particularly when

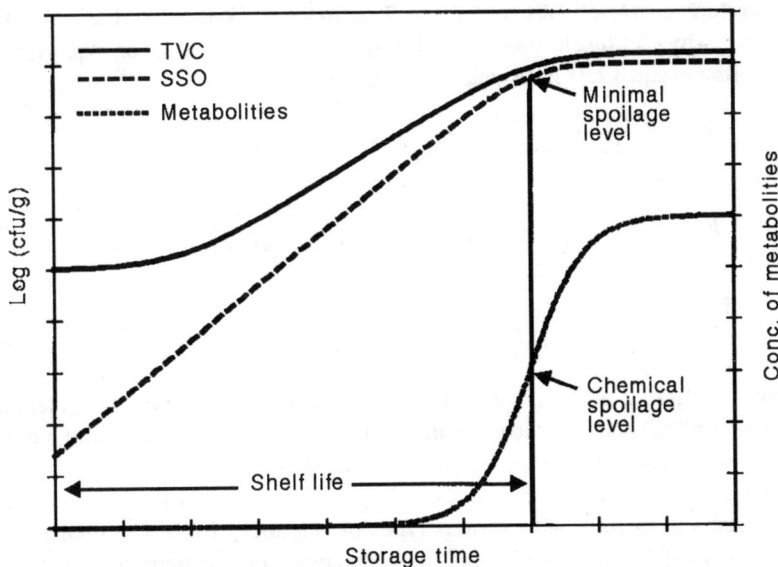

Fig. 12.1 Conceptual model of microbial seafood spoilage. The minimal spoilage level (MSL) and the chemical spoilage level (CSL) corresponds, respectively, to the concentration of specific spoilage organisms (SSO) and their metabolites at the time of sensory rejection (Dalgaard, 1993).

numerous trained assessors are used. Therefore, instrumental methods are needed to supplement or replace sensory assessments. This section describes models for the effect of temperature on shelf-life and models to relate responses of instrumental methods with sensory scores or remaining shelf-life of products.

12.2.1 Effect of temperatures on relative rates of spoilage

Shelf-life of seafoods varies considerably between fish species and between different types of preserved products. Nevertheless, the effect of storage temperature on shelf-life seems to be similar within groups of products and mathematical models have been developed that describe this effect. Fresh fish are often stored at close to 0°C in ice and the effect of temperature on shelf-life has been expressed by relative rates of spoilage (RRS) defined as the shelf-life at 0°C divided by shelf-life at T°C (Spencer and Baines, 1964; Nixon, 1971). With lightly preserved seafoods, however, a reference temperature (T_{ref}) of, e.g., 5°C instead of the 0°C may be more appropriate.

Models for the effect of temperature on RRS of a group of seafoods allow shelf-life of a product to be predicted at different storage temperatures. Shelf-life prediction by this type of RRS model is interesting because the only information required is the product shelf-life determined at one single known storage temperature. RRS models are typically developed on the basis of the shelf-life data determined by sensory evaluation and expressed as RRS values. The temperature characteristics a, E_A and T_{min} in eqns 12.1, 12.2, 12.3, are then estimated by fitting ln-transformed (eqns 12.1 and 12.2) or square root transformed (eqn 12.3) RRS data to the models.

$$RSS = \frac{Shelf\text{-}life\ at\ T_{ref}}{Shelf\text{-}life\ at\ T} = Exp[a \times (T - T_{ref})] \qquad 12.1$$

$$RSS = Exp\left[\frac{-E_A}{R} \times \left(\frac{1}{(T+273)} - \frac{1}{T_{ref}+273}\right)\right] \qquad 12.2$$

$$RSS = \left(\frac{T - T_{min}}{T_{ref} - T_{min}}\right)^2 \qquad 12.3$$

In eqns 12.1, 12.2 and 12.3, T and T_{ref} are the temperature and the reference temperature (°C) and R is the gas constant ($8.31\ J \times mol^{-1} \times K^{-1}$). Units of the temperature characteristics a, E_A and T_{min} are shown in Table 12.1.

The exponential (eqn 12.1) and the Arrhenius (eqn 12.2) RRS models have been used successfully with various types of seafoods (Table 12.1). Olley and Rathowsky (1973), however, found the Arrhenius model inappropriate for fresh seafood from temperate waters (Table 12.1). The square root spoilage model (eqn 12.3) was therefore suggested by Ratkowsky *et al.* (1982) on the basis of a simple model for growth of Gram-negative spoilage microorganisms (eqn 12.22). With a T_{min}-value of -10°C, which is typical for spoilage bacteria such as *Shewanella putrefaciens* and psychrotolerant pseudomonads, eqn 12.3 can be expressed as $\sqrt{RRS} = 1 + 0.1 \times T$. This simple square root spoilage model is applicable between -3°C and $+15$°C for both aerobically stored and modified atmoshere packed (MAP) fresh seafood from temperate waters (Dalgaard and Huss, 1997).

Chilled fresh seafood from tropical water usually keeps longer than fresh seafood from colder waters (Huss, 1995). At above ~15°C, however, RRS of tropical fish is higher than observed for cold-water fish and different RRS models are required for the two types of fresh seafoods (Table 12.1). Tropical fish are at 20–30°C when caught and delayed icing, or lack of icing, have a pronounced effect on shelf-life. Equations 12.1 and 12.2 can be used to evaluate the effect of temperature on shelf-life of fresh tropical seafood between 0°C–30°C (Table 12.1).

The effect of temperature on shelf-life of lightly preserved seafoods varies substantially between groups of products and it is most important to use the appropriate RRS model to predict shelf-life (Table 12.1). Some studies have used Q_{10}-values to characterize and evaluate various types of food deterioration. Q_{10}-values, defined as the change in reaction rate corresponding to a 10°C increase in temperature, are easily calculated from the exponential spoilage model (eqn 12.1) as $Q_{10} = \exp(a \times 10)$ (Taoukis *et al.*, 1997).

In the future it seems relevant to develop RRS models for frozen/thawed fish stored at chilled temperatures and for the effect of storage between -3°C and

Table 12.1. Relative rates of spoilage (RRS)

Type of seafood	Temperature characteristics			RRS at 25°C (T_{ret} = 0°C)[e]	References
	a in eqn. 12.1 (°C^{-1})	E_a in eqn. 12.2 (kJ mol^{-1})	T_{min} in eqn. 12.3 (°C)		
Fresh seafoods					
Fresh seafood from temperate or cold waters	nd[a]	~ 60[b]	−10	9–13	Spencer and Baines (1964), Olley and Ratkowsky (1973) Ratkowsky et al. (1982)
Fresh seafood from tropical waters	0.12	80	–[c]	~ 20	Dalgaard and Huss (1997)
Lightly preserved seafoods					
Smoked and packed cod and mackerel	0.025[d]	~ 20[d]	nd	~ 2[d]	Cann et al. (1983), Tinker et al. (1985)
Cold smoked and vacuum packed salmon, hot smoked and aerobically stored eel, halibut and herring	0.089	61	−10	9–13	Dalgaard et al. (2002a)
Cooked and brined MAP shrimps	> 0.15	> 100	–[c]	> ~ 50	Dalgaard and Jørgensen (2000)

(a) Not determined.
(b) E_A depends on the storage temperature (Olley and Ratkowsky, 1973).
(c) The square root model was less appropriate than eqns 12.1 and 12.2.
(d) Calculated from a limited number of experiments carried out between 0°C and 11.7°C.
(e) RRS values at 25°C were calculated to illustrate differences between the various models. The temperature range of applicability for each model is indicated in the text or in the references provided.

−60°C on the shelf-life of frozen seafood. In addition, the effect of temperature on shelf-life of smoked and packed seafood deserves further study.

12.2.2 Indices of quality and spoilage

Instrumental methods can be applied for shelf-life determination of a particular product when the response or a transformation of the response is expressed as a quality index (QI) that correlates with sensory scores or remaining shelf-life. As an example, Spencer and Baines (1964) applied a model equivalent to eqn 12.4 to predict shelf-life of aerobically stored cod as a function of both the storage temperatures and the initial values of microbiological, chemical and sensory quality indices ($QI_{Initial}$). See Ólafsdóttir et al. (1998) and Dalgaard (2000) for recent reviews of microbiological, biochemical, chemical or physical methods used as indices of seafood quality.

$$Remaining\ shelf\text{-}life\ (T^oC) = \frac{constant \times [QI_{Spoilage} - QI_{Initial}]}{RRS\ at\ T^oC} \qquad 12.4$$

where $QI_{Spoilage}$ is the value of a quality index determined at the time of sensory product rejection.

Single compound quality indices (SCQI)
The evolution of trimethylamine (TMA), trimethylamine-oxide (TMAO) and total volatile nitrogen (TVN), during storage of seafood, have been modelled by simple straight-line relationships (Ehrenberg and Shewan, 1955; Oehlenschläger, 1992). To obtain linear responses of microbial metabolites like TMA, data typically needs to be transformed into a TMA-index (eqn 12.5; Ehrenberg and Shewan, 1955). Levels of TMA corresponding to sensory rejection can be determined but microbial metabolites are produced in measurable concentrations only when high levels of spoilage microorganisms are reached (Figs 12.1 and 12.3). Consequently, TMA concentrations and TMA-index values cannot be used to predict the remaining shelf-life of products that are recently caught ('are fresh') but only with material in which the spoilage process is already under way and are neither 'fresh' nor spoilt.

$$TMA\text{-}index = log(1 + mg\ -\ N\ TMA/100mg) \qquad 12.5$$

In agreement with Fig. 12.1, levels of the SSOs in various fresh fish correlated closely with the remaining shelf-life of products (Table 12.2). Equation 12.4 can be used to predict shelf-life when close correlations, as shown in Table 12.2, have have been determined at different storage temperatures. This has been observed, e.g., with MAP cod fillets (Gibson 1985; Dalgaard, 1998).

During storage of seafood, sensory scores can change linearly with the time of storage (Shewan et al., 1953; Bremner, 1985) and eqn 12.4 can be used for shelf-life prediction (Branch and Vail, 1985). Sensory scores obtained by the quality index method (QIM) increase linearly during storage of various fresh fish in ice but further studies are needed to confirm the usefulness of QIM together with eqn 12.4 when seafood is stored at higher temperatures (Dalgaard, 2000).

Table 12.2 Relation between levels of specific spoilage organisms (SSO) and remaining shelf-life of fresh fish stored at 0°C

SSO and product	Atmosphere	Origin	Parameters in model[a]			References
			Intercept	Slope	Corr. coef.	
Shewanella putrefaciens						
Cod	Air	Denmark	15.0	−2.3	−0.974	Capell *et al.* (1998)
Herring	Air	UK	11.4	−1.9	−0.929	Capell *et al.* (1998)
Scad	Air	Portugal	15.7	−2.5	−0.937	Capell *et al.* (1998)
Trout	Air	UK	19.9	−3.3	−0.975	Capell *et al.* (1998)
Tropical fish	Air	Africa	28.2	−4.2	−0.950	Capell *et al.* (1998)
Photobacterium phosphoreum						
Cod	MAP	Denmark	18.5	−2.3	−0.954	Dalgaard (1998)
Brochothrix thermosphacta						
Red mullet	MAP	Greece	33.2	−3.9	−0.970	Koutsoumanis *et al.* (1998)
Sea bream	MAP	Greece	33.6	−3.8	−0.949	Koutsoumanis *et al.* (1998)

[a] Remaining shelf-life (days at 0°C) = Intercept + Slope × log(SSO/g).

Multiple compound quality indices (MCQI)

For many seafoods, particularly lightly preserved products, identification of a SCQI has not been successful. In some cases this problem has been overcome by development of quality indices relying on a combination of several compounds or measurements. Examples of this include the K-value, i.e. ratios of ATP degradation products, or ratios and sums of the concentrations of different biogenic amines (Saito *et al.*, 1959; Mietz and Karmas, 1977; Veciana-Nogués *et al.*, 1997; Duflos *et al.*, 1999). More recently, multivariate statistical methods have been used to identify MCQI. As an example, principal component analysis (PCA) of 44 volatile compounds resulted in a 93% correct allocation of samples into three quality grades, previously determined by sensory evaluation for Pacific pink salmon (Girard and Nakai, 1994). An MCQI relying on determination of 44 compounds, however, is difficult to use in practice. For cold-smoked salmon (CSS) partial least squares regression (PLSR) allowed a MCQI consisting of pH and the concentration of four biogenic amines to be developed. Values of this $MCQI_{CSS}$ corresponded to the percentage of sensory assessors that rejected a CSS-sample (eqn 12.6, Jørgensen *et al.*, 2000a). The PLSR approach to data analysis is capable of identifying a combination of responses that correlate with sensory scores or remaining shelf-life. In addition, PLSR can be used when the number of measured compounds or responses is higher than the number of samples and the techniques seem most valuable in MCQI development.

$$MCQI_{CSS} = 200 - 31 \times pH + 0.06 \times Tyramine(ppm) + 0.06$$
$$\times Cadaverine\ (ppm) + 0.04 \times Putrescine(ppm) + 0.15$$
$$\times Histamine\ (ppm) \qquad\qquad 12.6$$

$$Remaining\ shelf\text{-}life\ (weeks\ at\ 5^oC) = 4.78 - 0.34$$
$$\times Log(Lactobacillus/g)$$
$$- 0.06 \times TVN(mg - N/100g) \quad 12.7$$

Another MCQI, developed by multiple linear regression, related the remaining shelf-life of CSS at 5°C with the level of *Lactobacillus* and the content of TVN (eqn 12.7, Leroi *et al.*, 2001). Shelf-life of CSS depends strongly on the storage temperature and the content of salt and smoke components (Table 12.1, Leroi and Joffraud, 2000). It remains to be tested if the MCQI suggested for CSS (eqns 12.6 and 12.7) are valid at various storage temperatures and for products of different composition, e.g., with respect to salt and smoke components. Finally it needs to be noted that multivariate statistical methods have been used to identify relations between responses of various spectroscopic methods and quality attributes or time of seafood storage. This type of MCQI is discussed in Chapter 24.

Table 12.3   Kinetic parameters describing microbial growth and activity

Concentration of cells and metabolites	Absolute rate of growth and metabolite formation	Specific rate of growth or metabolite formation
Cell concentration (N) (cells × food mass^{-1})	$dN/dt = N \times \mu$ (cells × food mass^{-1} × time^{-1})	$\mu = dN/dt \times N^{-1}$ (time^{-1})
Metabolite concentration (M) (metabolite mass × food mass^{-1})	dM/dt (metabolite mass × food mass^{-1} × time^{-1})	$q_M = dM/dt \times N^{-1}$ (metabolite mass × cell^{-1} × time^{-1})

12.3   Modelling of microbial kinetics

Models of microbial growth and activity have been studied extensively, e.g., in relation to (i) fermentation technology, where models of microbial activity and transport of heat and metabolites have been used in control and up-scaling of bioreactors (Roels and Kossen, 1978; Nielsen and Villadsen, 1992), (ii) ecology where activity and interactions of microorganisms in various habitats including biofilms have been described (Lynch and Hobbie, 1988; Kreft *et al.*, 2001), and (iii) predictive food microbiology where the effect of product characteristics and storage conditions on growth and survival of pathogenic and spoilage microorganisms have been modelled and used to predict safety and shelf-life of foods.

Mathematical models of microbial growth and activity have often been developed based on the biomass of cells (Pirt, 1975). In foods, the biomass of microbial cells cannot easily be determined but if it is assumed that the average size of microorganisms remains constant during storage then the general principles of microbial kinetics can be applied (Table 12.3). As indicated in Fig. 12.1 models for growth of spoilage microorganisms can be used to predict the shelf-life of seafood. In addition to growth models, shelf-life prediction requires information about (i) the SSO and their spoilage domain (ii) the initial numbers of the SSO and (iii) a spoilage criterion. In some cases a level of 10^7 cells of the SSO per gram of product have been shown to correspond to the time of sensory product rejection and used as spoilage criterion, however, product may also spoil some time after the SSO reach their maximum cell concentration (Dalgaard *et al.* 1997; Koutsoumanis and Nychas, 2000). Within predictive food microbiology models for microbial responses at various conditions are usually developed by a two-step procedure with primary models describing changes in cell concentration as a function of time and secondary models describing the effect of environmental factors on the value of kinetic parameters like the lag time and the maximum specific growth rate (μ_{max}) (Buchanan, 1993; McClure *et al.*, 1994).

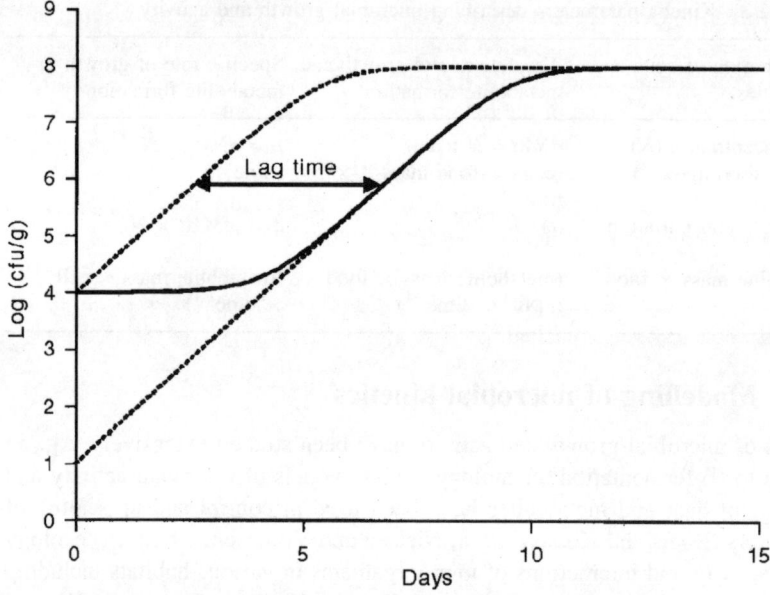

Fig. 12.2   Simulation of log-transformed Logistic models. Simple 3-parameter model (eqn 12.13) with N_0 equal to 10 and 10000 cfu/g, respectively (dotted lines). Four-parameter model (eqn 12.14) with N_{min} of 10000 cfu/g (solid line). For the three model simulations μ_{max}, was 0.070 h^{-1} and N_{max} of 1×10^8 cfu/g.

12.3.1   Primary models

Growth models can be used (i) to estimate values of kinetic parameters like lag-time, maximum specific growth rate (μ_{max}) and maximum cell concentration (N_{max}) from experimental data, and (ii) when values of kinetic parameters are known to predict the time required for a population to reach the cell concentration (N_t) at which the absolute rate of metabolite formation (dM/dt) is high and therefore may result in spoilage. The lag time has been defined as the time a culture in exponential growth ($\mu = \mu_{max}$) is delayed as compared to the culture growing from the same initial cell concentration with a constant specific growth rate equal to μ_{max} (Fig. 12.2). The lag time depends on the physiological condition of cells and of the substrates available. Therefore it is not a characteristic of a microbial population (Lodge and Hinshelwood, 1943; Monod, 1949; Pirt, 1975). The maximum specific growth rate (μ_{max}) (Table 12.3) is the specific growth rate a population can achieve under given conditions. μ_{max} is related to doubling time or generation time ($t_g = [Ln\ 2]/\mu_{max}$) and it is a remarkably reproducible characteristic of microbial populations (Monod, 1949).

Exponential growth

The simple exponential growth model (eqn 12.8) was suggested more than two centuries ago. Equation 12.9 shows the integrated and the log-tranformed form of this model. Within food microbiology, cell concentrations are often log-transformed, $\log(N_t)$. This stabilises the variance of viable counts data, allows kinetic parameters to be estimated more accurately when experimental data are fitted to growth models and finally it results in visually appealing growth curves. To retain the significance of the parameters in a growth model it is important to transform both sides of the equation (eqn 12.9).

The exponential growth model is used extensively to predict growth of microorganisms from known μ_{max} values (eqn 12.9) and to determine growth rates from experimental data (eqn 12.10). The simple exponential model, however, does not provide a realistic description of microbial growth curves which may include a lag phase and inevitably will show a maximum cell concentration has been reached.

$$\frac{dN}{dt} = N \times \mu_{max}; \quad N_t = N_0 \times \exp(\mu_{max} \times t) \tag{12.8}$$

$$Log(N_t) = Log(N_0 \times \exp(\mu_{max} \times t)) \tag{12.9}$$

$$\mu_{max} = \frac{(Log(N_{t_2}) - Log(N_{t_1})) \times Ln(10)}{t_2 - t_1} \tag{12.10}$$

Parameters in eqns 12.8, 12.9 and 12.10 are indicated in Table 12.3.

The exponential model can be expanded to include a lag phase (eqn 12.11). Values of the parameter t_{lag} in eqn 12.11 correspond to the classical definition of lag time and it can be estimated by piecewise regression (Lodge and Hinshelwood, 1943; Monod, 1949; Einarsson, 1994).

$$N_t = N_0 \text{ if } t < t_{lag}; N_t = N_0 \times \exp(\mu_{max}[t - t_{lag}] \text{ if } t \geq t_{lag} \tag{12.11}$$

Simple sigmoidal growth curves
The Logistic model (eqn 12.12, Verhulst, 1838) is an expansion of the exponential model where growth is dampened when N approaches N_{max}. The model is autonomous as the specific growth rate depends only on the cell concentration.

$$\frac{dN}{dt} = N \times \mu_{max}\left(1 - \frac{N}{N_{max}}\right) \tag{12.12}$$

$$N_t = \frac{N_{max}}{1 + \left[\dfrac{N_{max}}{N_0} - 1\right] \times \exp(-\mu_{max} \times t)} = \frac{N_{max}}{1 + \exp[-\mu_{max}(t - t_i)]} \tag{12.13}$$

In eqn 12.13 the parameter t_i corresponds to the inflection point i.e. the time when $N_t = N_{max}/2$.

The simple Logistic model (eqns 12.12 and 12.13) can be expanded to include a lag phase in the same way as shown above for the exponential models (Rosso, 1995; Rosso *et al.*, 1996). Lag time can also be estimated by another 4-parameter version of the Logistic model (eqns 12.14; 12.15; Fig. 12.2).

$$N_t = N_{min} + \frac{N_{max} - N_{min}}{1 + \exp[-\mu_{max}(t - t_i')]} \qquad 12.14$$

$$Lag\ time = t_i - \frac{1}{\mu_{max}} \times \ln\left(\frac{N_{max} + N_{max} \times \exp(\mu_{max} \times t_i)}{N_{max} + N_{max} \times \exp(\mu_{max} \times t_i)} - 1\right) \qquad 12.15$$

As for the simple Logistic model, the Gompertz model suggested in 1825 (eqn 12.16) contains no lag phase. The model is mentioned here only because modifed versions of it have been extensively used in predictive food microbiology (Gibson et al., 1988; Zwietering et al., 1990). The 'modified Gompertz' models have been obtained by a log-transformation of the left-hand side of eqn 12.16 and leaving the right-hand side non-transformed. Due to this awkward transformation, values of lag time and μ_{max} estimated from experimental data by the 'modified Gompertz' models do not correspond to the classical definition of these parameters. μ_{max}-values are typically overestimated by 10–20% (Baranyi et al., 1993; Dalgaard et al., 1994). It is unfortunate to use models that provide incorrect estimates of kinetic parameters and there seems to be no need to use the 'modified Gompertz' models in seafood microbiology.

$$\frac{dN}{dt} = N \times \mu' \times \ln\left(\frac{N}{N_{max}}\right) \qquad N_t = N_{max} \times \exp(-\exp(-\mu'(t - t_t)) \qquad 12.16$$

More flexible growth model
The Richards model (eqn 12.17) includes the parameter m to control the degree of growth dampening when N approaches N_{max}. As seen by comparison of eqn 12.12 with 12.17 the Logistic model is a special case of the Richards model where m is equal to 1.0 (Turner et al., 1976; Pruitt et al., 1979).

$$\frac{dN}{dt} = N \times \mu_{max} \times \left(1 - \frac{N^m}{N_{max}^m}\right) \qquad 12.17a$$

$$N_t = \frac{N_{max}}{[1 + \exp(-\mu_{max} \times m \times (t - t_i)]^{1/m}}$$

$$= \frac{N_{max}}{\left[1 + \left(\left(\frac{N_{max}}{N_0}\right)^m - 1\right) \times \exp(-\mu_{max} \times m \times t)\right]^{1/m}} \qquad 12.17b$$

The parameters in the Richards model, particularly m, have poor statistical properties (Ratkowsky, 1983, pp. 73–75) and the model it not usually appropriate for estimation of kinetic parameter values from experimental growth curve data. Nevertheless, by fixing values of the parameter m to 0.5; 1.0 or 2.0 the Richards model can be used to estimate μ_{max}-values from absorbance growth curves (Dalgaard and Koutsoumanis, 2001).

Flexible and non-autonomous models, with μ_{max} depending on both time (t) and cell concentration (N), have been suggested (Baranyi et al., 1993; Rosso,

1995). The Baranyi Growth Model has been popular. It contains the Richards model (eqn 12.17) and a delay function accounting for lag time. The model is complicated but the software 'MicroFit' (available free of charge at www.ifr.bbsrc.ac.uk/MicroFit/) and DMfit (available at a modest fee from www.ifr.bbsrc.ac.uk/safety/DMfit/) facilitates its practical use for curve fitting. When fitted to growth curves of seafood spoilage bacteria the Baranyi Growth Model and the log-transformed 4-parameter Logistic model (eqn 12.14) provided practically identical estimates of parameter values (Dalgaard, 1995b). The Baranyi Growth Model was developed specifically to predict growth under dynamic temperature conditions and a comparison of the two models under such conditions would be interesting.

Mathematical models that include an initial and/or a final decline in cell concentrations can be required, e.g., when studying superchilled seafood. Models that take into account these phenonoma have been suggested as simple expansions of the Logistic model (see, e.g., Pruitt and Kamau, 1993; Peleg, 1997).

In summary, primary models are available to describe the kinetics of spoilage microorganisms in seafood and to estimate μ_{max} and lag time values in agreement with the classical definitions of these parameters. Furthermore, model parameterisation that includes the initial cell concentration (N_0) are available and this is important to predict growth of microorganisms.

12.3.2   Modelling microbial activity and interactions

The yield concept suggests production of biomass of a microbial culture can be related to substrate consumption by a constant growth yield factor ($Y_{Biomass/Substrate}$) (Monod, 1942). In a similar way, the absolute rate of metabolite formation (dM/dt) can be related to the absolute growth rate (eqn 12.18; Pirt, 1975). Clearly, microorganisms use metabolic energy for production of biomass as well as for cell maintenance. However, in many situations the maintenance energy is low and integration of eqn 12.18 results in a simple model that allows formation of metabolites to be predicted (eqn 12.19). The yield concept has been a valuable tool to identify the SSO responsible for production of TMA, biogenic amines and volatile amines in different seafoods (Dalgaard 1995a, Jørgensen et al., 2000b, Koutsoumanis and Nychas, 2000). Prediction of metabolite formation in seafood, however, has been very little studied. Figure 12.3 shows examples of simulations where eqn 12.19 has been used in combination with the Logistic model (eqn 12.13) and the Richards model (12.17b). Figure 12.3 indicates that both the initial cell concentration (N_0) and the degree of growth dampening can influence formation of a metabolite like TMA substantially. Comparison of model simulations, like Figure 12.3, and experimental data deserves to be further studied as it is likely to increase our understanding of the factors that influence the production of spoilage matabolites in seafood.

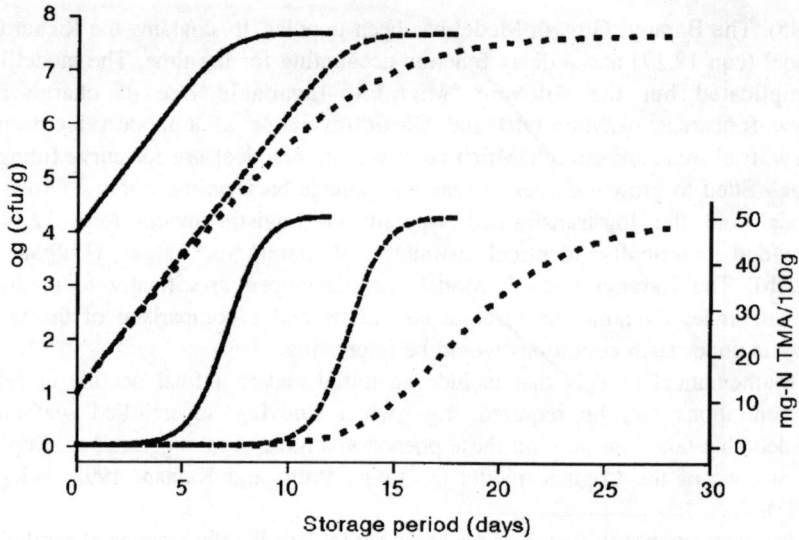

Fig. 12.3 Simulation of growth and trimethylamine (TMA) production by a specific spoilage organism. Simulations of all three growth curves were carried out by using a μ_{max} value of 0.050 h^{-1} and a N_{max}-value of $10^{7.7}$ cfu/g. TMA formation was simulated by using eqn 12.19 and a constant yield factor ($Y_{TMA/CFU}$) of $10^{-8.0}$ mg-N TMA/cfu. Logistic model (eqn 12.13) and N_0 of 10000 cfu/g. (solid lines), Logistic model (eqn 12.13) and N_0 of 10 cfu/g (dashed lines) and Richards model (eqn 12.17b) with N_0 equal to 10 10 cfu/g and m equal to 0.25 (dotted lines).

$$\frac{dM}{dt} = Y'_{M/N} \times \frac{dN}{dt} \qquad\qquad 12.18$$

$$M_t = M_0 + [Y_{M/N} \times (N_t - N_0)] \qquad\qquad 12.19$$

It has often been observed that groups of microorganisms present in low numbers in seafood stop growing at the time when the dominating microflora reaches its maximum cell concentration (Gram and Melchiorsen 1996; Dalgaard, 1999a, Koutsoumanis and Nychas, 2000; Koutsoumanis *et al.*, 2000; Emborg *et al.*, 2002). This phenomon, known as 'the Jameson effect' (Ross *et al.*, 2000) is not unique to seafood. The microbial interactions responsible for the Jameson effect can be due to substrate competition or be mediated by various metabolites including bacteriocines, siderophores or specific signal molecules. But for the time being the mechanisms remain poorly understood. In future studies the use of kinetic modelling may contribute to an increased understanding of the mechanisms behind the Jameson effect. Meanwhile microbial interactions can be predicted by a simple expansion of the Logistic model (eqn 12.20). This model has been used to predict how high levels of lactic acid bacteria inhibit growth of *L. monocytogenes* in cold-smoked salmon (Fig. 12.4; Dalgaard *et al.*, 2002a). The assumption behind eqn 12.20 is simply that growth of lactic acid bacteria inhibit growth of *L. monocytogenes* in the same way that the lactic acid

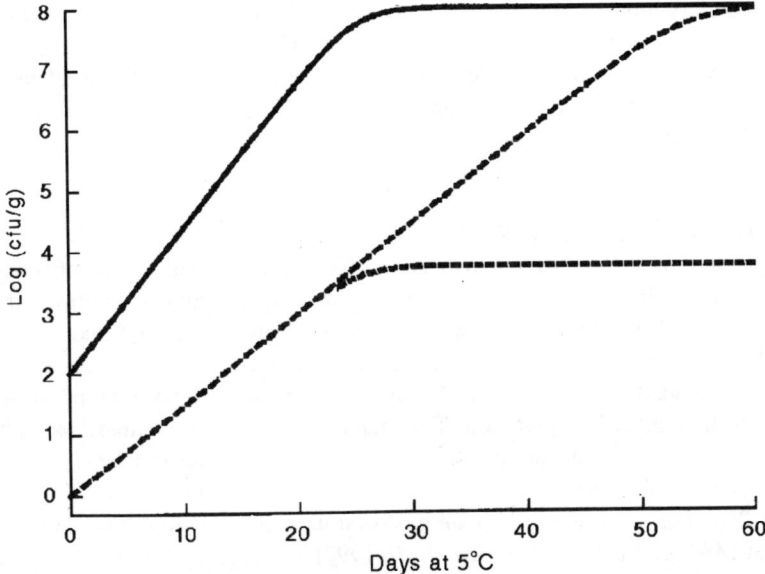

Fig. 12.4 Predicted growth of *Listeria monocytogenes* (Lm) and lactic acid bacteria at 5°C. LAB (solid lines), Lm alone (dashed lines) and Lm growing together with LAB (dotted line) (Jørgensen, 2000).

bacteria inhibit their own growth. When growth of *L. monocytogenes* in cold-smoked salmon is predicted it is important to take into account interaction with the spoilage microflora. In fact, the use of eqn 12.20 resulted in much more realistic risk estimates than the models typically used in predictive food microbiology (Fig. 12.4, Dalgaard *et al.*, 2002a).

$$\frac{dLm}{dt} = Lm_t \times \mu_{max}^{Lm} \times \left(1 - \frac{Lm_t}{Lm_{max}}\right) \times \left(1 - \frac{LAB_t}{LAB_{max}}\right) \qquad 12.20$$

where *Lm* and *LAB* signifies lactic acid bacteria and *L. monocytogenes*, respectively. *dLm/dt* is the absolute growth rates, Lm_t and LAB_t cell concentrations (cfu g^{-1}) at the time *t*, Lm_{max} and LAB_{max} the maximum cell concentrations (cfu g^{-1}) and μ_{max}^{Lm} the maximum specific growth rate (h^{-1}).

12.3.3 Secondary models
Lag time, μ_{max} and, in some cases, the maximum cell concentraion of spoilage bacteria depend on environmental factors like temperature, atmosphere/CO$_2$, a_w/NaCl, pH, lactate, preservatives and smoke compomnents. Within predictive food microbiology, secondary models are used to describe the effect of environmental factors on values of kinetic parameters. The separation of primary and secondary models has been extensively used but approaches to model growth responses in one step have been suggested (Membré *et al.*, 1997; Hajmeer *et al.*, 1997).

Secondary models covering wide ranges of several environmental factors can be developed but they may not be useful for shelf-life prediction if the spoilage domain of the SSO studied is less extensive than the model. This problem will be discussed later in relation to model validation. The present section focuses on models for the effect of environmental factors on μ_{max} values and on their application to SSOs from seafood.

Polynomial models and artificial neural networks
Simple polynomial models can account for the combined effect of several factors and their interactions. These models are easy to apply by multiple linear regression and they have been extensively used in predictive food microbiology (McClure *et al.*, 1994). Polynomial models have been criticised because some models included a very large number of parameters and because the parameters have no biological interpretation. This makes it difficult to compare parameter values from different studies (Baranyi *et al.*, 1996). Nevertheless, a simple polynomial model (eqn 12.21) for the effect of temperature (T) and CO_2 on the SSO *Photobacterium phosphoreum* has been used successfully to predict shelf-life of MAP cod fillets (Dalgaard *et al.*, 1997).

$$\sqrt{\mu_{max}} = 0.29 + 0.032 \times T - 1.16 \times 10^{-3} \times \%CO_2 - 9 \times 10^{-5}$$
$$\times T \times \%CO_2 + 9 \times 10^{-6} \times (\%CO_2)^2 \qquad\qquad 12.21$$

Artificial neural networks (ANN) comprise another type of empirical model suggested as secondary, or as primary as well as secondary, models in predictive food microbiology (Hajmeer *et al.*, 1997; Geeraerd *et al.*, 1998). ANN has not yet been used to predict shelf-life of seafood.

Kinetic models
Temperature is the most important factor influencing microbial growth and activity during seafood spoilage and numerous models for the effect of this environmental factor have been suggested. The Arrhenius equation (eqn 12.2), which is frequently used to model the effect of temperature on food deterioration in general, has also been applied to growth rates and lag times of seafood spoilage microorganisms. However, in many cases the activation energy (E_A) for the effect of temperature on growth rates has itself been found to be temperature dependent and this violates a major assumption inherent in the Arrhenius relationship, that E_A should be independent of temperature (Ratkowsky *et al.*, 1982; McMeekin *et al.*, 1993). For the seafood SSO *S. putrefaciens,* the average E_A-value was 85 kJ mol^{-1} between 0°C and 15°C but only 54 kJ mol^{-1} between 15°C and 27°C (Dalgaard, 1996). To substitute the Arrhenius equation, Ratkowsky *et al.* (1982) suggested a simple 2-parameter model for the effect of sub-optimal temperatures on growth rates of microorganisms (eqn 12.22). This model was later expanded to include the entire range of growth temperatures (eqn 12.23) (Ratkowsky *et al.*, 1983). These models, and their expansions described below, are referred to as 'square root' or 'Ratkowsky' type

models. Several studies have confirmed a square root transformation as appropriate to stabilise the variance of microbial growth rate data (McMeekin *et al.*, 1993; Zwietering *et al.*, 1994a).

$$\sqrt{\mu_{max}} = b(T - T_{min})$$ 12.22

$$\sqrt{\mu_{max}} = b(T - T_{min}) \times \{1 - \exp[c(T - T_{max})]\}$$ 12.23

With MAP fresh fish, growth of the SSO is controlled by temperatue and by the concentration of CO_2 in the modified atmosphere. To model the combined effect of these factors on *P. phosphoreum* a simple square root type model with three parameters has been used (eqn 12.24, Dalgaard *et al.*, 1997). Later, the usefulness of eqn 12.24 has been confirmed, e.g., for growth of *Brochothrix thermosphacta* in fresh MAP red mullet (Koutsoumanis *et al.*, 2000). μ_{max} values of spoilage bacteria growing in fresh MAP Mediterranean fish was also appropriately described by a combination of the Arrhenius model, for the effect of temperature, and the model suggested by Kalina (1993) for the effect of CO_2 (eqn 12.25; Koutsoumanis *et al.*, 2000).

$$\sqrt{\mu_{max}} = b(T - T_{min}) \times \frac{(\%CO_{2max} - \%CO_2)}{\%CO_{2max}}$$ 12.24

$$\ln(\mu_{max}) = \frac{E_a}{R} \times \left(\frac{1}{T_{ref}} - \frac{1}{T}\right) + \ln(\mu_{ref} - d_{CO_2} \times \%CO_2)$$ 12.25

where T and T_{ref} is temperature in K (eqn. 12.25).

In lightly and semi-preserved seafoods, a_w/NaCl content, pH, lactate content and various preservatives can influence growth of spoilage microorganisms and shelf-life of products. To model the combined effect of several environmental factors square root type models have been used extensively since McMeekin *et al.* (1987) expanded eqn 12.22 to include the effect of temperature and water activity. Later, several studies have documented that square root type models can be expanded by addition of terms for the effect of different environmental factors as shown in eqn 12.26 (Neumeyer *et al.*, 1997a; Devlieghere *et al.*, 1999; Wijtzes *et al.*, 2001). The range of the environmental factors relevant for shelf-life prediction of lightly preserved seafood will, in most practical situations, be limited and combinations of simple models like eqn 12.22 may be used (eqn 12.26). However, marine microorganisms and some seafood spoilage microorganisms have a substantial mineral requirement and more complex models like $\sqrt{\mu_{max}} = b(a_w - a_{wmax}) \times \{1 - \exp[c(a_w - a_{wmax})]\}$ can be required to predict the growth inihibiting effect of high water activities, i.e. low levels of salt (Miles *et al.*, 1997). As an example of this, growth of a spoilage bacteria from dried fish *Halobacterium salinarium* has been observed only at a_w values between 0.75 and 0.89 (Chandler and McMeekin, 1989).

$$\sqrt{\mu_{max}} = b$$
$$\times (T - T_{min})$$
$$\times \sqrt{(a_w - a_{wmin})}$$
$$\times \sqrt{(pH - pH_{min})}$$
$$\times (\%CO_{2max} - \%CO_2)/\%CO_{2max} \qquad\qquad 12.26$$

Square root type models have the advantage of being relatively simple and to contain parameters like T_{min}, a_{wmin}, pH_{min} and $\%CO_{2max}$ that have a biological interpretation. This facilitates comparison of data from different studies as shown in Table 12.4 where values of some 'cardinal' parameters have been collected for different SSOs from seafood. Reliable estimates of these parameters may also reduce the efforts required to develop a model for growth of a known SSO in novel or modified seafoods. In fact, a limited number of storage trials to estimate the value of 'b' in eqn 12.26 should be sufficient. This approach, or the very similar so-called gamma model approach (Zwietering *et al.*, 1992; Rosso, 1999; Wijtes *et al.*, 2001) has not yet been tried   for development of models to predict shelf-life of seafoods.

The structure of seafood is another factor that may be important when the aim is to predict growth of spoilage microorganisms. At 20°C growth rates of microorganisms can be substantially reduced in a medium solidified with gelatine as compared to growth in the same liquid medium without the gelatine. The growth inhibiting effect of texture/structure resulting from addition of gelatine has been observed particularly at low pH of 4.5–5.0 and with increased levels of NaCl (Robins and Wilson, 1994; Brocklehurst *et al.*, 1995). These data suggest predictive models developed in liquid media may overestimate microbial growth in lightly preserved seafoods. In fact, at 26–29°C growth of *Staphylococcus xylosus* was faster in broth than in salted and dried mackerel and tuna with the same water activity (Doe and Heruwati, 1988). Diffusion of substrates or metabolites may be related to the growth inhibiting effect of gelatine/structure. However, activiation energy of diffusion processes is much lower than that of microbial growth rates (Saguy and Karel, 1980). A possible growth inhibiting effect of seafood texture, therefore is likely to be markedly reduced with decreasing temperatures. Experiments with seafoods are needed to determine if growth of spoilage microorganisms is influenced by the texture of these products.

12.4   Validation of shelf-life models

Shelf-life models and models of microbial growth can be evaluated by comparison of predicted values with data from storage trials. The comparison of data can rely on (i) graphical methods where observed and predicted values are plotted against each other (McClure *et al.*, 1994) (ii) indices of performance like the bias and accuracy factors (Ross, 1996) where lag times, μ_{max}-values, doubling times, times for a 1000-fold increase in cell numbers or relative rates of

Table 12.4 Values of cardinal parameters for some seafood spoilage microorganisms

Spoilage microorganisms	Cardinal parameter values				References
	T_{min}	$a_{w\ min}$	pH_{min}	$\%CO_{2max}$	
Shewanella putrefaciens	−8.0 to −9.9	0.95	nd[a]	150–156	Dalgaard (1993); Dalgaard (1995b); Koutsoumanis et al. (2000)
Pseudomonads	−6.1 to −11.4	0.95	nd	121	Neumeyer et al. (1997a); Koutsoumanis et al. (2000)
Photobacterium phosphoreum	−9.0	0.95	4.3[b]	376	Dalgaard (1993); Dalgaard et al. (1997)
Lactobacillus curvatus	−3.3	0.93	4.2	nd	Wijtes et al. (2001)
Lactic acid bacteria	−3.6 to −11.4	nd	nd	232	Koutsoumanis et al. (2000); Dalgaard et al. (2002a)
Brochothrix thermosphacta	−10.9	nd	nd	187	Koutsoumanis et al. (2000)
Staphylococcus xylosus	+3.4	0.84	nd	nd	McMeekin et al. (1987)
Halobacterium salinarium	+8.3	0.70	nd	nd	Chandler and McMeekin (1989); Doe and Heruwati (1988)
Enterobacteriaceae	−0.7	nd	nd	nd	Dalgaard et al. (2002a)
Enterococcus faecalis	+3.6 to +7.7	nd	nd	nd	Thammavongs et al. (1996); Dalgaard et al. (2002a)

a Not determined.
b Unpublished data from DIFRES.

spoilage can be compared (eqn 12.27; 12.28; Neumeyer *et al.*, 1997b; Dalgaard et al., 2002a) and (iii) direct comparison of predicted and observed shelf-life (Dalgaard *et al.*, 1997).

$$\text{bias factor } (\mu_{max}) = 10^{(\Sigma \log \mu_{max \, predicted} / \mu_{max \, observed})/n)} \qquad 12.27$$

$$\text{accuracy factor } (\mu_{max}) = 10^{(\Sigma |\log \mu_{max \, predicted} / \mu_{max \, observed})|/n} \qquad 12.28$$

The bias factor indicates systematic over- or under prediction and a value of 1.2 indicates that predicted μ_{max}-values, on average, were 1.2 times higher than values observed, e.g., in storage trials. As compared to graphical methods for comparison of observed and predicted growth rates the bias factor has the advantage that results easily can be compared to reference values. It has been suggested that bias factor values between 0.75 and 1.25 can be used as a criterion for successful validation of a seafood shelf-life model (Dalgaard, 2000). Bias factor values may also be used to calibrate a model. As an example, the growth rate of lactic acid bacteria can be predicted as a function of temperature, water activity and pH (Wijtzes *et al.*, 2001). This model does not include the effect of smoke components and lactate and it overestimated the growth of lactic acid bacteria in cold-smoked salmon resulting in a bias factor of ~2. Until a more appropriate model is developed predicted μ_{max}-values can be divided by the bias factor. μ_{max}-values calibrated in this way can then be used to estimate growth responses of lactic acid bacteria in CSS as a function of temperature, water activity and pH (Dalgaard *et al.*, 2002a).

Although growth of a spoilage microorganism is accurately predicted over a range of conditions, shelf-life predictions may be strongly misleading if growth of the SSO is predicted outside its spoilage domain (Dalgaard, 1995b). Direct comparison of predicted and observed shelf-life instead of growth rates will overcome this problem. Furthermore, to obviate the generation of large amounts of growth data under conditions where a certain microorganism is of no importance for spoilage an iterative approach for development of spoilage models has been suggested (Dalgaard *et al.*, 1997). The basic idea behind the iterative approach is to validate the components used for model development, e.g., the SSO, the spoilage criterion, the spoilage domain, and the growth medium, before large amounts of data are generated and modelled. This philosophy is the same as in quality assurance where end-product testing is reduced and more efforts are placed on controlling critical steps in processing. The iterative approach was used for the development of a model including the effect of temperature and CO_2 on shelf-life of MAP cod fillets. Validation at constant and at varying temperatures showed shelf-life to be predicted with an average deviation of 9% from what was observed by sensory evaluation (Dalgaard *et al.*, 1997). The successful validation of the simple model, developed for growth of *P. phosphoreum* in a liquid medium, indicates food structure, microbial interactions and all factors other than temperature, $\%CO_2$ and the initial numbers of the SSO to be of minor importance for spoilage of MAP cod fillets.

To evaluate the effect of fluctuating temperatures during processing and distribution of seafood, actual temperatures must be recorded and used in combination with models developed under constant temperatures. The effect of temperature on shelf-life has been confirmed to be additive between ~0°C and +15°C for fresh fish and between +5°C and +15°C for cold-smoked salmon (Charm et al., 1972; McMeekin et al., 1988; Dalgaard et al., 2002a). Simple temperature profiles are therefore easy to evaluate (Ronsivalli et al., 1973). Shifts in temperature have very little, if any, lasting history effect on growth rates but lag times may be affected by fluctuating temperatures, especially at conditions close to the limits of growth (Fu et al., 1991; McMeekin et al., 1993; Zwietering et al., 1994b). More recent studies with fresh seafood stored between 0°C and 15°C concurred with these observations (Dalgaard et al., 1997; Koutsoumanis, 2001). Further studies are needed to evaluate the effect of more extreme temperatures,, e.g., as observed during superchilling and smoking, on growth of SSOs from seafood.

12.5   Application software

Various mathematical shelf-life models have been included in user friendly application software. These programmes are important as they allow a large number of people from industry, food inspection services, teaching, research and consumer organisations to conveniently use available shelf-life models. Several programmes can read temperature logger files and predict shelf-life as a function of a product temperature profile, e.g., measures during distribution in chill chains. Clearly, this type of electronic time-temperature integration increases the practical usefulness of shelf-life models substantially.

The Seafood Spoilage Predictor (SSP) software is available free of charge at www.dfu.min.dk/micro/ssp/ (Dalgaard et al., 2002b). SSP contains four models: (i) a RRS model for fresh seafood from temperate waters (ii) a RRS model for fresh tropical seafood (iii) a model for the effect of temperature and CO_2 on growth of P. phosphoreum in MAP cod fillets and (iv) a model for the effect of temperature on growth of Shewanella putrefaciens in different aerobically stored fresh fish. Temperature profiles recorded by different types of loggers can be read and evaluated by SSP. With more than 500 users, from most parts of the world, SSP has been relatively popular and an expanded version of this software is currently under development at the Danish Institute for Fisheries Research.

The Food Spoilage Predictor (FSP) contains a model for the effect of temperature and water activity on growth of psychrotolerant pseudomonads (Neumeyer et al., 1997a). The growth model included in FSP was not specifically developed to predict shelf-life of seafood. The model, however, has been successfully validated with a few seafoods (Neumeyer et al., 1997b; Dalgaard, 1999b). FSP is an integrated part of the Gemini Logger Manager that reads temperature profiles recorded, e.g., by Tinytag loggers. The software is commercially available from Gemini Data Loggers Ltd (www.gemini

dataloggers.com). A spreadsheet software which is in the process of becoming commercially available has been developed to predict shelf-life of Gilt-head Seabream, Seabass and Red Mullet from Greece (K. Koutsoumanis and G.-J. Nychas, personal communication). This software includes a model for the effect of temperature on growth of psychrotolerant *pseudomonads* spp. (Koutsoumanis and Nychas, 2000) and a model for the effect of temperature and atmosphere on *Brochothrix thermosphacta* (Koutsoumanis *et al.*, 2000). Models of some spoilage bacteria are included in the Food Micromodel software (Anon., 1997). It remains to be shown, however, if these models allow shelf-life of seafood to be predicted.

In recent years, tools for software development like Visual Basics, C# and Delphi have become both easier to use and substantially more powerful. For this reason and due to the fact that the number of shelf-life models keeps increasing it is most likely that new application software for seafood shelf-life prediction and electronic time-temperature integration will become available in the near future. These new programs may, for example, include stochastic models that allow variability in microbiological contamination and storage conditions to be taken into account when shelf-life is predicted (Giannakourou *et al.*, 2001; Rasmussen *et al.*, 2002).

12.6 Future trends

The concepts of relative rates of spoilage (RRS) and specific spoilage organisms (SSO) have allowed several seafood shelf-life models to be developed. The value of the SSO-concept has been acknowledged in other fields of food science where it is used increasingly, e.g., to develop models for the shelf-life of meat products. Nevertheless, shelf-life models are still lacking for a number of fresh, lightly preserved and semi-preserved seafoods. To address this problem and to expand the use of seafood shelf-life models the techniques and approaches listed below are likely to be applied in the future. The author is well aware that the list is by no means complete.

- Application of multivariate statistical methods to identify multiple compound quality indices (MCQI) that are applicable over a range of product characteristics and storage conditions.
- Identification of SSO and their spoilage domain in fresh and lightly preserved seafoods and development of new kinetic models for shelf-life prediction.
- Development of stochastic models to include variability in product characteristics and storage conditions in shelf-life prediction.
- Development of models and application software that allow changes in several types of spoilage reactions (e.g. both enzymatic and microbial activity) to be predicted.
- Development of application software to predict both spoilage and safety of seafood.
- Development of software that combines shelf-life models and traceability systems.

- Finally, kinetic models are likely to be used increasingly as a scientific tool to facilitate the identification of spoilage reactions and factors that influence the shelf-life of seafood.

12.7 References

ANONYMOUS (1997), *Food MicroModel – User Manual v. 2.5*, Surrey, UK, Food Micromodel Ltd.

BARANYI J, ROBERTS TA and MCCLURE P (1993), 'A non-autonomous differential equation to model bacterial growth', *Food Microbiol*, 10, 43–59.

BARANYI J, ROSS T, MCMEEKIN TA and ROBERTS TA (1996), 'Effects of parametrization on the performance of empirical models used in 'predictive microbiology', *Food Microbiol*, 13, 83–91.

BRANCH AC and VAIL AMA (1985), 'Bringing fish inspection into the computer age', *Food Technol Aust*, 37, 352–5.

BREMNER HA (1985), 'A convenient, easy to use system for estimating the quality of chilled seafoods', *Fish Processing Bulletin No.7*, Wellington, N.Z., Department of Scientific and Industrial Research, 59–70.

BROCKLEHURST TF, MITCHELL GA, RIDGE YP, SEALE R and SMITH AC (1995), 'The effect of transient temperatures on the growth of *Salmonalla typhimurium* LT2 in gelatin gel', *Int J Food Microbiol*, 27, 45–60.

BUCHANAN RL (1993), 'Predictive food microbiology', *Trends Food Sci Technol*, 41, 6–11.

CANN DC, SMITH GL and HOUSTON NC (1983), *Further studies on marine fish stored under modified atmosphere packaging*, Aberdeen, UK, Torry Research Station, MAFF.

CAPELL C, VAZ-PIRES P and KIRBY R (1998) 'Use of counts of hydrogen sulphide producing bacteria to estimate remaining shelf-life of fresh fish', in Ólafsdóttir *et al. Methods to determine the freshness of fish in research and industry*, Paris France, Int. Inst. Refrig., 175–82.

CHANDLER RE and MCMEEKIN TA (1989), 'Combined effect of temperature and salt concentration/water activity on growth rate of *Halobacterium* spp', *J Appl Bact*, 67, 71–6.

CHARM SE, LEARSON RJ, RONSIVALLI LJ and SCHWARTZ M (1972), 'Organoleptic technique predicts refrigeration shelf life of fish', *Food Technol*, 26, 65–8.

DALGAARD P (1993), *Evaluation and prediction of microbial fish spoilage (Ph.D-thesis)*, Lyngby, Denmark, Technological Laboratory, Danish Ministry of Fisheries.

DALGAARD P, ROSS T, KAMPERMAN L, NEUMEYER K and MCMEEKIN TA (1994), 'Estimation of bacterial growth rates from turbidimetric and viable count data', *Int J Food Microbiol*, 23, 391–404.

DALGAARD P (1995a), 'Qualitative and quantitative characterization of spoilage bacteria from packed fish', *Int J Food Microbiol*, 26, 319–333.

DALGAARD P (1995b), 'Modelling of microbial activity and prediction of shelf life for packed fresh fish', *Int J Food Microbiol*, 26, 305–17.

DALGAARD P (1996), 'Predictive modelling and time-temperature integration', *Refrig Sci Technol*, 409–19.

DALGAARD P AND HUSS HH (1997), 'Mathematical modelling used for evaluation and prediction of microbial fish spoilage', in Shahidi F, Jones Y and Kitts DD, *Seafood safety, processing and biotechnology*, Lancaster, PA, Technomic Publishing Co., Inc., 73–89.

DALGAARD P, MEJLHOLM O and HUSS HH (1997), 'Application of an iterative approach for development of a microbial model predicting the shelf-life of packed fish', *Int J Food Microbiol*, 38, 169–79.

DALGAARD P (1998) '*Photobacterium phosphoreum* – a microbial parameter for prediction of remaining shelf life in MAP cod fillets', in Ólafsdóttir *et al.* *Methods to determine the freshness of fish in research and industry*, Paris France, Int Inst Refrig, 166–74.

DALGAARD P (1999a), 'Importance and modelling of the interaction between specific fish spoilage bacteria', in Roberts TA, *COST 914 – Predictive modelling of microbial growth and survival in foods*, Wageningen, European Commission, 235–9.

DALGAARD P (1999b), 'Modelling of seafood spoilage', *Refrig Sci Technol*, EUR 18816, 143–51.

DALGAARD P (2000), 'Fresh and lightly preserved seafood', in Man CMD and Jones AA, *Shelf-Life Evaluation of Foods*, 2nd edn, London, Aspen Publishers, Inc., 110–39.

DALGAARD P AND JØRGENSEN LV (2000), 'Cooked and brined shrimps packed in a modified atmosphere have a shelf-life of > 7 months at 0°C, but spoil at 4–6 days at 25°C', *Int J Food Sci Technol*, 35, 431–42.

DALGAARD P AND KOUTSOUMANIS K (2001), 'Comparison of maximum specific growth rates and lag times estimated from absorbance and viable count data by different mathematical models', *J Microbiol Meth*, 43, 183–96.

DALGAARD P, MURILLO E and JØRGENSEN LV (2002a), 'Modelling the effect of temperature on shelf-life and on the interaction between the spoilage microflora and *Listeria monocytogenes* in cold-smoked salmon', in Proceedings of the 46th Atlantic Fisheries Technology Conference, 27–29 August 2001, Rimouski, Canada. submitted.

DALGAARD P, BUCH P and SILBERG S (2002b), 'Seafood Spoilage Predictor – development and distribution of a product specific application software' *Int J Food Microbiol*, 73, 227–33.

DEVLIEGHERE F, VAN BELLE B and DEBEVERE J (1999), 'Shelf life of modified atmosphere packed cooked meat products: a predictive model', *Int J Food Microbiol*, 46, 57–70.

DOE PE and HERUWATI E (1988), 'A model for the prediction of the microbial spoilage of sun-dried tropical fish', *J Food Eng*, 8, 47-72.

DUFLOS G, DERVIN C, MALLE P and BOUQUELET S (1999), 'Use of biogenic amines to evaluate spoilage in plaice (*Pleuronectes platessa*) and whiting (*Merlangus merlangus*)', *J AOAC Int*, 82 (6), 1357–63.

EHRENBERG ASC AND SHEWAN JM (1955), 'Volatile bases and sensory quality-factors in iced white fish', *J Sci Food Agric*, 6, 207–217.

EINARSSON H (1994), 'Evaluation of a predictive model for the shelf life of cod (*Gadus morhua*) fillets stored in two different atmospheres at varying temperatures', *Int J Food Microbiol*, 24, 93–102.

EMBORG J, LAURSEN BG, RATHJEN T and DALGAARD P (2002), 'Microbial spoilage and formation of biogenic amines in fresh and thawed modified atmosphere packed salmon (*Salmo salar*) at 2°C', *J Appl Microbiol*, 92, 790–799.

FU B, TAOUKIS P and LABOUZA TP (1991), 'Predictive microbiology for monitoring spoilage of dairy products with time-temperature integrators', *J Food Sci*, 56 (5), 1209–15.

GEERAERD AH, HERREMANS CH, CENENS C and VAN IMPE JF (1998), 'Application of artificial neural networks as a non-linear modular modeling technique to describe bacterial growth in chilled food products', *Int J Food Microbiol*, 44, 49–68.

GIANNAKOUROU MC, KOUTSOUMANIS K, NYCHAS GJE and TAOUKIS PS (2001), 'Development and assessment of an intelligent shelf life decision system for quality optimiztion of the food chill chain', *J Food Prot*, 64 (7), 1051–7.

GIBSON AM, BRATCHELL N and ROBERTS TA (1988), 'Predicting microbial growth: growth responses of salmonellae in a laboratory medium as affected by pH, sodium chloride and storage temperature', *Int J Food Microbiol*, 6, 155–78.

GIBSON DM (1985), 'Predicting the shelf life of packed fish from conductance measurements', *J Appl Bact*, 58, 465–470.

GIRARD B and NAKAI S (1994), 'Grade classification of canned pink salmon with static headspace volatile patterns', *J Food Sci*, 59 (3), 507–12.

GRAM L and MELCHIORSEN J (1996), 'Interaction between fish spoilage *Pseudomonas sp.* and *Shewanella putrefaciens* in fish extracts and on fish tissue', *J Appl Bact*, 80, 589–95.

HAJMEER MN, BASHEER IA and NAJJAR YM (1997), 'Computational neural networks for predictive microbiology II. Application to microbial growth', *Int J Food Microbiol*, 34, 51–66.

HUSS HH (1995), *Quality and quality changes in fresh fish*, Rome, FAO Fisheries Technical Paper No. 348.

JØRGENSEN, LV (2000), *'Spoilage and safety of cold-smoked salmon'* (Ph.D. thesis). Lyngby, Denmark, Danish Institute for Fisheries Research.

JØRGENSEN LV, DALGAARD P and HUSS HH (2000a), 'Multiple compound quality index for cold-smoked salmon (*Salmo salar*) developed by multivariate regression of biogenic amines and pH', *J Agric Food Chem*, 48 (6), 2448–53.

JØRGENSEN LV, HUSS HH and DALGAARD P (2000b), 'The effect of biogenic amine production by single bacterial cultures and metabiosis on cold-smoked salmon', *J Appl Microbiol*, 89, 920–34.

KALINA V (1993), 'Dynamics of microbial growth and metabolic activity and their control by aeration', *Antonie v Leeuw*, 63, 353-373.

KOUTSOUMANIS K (2001), 'Predictive modeling of the shelf life of fish under nonisothermal conditions', *Appl Env Microbiol*, 67 (4), 1821–9.

KOUTSOUMANIS K, TAOUKIS P, DROSINOS ES and NYCHAS GJE (1998) 'Lactic acid bacteria and *Brochothrix thermosphacta* – the dominant spoilage microflora of Mediterranean fresh fish stored under modified atmosphere packaging conditions', in Ólafsdóttir *et al. Methods to determine the freshness of fish in research and industry*, Paris France, Int Inst Refrig, 158–65.

KOUTSOUMANIS K AND NYCHAS G-JE (2000), 'Application of a systematic experimental procedure to develop a microbial model for rapid fish shelf life prediction', *Int J Food Microbiol*, 60, 171–84.

KOUTSOUMANIS KP, TAOUKIS PS, DROSINOS EH and NYCHAS G-JE (2000), 'Application of an Arrhenius model for the combined effect of temperature and CO_2 packaging on the spoilage microflora of fish', *Appl Env Microbiol*, 66 (8), 3528–34.

KREFT J-U, PICIOREANU C, WIMPENNY JWT and VAN LOOSDRECHT MCM (2001), 'Individual-based modelling of biofilms', *Microbiol*, 147, 2897–912.

LEROI F and JOFFRAUD JJ (2000), 'Salt and smoke simultaneously affect chemical and sensory quality of cold-smoked salmon during 5°C storage predicted using factorial design', *J Food Prot*, 63 (9), 1222–7.

LEROI F, JOFFRAUD JJ, CHEVALIER F and CARDINAL M (2001), 'Research of quality indices for cold-smoked salmon using a stepwise multiple regression of microbiological counts and physico-chemical parameters', *J Appl Microbiol*, 90, 578–87.

LODGE RM and HINSHLEWOOD CN (1943), 'Physiological aspects of bacterial growth. Part IX. The lag phase of *Bact. Lactis Aerogenes*', *J Chem Soc*, 213–19.

LYNCH JM and HOBBIE JE (1988), 'Microbial population and community dynamics', in Lynch JM and Hobbie JE, *Micro-organisms in action: Concepts and applications in microbial ecology*, second edn, Oxford, Blackwell Scientific Publications, 51–74.

MCCLURE PJ, BLACKBURN CD, COLE MB, CURTIS PS, JONES JE, LEGAN JD, OGDEN ID and PECK MW (1994), 'Modelling the growth, survival and death of microorganisms in foods: the UK Food Micromodel approach', *Int J Food Microbiol*, 23 (3,4), 265–75.

MCMEEKIN TA, CHANDLER RE, DOE PE, GARLAND CD, OLLEY J, PUTRO S and RATKOWSKY DA (1987), 'Model for the combined effect of temperature and salt concentration/water activity on growth rate of *Staphylococcus xylosus*', *J Appl Bact*, 62, 543–50.

MCMEEKIN TA, OLLEY J and RATKOWSKY DA (1988), 'Temperature effects on bacterial growth rates', in Bazin MJ and Prosser JI, *Physiological models in microbiology*, vol. 1, Boca Raton, Florida, CRC Press, Inc., 75–89.

MCMEEKIN TA, OLLEY J, ROSS T and RATKOWSKY DA (1993), *Predictive microbiology: Theory and application*, Taunton, UK, Research Studies

Press Ltd.

MEMBRÉ JM, THURETTE J and CATTEAU M (1997), 'Modelling the growth, survival and death of *Listeria monocytogenes*', *J Appl Bact*, 82, 345–50.

MIETZ JL and KARMAS E (1977), 'Chemical quality index of canned tuna as determined by high-pressure liquid chromatography', *J Food Sci*, 42, 155–8.

MILES DW, ROSS T, OLLEY J and MCMEEKIN TA (1997), 'Development and evaluation of a predictive model for the effect of temperature and water activity on the growth rate of *Vibrio parahaemolyticus*', *Int J Food Microbiol*, 38, 133–42.

MONOD J (1942), *Recherches sur la croissance des cultures bactériennes*, Paris, Hermann.

MONOD J (1949), 'The growth of bacterial cultures', *Annu Rev Microbiol*, 3, 371–94.

NEUMEYER K, ROSS T and MCMEEKIN TA (1997a), 'Development of a predictive model to describe the effect of temperature and water activity on the growth of spoilage pseudomonads', *Int J Food Microbiol*, 38, 45–54.

NEUMEYER K, ROSS T, THOMSON G and MCMEEKIN TA (1997b), 'Validation of a model describing the effect of temperature and water activity on the growth of psychrotrophic pseudomonads', *Int J Food Microbiol*, 38, 55–63.

NIELSEN J and VILLADSEN J (1992), 'Modelling of microbial kinetics', *Chem Eng Sci*, 47, 4225–70.

NIXON PA (1971), 'Temperature integration as a means of assessing storage conditions (Paper No. 6)', Report on quality in fish products, Seminar No. 3, Wellington, New Zealand, Fishing Industry Board, 34–44.

OEHLENSCHLÄGER J (1992), 'Evaluation of some well established and some underrated indices for the determination of freshness and/or spoilage of ice stored wet fish.', in Huss HH, Jacobsen M and Liston J, *Quality Assurance in the Fish Industry*, Amsterdam, Elsevier Science Publishers B.V., 339–50.

OLLEY J and RATKOWSKY DA (1973), 'Temperature function integration and its importance in the storage and distribution of fresh foods above the freezing point', *Food Technol Aust*, 25, 66–73.

ÓLAFSDÓTTIR G, LUTEN J, DALGAARD P, CARECHE M, VERREZ-BAGNIS V, MARTINSDÓTTIR E and HEIA K (1998), *Methods to determine the freshness of fish in research and industry* Paris, Int Inst Refrig.

PELEG M (1997), 'Modeling microbial populations with the original and modified versions of the continous and discrete Logistic equations', *Cri Rev Food Sci Nut*, 37, 471–90.

PIRT SJ (1975), *Principles of microbial and cell cultivation*, Oxford, Blackwell Scientific Publications.

PRUITT KM, DEMUTH RE and TURNER ME (1979), 'Practical application of generic growth theory and the significance of the growth curve parameters', *Growth*, 43, 19–35.

PRUITT KM and KAMAU DN (1993), 'Mathematical models of bacterial growth, inhibition and death under combined stress conditions', *J Ind Microbiol*, 12, 221–31.

RASMUSSEN SKJ, ROSS T, OLLEY J. and MCMEEKIN T (2002), 'A process risk model for the shelf-life of Atlantic salmon fillets', *Int J Food Microbiol*, 73, 47–60.

RATKOWSKY DA (1983), *Nonlinear regression modeling: A unified practical approach*, New York, USA, Marcel Dekker, Inc.

RATKOWSKY DA, OLLEY J, MCMEEKIN TA and BALL A (1982), 'Relation between temperature and growth rate of bacterial cultures', *J Bacteriol*, 149 (1), 1–5.

RATKOWSKY DA, LOWRY RK, MCMEEKIN TA, STOKES AN and CHANDLER RE (1983), 'Model for bacterial culture growth rate throughout the entire biokinetic temperature range', *J Bacteriol*, 154 (3), 1222–6.

ROBINS MM and WILSON PDG (1994), 'Food structure and microbial growth', *Trends Food Sci Technol*, 5, 289–93.

ROELS JA AND KOSSEN NWF (1978), 'On the modelling of microbial metabolism', in Bull MJ, *Progress in Industrial Microbiology*, Amsterdam, Elsevier, 95–203.

RONSIVALLI LJ, LEARSON RJ and CHARM SE (1973), 'Slide rule for predicting shelf life of cod', *Mar Fish Rev*, 35, 34–6.

ROSS T (1996), 'Indices for performance evaluation of predictive models in food microbiology', *J Appl Bact*, 81, 501–8.

ROSS T, DALGAARD P and TIENUNGOON S (2000), 'Predictive modelling of the growth and survival of *Listeria* in fishery products', *Int J Food Microbiol*, 62, 231–45.

ROSSO L (1995), *Modélisation et Microbiologie Prévissionnelle: Élaboration d'un nouvel outil pour l'Agro-alimentaire* (Ph.D. thesis), Lyon, Université Claude Bernard.

ROSSO L (1999), 'Models using cardinal values', *Refrig Sci Technol*, EUR 18816, 48–55.

ROSSO L, BAJARD S, FLANDOIS JP, LAHELLEC C, FOURNAUD J and VEIT P (1996), 'Differential growth of *Listeria monocytogenes* at 4°C and 8°C: Consequences for the shelf life of chilled products', *J Food Prot*, 59 (9), 944–9.

SAGUY I and KAREL M (1980), 'Modeling of quality deterioration during food processing and storage', *Food Technol*, 34 (2), 78–85.

SAITO T, ARAI K and MATSUYOSHI M (1959), 'A new method for estimating the freshness of fish', *Bull Jap Soc Sci Fish*, 24, 749–50.

SHEWAN JM, MACINTOSH RC, TUCKER CG and EHRENBERG ASC (1953), 'The development of a numerical scoring system for the sensory assessment of the spoilage of wet white fish stored in ice', *J Sci Food Agric*, 4, 283–98.

SPENCER R and BAINES CR (1964), 'The effect of tempetrature on the spoilage of wet white fish. I. Storage at constant temperatures between 1°C and 25°C', *Food Technol*, 18, 769-772.

TAOUKIS PS, LABUZA TP and SAGUY IS (1997), 'Kinetics of food deterioration and shelf-life prediction', in Valentas KJ, Rotstein E and Singh RP, *Handbook of Food Engineering Practice*, Boca Raton, CRC Press, 361–403.

THAMMAVONGS B, CORROLER D, PANOFF J-M, AUFFRAY Y and BOUTIBONNES P (1996), 'Physiological response of *Enterococcus faecalis* JH2-2 to cold shock: growth at low temperatures and freezing/thawing challenge', *Lett Appl Microbiol*, 23, 398–402.

TINKER BL, SLAVIN JW, LEARSON RJ and AMPOLA VG (1985), 'Evaluation of automated time-temperature monitoring system in measuring the freshness of chilled fish', *Refrig Sci Technol Proceedings*, 1985-4, 281–91.

TURNER ME, BRADLEY JrEL, KIRK KA and PRUITT KM (1976), 'A theory of growth', *Math Biosci*, 29, 367–73.

VECIANA-NOGUéS MT, MARINÉ-FONT A and VIDAL-CAROU MC (1997), 'Biogenic amines as hygienic quality indicators of tuna. Relationships with microbial counts, ATP-related compounds, volatile amines and organoleptic changes', *J Agric Food Chem*, 45, 2036–41.

VERHULST PF (1838), 'Notice sur la loi que la population suit dans son accroissement', *Corr Math Phys*, 41, 1–47.

WIJTZES T, ROMBOUTS FM, KANT-MUERMANS ML, RIET KV and ZWIETERING M (2001), 'Development and validation of a combined temperature, water activity, pH model for bacterial growth rate of *Lactobacillus curvatus*', *Int J Food Microbiol*, 63, 57–64.

ZWIETERING MH, JONGENBURGER I, ROMBOUTS FM and VAN'T REIT K (1990), 'Modeling of the bacterial growth curve', *Appl Environ Microbiol*, 56, 1875–81.

ZWIETERING MH, WIJTZES T, DE WIT JC and VAN'T RIET K (1992), 'A decision support system for prediction of the microbial spoilage in foods', *J Food Prot*, 55, 973–79.

ZWIETERING MH, CUPPERS HGAM, DE WIT JC and VAN'T RIET K (1994a), 'Evaluation of data transformations and validation of models for the effect of temperature on bacterial growth', *Appl Environ Microbiol*, 60, 195–203.

ZWIETERING MH, CUPPERS HGAM, DE WIT JC and VAN'T RIET K (1994b), 'Modelling of bacterial growth with shifts in temperature', *Appl Environ Microbiol*, 60, 204–13.

13

The role of enzymes in determining seafood color, flavor and texture

N. Haard, University of California, Davis

13.1 Introduction: the importance of enzymes in postmortem fish

Enzymes are ubiquitous components of the edible tissues from fish and other aquatic organisms and they influence seafood quality in many ways. It has been written that no other group of tissue components is more important to the seafood technologist than are enzyme systems.

13.1.1 Rigor mortis

Muscle tissue undergoes a shift from aerobic to anaerobic metabolism when oxygen is not available in sufficient amounts to fuel mitochondrial function. During anaerobic glycolysis, glycogen is converted to lactate with 2 or 3 moles adenosine triphosphate (ATP) per mole of glucose. ATP is gradually depleted by membrane and contractile ATPases resulting in the production of adenosine diphosphate (ADP), hydrogen ions and a decrease in pH. ATP depletion may occur very rapidly because of a loss of calcium sequestering ability by the sarcoplasmic reticulum or mitochondria. The rate and extent of pH decline has a profound influence on meat quality as well as chemical, enzymic and microbial spoilage reactions. After glycolysis ceases due to inhibition of regulatory enzymes, such as phosphofructokinase, or depletion of glycogen, the muscle passes into a state of 'death stiffening' known as rigor mortis. Subsequent enzyme catalyzed reactions by proteases result in the resolution of rigor. Normally it is best to delay processing fish until after the resolution of rigor to avoid serious quality problems. The post-rigor phase is characterized by loss of enzymic regulation with notable quality changes caused by nucleotide degrading enzymes, lipases, enzymic lipid oxidation and proteolytic activity, among others.

13.1.2 Loss of prime quality in fresh seafood

Several studies have shown that the initial rate of post-harvest quality deterioration in harvested aquatic organisms is essentially the same in sterile and non-sterile tissues (Haard, 1992d). This is because endogenous biochemical reactions catalyzed by enzymes are the main cause of quality loss in chilled fish during the first three or four days after harvest. Moreover, most methods developed to extend the shelf-life of chilled seafood, such as pasteurizing doses of ionizing radiation, chemical additives, and modified atmospheres are virtually ineffective in maintaining prime quality since they do not target enzymes that contribute to initial quality loss. Indeed, such treatments may promote enzyme-catalyzed reactions and associated loss of grade A quality.

13.1.3 Processed seafood quality

Enzyme catalyzed reactions may also play a key role in the quality loss of seafood preserved by techniques such as frozen storage, high pressure processing and sterilizing doses of ionizing radiation. Non-thermal preservation techniques such as these are normally ineffective in arresting enzyme activity. Thus it may be necessary to combine appropriate chemical or thermal pretreatments with non-thermal preservation techniques.

13.1.4 Enzymes as seafood processing aids

Industrial enzymes from a variety of sources are important in seafood technology because of their use as fish processing aids and as biosensors used in seafood quality control. Enzymes recovered from fishery products' process wastes have even been used as industrial enzymes in the seafood and other food and non-food industries.

13.2 Enzymes in fish myosystems

13.2.1 Homologous enzymes

Approximately 4,000 different enzymes have been classified by the International Commission on Enzymes, International Union of Biochemistry (Bairoch, 2000). For the most part, the same homologous enzymes are found in all food myosystems; i.e., muscles from aquatic organisms contain the same enzymes by name as those present in muscles from terrestrial organisms (Haard, 1990). Homologous enzymes show such great similarities in their structures and functions that a common name and enzyme classification (EC) number is given. Paralogous enzymes are homologous enzymes that have arisen by gene duplication and evolved side by side in a single line of descent. For example, the gene library of Atlantic salmon contains five distinct clones of trypsin that are presumably expressed at different times with respect to environment and ontogeny (Male *et al.*, 1995).

Orthologous enzymes are homologous enzymes in which structural differences arise from speciation. Such enzymes normally perform the same function within different species. Analogous enzymes have a similar catalytic function, but have large differences in structure. For example, some seafood trypsins are referred to as 'trypsin-like' because of their relatively low molecular mass (Simpson and Haard, 1987b). The term isoenzymes is used to describe multiple forms of an enzyme that catalyze the same reaction, and occur as genetically determined variants of the amino acid sequences within one species. Post-translational modification of enzymes that have a common amino acid sequence in an organism are called isoforms.

13.2.2   Interspecies differences in enzymes

Some enzymes are not found in all species of aquatic organisms. Whether the gene for the deficient enzyme is absent or it is not normally expressed is difficult to know without probing the gene library. It is difficult to establish phylogenetic-based differences in the presence of enzymes since their presence or absence is very sensitive to intraspecific factors (Rudneeva-Titova, 1997, Haard, 2000). Examples of enzymes that impact food quality and are present in only some seafood meat include gulonolactone oxidase (ascorbic acid biosynthesis), polyphenol oxidase (enzymic browning), amylase (glycogen hydrolysis), thiaminase (thiamine degradation), TMAO demethylase (texture toughening), inosine hydrolase and inosine phosphorylase (hypoxanthine accumulation), and the sequence of enzymes involved with conversion of carotenoids to oxycarotenoids (red color pigmentation). These and other examples of enzymes found in only some seafoods are discussed elsewhere (Simpson *et al.*, 2000, Haard, 2000).

13.2.3   Extrinsic enzymes

Seafood may also contain enzymes that originate from microorganisms (Norqvist *et al.*, 1990, Venugopal, 1990), parasites (Chang-Lee *et al.*, 1989), and feed (Haard, 1994). The existence of extrinsic enzymes that impact seafood quality is not always recognized. Examples include TMAO oxidoreductase from *Pseudomonas* and other bacteria, cathepsin L-like enzyme from myxosporidean protozoan (*Kudoa*) parasite infection, and nucleotide phosphorylase and inosine nucleosidase from *Proteus* bacteria.

13.3   Postmortem physiology

Food myosystems undergo profound changes in appearance, texture, flavor and functional properties due to endogenous biochemical reactions that are initiated by circulatory stoppage. These changes are collectively referred to as postmortem physiology. The edible quality of muscle at the time of harvest is normally quite different from that of the post-rigor meat. The conversion of muscle to meat is reviewed in detail elsewhere (Foegeding et al., 1996). One major factor that influences postmortem biochemistry is the biological function of the muscle. All muscles function to provide locomotion through the energy intensive process of contraction. Energy for muscle contraction is directly provided by adenosine triphosphate (ATP) which is hydrolyzed to adenosine diphosphate (ADP) by myosin ATPase during the contaction cycle. Individual muscles in the body of a given animal are specifically designed for biological function, e.g., for short-term, rapid and powerful contractions or for sustained, low amplitude contractions. The former design is called 'fast twitch' muscle and is characterized by being white in appearance and employing glycolysis for the re-synthesis of ATP. The latter design is called 'slow twitch' muscle and is typically red or dark in appearance and regenerates ATP by aerobic, mitochondrial metabolism. In fish, the 'red' and 'white' fiber types that make up white and dark muscles are more discretely segregated than is typical for terrestrial animals. Muscles of migratory pelagic fish, such as sardine, may contain more than 30% red fibers whereas those of demersal bottom feeding fish may contain less than 2% red fibers (Greer-Walker, 1970). Other types of adaptation in the biological function of muscle contraction and associated energy metabolism occur in fish and shellfish since they are poikilotherms that inhabit a wide range of environmental conditions with respect to temperature, water depth and pressure, water salinity, water pollution and oxygen tension (Haard, 2000).

A second class of factors that impact the biochemical changes in muscle after death are antemortem history (exercise, diet and nutritive status, fasting, stress, etc.) and method of slaughter prior to circulatory stoppage. Temperature of the carcass, processing, handling and species characteristics can profoundly influence the rate and extent of biochemical events leading to rigor mortis. The key events in postmortem muscle are anoxia and the depletion of ATP. Formation of ADP leads to ATP re-synthesis by glycolytic catabolism of glycogen for a limited time after death.

13.3.1   Onset of rigor mortis

Rigor mortis is the contractile process that occurs in postmortem muscle. The time of rigor onset and the extent of muscle shortening can have a profound effect on the edible quality of fish. The white muscles of fish are mainly responsible for rigor tension, while the red muscles do not significantly participate (Watabe et al., 1991). In the living animal, contraction occurs when the muscle is stimulated by an electrical impulse that depolarizes the muscle cell membrane called the sarcolemma. The stimulus is transmitted to the contractile organelles (myofibrils) located in the interior of the muscle cell by releasing Ca^{+2} from the sarcoplasmic reticulum. Muscle contraction involves interaction (shortening by a sliding mechanism) of the thick (myosin) and thin (actin + regulatory proteins) filaments that make up the myofibril. This interaction is triggered by myosin ATPase when the $[Ca^{+2}]$ increases from 0.1 to $10\mu M$. Relaxation occurs when the Ca^{+2} concentration in the sarcoplasm returns to $0.1\mu M$ and an ATP molecule re-associates with the myosin heads on the thick filaments. ATP is required for both contraction and relaxation. During contraction, ATP fuels the release of Ca^{+2} and the myosin ATPase reaction; and during relaxation ATP fuels the uptake of Ca^{+2} and causes relaxation by binding to the myosin heads.

Rigor mortis, an irreversible interaction of thick and thin filaments, occurs in postmortem muscle when the ATP declines significantly below the 5 mM concentration typical of resting muscle. Handling fish in rigor can cause physical damage such as 'gaping', the tearing apart of the muscle segments called myotomes. Filleting fish prior to the onset of rigor leads to a poor quality product because without the physical constraint imposed by the skeletal system extensive shortening during rigor can cause texture disorders and extensive loss of fluids ('drip'). In cod fish this may result in a 'sloppy' texture (Ang and Haard, 1985) whereas in sturgeon meat, with its high content of connective tissue, the product may exhibit a tough texture (Izquierdo-Pulido et al., 1992). Processing pre-rigor fish by techniques like cooking or freezing can also lead to excessive shortening and associated quality loss. The rapid depletion of ATP that occurs in pre-rigor fish that have been processed is caused by the loss of the ability of the sarcoplasmic reticulum and mitochondria to sequester Ca^{+2}. This, in turn, accelerates ATP hydrolysis, ATP resynthesis and the eventual depletion of ATP and decrease in pH.

13.3.2   Glycogen metabolism

In the anaerobic condition of postmortem fish muscle, glycogen is converted via phosphorylase (EC 2.4.1.1) to glucose-1-P and through a series of phosphorylated intermediates to pyruvate, which is then reduced to lactate (glycolysis). Some fish species also have a hydrolytic mechanism for converting glycogen to glucose (Burt, 1966). Glycolysis yields 2 or 3 moles of ATP per mole of glucose. ATP resynthesis occurs in postmortem flesh until glycogen is depleted or when pH declines to the point where key glycolytic enzymes, notably phosphofructokinase, are inhibited and glycolysis ceases. The concentration of glycogen in muscle depends on specie, biological condition, and antemortem stress. In some fish, free sugars and phosphorylated sugars generated by glycolysis or other reactions may persist in postrigor muscle. These sugars (especially sugar phosphates) may participate in browning reactions that lower the quality of processed (canned, frozen, dried) fishery products (see section 13.5.4).

13.3.3   ATP metabolism

The ATP concentration of rested fish muscle averages 7–10μmoles/g or about 5mM. The depletion of ATP normally occurs in postmortem fish muscle within 24h, although the process may take several days (Izquierdo-Pulido et al., 1992). The specific ATPase enzymes that catalyze hydrolysis of ATP in postmortem fish are incompletely understood with some coming from myosin ATPase and most coming from membrane ATPase pumps. Muscle contains about 20mM creatine phosphate and initial re-synthesis of ATP from ADP is catalyzed by creatine kinase (EC 2.7.3.2). The enzyme adenylate kinase (EC 2.7.4.3) catalyzes the formation 1 mole of ATP and 1 mole of AMP from 2 moles of ADP. For the most part, ATP regeneration from ADP occurs via anaerobic glycolysis. The terminal enzyme in the glycolytic sequence is lactate dehydrogenase (EC 1.1.1.27), which regenerates NADH from NAD^+ and forms lactate from pyruvate. The duration of ATP regeneration and consequently the rate of ATP depletion and rigor onset varies considerably in postmortem fish. Several key factors have been reviewed (Gill, 2000).

Storage temperature

For many fish from cold water, decreasing the storage temperature to 0°C decreases the rate of ATP depletion (Huynh et al., 1992). The temperature coefficient (Q_{10}) between 0° and 10°C may be as high as 10. However, for some species/muscles, the rate of ATP depletion increases and the time of rigor onset actually decreases with lower storage temperatures (Iwamoto et al., 1991). Studies with carp acclimated at different growing temperatures indicate that ATP depletion is faster when there is a greater difference between habitat temperature and storage temperature (Abe and Okuma, 1991). For most species, holding pre-rigor fish at or below the freezing point causes 'thaw rigor' associated with very rapid ATP degradation upon thawing (Hiltz et al., 1973). Accelerated ATP degradation resulting from chilling tropical fish or after freeze-thawing of some species appears to be caused by membrane damage that leads to the inability of the sarcoplasmic reticulum and or mitochondria to sequester Ca^{+2}.

Harvest and handling

Struggling by fish during capture may result in little or no muscle ATP or glycogen at the time they are landed. Trawling, especially for deep-water fish, can deplete muscle energy reserves causing rapid onset of rigor (Fraser et al., 1965). Likewise, farmed fish that are not rapidly anesthetized exhibit rapid ATP depletion and onset of rigor (Izquierdo-Pulido et al., 1992, Sigholt et al., 1997).

In some fresh fish markets pre-rigor fish command a high price. Brain spiking or severing the spinal cord immediately after capture can delay the depletion of ATP and onset of rigor by several fold (Mochizuki and Sato, 1996). The efficacy of brain spiking appears to be related to lowering the blood level of the stress hormone cortisol (Lowe et al., 1993).

13.3.4   Physiological disorders

Rapid and extensive hydrolysis of ATP may lead to poor quality meat because of the denaturation of contractile and/or sarcoplasmic proteins. Such muscle may have soft texture, poor water-holding capacity, and a pale color. ATP hydrolysis results in the formation of a H^+ and this is responsible for the decrease in muscle pH.

$$ATP + H_2O \Rightarrow ADP + PO_4^{-3} + 3\ H^+ + 11.6\ kcal$$

A rapid decline in pH thus causes disorders such as 'chalky halibut', 'burnt tuna' and 'sloppy cod'. It would appear that rapid and extensive pH decline is caused by stress hormone activity prior to death. Normally such disorders occur when the carcass has a relatively high temperature at the time that acid is produced from ATP hydrolysis.

13.4   Biochemical changes in post-rigor muscle

13.4.1   pH

With the cessation of ATP regeneration and hence ATP hydrolysis the pH of postmortem muscle reaches a stable value referred to as 'ultimate pH'. The ultimate pH varies with species of fish. Tuna, mackerel and sturgeon may have a pH as low as 5.5 whereas most other fish have a pH of 6.2–6.6. Factors affecting the extent of ATP hydrolysis, as described above, and buffering capacity will affect the ultimate pH. The pH relative to that which is normal for the specie appears to be more important than the absolute pH. For example, cod that have resumed feeding after a lengthy period of fasting may have a relatively low pH of 6.0 which is associated with 'gaping' due to weakening of the connective tissues and 'sloppy' texture (Ang and Haard, 1985). In some species, low ultimate pH is associated with tough cooked texture. A relatively low pH may also decrease water binding to the myofibrils and thereby influence light scattering and the appearance of fish. Low pH may also promote oxidation of myoglobin and lipids. Since the activity of enzymes is pH dependent it is expected that ultimate pH will influence a wide range of reactions.

13.4.2   Nucleotide catabolism

A series of enzymes convert ADP to hypoxanthine or uric acid. These reactions have been reviewed in detail (Gill, 2000). Together with ATP regeneration and hydrolysis, continued degradation of ADP is the most important factor responsible for loss of prime quality (Tarr, 1966). Adenosine monophoshate (AMP) is formed from ADP by adenylate kinase (also known as myokinase, EC 2.7.4.10):

$$[2ADP \leftrightarrow ATP + AMP]$$

In some species, such as squid (Sagedhal et al., 1997), AMP may accumulate and is believed to have significant influence on the flavor of seafood from some marine invertebrates. Conversion of AMP to adenosine (rather than IMP) has been proposed for invertebrate tissues like scallop adductor muscle (DeVido de Mattio *et al.*, 1992). Rapid deamination of AMP and adenosine is believed to contribute to ammonia formation in shrimp (Finne, 1982). However, in most species AMP is rapidly degraded to inosine monophosphate (IMP) by a reaction catalyzed by AMP deaminase (EC 3.5.4.6)

$$[AMP \leftrightarrow IMP + NH_3]$$

The conversion of IMP to inosine (Ino) is normally the rate limiting reaction of nucleotide catabolism. IMP may persist in chilled fish for several days or even for weeks in durable species like tuna. Accumulation of IMP is desirable since this molecule has flavor enhancing properties. Three enzymes are known to catalyze the conversion of IMP to Ino and phosphate (Pi), i.e., 5'-nucleotidase (EC 3.1.3.35), alkaline phosphatase (EC 3.1.3.1) and acid phosphatase (EC 3.1.3.2). The rate of IMP disappearance in chilled fish appears to be most closely related to the activity of 5'-nucleotidase.

$$[IMP \leftrightarrow Ino + Pi]$$

The importance of this reaction is seen by the equation for the useful freshness indicator called the 'K value' (Saito *et al.*, 1959).

$$K = \frac{[\text{Ino}] + [\text{Hx}]}{[\text{ATP}] + [\text{ADP}] + [\text{AMP}] + [\text{IMP}] + [\text{Ino}] + [\text{Hx}]} \times 100$$

In practice, the '*Ki* value' (Karube *et al.*, 1984) is also found to correlate well with fish freshness because of the rapid depletion of ATP, ADP and AMP:

$$\left[Ki = \frac{[\text{Ino}] + [\text{Hx}]}{[\text{IMP}] + [\text{Ino}] + [\text{Hx}]} \times 100 \right]$$

The degradation of Ino to hypoxanthine (Hx) is catalyzed by nucleoside phosphorylase (EC 2.4.2.1) or inosine nucleosidase (EC 2.4.2.2).

$$\text{Ino} + \text{Pi} \leftrightarrow \text{Hx} + \text{ribose-1-P} \quad \text{(nucleoside phosphorylase)}$$

$$\text{Ino} + \text{H}_2\text{O} \rightarrow \text{Hx} + \text{D-ribose} \quad \text{(inosine nucleosidase)}$$

Unlike the other reactions in nucleotide degradation, spoilage bacteria may contribute to this reaction (Surette *et al.*, 1988). The nucleoside phosphorylase reaction appears to be more important during the early stages of spoilage. Hx has been reported to have a bitter flavor and is believed to adversely affect the flavor of fish (Gill, 2000).

Xanthine oxidase (EC 1.1.3.22) catalyzes the conversion of Hx to xanthine (Xa) and uric acid.

$$\text{Hx} + \text{H}_2\text{O} + \text{O}_2 \rightarrow \text{Xa} + \text{H}_2\text{O}_2$$

$$\text{Xa} + \text{H}_2\text{O} + \text{O}_2 \rightarrow \text{Uric acid} + \text{H}_2\text{O}_2$$

These reactions would be promoted by the reintroduction of O_2 in spoiled fish and would be expected to promote lipid oxidation via H_2O_2 and hydroxyl radicals (Hultin, 1992).

13.4.3   Lipid hydrolysis and oxidation

Lipid hydrolysis

Lipases and phospholipases catalyze hydrolysis of glycerol-fatty acid esters resulting in the formation of free fatty acids (FFA). The role of these enzymes in seafood quality has been reviewed (Lopez-Amaya and Marangoni, 2000a, Lopez-Amaya and Marangoni, 2000b). Hydrolysis of acyltriglycerides may predominate in red muscle because of the higher fat content. Phospholipid (PL) hydrolysis predominates in white muscle where PL is the principal lipid class. Regardless of the source, FFA may be directly responsible for off-flavors, the destruction of some vitamins and amino acids, changes in texture, and changes in water holding capacity. The latter two results are particularly important in frozen fish and are related to the propensity of fatty acids to bind, denature and cross-link myofibrillar proteins (Haard, 1992a). Triglyceride hydrolysis by lipase (s) may promote lipid oxidation whereas PL hydrolysis by phospholipase (s) may inhibit lipid oxidation (Schewfelt, 1981). Since fatty acids appear to be more susceptible to oxidation than fatty acid esters (Toyomizu *et al.*, 1981), it appears that the stabilizing influence of phospholipase action in fish muscle is related to membrane location of PL (Schewfelt and Hultin, 1983).

Lipid oxidation

Oxidation of the highly unsaturated fatty acids associated with fish lipids is a primary cause of fish spoilage. The extent of lipid oxidation in fish meat is a function of natural and added antioxidants as well as enzymic and non-enzymic pro-oxidants. Microsomal enzymes can convert Fe^{+3} to Fe^{+2} which is a powerful pro-oxidant (Hultin, 1992, Haard, 1992a). Lipoxygenase (LOX; EC 1.13.11.12) can also accelerate oxidative rancidity but this group of enzymes have mainly been implicated in the formation of volatile alcohols and aldehydes (section 13.4.1) that contribute to fresh fish aroma (Josephson and Lindsay, 1986, Sun Pan and Kuo, 2000). Glutathione peroxidase (EC 1.11.1.12) has also been implicated in the antioxidant activity of meat (Chan and Dekker, 1994).

13.4.4   Protein hydrolysis

Resolution of rigor

When 'in rigor' fish is held for a period of time the muscle becomes extensible and more tender after cooking. This process, called the 'resolution of rigor', is not due to a sliding apart of thick and thin filaments, as occurs during relaxation, but is mainly due to the subtle action of endogenous proteases on key link proteins with a resulting loss in integrity of the myofibril structure. Proteases found in seafood have been reviewed (Haard, 1994). Two groups of muscle proteases have been implicated in the resolution of rigor. The calpains (EC 3.4.22.17) are calcium-activated neutral proteinases that cause separation of the myofibril units called 'sarcomeres'. Calpains have also been implicated in the physiological disorder ('burnt' tuna) (Watson *et al.*, 1988). The second group is the cathepsins, a class of lysosomal proteinases that are most active between pH 3 and 6.

Collagen degradation

Although type I collagen is not degraded in postmortem rockfish muscle held below critical temperatures (Cepeda *et al.*, 1990), disruption of myocommata junctions may occur during early postmortem stages (Bremner, 1992). These changes appear to involve hydrolysis of basement membrane type IV and V collagens (Sato *et al.*, 1994). Two muscle collagenases, classified as matrix metalloproteinases (EC 3.4.24.35), have been identified in fish (Bracho and Haard, 1995). Recently cDNA clones were obtained for a gelatinase type matrix metalloproteinase (EC 3.4.24.24) and a collagenase capable of degrading type I collagen (EC 3.4.24.-) from rainbow trout fibroblast (Saito *et al.*, 2000b, Saito *et al.*, 2000a).

Digestive proteinases

If fish or shellfish are not promptly eviscerated, seepage of digestive proteinases to the muscle can lead to extremely rapid quality deterioration (Almy, 1926). The problem can be extreme in heavily feeding fish and in crustaceans where the midgut gland is in close proximity to the meat (Nip and Moy, 1988).

13.4.5   Other enzymic reactions in post-rigor seafood

Examples of other enzymes that are known to influence the quality of post-rigor seafood include thiaminase (EC 3.5.99.2), urease (EC 3.5.1.5), trimethylamine oxide (TMAO) demethylase (EC 4.1.2.32), and polyphenol oxidase (PPO; EC 1.10.3.1). Melanosis caused by polyphenol oxidase is a common problem in crustacean species such as lobster, shrimp and crab. Seafood PPO is reviewed elsewhere (Kim *et al.*, 2000). TMAO demethylase catalyzes formation of dimethylamine and formaldehyde from TMAO (Sotelo and Rehbein, 2000). Formaldehyde is believed to contribute to protein cross-linking of muscle proteins causing an increase in toughness. Normally this reaction is more problematic in frozen fish (Haard, 1992a), however, the reaction may occur in species like red hake at temperatures above the freezing point in the absence of oxygen. Thiaminase is a transferase that catalyzes reaction of thiamine with amines. It appears to be associated with the viscera in several fresh-water and salt-water species (Haard, 1990). Urease activity is mainly a problem with elasmobranch species that contain high levels (e.g., 1.0–2.5% in shark meat) of urea. It is not clear whether or not endogenous or bacterial urease cause the rapid ammonia formation from urea in these species (Finne, 1992).

13.5   Enzymes and seafood color and appearance

Pigments in seafoods include those of natural origin as well as those resulting from postmortem changes. The principal pigments in seafood that normally impact quality are heme proteins, carotenoids, and melanins from enzymic and non-enzymic reactions. The biochemistry of seafood color has been reviewed (Haard, 1992b). Examples of enzymes that may influence the color and appearance of seafoods are summarized in Table 13.1.

13.5.1   Flesh opacity

Random light scattering of fish flesh normally increases resulting in a change of flesh appearance from a bluish translucent to an opaque, cooked appearance with post-harvest storage. Whiteness increases when the incident light is scattered reflectively by structures that exceed 0.7μ. A characteristic of post-rigor muscle is an increase in intracellular and extracellular space, a more random spacing of myofibrils, and dense packing of myofilaments within the myofibril. Changes in the interaction of thin and thick filaments and/or the network of cytoskeletal proteins (desmin, nebulin and titin) by proteases appear to give rise to disorganization of the alignment of intracellular myofibrils in post-harvest fish muscle. Both calpains and cathepsins (13.5.4) have been implicated in the degradation of cytoskeletal and myofibril proteins. Antemortem protein depletion brought on by starvation of fish can also result in flesh opacity (Haard, 1987).

Table 13.1   Enzymes that may influence seafood appearance

Enzyme(s)	Component(s)	Result
ATPases	H^+, Myoglobin	⇓ Low pH favors oxidation of myoglobin
Calpain, cathepsins	Cytoskeletal proteins	⇓ Increased flesh opacity
Glycolytic enzymes	Glycogen	⇓ Formation of reducing hexose sugars contributes to Maillard browning
Lipoxygenase-like enzyme; meyeloperoxidase	Carotenoids	⇓ Bleaching of yellow-red epithelial and flesh pigments
Metmyoglobin reductase	Metmyoglobin	⇑ Rate of brown heme discoloration reduced
Nucleoside phosphorylase, inosine nucleosidase	Ino	⇓ Formation of reducing pentose sugars contributes to Maillard browning
Polyphenoloxidase	Tyrosine	⇓ Melanin

13.5.2   Heme proteins

The heme proteins associated with the bright red color of fresh fish meat include oxymyoglobin and oxyhemoglobin. The oxidation products, metmyoglobin and methemoglobin, are responsible for the brown hue of spoiled meat. These oxidation reactions are non-enzymic and outside the scope of this chapter. However, enzymic reduction of metmyoglobin to myoglobin has been demonstrated in various fish muscles including dolphin, bluefin tuna, mackerel and blue white dolphin (Matsui *et al.*, 1975) and metmyoglobin reductase has been purified from tuna (Pong *et al.*, 2000, Al-Shaibani, 1977). The reduction of metmyoglobin in postmortem meat appears to require reduced pyridine nucleotides (MacDougall, 1982) and NADH-cytochrome b5 reductase has also been implicated in the reduction of metmyoglobin (Arihara *et al.*, 1991). Metmyoglobin reducing systems are inactivated by low pH (Foegeding *et al.*, 1996). Active mitochondria may also compete with myoglobin for oxygen thus increasing the rate of color change from bright red oxymyoglobin to blue-red myoglobin and its subsequent oxidation to brown metmyoglobin. Green discoloration of tuna is also an important color disorder that involves non-enzymic oxidation of myoglobin with TMAO and a sulfhydryl compound (Haard, 1992b).

13.5.3   Carotenoids

The carotenoids contribute to the striking yellow, orange and red hues of the integument and flesh of several important fish and shellfish products. In some cases, the grading and pricing of these species is directly related to the intensity of red color (Sackton, 1986). Animals obtain carotenoids from their diet, and species like salmonids are able to accumulate oxycarotenoids in the flesh as well as epithelial tissue. Fish species also differ in their ability to convert dietary carotenoids, e.g., crustacean species convert β-carotene to bright red oxycarotenoids (Fox, 1979).

Most animals have the enzyme carotene 15,15′-dioxygenase (EC 1.13.11.21) which catalyzes cleavage of carotenoids to vitamin A aldehydes. The pink or red color of skin or flesh tends to fade during storage on ice or freezing (Tsukuda and Amano, 1966). The rate of fading differs with species (Scott *et al.*, 1986). Cellular disruption by freezing and partial freezing normally accelerates the oxidation of carotenoids and this suggests the decompartmentation of catalysts. Tsukuda and associates have characterized a lipoxygenase-like enzyme that can catalyze oxidation of astaxanthin and other oxycarotenoids (Tsukuda, 1970). Fish leukocytes also contain a myeloperoxidase (EC 1.11.1.7) that can catalyze de-coloration of β-carotene in the presence of hydrogen peroxide and halogen ions (Kanner and Kinsella, 1983). Since carotenoids may be co-oxidized with lipids the importance of specific carotenoid oxidizing systems in stored seafood remains unclear.

13.5.4 Postharvest pigmentation

Enzymic browning

Melanosis in crustacean species like krill, lobster, shrimp and crab is one of the most studied enzyme problems in seafood technology. The occurrence of 'blackspot' gives the product an unacceptable appearance and thus lowers market value. Polyphenol oxidase (PPO, EC 1.10.3.1) in these species catalyzes a two-step reaction: (i) hydroxylation of monophenols like tyrosine (mono phenol oxidase reaction) to o-diphenols (Dopa) and (ii) oxidation of o-diphenols to di-quinones (diphenol oxidase reaction). A series of non-enzymic reactions are involved with the conversion of di-quinones to black melanin. Zymogens (Pre-PPOs) from crustacean species are activated by proteolytic enzymes such as trypsin-like enzymes in the tissue (Savagaon and Sreenivasan, 1978). Tyrosine appears to be the principal substrate for crustacean PPOs. The interaction of tyrosine with the copper protein hemacyanin in crab blood and other metal ions has also been implicated in blue discoloration of crab meat (Boon, 1975). The characterization of seafood PPOs and their control has been recently reviewed (Kim *et al.*, 2000).

Non-enzymic browning

The Maillard reaction has been identified as the cause of discoloration in a number of seafood products including frozen, dried, salted and canned fish and shellfish. Enzyme catalyzed reactions may be indirectly involved in this reaction. The presence of reducing sugars normally limits the occurrence of non-enzymic browning in fishery products. The reaction involves the formation of a Schiff base from a reducing sugar and an amino bearing compound. Amadori products slowly dehydrate and rearrange into yellow, brown fluorescent products that are able to cross-link with proteins, etc. Glucose and ribose are normally the main free sugars in fish and shellfish but phosphorylated sugars (e.g., ribose-5-P, glucose-6-P, fructose-6-P) are also present and very active in browning reaction (Haard and Arcilla, 1985).

Orange discoloration in canned tuna involves the reaction of glucose-6-P with the amino groups of histidine and dipeptides (Nakayama *et al.*, 1976). Conditions which lead to the accumulation of glucose-6-P and other sugars appear to relate to the occurrence of glycolysis and glycogen hydrolysis (14.2.2) for hexose sugars and nucleotide degradation (14.3.2) for pentose sugars under appropriate conditions such as frozen storage (Burt, 1971). Browning in surimi products may involve the enzymic conversion of glucose to 2,5-diketogluconic acid by bacterial enzyme (s). The glucose derivative is a very active reducing sugar in the Maillard reaction (Fujita *et al.*, 1978).

Protein-lipid browning may also be an important reaction in fishery products (Haard, 1992b). Unsaturated carbonyls, such as 2-hexenal and acetaldehyde, appear to react by aldol condensation and dehydration to form crotonaldehyde and 2-(1-butenyl)-octa-2,4-dienal) that react with amino compounds during the early stages of this reaction (Fujimoto and Kaneda, 1973b). While the generation of reactive carbonyls appears to arise from non-enzymic lipid oxidation, volatile amines that arise from endogenous and bacterial enzymic reactions (e.g., NH_3, TMA) are the main amino compounds in this reaction (Fujimoto and Kaneda, 1973a).

13.6   Enzymes and seafood flavor

The aroma of raw fish is an important indication of quality (Lindsay, 1990). Seafood flavor is complex because of the broad range of taxa including both vertebrate and invertebrate organisms. One can distinguish between (i) the pleasant, plant-like, melon-like aroma of very fresh raw fish, (ii) the strong and objectionable aroma of spoiled raw fish, and (iii) the flavor of cooked fish and shellfish. Endogenous enzymes are directly responsible for the first type of aroma. Endogenous enzymes together with spoilage bacteria are responsible for the 'fishy' odor characteristic of stale or rancid fish. Cooked flavor also includes taste active compounds that are partly the result of endogenous enzyme activity. Examples of enzymes that may influence the flavor of seafoods are summarized in Table 13.2.

13.6.1   Fresh fish flavor

The role of enzymes in fresh aroma biogenesis in fish has been recently reviewed (Cadwallader, 2000). The delicate aroma and taste of very fresh prime quality fish is quite different from that in most commercially available seafood.

Table 13.2   Enzymes that may influence seafood flavor

Enzyme(s)	Component(s)	Result
AMP deaminase	AMP	⇑ IMP formation contributes to delicious taste
Bromoperoxidase	Haloforms	⇑ Characteristic aroma of some seaweeds
Dethiomethylase	DMPT	⇓⇑ DMS may positively or negatively impact aroma
LAFE	Fatty acids	⇑ Long chain aldehydes in seaweed
Lipoxygenase, hydroperoxide lyase	Polyunsaturated fatty acids	⇑ Plant-like fresh aroma
Nucleoside phosphorylase, inosine nucleosidase	Ino	⇓ Hypoxanthine contributes to bitter taste
Phospholipase, lipase	Phospholipids, acyltriglycerides	⇓ hydrolytic and oxidative rancidity
TMAO reductase	TMAO	⇓ 'fishy' off-odor
Urease	Urea	⇓ Ammonia off-odor
Various deaminases	Amino acids, nucleotides	

Lipoxygenase (LOX; EC 1.13.11.12) catalyzes the formation of fatty acid hydroperoxides that may undergo non-enzymic or enzymic degradation to form low molecular weight volatile, odorous compounds. Hydroperoxide lyase (HOL, EC 4.2.1.92) catalyzes the hydrolysis of hydroperoxides to form Z-aldehydes which can undergo isomerization to form E-aldehydes. Reduction of aldehydes to alcohols may also occur by the action of alcohol dehydrogenase (EC 1.1.1.2). The spectra of products formed depend on the structure of the fatty acid substrate, the fatty acid specificity of the lipoxygenase, and the positional specificity of the LOX. Lipoxygenases identified in different seafoods include specificity for the 8, 11 (R), 12 (R),13 (S), 15, 5 + 12 + 15, 12+15, 5+ 12, 8 + 12 + 15, and 9 + 13 positions of arachidonic acid (Sun Pan and Kuo, 2000). Hence a wide variety of carbonyls and alcohols may be formed in different species. The principal compounds contributing to fresh fish aroma are C_6, C_8, and C_9 carbonyls and alcohols. It has also been suggested that co-oxidation of carotenoids can also give rise to contributory aroma compounds (Winterhalter, 1996).

The taste of seafood from coldwater organisms is normally optimal at around 24 h ice time post harvest when IMP levels are at a maximum. The enzymology of IMP formation has already been discussed in section 13.5.2. IMP is a flavor enhancer, and this nucleotide is responsible for the delicious taste of prime quality seafood. As mentioned, IMP may persist for several days in some species such as tuna. There is also evidence that AMP may contribute to the characteristic fresh taste of species that accumulate this nucleotide, such as some species of scallop and squid. The taste of fresh shellfish is also strongly influenced by pre-formed, free amino acids. For example, sweet tasting L-glycine is present in crab meat at concentrations as high as 2% on a fresh weight basis. Dimethyl-β-propriothetin (DMPT) from zooplankton feed is an important precursor of volatile sulfides such as dimethyl sulfide (DMS) in some finfish. DMS is particularly important to the top-note aroma of fresh cooked clams and oysters. However, fresh fish which ingest excessive quantities of feed with sulfide precursor may have an objectionable off-flavor (Motohito, 1962). Sulfides may arise by thermal degradation of precursors and it is not clear whether enzyme catalyzed reactions are involved with their formation.

13.6.2   Seaweed flavor

The enzymology of seaweed flavor biogenesis has been reviewed (Fujimura and Kawai, 2000). It appears seaweeds would lack most of their delicate and distinctive flavor without the occurrence of flavor precursors and enzymes. DMS is a major component of green seaweed aroma and green seaweed contains a greater content of an enzyme (dethiomethylase) that decomposes DMPT to DMS (Iida et al., 1985, Cantoni and Anderson, 1956). DMS also contributes to the aroma of red and brown seaweeds. Long chain, (un)saturated aldehydes ($C_{17}CHO$) are formed enzymatically from stearic, oleic, linoleic, and linolenic acids. Long chain aldehyde-forming enzyme (LAFE) appears to be present in seaweeds but not terrestrial plants (Kajiwara *et al.*, 1996). The reaction involves formation of a hydroperoxide adjacent to the fatty acid carboxyl followed by decarboxylation to form a long chain aldehyde. LOX and HOL activities also form relatively short chain aldehydes (e.g., (2E)-nonenal) that contribute to seaweed flavor. Halogen containing compounds are important to the aroma of red seaweeds. Bromoperoxidases (EC 1.11.1.8) have been identified in several species of red algae (Yamada *et al.*, 1985) but it is not clear whether they are involved in postharvest transformation of haloforms. Red seaweed also contain greater quantities of sulfate-adenyltransferase (EC 2.7.7), the initial enzyme in the inorganic sulfate assimilation pathway. A brown seaweed (*Desmarestia*) contains an unidentified enzyme that catalyzes release of large quantities of sulfuric acid and an offensive odor after harvest (Chihara, 1982).

Stale fish flavor
Much of the spoiled odor and taste of postharvest fish is due to bacterial enzymes. However, endogenous enzymes are important catalysts of spoiled fish aroma in some species. Meat of prerigor elasmobranches (rays and sharks) may develop a strong ammoniacal odor and bitter taste. This is associated with the hydrolysis of urea (1–2.5% fwb) to ammonia and carbon dioxide by urease (EC 3.5.1.5). The problem can be avoided by washing the flesh in water to remove urea. Stringent aftertaste and other quality defects in the flesh of struggling tuna (burnt tuna) has been linked to endogenous calpains (Watson *et al.*, 1988). Blackberry odor in herring, a flavor defect caused by DMS, has been linked to the rapid degradation of DMPT, which as mentioned above, may be enzyme catalyzed. Rapid formation of ammonia in shrimp has also been partly attributed to endogenous deaminases. Varying amounts of TMAO have been found in all tissues of saltwater fish as well as in some freshwater species. Trimethylamine oxide reductase (EC 1.6.6.9) catalyzes formation of trimethylamine (TMA) from TMAO. The enzyme has been purified from spoilage bacteria (Clark and Ward, 1988). Although there are claims for the presence of endogenous fish TMAO reductase the most important source of this enzyme is spoilage bacteria. TMA has a very low odor threshold and is the main indication of spoilage of marine fish. Endogenous phospholipases and possible lipases are clearly involved with the deterioration of flavor in some seafood products during frozen storage. Free fatty acids from phospholipids may be responsible for off-flavors and the hydrolytic reaction may facilitate enzymic or non-enzymic oxidative rancidity (Lopez-Amaya and Marangoni, 2000b).

13.7   Enzymes and seafood texture

Enzymes in seafood can cause either excessive softening or toughening of seafood products (Table 13.3). The enzymes involved in the softening of chilled fish muscle are mainly active on myofibrillar and connective tissue proteins.

Table 13.3   Enzymes that may influence seafood texture

Enzyme(s)	Component(s)	Result
ATPases	ATP	⇓ Onset of rigor, physiological disorders
Calpains, cathepsins	Myofibrillar proteins	⇑ Resolution of rigor
Cathepsin L, alkaline proteases	Myosin, etc.	⇓ Surimi gel weakening
Matrix metalloproteinases	Collagen	⇓ Gaping, raw fish texture
Phospholipases, lipases	Phospholipids, acyltriglycerides	⇓ Toughening of frozen fish
TMAO demethylase	TMAO	⇓ Toughening of frozen fish
Transglutaminase	Myosin	⇑ Improved surimi gel strength

Heat activated proteases are responsible for surimi gel softening (Kang and Lanier, 2000). More recently, true collagenases have been purified from seafood muscle (Bracho and Haard, 1995, Saito *et al.*, 2000a, Saito *et al.*, 2000b, Sato *et al.*, 1994) and these enzymes may contribute to 'gaping' and loss of muscle integrity during filleting. Transglutaminase (Tgase, EC 2.3.2.13) catalyzes formation of covalent crosslinks and is responsible for improving the gel strength of surimi pastes at low temperatures (5–40°C) (Ashie and Lanier, 2000). TMAO demethylase (EC 4.1.2.32) contributes to the loss of succulence and tough texture of some fish species during frozen storage by formation of the cross-linking agent formaldehyde.

13.7.1   Tenderization and softening

Resolution of rigor

As mentioned in section 13.4.4, proteases implicated in the softening of post-rigor muscle include the cathepsins, calpains and collagenases. Tenderization during the resolution of rigor is not normally important in seafood products although there are exceptions (Izquierdo-Pulido *et al.*, 1992). Early studies revealed that muscle proteolytic activity is about ten times greater in fish than in land animals (Siebert, 1958). Of the 13 lysosomal cathepsins, B, D, H, L, L-like and X have been purified from fish and shellfish muscles (Jiang, 2000). The neutral proteinase calpain has also been purified from fish and shellfish (Wang *et al.*, 1993). The characteristics of these endogenous fish proteases have been recently reviewed (Jiang, 2000). Most research on the postrigor tenderization of muscle has been done on terrestrial myosystems. In general, tenderization involves proteolytic degradation of inter-(desmin, vinculin) and intra-myofibril (titin, nebulin, troponin-T) link proteins. Increased fragility of myofibrils at the Z-disk (calpains) and at the junctions of A-bands and I-bands (lysosomal cathepsins) are primarily responsible for tenderization.

Connective tissue degradation
Available evidence indicates that enzymic collagen degradation and associated tissue softening is much more important in fish than it is in terrestrial myosystems. In fish, flesh softening has been related to the increased solubility of type V collagen (Sato *et al.*, 1997, Sato *et al.*, 1991, Sato *et al.*, 1994). Rapid degradation of connective tissues is responsible for the soft texture of spawning ayu, *Plecoglossus altivelis* (Itoh *et al.*, 1992). Very soft raw flesh texture has also been observed in other species such as post-spawning cod, *Gadus morhua* (Ang and Haard, 1985). Scanning electron microscopy has revealed the degradation of connective tissues in iced freshwater prawns (Nip and Moy, 1988) and it has been suggested that enzymic degradation of the non-helical domains of collagen contributes to the softening of kumara prawn (Mizuta *et al.*, 1997).

13.7.2   Heat induced proteases
Cooking at intermediate temperatures (e.g., 60°C) may cause activation of endogenous proteases that can cause severe texture breakdown of fish muscle or surimi-based products. Unlike chilled muscle, heat induced proteases frequently cause extensive breakdown of the main muscle protein myosin. The occurrence of heat-induced proteinases is highly variable with species. In some species, such as Pacific whiting (An *et al.*, 1994), Arrowtooth flounder (Izquerdo-Pulido *et al.*, 1994), chum salmon (Yamashita and Konagaya, 1991) and mackerel (Jiang *et al.*, 1997), the enzyme responsible for excessive softening is the cysteine proteinase cathepsin L (EC 3.4.22.15) or L-like enzyme. In many other fish species, serine proteinases (so-called alkaline proteinases) have been found to be responsible for gel weakening in surimi (Kang and Lanier, 2000). Several of these alkaline proteases have a very high molecular weight (Iwata *et al.*, 1973). Some of the alkaline proteases that are responsible for gel weakening are also salt activated and not removed by the water washing step of surimi manufacture and are thus especially problematic in surimi products (Kinoshita *et al.*, 1990, Haard, 1994, An *et al.*, 1994).

13.7.3   Transglutaminase
Endogenous or added Tgase can form isopeptide bonds between myosin molecules and this reaction is responsible for improvement in gel strength during the 'setting' stage of surimi processing (Jiang, 2000, Ashie and Lanier, 2000). Gels formed with a low temperature setting stage have much greater strength after cooking than those made by direct cooking. Tgase occurrence in fish species is variable. The enzyme has been identified in several fish species including carp, rainbow trout, chum salmon, atka mackerel and white croaker (Ashie and Lanier, 2000). Tgase has been purified from the muscle and surimi of Alaska pollack (Seki *et al.*, 1990). Other enzymes may also contribute to protein crosslinking but they have not been studied in seafoods (Haard, 2001).

13.7.4   Protein aggregation in frozen fish

Frozen fish may develop a dry, stringy texture during frozen storage. The process is mainly due to the denaturation and aggregation of myofibrillar proteins and perhaps collagen (Haard, 1992a). Enzymes appear to contribute to this quality defect in several ways. Free fatty acids resulting from lipid or phospholipid hydrolysis can bind to myofibrillar proteins causing denaturation and aggregation. Products of lipid oxidation, such as malonaldehyde or butyraldehyde, can react with protein to form fluorescent cross-linked proteins (Leake and Karel, 1985). TMAO demethylase forms formaldehyde (with dimethylamine) in fish and the presence of formaldehyde is related to texture deterioration during frozen storage, especially gadoid species (Castell *et al.*, 1971, Sotelo and Rehbein, 2000). The occurrence and purification of this membrane protein from various tissues and species has been reviewed (Sotelo and Rehbein, 2000).

13.8   The use of enzymes in seafood processing and quality control

The use of enzymes as processing aids and as analytical tools in the seafood industry has increased greatly in the past three decades. Enzyme processing aids offer numerous advantages over most other process strategies for several reasons. Normally, enzyme catalyzed reactions are less energy intensive, more environmentally friendly, more selective, result in less undesirable side reactions, can be conducted under moderate conditions of temperature and pH, and are readily subject to control. Analytical techniques, such as biosensors utilizing enzymes, are also being used by the seafood industry for quality control, species identification and freshness testing. Finally, enzymes from fish processing wastes are being recovered for use as additives in a variety of industrial applications. This topic has been the subject of numerous reviews and the reader is referred to the following for details and references (Haard and Simpson, 1985, Reece, 1988, Stefansson and Steingrimsdottir, 1990, Haard, 1992c, Gildberg, 1993, Haard and Simpson, 1994, Kolodziejska and Sikorski, 1995, de Vecchi and Coppes, 1996, Raa, 1997, Vilhelmsson, 1997, Haard, 1998, Diaz-Lopez and Garcia-Carreno, 2000, Gildberg *et al.*, 2000, An and Visessanguan, 2000, Venugopal *et al.*, 2000). That these reviews have originated from so many countries around the world, including Canada, Denmark, Iceland, India, Mexico, Norway, Poland, Spain, USA, and Uruguay, is testimony to the global importance of this topic.

13.9   Enzymes as seafood processing aids

13.9.1   Proteolytic enzymes

The main group of enzymes used as seafood processing aids are the proteases. Examples of the use of proteases as seafood processing aids are given in Table

13.4. Some of the applications shown (e.g., fish sauce, protein isolates) require a protease or mixture of proteases with broad specificity that provide a high degree of peptide bond hydrolysis. Others (e.g., carotenoprotein isolation, Matjes herring) require an enzyme with relatively narrow specificity such as trypsin. A problem often encountered with protein hydrolysis is the formation of bitter peptides (Haard, 2001). Many of the examples summarized in Table 13.4 utilize digestive proteases from fish processing wastes (Simpson, 2000) although microbial, plant and terrestrial animal proteinases have also been used in these studies.

Other enzymes
Examples of other enzymes including a transferase (Tgase), oxidoreductases (glucose oxidase, superoxide dismutase) and hydrolases (lipase, phospholipase, lysozyme, urease, carbohydrases) are summarized in Table 13.5. Of the examples shown, Tgase has received the most attention. Uses of Tgase in the fish processing industry include improving the gel strength of surimi and surimi-based analogues, binding of other protein ingredients with fish meat proteins, fish meat sheet formation, raw and processed fish egg products, fin fish texture modification with gelatin, shark fin processing, reduction of drip in thawed fish, freeze-texturization of fishery products and bonding of fish pieces into uniform shapes (Venugopal *et al.*, 2000).

Table 13.4 Application of proteolytic enzymes in seafood processing

Application	Reference
Carotenoprotein from crustacean waste	(Simpson and Haard, 1985)
Caviar/roe production	(Sugihara *et al.*, 1973)
Clam shucking	(Fehmerling, 1970)
Fish meal stickwater viscosity	(Jacobsen and Rasmussen, 1984)
Fish protein hydrolysate	(Barzana and Garcia-Garibay, 1994)
Fish sauce	(Raksakulthai *et al.*, 1986)
Matjes herring	(Simpson and Haard, 1984)
Peptones from whole fish	(Gildberg *et al.*, 1989)
Prevention of curd in canned salmon	(Yamahoto and Mackey, 1981)
Protein recovery from frame waste	(Kim *et al.*, 1997)
Ripening of salt fish	(Gora, 1972)
Scale removal	(Stefansson and Steingrimsdottir, 1990)
Seafood flavors	(In, 1990)
Shiokara squid	(Lee *et al.*, 1982)
Shrimp flavor extract	(Gildberg *et al.*, 2000)
Shrimp peeling and deveining	(Fehmerling, 1970)
Skin removal	(Stefansson, 1988)
Squid tenderization	(Kolodziejska *et al.*, 1992)
Swim bladder membrane removal	(Stefansson and Steingrimsdottir, 1990)

Table 13.5   Application of other enzymes in seafood processing

Enzyme/application	Reference
Amylase, Cellulase/Clam shucking	(Scott, 1975b)
Carbohydrases/Skin removal	(Stefansson, 1988)
Glucose oxidase/Color retention of shrimp	(Scott, 1975a)
Glucose oxidase/Color retention of tuna	(Koczot et al., 1985)
Glucose oxidase/Preservation	(Field et al., 1986)
Lipase/Marine oil modification	(Wanasundra and Shahidi, 1997)
Lysozyme/Preservation	(Ramesh and Lewis, 1980)
Phospholipase A2/Marine oil modification	(Tocher et al., 1986)
Superoxide dismutase/Preservation	(Ashie et al., 1996)
Transglutaminase/Texture improvement, etc.	(Ashie and Lanier, 2000)
Urease/Off-odor reduction	(Ghosh, 1989)

13.9.2   Enzymes as quality indices and analytical tools

Endogenous enzymes as quality indices

Endogenous enzyme activity in fish and fish products can be used as a measure of freshness, storage history or quality (Gopakumar, 2000). Some examples of enzymes used to assess fish quality are summarized in Table 13.6. In measuring fish freshness the marker enzyme normally shows a progressive decrease in activity with storage time. In some markets, such as Europe, fresh seafood commands a higher price than previously frozen seafood and the activation of some enzymes due to de-compartmentation from freeze-thawing is used as a measure of storage history.

Exogenous enzymes as biosensors

The assessment of fish quality may also involve using exogenous enzymes as analytical tools. Some examples of enzyme-linked reactions used to evaluate seafood quality are summarized in Table 13.7. Initial studies utilized enzyme test-tube assays while most recent developments have used immobilized

Table 13.6   Endogenous enzymes used as fish quality indices

Enzyme	Measurement	Reference
Ca^{+2} ATPase	Quality of frozen fish	(Nambudiri and Gopakumar, 1990)
Ca^{+2} ATPase	Quality of surimi	(Katoh et al., 1979)
Lactate dehydrogenase	Quality of iced and frozen fish	(Nambudiri and Gopakumar, 1990)
Sarcoplasmic Mg^{+2} ATPase	Freshness during ice storage	(Yamanaka and Mackie, 1971)
Transglutaminase	Gel strength of surimi	(An et al., 1996)
Various lysosomal and mitochondrial enzymes	Freeze-thaw history of fish and shellfish	(Chhatbar and Velankar, 1977)

Table 13.7   Exogenous enzymes used as fish quality indices

Enzyme	Measurement	Reference
5′-Nucleotidase, nucleoside phosphorylase, xanthine oxidase	K-value	(Lou, 1998)
Alcohol dehydrogenase	Ethanol in canned fish	(Gill, 1990)
Diamine oxidase, peroxidase	Histamine	(Lopez-Sabater et al., 1993)
ELISA	Species identification	(Carrera, 1996)
Glutamate dehydrogenase	Ammonia	(Huss, 1995)
Octopine dehydrogenase, pyruvate oxidase	Scallop freshness	(Shin et al., 1998)
Purine nucleoside phosphorylase, xanthine oxidase	ATPase	(Cheun et al., 1996)
Putrescine oxidase	Polyamines	(Chemnitius et al., 1992)
Trimethylamine dehydrogenase	Trimethylamine	(Wong and Gill, 1987)
Urease	Urea	(Sheppard et al., 1996)
Xanthine oxidase	Hypoxanthine	(Shen et al., 1996)

enzymes as re-useable components of biosensors. The hypoxanthine biosensor developed by Watanabe et al. can be used for more than 100 assays and is stable for one month at 5°C (Watanabe et al., 1983). The next level of sophistication in biosensor technology might be the ability to measure a wide spectrum of freshness indices at the same time and a reference data base for these indices for individual species.

13.9.3   Enzymes as by-products from seafood processing wastes

Although most industrial enzymes are obtained from microorganisms, plants and terrestrial animal by-products there is a growing market for specialty enzymes from aquatic organisms (Haard, 1998). Homologous enzymes from aquatic organisms may be advantageous for selective niche applications over conventional sources because differences in catalytic and physio-chemical properties. For example, some differences between trypsin from bovine aquatic organisms are summarized in Table 13.8. The key properties of homologous enzymes from coldwater aquatic organisms that have been exploited are lower thermal stability and higher molecular activity at low reaction temperatures (Simpson and Haard, 1987a). Thus, such enzymes can be used at reduced concentration at low reaction temperatures compared to sources from warm-blooded organisms and can be readily heat inactivated at moderate temperatures thus avoiding side reactions sensitive to high temperatures. Some examples of the use of cold-adapted enzymes from aquatic organisms are summarized in Table 13.9. Various methods have been used to economically recover enzymes

Table 13.8   Comparison of some properties of terrestrial and aquatic trypsins[1]

Property	Bovine	Aquatic
Acid pH stability	Very high	Very low
Alkaline pH stability	Moderate	High
Catalytic specificity	Same	Same
Michaelis constant	Lower	Higher
Molecular activity at 0–5°C	Low	High
pH optimum	Same	Same
Soybean inhibitor sensitivity	Moderate	Very high
Temperature coefficient	2.0	1.4–1.6
Thermal denaturation	80°C	35–45°C

[1] Comparisons are mostly representative of data for trypsin from coldwater aquatic organisms.

from fish processing wastes including ensilage, membrane technology, ohmic heating, selective heat denaturation, aqueous two-phase systems, and scale-up chromatography (An and Visessanguan, 2000).

Table 13.9   Applications of cold adapted enzymes from aquatic organisms

Enzyme	Source	Application	Reference
Alkaline phosphatase	Northern pink shrimp	r-DNA kit	(Olsen et al., 1990)
Aminopeptidases	Illex Squid	Cheddar cheese ripening	(Raksakulthai et al., 2001)
Collagenase	Crab	Caviar production	(Gildberg, 1993)
Digestive proteases	Illex squid	Acceleration of fish sauce	(Raksakulthai et al., 1986)
Lipases	Atlantic cod	Oil modification	(Lie and Lambertson, 1985)
Lysozyme	Arctic scallop	Anti-microbial	(Myrnes and Johansen, 1994)
Pepsin	Atlantic cod, Arctic cod	Cold renneting	(Brewer et al., 1984)
Pepsin	Atlantic cod	Skin removal	(Haard, 1992c)
Trypsin	Greenland cod	Prevention of oxidized flavor in milk	(Simpson and Haard, 1984)
Trypsin	Greenland cod	Acceleration of herring fermentation	(Simpson and Haard, 1984)
Trypsin	Atlantic cod	Carotenoprotein isolation from shrimp waste	(Cano-Lopez et al., 1987)
Trypsin/ chymotrypsin	Turbot	Shrimp flavor extract	(Gildberg et al., 2000)

13.10   References

ABE, H. and OKUMA, E. (1991) Rigor-mortis progress of carp acclimated to different water temperatures. *Nippon Suisan Gakk*, **57**, 2095–100.

AL-SHAIBANI, K. A. (1977) Purification of metmyoglobin reductase from bluefin tuna. *J Food Sci*, **42**, 1013.

ALMY, L. H. (1926) The role of the proteolytic enzymes in the decomposition of the herring. *J Am Chem Soc*, **48**, 2136–46.

AN, H., PETERS, M. Y. and SEYMOUR, T. A. (1996) Roles of endogenous enzymes in surimi gelation. *Trends Food sci Technol*, **7**, 321–7.

AN, H. and VISESSANGUAN, W. (2000) Recovery of enzymes from seafood-processing wastes. In *Seafood Enzymes* (eds, Haard, N. F. and Simpson, B. K.) Marcel Dekker, Inc., New York, pp. 641–64.

AN, H., WEERASINGHE, V. and SEYMOUR, T. (1994) Cathepsin degradation of Pacific whiting surimi proteins. *J Food Sci*, **59**, 1013–17.

ANG, J. and HAARD, N. F. (1985) Chemical composition and postmortem changes in soft textured muscle from intensely feeding Atlantic cod, *Gadus morhua*. *J Food Biochemistry*, **9**, 49–64.

ARIHARA, K., ITOH, M. and KONDO, Y. (1991) Biochemical basis for meat color control systems. *Rakuno Kagaku Shokuhin no Kenkyu*, **40**, 317–22.

ASHIE, I. N., SIMPSON, B. K. and SMITH, J. P. (1996) Spoilage and shelf life extension of fresh fish and shellfish. *Crit Rev Food Sci Nutr*, **36**, 87–121.

ASHIE, I. N. A. and LANIER, T. C. (2000) Transglutaminases in seafood processing. In *Seafood Enzymes* (eds, Haard, N. F. and Simpson, B. K.) Marcel Dekker, Inc., New York, pp. 147–90.

BAIROCH, A. (2000) The enzyme database in 2000 *Nucleic Acids Res.*, **28**, 304–5.

BARZANA, E. and GARCIA-GARIBAY, M. (1994) Production of fish protein concentrates. In *Fisheries Processing: Biotechnological Applications* (ed. Martin, A. M.) Chapman & Hall, London, pp. 206–22.

BOON, D. D. (1975) Discoloration in processed crab meat. A review. *J Food Sci*, **40**, 756–61.

BRACHO, G. and HAARD, N. F. (1995) Identification of two matrix metalloproteinases in the skeletal muscle of Pacific rockfish (*Sebastes sp.*). *J Food Biochemistry*, **19**, 299–319.

BREMNER, H. A. (1992) Fish flesh structure and the role of collagen-its postmortem aspects and implications for fish processing. In *Quality Assurance in the Fish Industry* (eds, Huss, H. H., Jacobsen, M. and Liston, J.) Elsevier, London, pp. 39–62.

BREWER, P., HELBIG, N. and N.F., H. (1984) Atlantic cod pepsin. Characterization and use as a rennet substitute. *Can Inst Food Sci Technol J*, **17**, 38–43.

BURT, J. R. (1966) Glycogenolytic enzymes of cod (*Gadus callaria*) muscle. *J Fish Res Bd Can*, **23**, 527.

BURT, J. R. (1971) Changes in sugar phospate and lactate concentration in trawled cod (*Gadus callarias*) muscle during frozen storage. *J Sci Food Agric*, **22**, 536–9.

CADWALLADER, K. R. (2000) Enzymes and flavor biogenesis in fish. In *Seafood Enzymes* (eds, Haard, N. F. and Simpson, B. K.) Marcel Dekker, Inc., New York, pp. 365–83.

CANO-LOPEZ, A., SIMPSON, B. K. and HAARD, N. F. (1987) Extraction of carotenoprotein from shrimp process waste with the aid of trypsin from Atlantic cod. *J Food Sci*, **52**, 503–4.

CANTONI, G. L. and ANDERSON, D. G. (1956) Enzymatic cleavage of dimethylpropriothetin by *Polysiphonia lanosa*. *J Biol Chem*, **222**, 171–7.

CARRERA, E. (1996) Development of an enzyme-linked immunosorbent assay for the identification of smoked salmon (*Salmo salar*), trout (*Oncrhynchus mykiss*) and bream (*Brama raii*). *J Food Protect*, **59**, 521–4.

CASTELL, C. H., SMITH, B. and NEAL, W. (1971) Production of dimethylamine in muscle of several species of gadoid fish during frozen storage, especially in relation to presence of dark muscle. *J Fish Res Board Can*, **28**, 1–5.

CEPEDA, R., CHOU, E., BRACHO, G. and HAARD, N. F. (1990) An immunological method for measuring collagen degradation in muscle of fish. In *Advances in Fisheries Technology and Biotechnology for Increased Profitability* (eds, Voigt, M. N. and Botta, J. R.) Technomic Publishing Co., Lancaster, pp. 487–506.

CHAN, K. M. and DEKKER, E. A. (1994) Endogenous skeletal muscle antioxidants. *Crit Rev Food Sci Nutri*, **34**, 403–26.

CHANG-LEE, M. V., PACHECO-AGUILAR, R., CRAWFORD, D. L. and LAMPILA, L. E. (1989) Proteolytic activity of surimi from Pacific whiting (*Merluccius productus*) and heat-set gel texture. *J Food Sci*, **54**.

CHEMNITIUS, G. C., SUZUKI, M., ISOBU, K., KIMURA, J., KARUBE, I. and SCHMID, R. D. (1992) Thin-film polyamine biosensor: Substrate specificty and application to fish freshness determination. *Anal Chimica Acta*, **263**, 93–100.

CHEUN, B., ENDO, H., HAYSHI, T. and WATANABE, E. (1996) Development of a sensor for ATPase activity. *Fisheries Sci Tokyo*, **62**, 950–4.

CHHATBAR, S. K. and VELENKAR, N. K. (1977) A biochemical test for the distinction of fresh fish from frozen and thawed fish. *Fish Technol*, **14**, 131–6.

CHIHARA, M. (1982) *Common Seaweeds of Japan in Color,* Hoikusha Publishing Co., Ltd., Osaka.

CLARK, G. J. and WARD, D. R. (1988) Purification and properties of trimethylamine N-oxide reductase from *Schewanella* sp. NCMB 400. *J Gen Microbiol*, **133**, 379–86.

DE VECCHI, S. and COPPES, Z. (1996) Marine fish digestive enzymes. Relevance to food industry and the South-West Atlantic region – a review. *J Food Biochemistry*, **20**, 193–214.

DEVIDO DE MATTIO, N., PAREDI, M. E. and CRUPKIN, M. (1992) Post mortem changes in glycogen, ATP, hypoxanthine and 260/250 absorbance ratio in extracts of adductor muscles from *Aulacomya ater ater* (Molina) at different biological conditions. *Comp Biochem Physiol*, **103A**, 605–8.

DIAZ-LOPEZ, M. and GARCIA-CARRENO, F. L. (2000) Applications of fish and shellfish enzymes in food and feed products. In *Seafood Enzymes* (eds, Haard, N. F. and Simpson, B. K.) Marcel Dekker, Inc., New York, pp. 571–618.

FEHNERLING, G. B. (1970) Separation of edible tissue from edible flesh of marine creatures. US Patent No 3513071.

FIELD, C. E., PIVARNIK, L. F., BARNETT, S. M. and RAND, A. G. (1986) Utilization of glucose oxidase for extending the shelf life of fish. *J Food Sci*, **51**, 66–70.

FINNE, G. (1982) Enzymatic ammonia production in penaied shrimp held on ice. In *Chemistry and Biochemistry of Marine Food Products* (eds, Flick, G. E., Hebard, C. E. and Ward, D. W.) AVI, Westport, pp. 323–31.

FINNE, G. (1992) Non-protein nitrogen compounds in fish and shellfish. In *Advances in Seafood Biochemistry* (eds, Flick, G. J. and Martin, R. E.) Technomic Publishing Co., Lancaster, pp. 393–401.

FOEGEDING, E. A., LANIER, T. C. and HULTIN, H. O. (1996) Characteristics of edible muscle tissues. In *Food Chemistry* (ed. Fennema, O. R.) Marcel Dekker, Inc., New York, pp. 879–942.

FOX, D. L. (1979) *Biochromy: Natural Coloration of Living Things.*, University of California Press, Berkeley.

FRASER, D. I., WEINSTEIN, W. J. and DYER, W. J. (1965) Post-mortem glycolytic and associated changes in the muscle of trap- and trawl-caught cod. *J Fish Res Bd Can*, **22**, 83–100.

FUJIMOTO, K. and KANEDA, T. (1973a) Studies on the brown discoloration of fish products. IV. Nitrogen content of the browning substances. *Bull Jpn Soc Sci Fish*, **39**, 179–83.

FUJIMOTO, K. and KANEDA, T. (1973b) Studies on the brown discoloration of fish products. V. Reaction mechanism in the early stage. *Bull Jpn Soc Sci Fish*, **39**, 185–90.

FUJIMURA, T. and KAWAI, T. (2000) Enzymes and seaweed flavor. In *Seafood Enzymes* (eds, Haard, N. F. and Simpson, B. K.) Marcel Dekker, Inc., New York, pp. 385–409.

FUJITA, Y., MATSUMOTO, T. and MATSUDA, T. (1978) On the browning discoloration in fish jelly products. VII. The mechanism of browning precursor production by *Enterobacter cloacae* UFF-107. *Bull Jpn Soc Sci Fish*, **44**, 643–51.

GHOSH, S. (1989) A simple method to remove urea from shark. *J Food Sci Technol*, **26**, 164–5.

GILDBERG, A. (1993) Enzymic processing of marine raw materials. *Process Biochem*, **28**, 1–15.

GILDBERG, A., BATISTA, I. and STROM, E. (1989). Preparation and characterization of peptones obtained by two-step enzymatic hydrolysis of whole fish. *Biotechnology and Applied Biochemistry*, **11**, 413–23.

GILDBERG, A., SIMPSON, B. K. and HAARD, N. F. (2000) Uses of enzymes from marine organisms. In *Seafood Enzymes* (eds, Haard, N. F. and Simpson, B. K.) Marcel Dekker, Inc., New York, pp. 619–39.

GILL, T. (2000) Nucleotide-degrading enzymes. In *Seafood Enzymes* (eds, Haard, N. F. and Simpson, B. K.) Marcel Dekker, Inc., New York, pp. 37–68.

GILL, T. A. (1990) Objective analysis of seafood quality. *Food Rev Int*, **6**, 681–714.

GOPAKUMAR, K. (2000) Enzymes and enzyme products as quality indices. In *Seafood Enzymes* (eds, Haard, N. F. and Simpson, B. K.) Marcel Dekker, Inc., New York, pp. 337–3.

GORA, A. (1972) The technology of light salting of herring and mackerel and of enzymatically salted herring fillets. Sea Fisheries Institute, Gdynia, pp. 1–37.

GREER-WALKER, M. (1970) Growth and development of the skeletal muscle fibers of the cod (*Gadus morhua*, L.). *J. Cons. Int. Explor. Mer.*, **33**, 228–44.

HAARD, N. F. (1987) Protein and non-protein constituents in jellied American plaice, *Hippoglossoides platessoides*. *Can Inst Food Sci and Technol J*, **20**, 98–101.

HAARD, N. F. (1990) Enzymes from food myosystems. *J Muscle Foods*, **1**, 293–338.

HAARD, N. F. (1992a) Biochemical reactions in fish muscle during frozen storage. In *Seafood Science and Technology* (ed. Bligh, E. G.) Fishing News Books, Oxford, pp. 176–209.

HAARD, N. F. (1992b) Biochemistry and chemistry of color and color changes in seafoods. In *Advances in Seafood Biochemistry* (eds, Flick, G. J. and Martin, R. E.) Technomic Publishing Co., Lancaster, pp. 305–60.

HAARD, N. F. (1992c) A review of proteolytic enzymes from marine organisms and their application in the food industry. *J Aquatic Food Product Technol*, **1**, 17-35.

HAARD, N. F. (1992d) Technological aspects of extending prime quality of seafoods: a review. *J Aquatic Food Product Technol*, **1**, 9–27.

HAARD, N. F. (1994) Protein hydrolysis in seafood. In *Seafoods: Chemistry, Processing Technology and Quality* (eds, Shahidi, F. and Botta, R.) Chapman and Hall, London, pp. 10–33.

HAARD, N. F. (1998) Specialty enzymes from marine organisms. *Food Tech*, **57**, 64–7.

HAARD, N. F. (2000) Seafood enzymes: The role of adaptation and other intraspecific factors. In *Seafood Enzymes* (eds, Haard, N. F. and Simpson, B. K.) Marcel Dekker, Inc, New York, NY, pp. 1–36.

HAARD, N. F.. (2001) Enzymic modification of proteins. In *The Chemical and Functional Properties of Food Proteins* (ed. Sikorski, Z.) Technomic Publishing Co., Lancaster, pp. 155–90.

HAARD, N. F. and ARCILLA, R. (1985) Precursors of Maillard browning in dehydrated squid, *Illex illecebrosus*. *Can Inst Food Sci Technol J*, **18**, 326–31.

HAARD, N. F. and SIMPSON, B. K. (1985) Cold-adapted enzymes from fish. In *Food Biotechnology* (ed. Knorr, D.) Marcel Dekker, Inc., New York, pp. 495–527.

HAARD, N. F. and SIMPSON, B. K. (1994) Proteases from aquatic organisms and their uses in the seafood industry. In *Fisheries Processing: Biotechnological applications* (ed. Martin, A. M.) Chapman and Hall, London, pp. 132–54.

HILTZ, D. F., BISHOP, L. J. and DYER, W. J. (1973) Accelerated nucleotide degradation and glycolysis during warming to and subsequent storage at $-5°C$ in pre-rigor cod muscle frozen at various rates. *J Fish Res Bd Can*, **31**, 1181–7.

HUŁTIN, H. O. (1992) Biochemical deterioration of fish flesh. In *Quality Assurance in the Fish Industry* (eds, Huss, H. H., Jakobsen, M. and Liston, J.) Elsevier, Oxford, pp. 125–38.

HUSS, H. H. (1995) *Quality and quality changes in fresh fish* FAO, Rome.

HUYNH, M. D., MACKEY, J. and GAWLEY, R. (1992) Freshness assessment of Pacific fish species using K-value. In *Seafood Science and Technology* (ed. Bligh, E. G.) Fishing News Books, Oxford, pp. 258–68.

IIDA, H., NAKAMURA, K. and TOKUNAGA, T. (1985) Dimethyl sulfide and dimethyl-β-propriothetin in sea algae. *Bull Japn Soc Sci Fish*, **51**, 1145–50.

IN, T. (1990) Seafood flavourants produced by enzymatic hydrolysis. In *Advances in Fisheries Technology and Biotechnology for Increased Profitability* (eds, Voigt, M. N. and Botta, J. R.) Technomic Publishing, Lancaster, pp. 425–36.

ITOH, K., TOYOHARA, H., ANDO, M. and SAKAGUCHI, M. (1992) Disintegration of the pericellular connective tissue of ayu muscle in the spawning season relevant to softening. *Nippon Suisan Gakkaishi*, **58**, 1553.

IWAMOTO, M., YAMNAKA, H., WATABE, S. and HASHIMOTO, K. (1991) Changes in ATP and related breakdown compounds in the adductor muscle of Itayagai scallop *Pecten albicans* during storage at various temperatures. *Nippon Suisan Gakk*, **57**, 153–6.

IWATA, K., KOBASHI, K. and HASE, J. (1973) Studies on muscle alkaline protease. I. Isolation, purification and some physico-chemical properties of an alkaline protease from carp muscle. *Bull Jpn Soc Sci Fish*, **39**, 1325–37.

IZQUIERDO-PULIDO, M. L., HAARD, T. A., HUNG, J. and HAARD, N. F. (1994) Oryzacystatin and other proteinase inhibitors in rice grain: Potential use as fish processing aid. *J Agric Food Chem*, **42**, 616–22.

IZQUIERDO-PULIDO, M., HATAE, K. and HAARD, N. F. (1992) Nucleotide catabolism and changes in texture during ice storage of cultured sturgeon, *Acipenser transmontanus*. *J Food Biochemistry*, **16**, 173–92.

JACOBSEN, F. and RASMUSSEN, O. L. (1984) Energy savings through enzymatic treatment of stickwater in the fish meal industry. *Process Biochem*, **19**, 165–9.

JIANG, S.-T. (2000) Enzymes and their effects on seafood texture. In *Seafood Enzymes* (eds, Haard, N. F. and Simpson, B. K.) Marcel Dekker, Inc., New York, pp. 411–50.

JIANG, S. T., LEE, J. J. and CHEN, H. C. (1997) Mackerel cathepsin B and L effects on thermal degradation of surimi. *J Food Sci*, **62**, 1–6.

JOSEPHSON, D. B. and LINDSAY, R. C. (1986) Enzymatic generation of volatile aroma compounds from fresh fish. In *Biogeneration of Aromas* (eds, Parliment, T. H. and Croteau, R.) American Chem Soc, Washington, D.C., pp. 201–21.

KAJIWARA, T., MATSUI, K. and AKAKABE, Y. (1996) Biogeneration of volatile compounds via oxylipins in edible seaweeds. In *Biotechnology for Improved Foods and Flavors*, Vol. 637 (eds, Takeoka, G. R., Teranishi, R. and Williams, P. J.) American Chemical Society, Washington, D.C., pp. 146–66.

KANG, I.-S. and LANIER, T. C. (2000) Heat-induced softening of surimi gels by proteinases. In *Surimi and Surimi Seafood* (ed. Park, J. W.) Marcel Dekker, Inc., New York, pp. 445–74.

KANNER, J. and KINSELLA, J. E. (1983) Lipid deterioration initiated by phagocytic cells in muscle foods; beta carotene destruction by myeloperoxidase-halide system. *J Agric Food Chem*, **31**, 370–6.

KARUBE, H., MATSUOKA, H., SUZUKI, S., WATANABE, E. and TOYAMA, K. (1984) Determination of fish freshness with an enzyme sensor. *J Agric Food Chem*, **32**, 314–19.

KATOH, N., NOZAKI, N., KOMATSU, K. and ARAI, K. (1979) A new method for evaluation of quality of frozen surimi from Alaska pollack. Relationship between myofibrillar ATPase activity and kamaboko forming ability of frozen surimi. *Bull Jpn Soc Sci Fish*, **48**, 1027–32.

KIM, J., MARSHALL, M. R. and WEI, C.-I. (2000) Polyphenoloxidase. In *Seafood Enzymes* (eds, Haard, N. F. and Simpson, B.K.) Marcel Dekker, Inc., New York, pp. 271–315.

KIM, S. K., JEON, Y. J., BYENN, H. G., KIM, Y. T. and LEE, C. K. (1997) Enzymatic recovery of cod frame proteins with crude proteinases from tuna pyloric caeca. *Fisheries Sci (Tokyo)*, **63**, 421–7.

KINOSHITA, M., TOYOHARA, H. and SCHIMIZY, Y. (1990) Diverse distribution of four distinct types of modori (gel degrading)-inducing proteinases among fish species. *Nippoon Suisan Gakkaishi*, **56**, 1485–92.

KOCZOT, A. B., , SZEWCZUK, M. A. and MALDER, J. (1985) Protecting fish against color changes during refrigeration *Poland*, pp. 2.

KOLODZIEJSKA, I., PACANA, J. and SIKORSKI, Z. E. (1992) Effect of squid liver extract on proteins and on cooked squid mantle. *J Food Biochemistry*, **16**, 141–50.

KOLODZIEJSKA, I. and SIKORSKI, Z. E. (1995) The properties and utilization of proteases of marine fish and invertebrates. *Pol J Food Nutr Sci*, **45**, 5–12.

LEAKE, L. and KAREL, M. (1985) Nature of fluorescent compounds generated by exposure of proteins to oxidized lipids. *J Food Biochemistry*, **9**, 117.

LEE, Y. Z., SIMPSON, B. K. and HAARD, N. F. (1982) Supplementation of squid fermentation with proteolytic enzymes. *J Food Biochemistry*, **6**, 127–34.

LIE, O. and LAMBERTSON, G. (1985) Digestive enzymes in cod (*Gadus morhua*): Fatty acid specificity. *Comp Biochem Physiol*, **80B**, 447–50.

LINDSAY, R. C. (1990) Fish flavors. *Food Rev Int*, **6**, 437–55.

LOPEZ-AMAYA, C. and MARANGONI, A. G. (2000a) Lipases. In *Seafood Enzymes* (eds, Haard, N. F. and Simpson, B. K.) Marcel Dekker, Inc., New York, pp. 121–46.

LOPEZ-AMAYA, C. and MARANGONI, A. G. (2000b) Phospholipases. In *Seafood Enzymes* (eds, Haard, N. F. and Simpson, B. K.) Marcel Dekker, Inc., New York, pp. 91–119.

LOPEZ-SABATER, E. I., RODRIGUEZ-JEREZ, J. J., ROIG-SAQUES, A. X. and MORA-VENTURA, M. T. (1993) Determination of histamine in fish using enzymic method. *Food Additives and Contaminants*, **19**, 593–602.

LOU, S. (1998) Purine content in grass shrimp during storage as related to freshness. *J Food Sci*, **63**, 442–9.

LOWE, T., RYDER, J. M., CARREGER, J. F. and WELLS, R. M. G. (1993) Flesh quality in snapper, *Pagrus auratus*, affected by capture stress. *J Food Sci*, **58**, 770–3, 796.

MACDOUGALL, D. B. (1982) Changes in colour and opacity of meat *Food Chem*, **9**, 75.

MALE, R., LORENS, J., SMALAS, A. and TORRISEN, K. (1995) Molecular cloning and characterization of anionic and cationic variants of trypsin from Atlantic salmon. *Eur J Biochem*, **232**, 677–85.

MATSUI, T., SHIMIZU, C. and MATSUURA, F. (1975) Studies on metmyoglobin reducing systems in the muscle of blue white dolphin. II. Purification and some physico-chemical properties of ferrimyoglobin reductase. *Bull Jpn Soc Sci Fish*, **41**, 771.

MIZUTA, S., YOSHINAKA, R., SATO, M. and SAKAGUCHI, M. (1997) Histological and biochemical changes of muscle collagen during chilled storage of the kuruma prawn *Panaeus japonicus*. *Fish Sci*, **63**, 784–95.

MOCHIZUKI, S. and SATO, A. (1996) Effects of various killing procedures on post-mortem changes in the muscle of chub mackeral and round scad. *Nippon Suisan Gakk*, **62**, 453–7.

MOTOHITO, T. (1962) Studies on the petroleum odor in canned chum salmon. *Mem Fac Fisheries, Hokkaido University*, **10**, 1–5.

MYRNES, B. and JOHANSEN, A. (1994) Recovery of lysozyme from scallop waste. *Prep Biochem*, **24**, 69–80.

NAKAYAMA, H., BITO, M. and YOKOSEKI, M. (1976) Orange discoloration of canned skipjack. VIII. Effects of G-6-P accumulation and raw material treatments and storage conditions. *Bull Jpn Soc Sci Fish*, **42**, 1299–308.

NAMBUDIRI, D. O. and GOPAKUMAR, K. (1990) *Effect of freezing and thawing on press juice and enzyme activity in the muscle of farmed fish and shellfish*, I.I.F.-I.I.R Commission, Aberdeen.

NIP, W. K. and MOY, J. H. (1988) Microstructural changes of ice-chilled and cooked freshwater prawn, *Macrobrachium rosenbergii*. *J Food Sci*, **53**, 319–22.

NORQVIST, A., NORMAN, B. and WOLF-WATZ, H. (1990) Identification and characterization of a zinc metalloproteinase associated with invasion by the fish pathogen *Vibrio anguiillarum*. *Infect Immun*, **58**, 3731–6.

OLSEN, R. L., JOHANSEN, A. and MYRNES, B. (1990) Recovery of enzymes from shrimp waste. *Process Biochem*, **25**, 67–8.

PONG, C.-Y., CHIOU, T.-K., NIEH, F.-P. and JIANG, S.-T. (2000) Purification and characterization of metmyoglobin reductase from ordinary muscle of blue-fin tuna. *Fish Sci*, **66**, 599–604.

RAA, J. (1997) New commercial products based on waste from the fish processing industry. In *Making the Most of the Catch* (eds, Bremner, A., Davis, C. and Austin, B.) AUSEAS, Brisbane, pp. 1–4.

RAKSAKULTHAI, N., LEE, Y. Z. and HAARD, N. F. (1986) Effect of enzyme supplements on the production of fish sauce from male capelin (*Mallotus villosus*). *Can Inst Food Sci Technol J*, **19**, 28–33.

RAKSAKULTHAI, R., ROSENBERG, M. and HAARD, N. F. (2001) Accelerated Cheddar cheese ripening with an aminopeptidase fraction from squid hepato-pancreas. *J Food Sci*, in press.

RAMESH, C. and LEWIS, N. F. (1980) Effect of lysozyme and sodium EDTA on shrimp microflora. *J Appl Microbiol Biotechnol*, **10**, 253–8.

REECE, P. (1988) Recovery of proteinase from fish waste. *Process Biochem*, **6**, 62–6.

RUDNEEVA-TITOVA, I. (1997) Ecological and phytogenetic features of some antioxidant enzymes activity and antioxidant content in Black Sea cartilaginous and teleost fish. *Z Evol Biokh Fiziol*, **33**, 29–37.

SACKTON, J. (1986) *The Seafood Handbook*, Seafood Business, Seattle.

SAGEDHAL, A., BRUSALMEN, J. P., ROLDMAN, H. A., PAREDI, M. E. and CRUPKIN, M. (1997) Post-mortem changes in adenosine triphosphate and related compounds in mantle of squid (*Illex argentinus*) at different stages of sexual maturation. *J Aquat Food Prod Technol*, **6**, 43–56.

SAITO, M., KUNISAKI, N., URANO, N. and KIMURA, S. (2000a) Characterization of cDNA clone encoding the matrix metalloproteinase 2 from rainbow trout fibroblast. *Fisheries Science*, **66**, 334–42.

SAITO, M., SATO, K., KUNISAKI, N. and KIMURA, S. (2000b) Characterization of a rainbow trout matrix metalloproteinase capable of degrading type I collagen. *Eur J Biochem*, **267**, 6943–50.

SAITO, T., ARAI, K. and MATSUYOSHI M. (1959) A new method for estimating the freshness of fish. *Bull Jpn Soc Sci Fish*, **24**.

SATO, K., ANDO, M. and KUBATA, S. (1997) Involvement of type V collagen in softening of fish muscle during short-term chilled storage. *J Agric Food Chem*, **45**, 343–8.

SATO, K., KOIKE, A., YOSHINAKA, R., SATO, M. and SHIMIZU, Y. (1994) Postmortem changes in type I and V collagens in myocommatal and endomysial fractions of rainbow trout (*Oncorhynchus mykiss*) muscle. *J Aquat Food Prod Technol*, **3**, 5–11.

SATO, K., OHASHI, C., OHTSUKI, K. and KAWABATA, M. (1991) Type V collagen in trout (*Salmo gairdneri*) muscle and its solubility change during chilled storage of muscle. *J Agric Food Chem*, **39**, 1222–5.

SAVAGAON, K. A. and SREENIVASAN, A. (1978) Activation mechanism of pre-polyphenoloxidase in lobster and shrimp. *Fish Technol*, **15**, 49–55.

SCHEWFELT, R. L. (1981) Fish muscle lipolysis- a review. *J Food Biochemistry*, **5**, 79–100.

SCHEWFELT, R. L. and HULTIN, H. O. (1983) Inhibition of enzymic, non-enzymic lipid peroxidation of flounder muscle sarcoplasmic reticulum by pretreatment with phospholipase A_2. *Biochim Biophys Acta*, **751**, 432–8.

SCOTT, D. (1975a) Application of glucose oxidase. In *Enzymes in Food Processing* (ed. Reed, G.) Academic Press, New York, pp. 519–47.

SCOTT, D. (1975b) Miscellaneous applications of enzymes. In *Enzymes in Food Processing* (ed. Reed, G.) Academic Press, New York, pp. 493–547.

SCOTT, O. N., FLETCHER, G. C., HOGG, M. G. and RYDER, J. M. (1986) Comparison of whole with headed and gutted orange roughy stored in ice: Sensory, microbiology and chemical assessment. *J Food Sci*, **51**, 79–83, 86.

SEKI, N., UNO, H., LEE, N. H., KIMURA, I., TOYODA, K., FUJITA, T. and ARAI, K. (1990) Transglutaminase activity in Alaska pollack muscle and surimi, and its reactions with myosin B. *Nippon Suisan Gakk*, **56**, 125–32.

SHEN, L.-Q., YANG, L.-J. and PENG, T.-Z. (1996) Amperometric determination of fish freshness by a hypoxanthine biosensor. *J Sci Food Agric*, **70**, 298–302.

SHEPPARD, N. F., MEANS, D. J. and GINSEPPI, E. A. (1996) Model of an immobilized enzyme conductimetric urea biosensor. *Biosensors & Bioelectronics*, **11**, 967-979.

SHIN, S. J., YAMARIAKA, H., ENDO, H. and WATANABE, E. (1998) Development of an octopine biosensor and its application to the estimation of scallop freshness. *Enz Microbiol Technol*, **23**, 10–13.

SIEBERT, G. (1958) Protein-splitting enzyme activity of fish flesh. *Experientia*, **14**, 65–66.

SIGHOLT, T., ERIKSON, U., RUSTAD, T., JOHANSEN, S., NORDVEDT, T. S. and SELAND, A. (1997) Handling stress and storage temperature affect meat quality of farm-raised Atlantic salmon (*Salmo salar*). *J Food Sci*, **62**, 895–905.

SIMPSON, B. K. (2000) Digestive proteinases from marine animals. In *Seafood Enzymes* (eds, Haard, N. F. and Simpson, B. K.) Marcel Dekker, Inc., New York, pp. 191–213.

SIMPSON, B. K. and HAARD, N. F. (1984) Trypsin from Greenland cod as a food-processing aid. *J Appl Biochem*, **6**, 135–43.

SIMPSON, B. K. and HAARD, N. F. (1985) Extraction of carotenoprotein from shrimp processing offal with the aid of trypsin. *J Appl Biochem*, **7**, 212–22.

SIMPSON, B. K. and HAARD, N. F. (1987a) Cold-adapted enzymes from fish. In *Food Biotechnology* (ed. Knorr, D.) Marcel Dekker, Inc., New York, pp. 495–572.

SIMPSON, B. K. and HAARD, N. F. (1987b) Trypsin and trypsin-like enzymes from stomachless cunner. Kinetic and physical properties. *J Agric Food Chem*, **35**, 652–6.

SIMPSON, B. K., SEQUEIRA-MUNOZ, A. and HAARD, N. F. (2000) Marine enzymes In

Encyclopedia of Food Science and Technology, Vol. 3 (ed. Francis, F. J.) John Wiley & Sons, Inc., New York, pp. 1525–34.

SOTELO, C. G. and REHBEIN, H. (2000) TMAO-degrading enzymes. In *Seafood Enzymes* (eds, Haard, N. F. and Simpson, B. K.) Marcel Dekker, Inc., New York, pp. 167–90.

STEFANSSON, G. (1988) Enzymes in the fishing industry. *Food Technol*, **42**, 64–6.

STEFANSSON, G. and STEINGRIMSDOTTIR, U. (1990) Applications of enzymes for fish processing in Iceland – present and future aspects. In *Advances in Fisheries Technology and Biotechnology for Increased Profitability* (eds, Voigt, M. N. and Botta, J. R.) Technomic Publishing, Lancaster, pp. 237–50.

SUGIHARA, T., YASHIMA, C., TAMURA, H., KAWASAKI, M. and SHIMIZU, S. (1973) Process for preparation of ikura (salmon egg) US No 3759718.

SUN PAN, B. and KUO, J.-M. (2000) Lipoxygenases. In *Seafood Enzymes* (eds, Haard, N. F. and Simpson, B. K.) Marcel Dekker, Inc., New York, pp. 317–36.

SURETTE, M., GILL, T. A. and LEBLANC, P. J. (1988) Biochemical basis of postmortem nucleotide catabolism in cod (*Gadus morhua*) and its relationship to spoilage. *J Agric Food Chem*, **36**, 19–22.

TARR, H. L. A. (1966) Post-mortem changes in glycogen, nucleotides, sugar phosphates, and sugars in fish muscle – a review. *J Food Sci*, **53**, 6–11.

TOCHER, D. R., WEBSTER, A. and SARGENT, J. R. (1986) Utilization of porcine pancreatic phospholipase A for the preparation of marine fish oil enriched in n-3 polyunsaturated fatty acids. *Biotechnol Appl Biochem*, **8**, 657–79.

TOYOMIZU, M., HANAOKA, K. and YAMAGUCHI, K. (1981) Effect of release of free fatty acids by enzmatic hydrolysis of phospholipids on oxidation during storage of fish muscle at −5°C. *Bull Jpn Soc Sci Fish*, **47**, 615–20.

TSUKUDA, N. (1970) Studies on the discoloration of red fishes. VI. Partial purification and specificity of the lipoxygenase-like enzyme responsible for carotenoid discoloration of fish skin during refrigerated storage. *Bull Jpn Soc Sci Fish*, **36**, 725–33.

TSUKUDA, N. and AMANO, K. (1966) Studies on the discoloration of red fishes II. The discoloration of three species during ice and freeze storage. *Bull Jpn Soc Sci Fish*, **32**, 522–9.

VENUGOPAL, V. (1990) Extracellular proteases of contaminant bacteria in fish spoilage: a review *J Food Protection*, **53**, 341–50.

VENUGOPAL, V., LAKSHMANAN, R., DOKE, S. N. and BONGIRWAR, D. R. (2000) Enzymes in fish processing, biosensors and quality control. *Food Biotechnology*, **14**, 21–77.

VILHELMSSON, O. (1997) The state of enzyme biotechnology in the fish processing industry. *Trends Food Sci Technol*, **8**, 266–70.

WANASUNDRA, U. N. and SHAHIDI, F. (1997) Lipase assisted concentration of w-3 polyunsaturated fatty acids in acylglycerols from marine oils. *J Am Oil Chem Soc*, **75**, 945–51.

WANG, J. H., MA, W. C., SU, J. C., CHEN, C. S. and JIANG, S. T. (1993) Comparison of the properties of m-calpain from tilapia and grass shrimp muscles. *J Agric Food Chem*, **41**, 1379–84.

WATABE, S., KAMAL, M. and HASHIMOTO, K. (1991) Postmortem changes in ATP, creatine phosphate, and lactate in sardine muscle. *J Food Sci*, **56**, 151–3.

WATANABE, E., ANDO, K., KARUBE, I., MATSUOKA, H. and SUZUKI, S. (1983) Determination of hypoxanthine in fish meat with an enzyme sensor. *J Food Sci*, **48**, 496–500.

WATSON, C., BOURKE, R. E. and BRILL, R. W. (1988) A comprehensive theory on the etiology of burned tuna. *Fish Bull*, **86**, 367–72.

WINTERHALTER, P. (1996) Carotenoid-derived aroma compounds: biogenic and biotechnological aspects. In *Biotechnology for Improved Foods and Flavors*, Vol. 637 (eds, Takeoka, G. R., Teranishi, R., Williams, P. J. and Kobayashi, A.) American Chemical Society, Washington, D.C., pp. 295–308.

WONG, K. and GILL, T. A. (1987) Enzymatic determination of trimethylamine and its relationship to fish quality. *J Food Sci*, **52**, 1–3, 6.

YAMADA, N., ITOH, N., MURAKAMI, S. and IZUMI, Y. (1985) New bromoperoxidase from coralline algae that brominates phenol compounds. *Agric Biol Chem*, **49**, 2961–7.

YAMAHOTO, M. and MACKEY, J. (1981) An enzymatic method for reducing curd formation in canned salmon. *J Food Sci*, **46**, 656–7.

YAMANAKA, A. and MACKIE, I. M. (1971) Changes in the activity of a sarcoplasmic adenosine triphosphatase during iced-storage and frozen storage of cod. *Bull Jpn Soc Sci Fish*, **37**, 1105–9.

YAMASHITA, M. and KONAGAYA, S. (1991) Immunochemical localization of cathepsins B and L in the white muscle of chum salmon (*Oncorhynchus keta*) in spawning migration. Probable participation of phagocytes rich in cathepsins in extensive muscle softening of the mature salmon. *J Agric Food Chem*, **39**, 1402–5.

14

Understanding lipid oxidation in fish

I. P. Ashton, Unilever R&D, Sharnbrook

14.1 Introduction

Rancidity is a perennial problem in oily fish associated primarily with the frozen and dried storage state. Indeed, the shelf-life of frozen oily fish is usually terminated by the onset of rancid flavours. Although there are studies investigating off-lavour development during chill storage, this is not generally considered to be as much of a problem because the susceptibility of fish to microbiologically induced spoilage during extended chill storage will precede any perceivable changes caused by lipid oxidation. In canned fish, the total elimination of oxygen during processing is sufficient to give these products a shelf-life of many years.

This chapter discusses the factors affecting the development of rancid off-flavours in fish, particularly on frozen storage, and the measures that are reported to reduce the problem. The approaches discussed include the addition of antioxidants and chelators, the rise in interest in natural herb extracts, their efficacy and the legislation involved in the addition of antioxidants to fish products. The storage stability of products with elevated antioxidant content achieved via dietary supplementation (particularly tocopherol), modified atmosphere and vacuum packaging are also discussed.

In the living animal the ingestion and regeneration of antioxidants prevents excessive oxidative deterioration of important biological components. Post-mortem, the protective systems become depleted and are unable to regenerate. Thus the edible muscle tissues of fish are liable to react with oxygen when exposed to air (oxidation). Oxygen may react with many of the biochemical components of animal tissues including lipids. It is the oxidation of polyunsaturated fatty acids (PUFAs)-containing lipids that causes the development of off-flavours and aromas, often referred to as 'rancidity'. The primary products of fatty acid oxidation (lipid hydroperoxides) are generally considered not to have a flavour impact. The compounds giving rise to rancid flavours and aromas are volatile secondary oxidation products derived from the breakdown of these lipid hydroperoxides.

The main intrinsic factors that will determine the rate and extent of rancidity development in fish are:

- lipid level and fatty acid composition of the lipids
- levels of endogenous antioxidants and endogenous oxidative catalysts.

External influencing factors include:

- oxygen concentration
- surface area exposed to atmospheric oxygen
- storage temperature
- processing procedures that lead to tissue damage.

The levels of lipid in fish flesh vary depending on species, ranging from lean fish (<2% total lipid) such as cod, haddock and pollack, to high lipid species (8–20% total lipid) such as herring, mackerel and farmed salmon. A total lipid level of 5% has been suggested as a cut-off point between low and medium fat fish. In addition to species variability, lipid levels vary with sex, diet, seasonal fluctuation and tissue. For example the dark muscle of Mackerel was found to contain 20% lipid in comparison with 4% in the white muscle,[1] while Atlantic herring can have seasonal variation of 1–25% total fat.

It is well established that oily fish are particularly susceptible to lipid oxidation and rancidity development because of the high content of PUFAs in their lipid, particularly the nutritionally important *n-3* fatty acids eicosapentaenoic acid (20:5 n-3) (EPA) and docosahexaenoic acid (22:6 n-3) (DHA). Again, the fatty acid profile of fish varies quite considerably between and within species and are also influenced by the factors already mentioned.

Within edible muscle tissues there are two pools of lipids that may be involved in oxidation reactions. These are storage triacyl glycerides (TAGs), which constitute the majority of the total lipid in oily fish and the fraction that accounts for the variability in fat content, and phospholipids present in the cellular membranes. This lipid fraction represents around 1% of the total weight of the flesh. The contribution of these lipid pools to rancidity development is a matter of some debate. Although present in much smaller amounts, the physical form of phospholipids in the membrane bilayer present a large surface area for oxidation reactions to occur compared with 'droplets' of storage TAG present in the muscle associated storage depots of oily fish. The amount of phospholipid substrate available for oxidation by catalysts is proportional to the surface area exposed to the aqueous phase.[2] Potentially phospholipids are more important in the initial onset of lipid oxidation. Phospholipids also tend to contain a higher proportion of unsaturated fatty acids than TAG. It is interesting to note that it is generally accepted that white lean fish such as cod are not prone to rancidity development on storage even though they contain highly unsaturated phospholipid. These species contain virtually no TAG or myoglobin in the muscle tissue. This implies that one or both these components have important implications for storage stability.

The involvement of membranes as a principal substrate for lipid oxidation may also explain why dietary tocopherol supplementation offers such good protection for products on storage (see Section 14.5). This lipid soluble antioxidant is primarily involved in protecting membranes from oxidative damage. The structure of the tocopherol molecule is such that it is anchored in the hydrophobic part (fatty acyl chains) of the membrane bi-layer by its C13 chain while the chromanol nucleus lies at the membrane/water interface. It is the latter part of the molecule (the phenolic component) which is responsible for its antioxidant property. Indeed, a synergist effect has been reported between phospholipids and α-tocopherol in protecting lipids in a number of systems including sardine and mackerel lipids.[3]

14.2   The role of lipolysis in rancidity development

An increase in free fatty acid (FFA) (lipolysis) is a well established post-mortem feature in fish tissue[4,5] resulting from the enzymatic hydrolysis of esterified lipids.[6,7] The build up of FFA can be quite substantial. Ingemansson et al.[8] have shown that in frozen stored (−15°C) rainbow trout the FFAs increased after 34 weeks to between 5–15% of the total muscle lipids, but more so in the light muscle compared to dark. Tsukuda[9] investigated the change in lipids of skipjack tuna dark and light muscle tissue stored at −10, −20 and −30°C. This study showed that FFA increased in the dark and light muscle ca. 13- and 4-fold respectively after 80 days storage at −10°C with the total FFA content reaching more than 50% of the total lipid. This compared to 13% after 140 days at −30°C.

The important question with regard to the relationship between lipolysis and lipid oxidation/rancidity is, do free (polyunsaturated) fatty acids oxidise more readily than esterified lipid. There are a number of potential routes by which phospholipid and TAG can undergo oxidation (Fig. 14.1). However, the relationship between lipolysis and lipid oxidation is a matter of some debate. It is generally accepted that FFAs oxidise more readily than esterified FA, especially when enzymes such as lipoxygenase (LOX) can be involved in uncooked tissue.[10,11] Conceivably, it may be the rate-limiting step in fatty acid oxidation. However, there is evidence that indicates quite the contrary. In a study with isolated flounder muscle sarcoplasmic reticulum, preincubation with PLA2 prior to initiation of enzymatic and non-enzymatic oxidation caused a decrease in the production of lipid oxidation products. It was suggested that a structural rearrangement of the membrane fatty acids following hydrolysis caused a decrease in free radical chain propagation.[12]

The lipid origin of FFA has also been debated. Work by Dekoning et al.[13] showed that FFA formed during the storage of hake mince at −18°C originated

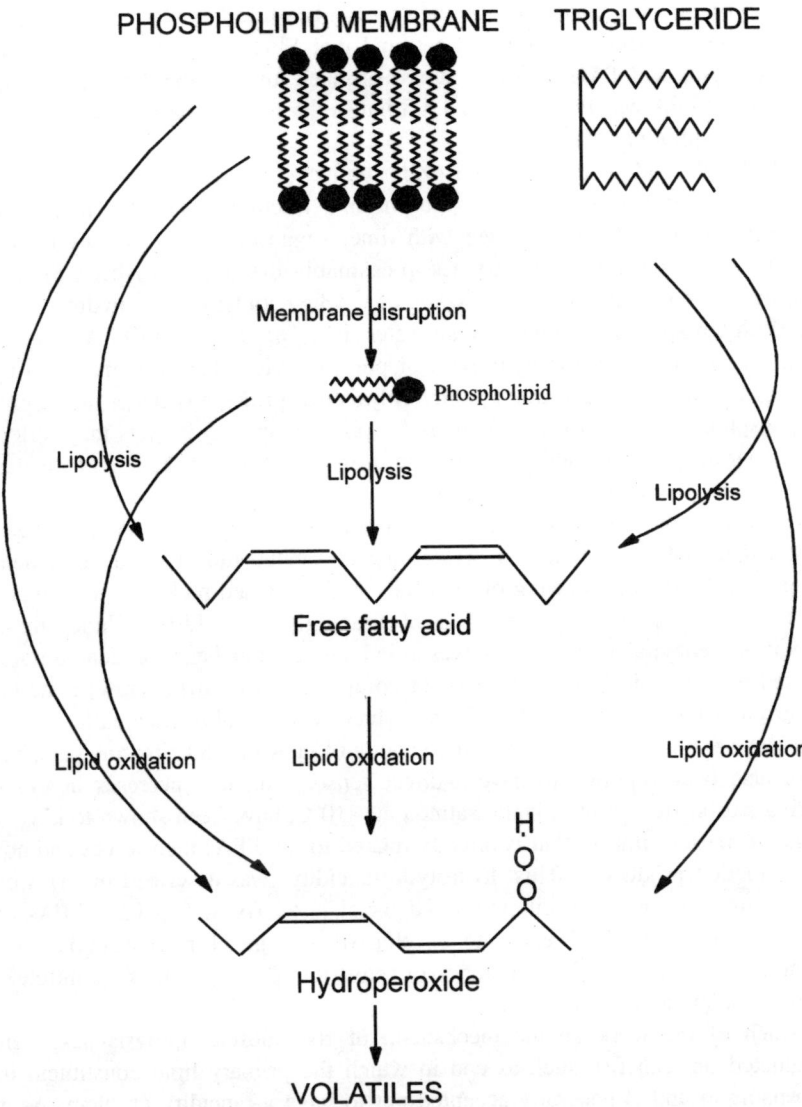

Fig. 14.1   Possible routes of lipid oxidation.

from both phospholipid and neutral lipid fractions. Dekoning and Theodora[14] carried out more detailed follow-up work with the same system in which they calculated the rates of enzymatic hydrolysis of phospholipid and neutral lipids over a range of temperatures. The decrease in rate of FFA formation with decreasing temperature was greater for the phospholipids than for neutral lipids. An initial rapid surge of FFA was observed where the hydrolysis of both lipid fractions was identical at −12°C. At higher temperature the phospholipid fraction hydrolysed faster than the neutral lipids while below this temperature the reverse was true. Tsukuda[9] also found that FFA were derived from both phospholipids and TAG in frozen stored skipjack tuna. In the dark muscle the majority of FFA was formed from TAG but in the light tissue FFA was mainly due to phospholipid hydrolysis.

Mackerel lipid hydrolysis on frozen storage showed an increase in FFA consistent with hydrolysis of the phospholipid fraction.[6] The FFA in cooked samples however did not change with time, suggesting lipid hydrolysis was mainly due to the action of enzymes, presumably inactivated by heat. Similar findings were obtained by Kaneniwa et al.[15] when studying lipid hydrolysis in the flesh ten species of chinese fresh water fish. The increase in FFA in silver carp was attributed to the hydrolysis of phospholipids. This was inhibited by heating the muscle, suggesting lipid hydrolysis was principally under the control of phospholipase. The overall findings for phospholipid hydrolysis may reflect the importance of substrate physical state in situ and the interfacial surface area provided by membrane phospholipids.

Evidence for non-enzymatic mechanisms of TAG hydrolysis have been reported in fish.[16] Comparison of frozen and canned tuna FFA development found that thermal processing of tuna for canning caused preferential cleavage of fatty acids at the sn-2 position of triacylglycerol (TAG). Under these conditions enzymatic hydrolysis was ruled out on the basis of denaturation caused by the rapid heating process. In comparison, the frozen (raw) material occurred at the sn-1 and sn-3 positions, which was probably enzymatic.

FFAs are not only important from the point of view of oxidation products, but have also been reported to have a direct sensory impact. Increases in FFAs during frozen storage of Atlantic salmon at $-10°C$, have been shown to have a negative sensory impact that is directly related to the FFAs themselves and not the oxidation products.[17] This 'hydrolytic rancidity' was described in terms of increasing train oil taste, bitterness and metal taste. By adding back FFAs to fresh mince at levels observed to develop on storage (1 mg/g salmon), the authors were able to rank the potency of sensory impact as DHA > palmitoleic acid > linoleic acid > EPA.

Much of the work on the mechanism of fish muscle lipolysis has been conducted on lean fish such as cod in which the primary lipid constituent is phospholipid and is generally accepted not to have a rancidity problem. As a result, work has tended toward phospholipid hydrolysis with little work being conducted on TAG hydrolysis. Indeed, findings have not been conclusive with regard to the contribution of phospholipids and TAGs to oxidative rancidity in fish. There is a view that phospholipids are the major lipid fraction responsible for oxidative deterioration due to the high levels of PUFA.[9]

14.2.1   Phospholipases

The findings to date indicate that catalytic hydrolysis of phospholipids in fish flesh is principally under the control of phospholipase A2 (PLA2). This enzyme hydrolyses the ester bond at position 2 of phospholipids to yield FFA and lysophospholipid. Both cytosolic and microsomally associated forms of PLA activity have been reported in different species of fish.[18,4] A microsomal PLA activity has been reported in the dark muscle of rainbow trout.[19] This enzyme showed optimum activity at pH 7.00 and did not require Ca^{2+} for activity. Microsomally associated PLA2 activity has also been reported in the muscle tissue of flounder[12] and the results from the study of Chawla and Ablett[20] demonstrated the presence of a similar phospholipase in the myotomal microsomes of Atlantic cod. The capacity for phospholipid hydrolysis was apparently enhanced in cod during an initial period of frozen storage at $-30°C$ up to eight weeks and then decreased on further storage to 12 weeks.[21] Subsequently a calcium independent, 50kDa PLA2 was partially purified from cod muscle cytosol with an acidic pH optimum.[22] It is worth noting that the action of phospholipase in lean fish including cod has been associated with the development of toughened texture. This has been attributed to the interaction of FFA and muscle protein.[23]

A well documented property of PLA2 is its obligatory requirement for $Ca.^{2+}$ The enzyme is inactive in the absence of Ca^{2+} or in the presence of excess chelating agent such as EDTA. This requirement was demonstrated with the 13.7kDa PLA2 isolated from the cytosolic fraction of pollack muscle.[18] In contrast to the cod, the enzyme from pollack exhibited an alkaline pH optimum. Earlier work described a PLA activity occurring in rainbow trout muscle tissue. Activity was greater in the dark muscle compared with white tissue and had no Ca^{2+} requirement.[24] In addition, this study found no evidence of PLC or PLD activity. Follow-up work[19] on the subcellular location indicated that the PLA activity occurs in various sub-cellular fractions of rainbow trout muscle tissue with the majority in the microsomal fraction.

One of the important functions of PLA2 in mammals is its involvement in the inflammatory response where it releases fatty acids that are subsequently metabolised to form pro-inflammatory eicosanoids.[25] There is clear evidence for the activation and expression of a number of PLA2 isozymes in mammals in response to proinflammatory initiators. In piscines such as trout and salmon, there is strong circumstantial evidence for the involvement of PLA2 in the inflammatory response,[26,27] but the mechanisms of control for PLA2 in this process are unclear. If there are similarities to mammalian systems, then tissue damage sustained during capture struggle and perhaps even stress, could cause an increase in PLA2 activity. This may have important consequences for the shelf-life during frozen storage of the muscle.

Other phospholipid acyl hydrolases have been reported in fish muscle. The observed accumulation of large amounts of lyso phospholipid in the muscle of the tuna species *Euthynnus pelamis* have been attributed to PLA1.[28] Follow up work confirmed the presence of PLA1 in the white muscle of this species.[29] The enzyme was found to be Ca^{2+}-independent and had a neutral pH optimum of 6.5–7.0 for activity The enzyme was not stereospecific but was position specific. The authors, however, did not explore the possibility that the enzyme may be a lipase acting at the 1-position of phospholipids. Similar findings were obtained in the recently completed EU FAIR project (FAIR CT 97 3228) where the main lipolytic activities present in Atlantic mackerel muscle tissue was a 66kDa cytosolic, calcium independent PLA1 with an acid pH4 optimum. This enzyme did not display any TAG lipase activity *in vitro* even under phosphorylating conditions (I. Ashton, unpublished data).

14.2.2   Lipases

Most of the work reported on lipases from fish has concentrated on enzymes occurring in the intestinal tract.[30] In terms of frozen storage, the intestinal lipases will pose a problem in the muscle tissue only if the intestinal contents are allowed to mix with the flesh during processing, as can occur in whole frame mince. In filleted material, the intrinsic lipases that are present in the muscle tissue itself are potentially more of a problem. Lipid profile data from storage studies point to the possible importance of TAG hydrolysis in the development of FFAs in fish muscle tissue.[7]

Early work described a short-chain lipase from a pelagic source (mackerel).[31] Comparison of activity from light and dark muscle revealed greater activity in the latter. This may be a contributory factor as to why lipid oxidation in mackerel occurs at an accelerated rate in the dark tissue.[32] Bosund and Ganrot[6] found that TAG were hydrolysed more readily than phospholipids. This may have important implications especially when considering the fact that the majority of lipid present in pelagic species is in the form of storage triglyceride.

Bilinski *et al*.[33] have described a lipase from the dark lateral line of rainbow trout associated with the mitochondrial fraction which is specific for TAG containing long-chain fatty acid. This appeared to be a different enzyme from that previously demonstrated from this tissue which had a higher optimum pH and was stimulated by triton x-100.[34] Further work with the acid lipase from rainbow trout dark lateral line muscle revealed that activity was increased in fasting fish the majority of which (60%) was located in the lysomal fraction.[35] Fast and intermediate rates of freezing did not cause the release of lipase. However, slow freezing and fluctuation of temperature of frozen fillets resulted in appreciable release from the lysosomes. In frozen storage most of the lipase activity was released within the first month. Lower storage temperatures decreased the release rate. As yet there has been no purification of a fish muscle derived TAG lipase. There has been a partial purification of TAG lipase from steelhead trout adipose tissue.[36] The cytosolic enzyme was partially purified ca. 71-fold and had a molecular weight of 48kDa.

Evidence indicates that the lipases which have been detected in some fish muscle tissues are hormone-sensitive (HSL). This type of lipase is involved in the mobilisation of fatty acids from storage TAG. Therefore, the nutritional status of the fish before capture could be an important factor in determining the level of lipase activity in the muscle tissue and hence post-mortem FFA generation. Stress too could also have an effect through the hormones released, as is the case in mammals where catecholamines such as adrenaline cause activation of hormone sensitive lipase.[37]

HSL is activated via signal transduced protein kinase phosphorylation upon stimulation of adipocyte hormone receptors.[38,39] Fish store TAG in several compartments, including mesentric fat, liver, and dark muscle. Studies with rainbow trout have shown that fasting the fish prior to sacrifice results in higher lipase activity in the dark muscle compared to non-fasted fish.[35] It is probable that there was an increase in lipase phosphorylation and or expression in the fasted fish indicating that the nutritional status of the fish upon capture could have an effect on post-mortem lipolytic events. This indirect evidence for the hormone-sensitive nature of lipase involved in the mobilisation of storage TAG in fish dark muscle has been confirmed.[40] Lipase activity from the adipose tissue of rainbow trout was found to be modulated by phosphorylation,[41] which is characteristic of HSLs. This enzyme was activated up to 137% by cAMP dependent phosphorylation. Migliorini et al.[42] have demonstrated that cAMP and known adipokinetic agents significantly stimulated fatty acid release from tigerfish adipocytes in vitro.

14.3   Lipid oxidation reactions

The oxidation of PUFAs requires an active form of oxygen because the reaction of PUFAs with ground state oxygen is spin restricted. The spin restriction is overcome by an activation reaction (initiation) involving a catalyst to initiate free radical chain reactions (propagation). The lipid hydroperoxides formed are unstable and subsequently break down to form volatile compounds which are associated with rancidity development. The volatiles produced during storage have been thoroughly characterised for a number of oily fish species including anchovy,[43] Atlantic salmon[44] and mackerel[45] and the mechanism of breakdown of lipid hydroperoxides to produce volatiles reviewed.[46]

In fish muscle there are a number of potential catalysts and mechanisms that may be involved in the activation reaction to generate active oxygen species for lipid oxidation. These include non-enzymatic mechanisms involving haem proteins such as myoglobin, haemoglobin and cytochrome P450, free iron and enzyme initiators such as lipoxygenase (LOX) and cyloxygenase (COX), all of which have been extensively reviewed.[47,48,49] In reality, some of these mechanisms are of minor importance. For example, although enzymes such as LOX have been reported in post-mortem flesh, the overall impact on rancidity development is still questionable. Also there is no strong evidence for the role played by singlet oxygen. This is especially true when light is not a factor during storage.

14.3.1   Non-enzymatic lipid oxidation

Haem proteins and free iron ions play a key role in non-enzymatic lipid oxidation. Iron ions can take part in electron transfer reaction with molecular oxygen and thus remove the spin restrictions imposed on the unpromoted reaction (reaction 1).

$$Fe^{2+}+O^2 \rightarrow [Fe-O_2]^{2+} \rightarrow Fe^{3+} + O_2^- \qquad 14.1$$

$$O_2^- + H^+ \rightarrow HOO^{\cdot} \qquad 14.2$$

$$2HOO^{\cdot} \rightarrow H_2O_2 + _2 \qquad 14.3$$

Neither superoxide (O_2^-) or hydrogen peroxide (H_2O_2) is thought to cause lipid oxidation directly. This requires a more reactive oxygen species such as the hydroxy radical. Superoxide can dismute to form hydrogen peroxide(reactions 2 and 3) from which hydroxy radicals can be formed by further reaction of iron complexes with hydrogen peroxide via the Fenton-Haber-Weiss reaction (reaction 4):

$$Fe^{2+} + H_2O_2 \rightarrow Fe^{3+} + HO^{\cdot} + HO^- \qquad 14.4$$

This form of oxidation is inhibitable by peroxidases, scavengers of HO˙ and chelators that bind iron, although in some cases, sequestration of iron by chelators such as EDTA has been reported to stimulate oxidation, work on which has been reviewed.[49]

Iron also has another role in lipid oxidation. As in the Fenton reaction with hydrogen peroxide, iron can also cause the decomposition of lipid peroxides (reaction 5).

$$Fe^{2+} + L-O_2H \rightarrow Fe^{3+} + LOO^{\cdot} + HO^- \qquad 14.5$$

Haem proteins are an abundant source of iron in pelagic oily fish, especially myoglobin and blood haemoglobin. These iron-containing proteins are able to propagate lipid oxidation via redox reaction with lipid hydroperoxide or hydrogen peroxide already present. This redox cycle drives the formation of further lipid oxidation products that break down to form the volatiles.

It is well established that ferryl myoglobin (Fe^{4+}), or more precisely the porphyrin cation radical ([P−FeIV=O] resulting from the reaction of met (Fe^{3+}) or ferrous (Fe^{2+}) myoglobin with hydrogen peroxide can cause membrane lipid oxidation.[50] The ferryl isoform is able to abstract a hydrogen from a lipid (LH) to form an alkyl radical (L˙). In the presence of oxygen a hydroperoxide radical is formed (LOO˙). This can abstract hydrogen from another lipid to form a lipid hydroperoxide (LOOH) thus propagating the cycle of oxidative damage. The lipid hydroperoxide is also able to oxidise both ferric and ferryl myoglobin to form a redox cycle.[51,52] This process was significantly affected by pH when haemoglobin-mediated lipid oxidation was studied by adding hemolysate to washed cod muscle. The lag time prior to rancidity and thiobarbituric acid reactive substance development decreased greatly as the pH was reduced.[53] Using the same model Richards and Hultin showed that rancid odours and extensive lipid oxidation could occur even in cod, which has very low levels of fish lipid, when blood haem catalysts are present.[54]

The recently completed EU FAIR project (FAIR CT 97 3228) investigated the importance of interfacial oxidation at membranes in the onset of lipid oxidation in Atlantic mackerel. The major catalyst for lipid oxidation was the haem protein myoglobin, the major isoform of which was metmyoglobin (Fe^{3+}). The oxidative activity of the purified mackerel myoglobin was tested using two model systems, a phospholipid monolayer and fluoresence studies in liposomes. Monolayer studies using dilinoleoyl-PC, dioleoyl-PC and dipalmitoyl-PC (K. Abousalham and R. Verger, CNRS, Marseilles, France unpublished data.) at various initial surface pressures have shown that metmyoglobin is able to catalyse oxidation of intact phospholipid at three initial (applied) surface pressures (Fig. 14.2). Oxidation rates and/or penetration (as measured by increase in surface pressure) were greater in the expanded state, i.e., low initial compression ($2mN.m^{-1}$) in comparison with higher initial surface pressures up to and including $25mN.m^{-1}$ which are exhibited in natural membrane systems. Even at the latter initial pressure significant oxidation of both dilinoleoyl-PC and dioleoyl-PC was observed, with greater rates of oxidation occurring with the former substrate. These data suggest a steric effect that reduces the access of myoglobin in a more ordered membrane to the fatty acid moieties. This is borne out by the fact that no pressure increase was observed at the high initial pressures when using dipalmitoyl-PC. This is inert to oxidation and thus indicates virtually no penetration of the protein into the monolayer. The final results show that oxidation rates were higher in membranes (i) containing polyunsaturated phospholipids and (ii) with lower applied pressures, i.e., the lipid monolayer is in the expanded state. Under these conditions the fatty acid moieties are far more exposed to the action of the prooxidative myoglobin and thus undergo oxidation far more readily.

Using L-α-phosphatidylcholine, β-arachidonyl-γ-stearoyl (20:4(n-6)/18:0) liposomes containing the C11 fluorescent probe BODIPYn581/591 to monitor oxidation confirmed that myoglobin can cause oxidation of esterified phospholipid. Oxidation was markedly increased in the presence of cumenehydroperoxide. Similar studies have also shown that oxidation increased as unsaturation of phospholipids increased.[55] Size measurements of the single unilamellar vesicles have demonstrated that an increase of lipid unsaturation results in smaller SUVs and therefore a larger curvature of the outer bilayer leaflet. This suggests that the lipid-lipid spacing has increased and that the unsaturated fatty acyl chains are more accessible for oxidation. This can be

Fig. 14.2 (opposite) The effect of Atlantic mackerel metmyoglobin (oxidative catalyst) on the surface pressure of phosphatidyl choline films (diC18:2PC, diC18:1PC and diC16:0PC). Monolayers of each PC molecular species were spread at various initial surface pressures (A-$2mN.m^{-1}$, b-$15mN.m^{-1}$, C-$25mN.m^{-1}$) and constant area. Metmyoglobin (0.064 $\mu g.ml^{-1}$, final concentration) was injected under the monolayer and the subsequent change in monolayer surface pressure was recorded as a function of time. (By kind permission of Dr Karim Abousalham and Dr Robert Verger CNRS, Marseilles, France.)

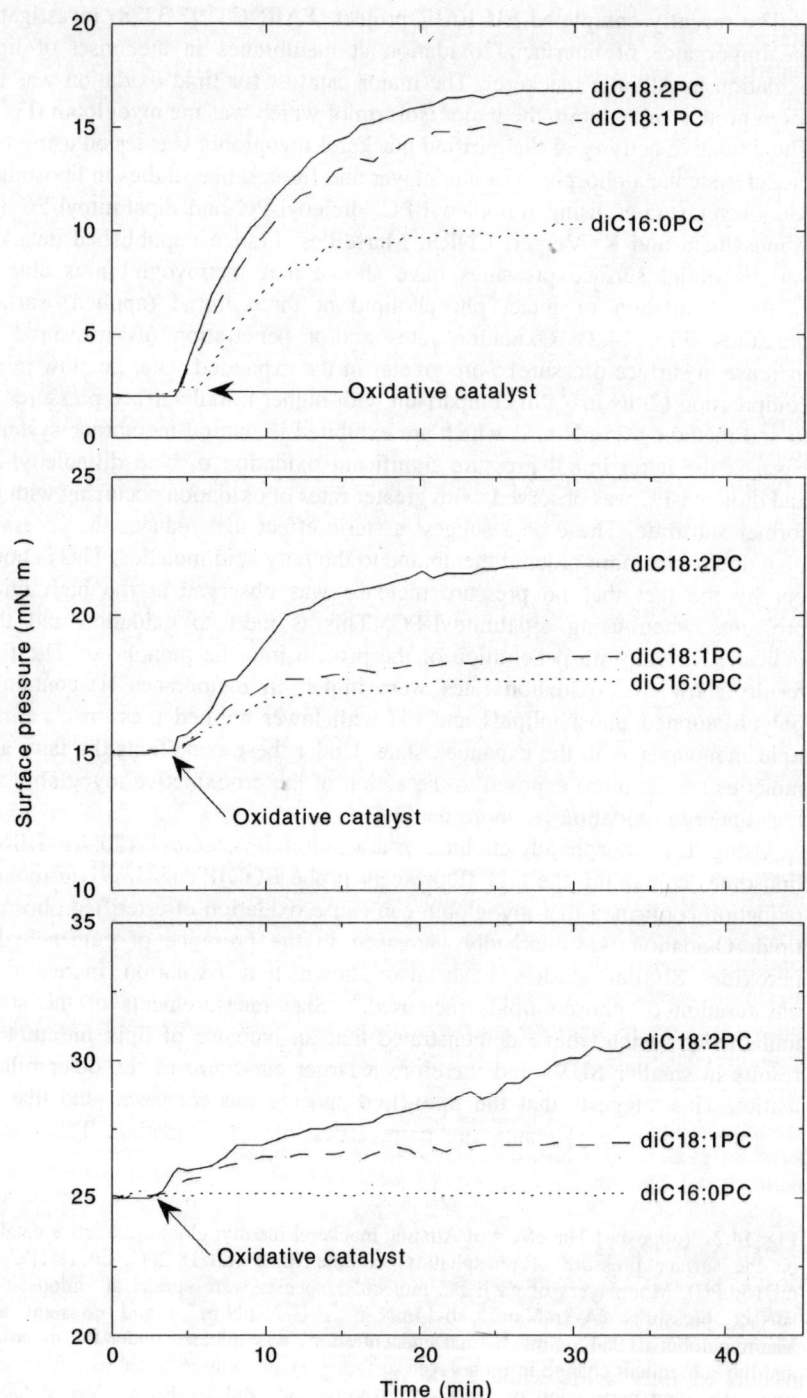

interpreted as a change in density of the lipid packing in the outer leaflet and confirms results from monolayer studies. Thus the high degree of unsaturation of fish membrane lipids make them prone to oxidation. In addition, fluidity measurements of oxidised arachidonic acid-PC liposomes showed a decrease in fluidity upon oxidation, i.e., increased rigidity.

The monolayer and liposome studies indicate that it is not absolutely necessary for FFA to be available in order for oxidation to occur. Nevertheless, it is possible that lipolysis prior to oxidation may cause expansion of the membrane and therefore increased oxidation as the result of increased access to the remaining fatty acid moiety and the FFA. However, work with flounder muscle microsomes showed that lipolysis using exogenous PLA2 prior to oxidation caused inhibition.[12] Studies with microsomal model systems have shown that when peroxidation preceded PLA2 activity, hydrolysis facilitated the propagation of the peroxidative process. With such high oxidative activity particularly in the dark muscle, one might expect high levels of lipolytic activity, e.g., PLA2, which would preferentially hydrolyse oxidised phospholipids and thus maintain membrane integrity by facilitating acyl exchange.[56,57] The major lipolytic activity present in mackerel muscle was a Ca^{2+} independent PLA1 with an acid pH optimum (pH4). The pH optimum for this enzyme is surprisingly low but may serve to maintain membrane integrity at times of high oxidative stress. It is possible that the build up of lactic acid in the fish muscle tissues during capture may cause localised drops in pH that effectively activate the enzyme. Post-mortem this could effectively increase the level of FFAs present in the tissue depending on the level of lactic acid build up.

14.3.2 Enzymatic lipid oxidation
In cooked fish products lipid oxidation is non-enzymatic as any enzymes present have been denatured at the cook temperature. Thus the warmed-over flavour that can be associated with cooked and reheated meat and fish is the result of non-enzymatic reactions. In raw fish there is the potential for lipid oxidation to be catalysed by one or more enzymes. There are two broad groups of enzymic lipid oxidation. Enzymes such as LOX and COX directly introduce oxygen to the fatty acid moiety. In the second type of oxidation, the enzymes are involved in reducing iron complexes such as haem proteins that can react with hydrogen peroxide as described above, forming hydroxy radicals that are in turn able to oxidise lipid. This group includes the microsomal lipid oxidation enzymes.

Lipoxygenase (LOX)
Most LOX and COX enzymes have a strong preference for FFAs. Both enzymes directly introduce oxygen onto the acyl chain(s) of the lipid. LOX enzymes stereospecifically incorporate one molecule of oxygen at a position on unsaturated fatty acids while COX incorporates two molecules of oxygen. There is evidence suggesting that enzymatic fatty acid oxidation by LOX is involved in the production of aroma compounds and initiation of rancidity in

fish.[58,59] LOX activity has been reported in tissues from a number of fish species.[46] German and Creveling[60] have identified a 12-LOX from rainbow trout skin that produced 12-HETE as the major monohydroxy product. 12-LOX has also been detected in trout gill tissue[61] and the activity related to the generation of oxidative volatile carbonyl compounds in a model system.[62] This exhibited activity toward EPA, arachidonic acid (AA) and DHA but low activity toward linoleic acid. 12-LOX activity was also detected in the skin and gill tissue of sixteen other species of fish during this study. Comparison of the relative activities in both tissues showed species variation. Further work on the trout gill 12-LOX indicated the potencies of flavonoids as inhibitors of this activity.[63]

Detailed study of trout gill 12-LOX related the generation of oxygenated products from AA and DHA to the availability of glutathione.[64] The absence of glutathione was found to result in the formation of non-enzymatic trihydroxy products of the initial hydroperoxides. The presence of added glutathione increased the conversion of hydroperoxides to the corresponding monohyydroxy derivative. This was consistent with glutathione's role as a co-substrate in the peroxidase reaction

LOX activity found in Menhaden gill homogenate has been directly linked to the production of volatiles.[65] Sensory analysis judged the smell of n-3 ethyl esters that had been incubated with the enzyme extract to be significantly stronger than untreated controls. However, a pro-oxidant other than LOX could have been responsible for the oxidation.

A 15-LOX activity was discovered in the gill of trout during the purification of the previously detected 12-LOX activity.[66] On purification using hyroxy-apatite the activity was significantly greater than in the crude homogenate. The 15-LOX was active toward C18, C20 and C22 PUFA producing hydroxylated metabolites. The generation of the eicosanoid leucotriene C from ionophore challenged rainbow trout gill tissue has implied the presence of 5-LOX activity.[67] This finding in conjunction with previous work[64] suggests the presence of three LOX activities.

LOX has also been detected in the tissues of pelagic species including sardine[68,69] and mackerel flesh.[10] The partially purified LOX activity from sardine skin[69] was readily oxidized esterified fatty acids such as methyl linoleate and trilinolein. When tested, the specificity of LOX in a crude homogenate from mackerel flesh oxidation of linoleic acid and DHA was found to be more efficient than EPA or linolenic acid.[10] Complex acyl lipids such as TAG and lecithin were not effective substrates whereas 1,3-dilinolein was utilised as substrate to an appreciable extent. A recent report on the detection and purification of 12-LOX from Atlantic mackerel muscle indicates that this enzyme may also have a role in the onset of lipid oxidation[70] although it is possible that the enzyme was derived from blood left in the capillaries and not the muscle itself.

Relatively little is known about the variation in LOX expression in different species, tissues and cell types. In their study, Knight et al.[67] have demonstrated a plethora of LOX metabolites from a number of rainbow trout tissues. All the tissues examined generated 12-LOX products implying this enzyme is widely distributed in this species. Considerable variation in the profile of LOX products was observed indicating differences in tissue LOX expression. *In vivo*, these enzymes are involved in the biosynthesis of C20 fatty acid oxidation products known as eicosanoids, which are involved in modulating a number of physiological processes including the piscine inflammatory response.[71] It is interesting to speculate that the activities of these enzymes, as suggested for phospholipase A_2, may be induced by tissue damage during capture that may impact on post-mortem storage.

14.4   Methods to control lipid oxidation and off-flavour development in fish

To reduce the formation of the volatile compounds associated with off-flavour, the oxidation process has to be stopped or slowed down. In frozen products, a number of potential strategies exist which may help achieve this. None of these methods is able to improve the flavour status of an already rancid product. They are preventative measures for reducing lipid oxidation and/or the breakdown of lipid oxidation products and thus volatile formation. Many of the strategies discussed have been extensively studied. Other strategies have had less attention but still warrant consideration.

14.5   The direct application of antioxidant(s) to fish

This is the most widely studied method for reducing off-flavour development in fish. There are several categories of antioxidants that exist for application. These compounds must meet several important criteria. They must be effective at low concentration, not have too much of a sensory impact and not be toxic.

Antioxidants have a number of modes of action. These include:

- sequestration (chelation) of catalytic metal ions thus preventing propagation
- decreasing oxygen concentration
- quenching singlet oxygen and superoxide anion
- decomposing primary oxidation products to non-volatile compounds
- preventing first chain initiation by scavenging initial radicals such as hydroxyl radical
- chain breaking or free radical interceptor antioxidants (AO) that can donate a hydrogen to lipoperoxyl radical forming an AO radical. The AO radical (A·) formed is unable to abstract a hydrogen from a fatty acid. The antioxidant radical is stabilised by the conjugated system that exists in this type of AO. The mechanism of free radical interceptor antioxidants can be represented by equation 14.6.

$$L + AH \rightarrow LH + A \qquad\qquad 14.6$$

However, this equation is accurate only in situations where the levels of oxygen are extremely low. The more realistic mechanism is given by equation 14.7.

$$LOO + AH \rightarrow LOOH + A \qquad\qquad 14.7$$

As this equation suggests, the addition of antioxidants will not stop oxidation altogether as lipid hydroperoxides are produced. In addition peroxy radicals also react with other PUFAs to produce lipid free radicals. So at best, the use of antioxidants can serve only to reduce the rate at which lipid oxidation occurs. It is also worth noting that antioxidants will not prevent the formation of FFA either.

It is also true that a given AO of this type is able to react only with certain types of free radicals for reasons of solubility, steric hindrance, etc. Thus, an antioxidant may not intercept all the radical species generated. This is why in some cases synergism can be observed from the use of AO combinations. Where synergism occurs a wider array of free radical species are intercepted.[72]

14.5.1   Legal issues relating to the direct addition of antioxidants in fish

The use of antioxidants in fish in the EU is strictly controlled by legislation. Fish blocks are considered to be unprocessed and therefore antioxidants, apart from ascorbic acid and citric acid, cannot be used in their manufacture. Any other antioxidants would have to be cleared for use and this can take between 2–3 years. Antioxidant blends can be used which contribute to flavouring, but the addition would need to have a discernible flavour impact for valid declaration. Relevant governing legislation is EC directive No. 95/2/EC on food additives other than odours and sweeteners.

If the final product is to be a composite product then antioxidants allowed for use under legislation include those for general food use. However, the issue is always one of naturalness (as perceived by the consumer) and label declaration. In terms of further guidance on the use of antioxidants in the EC in composite meat and fish products, information on the exact type of product is needed because all have slightly different nuances in terms of EC legislation.

14.5.2   Metal ion sequestrants

This group of compounds acts by binding pro-oxidative metal ions such as iron and copper. Examples include citric acid, citrate esters, EDTA, phosphoric acid, seed-derived phytate, tartaric acid, and polysaccharides. The use of chelators usually relates to sequestration of free Fe, i.e., the storage of cooked meats. Chelators are ineffective with the Fe associated with haem protein such as myoglobin in raw meat as the chelator presumably cannot access the Fe centre.[73] EDTA has been shown to be an effective chelating antioxidant in many studies on lipid oxidation in oils and fats[74] and meat.[75] However, some work in model systems indicates that EDTA can have a pro-oxidative effect when chelating iron.[76]

Another naturally occurring chelator, phytate, is the hexaphosphate ester of myoinositol and is an abundant phosphorous storage constituent of most cereals, legumes and oilseeds. This compound is an effective chelator of many transition metal ions including Mg^{2+}, Ca^{2+}, Fe^{3+}, Cu^{2+} and Zn^{2+}.[77] It is a powerful inhibitor of iron driven hydroxyl radical formation because of this chelation ability.[78] In a study carried out by Graf *et al.*[79] phytate was found to be one of the most effective agents for inhibiting lipid oxidation. Its effectiveness at providing protection against lipid oxidation and warmed over flavour in refrigerated chicken has also been described.[80]

14.5.3 Synthetic antioxidants

There are a limited number of these antioxidants available for application in food. Oil soluble synthetic antioxidants include:

- Butylated hydroxyanisole (BHA – E320)
- Butylated hydroxytoluene (BHT – E321)
- Propyl, octyl and dodecyl gallate (E 310-12)
- Tert-butyl hydroquinone (TBHQ).

The antioxidant properties of TBHQ have been assessed in comparison with natural antioxidants, rosemary extract, and ascorbic acid alone and in combination. These were added as solutions to cooked trout fish flakes that were then stored at −20°C. Oxidation was measured over 90 days using thiobarbituric acid and sensory evaluation. The effectiveness was in the order TBHQ + ascorbic acid > TBHQ + ascorbic acid + rosemary > rosemary > untreated control −70°C > untreated control −20°C.[81]

Synthetic versions of naturally occurring antioxidants are also available cheaply and in large quantities. Among these are the tocopherols (α, β, γ, δ), which are used in the diets of farmed fish to improve appearance and potentially prolong shelf-life, tocotrienols (α, β, γ, δ), ascorbic acid and ascorbyl palmitate (see Section 14.5).

Maillard reaction products have been shown to possess antioxidant activity in model systems.[82] However, crude and partially purified Maillard reaction products were found to be ineffective at reducing lipid oxidation in minced herring frozen stored at −20°C for 28 weeks.[83]

14.5.4 Intact herbs and spices

A number of herbs and spices have been identified as having substantial antioxidant activity. A general problem encountered with these sources of antioxidants is their characteristic odour and taste. This may make them unsuitable for use in many foods. The antioxidant activity of these ingredients is also less than the corresponding extracts or purified compounds. Moreover, the antioxidant accessibility to the meat or fish tissue may also be limited because much of the active component(s) may be held within the herb or spice used. Ramanathan and Das[84] tested a number of polyphenols, dried and fresh herbs for their antioxidant effect in salted minced mackerel as assessed by TBARS. The antioxidant efficacy for the dried herbs was cloves > cinnamon > cumin = black pepper > fennel = foenugreek. The dried spices were more effective than the fresh. The polyphenols were the most potent group of natural antioxidants tested. At 0.01% the order was ellagic acid > tannic acid > myricetin > quercetin.

14.5.5   Natural antioxidant extracts

Natural products are in growing demand by consumers and additives of any sort are increasingly being questioned. This has led to trends in replacing synthetic antioxidants with natural products of comparable efficacy. Wide interest in the natural antioxidant extracts is indicated by the recent extensive patent activity in this area.

Herb extracts

In another study the antioxidant properties of rosemary extract, TBHQ and ascorbic acid alone and in combinations were determined by their addition to cooked trout fish flakes stored at −20°C.[82] Oxidation was measured over 90 days using thiobarbituric acid and sensory evaluation. The effectiveness was in the order TBHQ + ascorbic acid > TBHQ + ascorbic acid + rosemary > rosemary > untreated control −70°C > untreated control −20°C. Sensory evaluation indicated that a green aroma and flavour was associated with the rosemary treated samples. Other reports have indicated a synergistic effect of α-tocopherol with rosemary extract in a sardine oil model system and dark muscle tissue.[85]

Plant extract

Some plant tissues such as leaf tea contain a class of phenolic or polyphenolic compounds known as flavonoids that display potent antioxidant activity. Flavonoids are highly effective antioxidants because of their multifunctional mode of action. Apart from reacting with hydroxy radicals, superoxide anions and singlet oxygen, these compounds have strong sequestration properties. Flavonoids also have high TEAC values, a measure of total antioxidant efficacy, as assessed by the ease with which they can donate a hydrogen atom. Green tea extract also has a well-documented deodorising effect that may help reduce rancidity.

A number of flavonoid compounds have been tested in studies to control lipid oxidation in ground mackerel.[86] These included quercetin, myricetin, tannic acid and ellagic acid at levels of 200ppm. All the polyphenols tested appeared to be effective at inhibiting lipid oxidation in ground fish stored at −20°C as indicated by TBAR measurement. However, the storage period for the trial was only one week and there was no correlation with sensory analysis. It is, therefore, difficult to assess the real effectiveness of any of the compounds tested for use over longer storage periods. Flavonoids from tea extract have also been shown to modify the most potent pro-oxidant form of myoglobin in meat, i.e., ferryl-myoglobin, reducing it to inactive metmyoglobin.[87]

14.5.6   Endogenous antioxidants

Skeletal muscle contains an endogenous multi-component, bi-phasic antioxidant system, being found both in the aqueous and lipid environments. The purpose of the system is to control the activity of prooxidants and free radicals in the aqueous, lipid and interfacial environments. Among the components in fish muscle are anserine, ascorbic acid, α-tocopherol and glutathione.[88,89] The components and their modes of action in muscle systems have been extensively reviewed.[90]

The levels of endogenous ascorbic acid, tocopherols and glutathione were monitored in minced and fillets of channel catfish over a six-month period of frozen storage at $-6°C$. The rate of decline for the antioxidants for both sample sets was glutathione >ascorbic acid > α-tocopherol. The rate of decline of ascorbic acid and glutathione was greater in the minced samples (87% and 84% drop respectively at four months) compared to fillet (71% and 80% at four months). The increase in the lipid oxidation volatile hexanal was similar in both sets.[91]

Glutathione depletion was measured in mince and fillets of mackerel and blue fish muscle during frozen storage at $-20°C$. The mackerel system lost glutathione faster than blue fish at $-20°C$. Sensory scores declined in both species after approximately two-thirds of the glutathione was lost. Model system studies indicated the effective antioxidant activity of glutathione against lipid oxidation in the early stages.[89] In order for glutathione to reduce lipid hydroperoxides it usually works in tandem with the enzyme glutathione peroxidase. This had been detected in a number of fish species including sardine, coho salmon, and rainbow trout.[92] However, Bell et al.[93] showed that the two components together provided no additional protection in a trout liver microsmol lipid peroxidation model above that provided by reduced glutathione alone.

Carnosine and anserine are aqueous dipeptide antioxidants found in muscle tissue including fish which are able to inhibit the oxidation of lipid catalysed by haem proteins, LOX and singlet oxygen in vitro.[90] The pure dipeptide is an effective antioxidant in model systems of muscle phospholipid liposomes. Carnosine inhibits the initiation step of oxidation, it decreases the amount of pre-formed peroxides and it reacts with some secondary products.[94]

14.6   Modification of the diet of farmed fish

This area has been more extensively studied in the production of meat from farmed animals[95] and to a lesser extent in aquaculture. The work in meat may provide useful learning applicable to the farming of oily fish such as salmon. The composition and stability of animal tissues after slaughter is influenced by dietary composition, including antioxidants and fatty acid content. Many studies have shown that elevating antioxidant content of muscle tissue via increased consumption improves the stability of stored meat and fish. Antioxidant supplementation studies of animal diets have included tocopherols, ascorbic acid and carotenoids.[96]

The assimilation of feed antioxidants by farmed animals has the clear advantage over applying antioxidants in that it gets the active molecules to the site where they can act most effectively. For example, dietary α-tocopherol offers greater protection for meat compared to postmortem applied tocopherol.[95] The reason is that dietary α-tocopherol is incorporated into the membranes whereas the applied antioxidant is exogenous to the membrane. Dietary tocopherol has been one of the most studied dietary antioxidant supplements and has been reported as one of the most potent stabilisers of lipids in meat.[97] In some studies, beneficial effects were observed only when the level of antioxidant supplementation was very high. For example, in the study by Raskin et al.[98] tocopherol incorporation was at 10–20 times the normal dietary requirement, causing a significant increase in feeding costs. Dietary supplementation strategies do have the added advantage in that they pose no legislative issues regarding addition antioxidants.

Dietary fat and vitamin E content may have an interactive effect on oxidative stability of meat. Increased PUFA content of meat is generally considered to increase the rate of off-flavour development. However, recent studies have successfully improved the nutritional value of pork meat by increasing the n-3 PUFA levels using linseed oil in the diets. The improvements were achieved without affecting the sensory qualities and oxidative stability of products under storage conditions including freezing.[99,100] Boggio et al.[101] found that dietary lipid and α-tocopherol had no significant effect in preventing oxidative deterioration in trout during frozen storage. However, Frigg et al.[102] found that dietary supplementation with dietary α-tocopherol at 100 and 200mg/kg effectively stabilised against oxidation in trout fillet at −28°C as determined organoleptically and chemically.

The impact of oxidation of fishmeal lipids and feed α and γ tocopherol content has been investigated on quality of frozen fillets from farmed Atlantic Salmon.[103] Feeding with supplements resulted in good uptake of the antioxidants with α-tocopherol found in phospholipid-rich organ tissues and γ-tocopherol in the adipocyte TAG storage in the muscle. Feeding with oils that were oxidised to different extents did not show any significant sensorical initial differences, and post-slaughter storage of salmon fed tocopherol did not show significant differences in TBAR value following six months storage at −40°C.

The storage ($-40°C$) conditions for this trial were probably too low to observe any changes.

Dietary α-tocopherol was reported not to protect the white muscle lipids of rainbow trout from rancidity, although a slight protective effect was reported for the dark muscle lipids.[104] Hamre et al.[105] investigated the effect of enriching Atlantic salmon fillet with α, δ and γ tocopherol via diet on their oxidative stability during ice and frozen storage. They found that non α-tocopherol was prooxidative during storage on ice but effective in a dose dependent manner at reducing oxidation during frozen storage at $-30°C$ over 48 weeks storage. It was concluded that α-tocopherol may be better suited than tocopherol mixes as a tool to optimise the oxidative stability of salmon fillet.

The use of carotenoids such as astaxanthin in aquaculture has been widely studied mainly in rainbow trout and Atlantic salmon. Lipid oxidation was less significant after 18 months of frozen storage of rainbow trout fed high astaxanthin diet (100ppm) in comparison to 40ppm diets.[106] The levels of α tocopherol tested (1000 and 600ppm) were found to have no significant effect. The converse was true with the same but more processed fish, where lipid oxidation was more advanced. It was concluded that the carotenoid was important in controlling early stages of lipid oxidation.

14.7   Modified atmosphere and vacuum packaging

Modified atmosphere packaging (MAP) involves the replacement of air by other gases, usually CO_2 or N_2, whereas vacuum packaging (VP) involves the removal of atmospheric air and reducing the headspace as much as possible. Both techniques are designed to extend shelf-life by eliminating oxygen and thereby reducing oxidative deterioration during storage. Published studies reporting the use of modified atmosphere/vacuum packaging in a number of susceptible meat products such as pork and beef suggest that these methods significantly extend shelf-life of frozen meat products, in some cases more successfully than antioxidants.[107] However, there are fewer reported studies with frozen oily fish.

In order to maintain the correct atmospheric composition around the product, packaging material with low gas and moisture permeability ($<50cm^3$. $m^{-2}day^{-1}atm^{-1}$) must be used. This is to prevent oxygen diffusing in and, in the case of modified atmosphere, the applied inert gases diffusing out. The influence of packaging materials with high, medium and low oxygen permeability (OTR) on the development of TBARS and astaxanthin retention in rainbow trout fillets was investigated during 36 weeks frozen storage.[108] Rancidity developed more rapidly in packages with high oxygen permeability. This effect was more pronounced in illuminated samples, also verified by sensory evaluation. Fillets with highest astaxanthin content reached maximum TBARS after 29 weeks. The two less pigmented sample groups reached maximum after 17 weeks frozen storage, indicating the antioxidative effects of astaxanthin. In a similar study, Anelich et al.[109] monitored the quality of frozen catfish fillets over an 11-month period at $-20°C$, in two types of packaging materials, namely vacuum packaging and oxygen-permeable packaging. Lipid oxidation, irrespective of packaging, showed no fixed trends throughout the study. However, the use of vacuum packaging alone and in combination with the antioxidant erythorbic acid was found significantly to protect frozen Atlantic mackerel stored at $-7°C$ as indicated by TBA test and sensory panel.[110]

Christophersen et al.[111] characterised the combined effect of oxygen permeability of packaging material and exposure to light on the oxidative degradation of lipids in frozen steaks of rainbow trout during storage. Lipid oxidation was found to be dependent on the accessibility of oxygen rather than on exposure to UV light.

Oxidation can also cause undesirable deterioration in the appearance of fish tissues with high myoglobin content via the formation of metmyoglobin. Temperature abuse during frozen storage has been shown to exacerbate this problem in tuna.[112] Subsequent altered atmosphere packaging studies with tuna using carbon monoxide were shown to eliminate this problem over a six-month period at $-20°C$, with the percentage of metmyoglobin remaining virtually constant.[113]

Chemical oxygen absorption is another potentially interesting solution in attempting to reduce oxygen levels. An example of a chemical oxygen scavenger is that patented and produced by Mitsubishi Gas Chemical Company and sold under the brand name 'Ageless' and available as small sachets. Oxygen is absorbed chemically via reaction with powdered iron oxide to form stable oxidation products. The product is recommended for use at temperatures as low as $-40°C$. The treatment is claimed to able to reduce levels of oxygen in a closed container down to 0.01%.

14.8   The effects of freezing

The impact of freezing practice on rancidity development remains poorly understood and many of the reports appearing to date are inconclusive. Good freezing practices may reduce rancidity during storage by reducing ice crystal damage, particularly to membrane lipids where the initial onset of lipid oxidation may occur. Although the rate of chemical reactions diminishes with falling temperature, oxidation in meat and fish lipid is accelerated below freezing point, with a maximum around $-10°C$. The temperature deceleration effect is more than offset by the accelerating effect of freeze concentration of reactants and catalysts. The acceleration of rancidity upon freezing is related to ice formation, especially in the oily regions of the fish. As the temperature is progressively reduced below $0°C$ an increasing proportion of pure ice is formed. The remaining liquid water forms a solution of increased concentration and depressed freezing point (see Chapter 20). Therefore, even at temperatures as low as $-20°C$ some liquid water remains in which reaction can occur.

The formation of ice crystals and their expansion may cause freeze-damage to the ultra-structure of the fish tissue. For example, membrane may be disrupted allowing lipid substrates readily to mix with pro-oxidants. Recent work with catfish has indicated the deleterious, pro-oxidative effect of freeze thaw cycling.[114] One important factor was the release of iron from haem protein which is reported to be more pro-oxidative. One of the reasons why free iron displays higher oxidative activity may relate to its increased penetration of membranes compared to the parent haem-protein. In addition, it was found that freeze thaw cycling caused the appearance of what appeared to be peroxide-activated haem-proteins (Fe^{4+}) with a concomitant decrease in metmyoglobin. Prefreeze holding time also has a significant impact on subsequent frozen storage stability of oily fish. This has been demonstrated in herring fillet where changes that negatively impact the oxidation process took place between three and six days after catch relating to endogenous antioxidant depletion.[115]

14.8.1   Rapid freezing and cryoprotection

Rapid freezing practice may reduce freezing-induced damage to membranes and potentially reduce rancidity development. It is possible that the formation and even distribution of small ice crystals via a rapid freezing process could reduce initial damage. It is important to note that even when starting with rapidly frozen material, temperature fluctuation during frozen storage will still cause ice crystal growth. There is little work reported in the literature regarding the effect of ice crystal size and location on rancidity development. Some published results indicate that different freezing regimes have no effect on lipid oxidation.[116] Chicken breast and salmon muscle were frozen either slowly or rapidly at −25°C, and then stored at −5°C for up to 83 and 47 days respectively. TBA values increased significantly with storage time, but freezing rate had no effect. It should be noted, however, that the freezing method used in this investigation might not have been optimal for the material used. Even though one of the freezing methods (plate freezing at −35°C) was classed as rapid, temperature gradients may have existed in the muscle tissue, and if this were the case then uneven freezing would have resulted. In addition, ice crystal morphology, location and distribution were not determined.

The use of cryoprotectants to reduce freezing damage could also be a potential route to prevention of off-flavour development. This would have the added advantage over a cryo freezing process in that it may be able to withstand subsequent temperature abuse more readily, although to date there is no evidence for this.

14.9   Conclusion and future trends

Rancidity is clearly a perennial problem associated with frozen storage of oily fish. Measures have to be taken in order to extend shelf-life with regard to flavour deterioration. It is apparent that there is no 'all-encompassing' solution to the generic problem of rancidity in fish. Indeed, the evidence can sometimes be conflicting for a number of reasons. This does not mean that solutions cannot be formulated. It means that potential strategies have to be tailored, systematically investigated and continually checked when implemented with the particular raw material of interest.

It is clear that consumers are becoming increasingly aware of and selective against foods that are perceived by them to be unnatural and containing additives. This means that controlling rancidity in fish with applied antioxidants has to be carefully considered. This would account for the trend toward investigating the use of more natural antioxidants such as herb extracts, e.g., rosemary and the recent patent activity in this area. Here there is scope to screen for new and perhaps more efficacious natural extracts.

The huge growth in aquaculture is already providing an important and widely consumed source of oily fish in the form of salmonids. This may extend in the future to the farming of other oily fish species. The use of dietary antioxidant and manipulating dietary lipids is clearly an important area to investigate further in minimising lipid oxidation in these raw materials.

Treatments to minimise rancidity development are often unpredictable and the assessments of these treatments time consuming, involving long storage trials and lengthy sensory analyses. The use of accelerated storage conditions particularly for frozen conditions can go some way toward reducing the study times involved, but caution has to be exercised when extrapolating data from higher frozen storage temperatures to lower temperatures. Having reliable and correlated accelerated storage trials or even more rapid model systems would be a great benefit to the area. Also having reliable and rapid analytical techniques that correlate with and possibly even replace sensory analysis would be advantageous. Many of the current techniques such as volatile analysis although powerful, are still quite lengthy procedures requiring rigorous optimisation. Thus the development of a rapid, reliable analytical tool to determine the extent of a product's oxidative deterioration would be invaluable to screen treatments rapidly.

14.10   Sources of further information

'Antioxidants in Muscle Foods; Nutritional Strategies to Improve Quality' (2000) (E.A. Decker, C. Faustman and C.J. Lopez-Bote eds) Wiley Interscience, New York.
'Free Radicals, Oxidative Stress, and Antioxidants – Pathological and Physiological Significance' (1998) (T. Ozben ed.) Plenum Press, New York and London.
'Free Radicals in Biology and Medicine' 2nd edition (1989) (B. Halliwell and J.C. Gutteridge eds) Clarendon Press, Oxford.

'Natural Antioxidants – Chemistry, Health effects and Applications' (1997) (F. Shahidi ed.) AOCS Press, Champaign, Illinois.

'Rancidity in Foods' 3rd edition (1999) (J.C. Allen and R.J. Hamilton eds) Aspen Publications, Gaithburg, Maryland.

'Muscle as Food' (1986) (P.J. Becktel ed.) Academic Press, New York.

Aust S.D. and Svingen B.A. (1982) 'The role of iron in enzymatic lipid per oxidation' in Free Radical Biology vol. 5 (Pryor, W. ed).

Decker, E.A. and Hultin, H.O. (1992) 'Lipid Oxidation in Muscle Foods via Redox Iron' ACS Symposium Series 500, p33–54.

Frankel, E.N. (1985) 'Chemistry of free radical and singlet oxidation of lipids' Prog. Lipid Res. 23, p197–221.

Hsieh, R.J. and Kinsella, J.E. (1989) 'Oxidation of PUFAs, Mechanisms, Products and Inhibition with Emphasis on Fish' Adv. Food Nutr. Res. 33, p233–41.

Kanner, J. (1992) 'Mechanisms of Nonenzymatic Lipid Peroxidation in Muscle Foods in Lipid Oxidation In Foods' ACS Symposium Series 500, p55–73.

14.11 References

1. BODY, D.R. and VLIEG, P. (1989) 'Distribution of the lipid classes and eicosapentaenoic and docosahexaenoic acids in different sites in blue mackerel Fillets.' J. Food Sci. 54, 569–72.

2. DEEMS, R.A., EATON, B.R. and DENNIS, E.A. (1975) 'Kinetic analysis of phospholipase A2 activity toward mixed micelles and its implications for the study of lipolytic enzymes' J. Biol. Chem., 250, 9013–20.

3. OHSIMA, T., FUJITA, Y. and KOIZUMI, C. (1993) 'Oxidative stability of sardine and mackerel lipids with a reference to synergism between phospholipids and alpha tocopherol. J. Amer. Oil Soc. 70, 269–76.

4. SHEWFELT, R.L. (1981) 'Fish Muscle Lipolysis' J. Food Biochem. 5, 79–100.

5. OHSHIMA T., WADA S. and KOIZUMI, C. (1985) 'Accumulation of lyso phospholipids in several species of fish flesh during storage at −5°C.' Bulletin Jap. Soc. Sci. Fisheries 51, 965–71.

6. BOSUND, I. and GANROT, B. (1969) 'Lipid hydrolysis in frozen Baltic herring' J. Food Sci., 34, 13–17.

7. HWANG, K.T. and REGENSTEIN J.M. (1993) 'Characteristics of mackerel mince lipid hydrolysis.' J.Food Sci. 58, 79–83.

8. INGEMANSSON, T., KAUFMANN, P. and EKSTRAND, B. (1995) 'Multivariant evaluation of lipid hydrolysis and oxidation data from light and dark muscle of frozen stored Rainbow trout.' J. Agri. Food Chem. 43, 2046–52.

9. TSUKADA, N. (1976) 'Changes in the Lipids of Skipjack during Frozen Storage' Bull. Tokai Reg. Fish. Res. Lab. 84, 31–41.

10. HARRIS, P. and TALL, J. (1994b). 'Substrate Specificity of Mackerel Flesh Lipoxygenase'. J. Food Sci. 59, 504–6.

11. MIYASHITA, K. and TAKAGI, T. (1986) 'Study on the oxidative rate and prooxidant activity of free fatty acids' *JAOCS* **63**, 1380–4.

12. SCHEWFELT, R.L and HULTIN, H.O. (1983) 'Inhibition of Enzymatic and Non-Enzymatic Lipid Peroxidation of Flounder Muscle Sarcoplasmic Reticulum by Pretreatment with Phospholipase A2' *Biochim. Biophys. Acta* **751**, 432–8.

13. DEKONING A.J., MILKOVITCH, S. and THEODORA, H.M. (1987) 'The origin of free fatty acids formed in frozen cape hake mince during cold storage at −18°C.' *J. Sci. Food Agric.*, **39**, 79–84.

14. DEKONING A.J. and THEODORA, H.M. (1990) 'Rates of free fatty acid formation from phospholipids and neutral lipids in frozen cape hake mince at various teperatures.' *J. Sci. Food Agric.*, **50**, 391–8.

15. KANENIWA, M., MIAO, S., YUAN. C.H., IIDA, H., and FUKUDA, Y. (2000) 'Lipid components and enzymatic hydrolysis of lipids in muscle of Chinese freshwater fish' *J. Amer. Oil Chem. Soc.* **77**, 825–30.

16. MEDINA I., SACCHI R. and AUBOURG, S. (1994) '[13]C Nuclear magnetic resonance monitoring of free fatty acid release after fish thermal processing' *JAOCS* **71**, 479–82.

17. REFSGAARD H.H.F., BROCKHOFF P.M.B. and JENSEN B. (2000) 'Free polyunsaturated fatty acids cause taste deterioration of salmon during frozen storage.' *J. Agri Food Chem.* **48**, 3280–5.

18. AUDLEY, A., SHETTY, K.J. and KINSELLA, J.E. (1978) 'Isolation and properties of phospholipase A from pollock .' *J. Food Sci.* **43**, 1771–5.

19. BILINSKI, E. and JONAS, R.E.E. (1966b) 'Distribution of Lecithinase in the subcellular fractions of rainbow trout lateral line' *J. Fish Res. Bd. Canada* **23**, 1811–3.

20. CHAWLA, P. and ABLETT, R.F. (1987) 'Detection of microsomal phospholipase activity in myotomal tissue of Atlantic cod.' *J. Food Sci.* **52**, 1194–7.

21. CHAWLA P., MACKEIGAN B., GOULD S.P. and ABLETT R.F. (1988) 'Influence of frozen storage on microsomal phospholipase activity in myotomal tissue of Atlantic cod' *Canadian Institute of Food Science and Technology Journal* **21**, 399–402.

22. AAEN, B., JESSEN, F. and JENSEN, J. (1995) 'Partial purification and characterisation of a cellular acid phospholipase A(2) from cod.' *Comp. Biochem. Physiol.* **110B** (3) 547.

23. SIKORSKI, Z., OLLEY, J. and KASTUCH, S. (1976) 'Protein changes in frozen fish.' *Food Sci. and Nutr.* **8**, 97–104.

24. BILINSKI, E. and JONAS, R.E.E. (1966a) 'Lecithinase activity in the muscle of rainbow trout' *J. Fish Res. Bd. Canada* **23**, 207–20.

25. LOCATI, M., LAMORTE, G., LUINI, W., INTRONA, M., BERNASCONI, S., MANTOVANI, A. and SOZZANI, S. (1996) 'Inhibition of monocyte chemotaxis to C-C chemokines by antisense oligonucleotide for cytosolic phospholipase A(2)' *J. Biol. Chem.* **271**, 6010.

26. PETTIT, T.R., ROWLEY, A.F. and BARROW, S.E. (1989) Synthesis of

leukotriene-B and other conjugated triene lipoxygenase products by blood-cells of the rainbow trout (Salmo-Gairdneri) *Biochem. Biophys. Acta* **1003**, 397–402.

27. BELL, J.G., SARGENT, J.R. and RAYNARD, R.S. (1992) 'Effects of increasing dietary linoleic acid on phospholipid fatty acid composition and eicosanoid production in leucocytes and gill cells of Atlantic Salmon' *Prostaglandins, Leukotriens and Essential Fatty Acids* **45**, 197–206.

28. SATOUCHI K., SAKAGUCHI M., SHIRAKAWA M., HIRANO K. and TANAKA T. (1994) 'Lyso-phosphatidylcholine from white muscle of Bonito euthynnus-Pelamis – Involvement of phospholipase A(1) activity for its production' *Biochim Biophys Acta* **1214**, 303–8.

29. HIRANO K., TANAKA A., YOSHIZUMI K., TANAKA T. and SATOUCHI K (1997) 'Properties of phospholipase A(1) transacylase in the white muscle of bonito Euthynnus pelamis (Linnaeus)' *J. Biochem.* **122**, 1160–6.

30. GREENE, D.H.S. and SELIVONCHICK, D.P. (1987) 'Lipid metabolism in fish' *Prog. Lipid Res.* **26**, 53–85.

31. GEORGE, J.C. (1962) 'A histophysiological study of the red and white muscle of mackerel' *Am. Midl. Nat.* **68**, 487–94.

32. KE, P.J., ACKMAN, R.G., LINKE, B.A. and NASH, D.M. 'Differential lipid oxidation in various parts of frozen mackerel' (1977) *J. Food Technol.* **12**, 37–42.

33. BILINSKI, E.R., JONAS, R.E.E. and LAU, Y.C. (1971) 'Lysosomal Triglyceride lipase from the lateral line tissue of rainbow trout.' *J. Res. Bd. Canada*, **28**, 1015–18.

34. BILINSKI, E.R. and LAU, Y.C. (1969) 'Lipolytic activity toward long-chain triglycerides in lateral line muscle of rainbow trout.' *J. Fish Res. Bd. Canada* **26**, 1857–66.

35. GEROMEL, E.J. and MONTGOMERY, M.W (1980) 'Lipase release from lysosomes of rainbow trout (*Salmo-Gairdneri*) muscle subjected to low temperature' *J. Food Sci.* **45**, 412–15.

36. SHERIDAN, M.A. and ALLEN, W.V. (1984) 'Partial-purification of a triacylglycerol lipase isolated from steelhead trout (*Salmo-Gairdneri*) adipose tissue' *Lipids* **19**, 347–52.

37. GREENBERG A.S., SHEN W.J., MULIRO K., PATEL S., SOUZA S.C., ROTH R.A. and KRAEMER F.B. (2001) 'Stimulation of lipolysis and hormone-sensitive lipase via the extracellular signal-regulated kinase pathway' *J. Biol. Chem.* **276**, 45456–61.

38. FREDERIKSSON, G., TORQUVIST, H. and BELFRAGE, P. (1986) 'Hormone sensitive lipase and monoacylglycerol lipase are both required for complete degradation of adipocyte triacylglycerol' *Biochem. Biophys. Acta*, **876**, 288–93.

39. SHERIDAN, M.A. (1988) 'Lipid dynamics in fish – aspects of absorption, transport, deposition and mobilization' *Comp. Biochem. Physiol.* **90B**, 679–90.

40. HENDERSON, R.J. and TOCHER, D.R. (1987) 'The lipid composition and

biochemistry of fresh water fish' *Prog. Lipid Res.* **26**, 281–347.

41. MICHELSEN, K.G., HARMON, J.S. and SHERIDAN, M.A. (1994) 'Adipose-tissue lipolysis in rainbow trout, *Onorhynchus mykiss*, is modulated by phosphorylation of triacylglycerol lipase' *Comp. Biochem. Physiol.* **107B**, 509–13.

42. MIGLIORINI, R.H., LIMA-VERDE, J.S., MAVHADO, C.R., CARDONA, G.M.P., GAROFALO, M.A.R. and KETTLEHUT, I.C. (1992) 'Control of adipose-tissue lipolysis in ectotherm vertebrates' *Am. J. Physiol.* **263**, 857–62.

43. TRIQUI R. and REINECCIUS G.A. (1995) 'Changes in Flavour Profiles with Ripening Anchovy' *J. Agric Food Chem*, **43**, 1883–9.

44. REFSGAARD H.H.F., BROCKHOFF, P.B. and JENSEN B. (1998) 'Sensory and chemical changes in farmed Atlantic salmon (*Salmo salar*) during frozen storage.' *J. Agri. Food Chem.* **46**, 3473–9.

45. REFSGAARD H.H.F., HAAHR A.M. and JENSEN B. (1999) 'Isolation and quantification of volatiles in fish by dynamic headspace sampling and mass spectrometry.' *J. Agri. Food Chem.* **47**, 1114–18.

46. HSIEH, R.J. and KINSELLA, J.E. (1989). 'Oxidation of PUFAs, Mechanisms, Products and Inhibition with Emphasis on Fish' *Adv. Food Nutr. Res.* **33**, 233–41.

47. HARRIS, P. and TALLI, J. (1994) in *'Rancidity in Foods'* (J.C. Allen and R.J. Hamilton eds) Blackie Academic & Professional 256–70.

48. KANNER, J. (1992) 'Mechanisms of Nonenzymatic Lipid Peroxidation in Muscle Foods in Lipid Oxidation In Foods' *ACS Symposium Series* **500**, 55–73.

49. AUST S.D. and SVINGEN B.A. (1982) 'The role of iron in enzymatic lipid peroxidation' in *Free Radical Biology* vol. 5 (Pryor, W.A., ed) Academic Press, pp 1–28.

50. KANNER, J. and HAREL, S. (1985) 'Initiation of membranal lipid peroxidation by activated metmyoglobin and methemoglobin' *Arch. Biochem. Biophys.* **237**, 314–21.

51. REEDER, B.J. and WILSON, M.T. (1998) 'Mechanism of reaction of myoglobin with the lipid hydroperoxide hydroperoxyoctadecadienoic acid' *Biochem. J.* **330**, 1317–23.

52. CHAN, W.K.M. FAUSTMAN C. and DECKER E.A. (1997) Oxymyoglobin Oxidation as affected by Oxidation Products of Phosphatidylcholine Liposomes, *J. Food Sci.* **62**, 709–12.

53. RICHARDS M.P. and HULTIN H.O. (2000) Effect of pH on lipid oxidation using trout hemolysate as a catalyst: A possible role for deoxyhemoglobin.' *J. Agri. Food Chem.* **48**, 3141–47.

54. RICHARDS M.P. and HULTIN H.O. (2001) 'Rancidity development in a fish model system as affected by phospholipids' *J. Food Lipids* **8**, 215–30.

55. BORST J.W., VISSER N.V., KOUPTSOVA O. and VISSER A.J.W.G. (2000) 'Oxidation of unsaturated phospholipids in membrane bilayer mixtures is accompanied by membrane fluidity changes' *Biochim. Biophys. Acta* **1487**, 61–73.

56. HAN, T.J. and LISTON, J. (1987) 'Lipid peroxidation and phospholipid hydrolysis in fish muscle microsomes and frozen fish' *J. Food Sci.* **52**, 294–301.

57. TOYOMIZU, M., HANAOKA, K. and YAMAGUCHI, K. (1981) 'Effect of release of free fatty acids by enzymatic hydrolysis of phospholipids on lipid oxidation during storage of fish muscle' *Bulletin of the Japanese Soc. Sci. Fish* **47**, 615–20.

58. JOSEPHSON, D.B., LINDSAY, R.C. and STUIBER, D.A. (1984) 'Biogenesis of lipid-derived volatile aroma compounds in the emerald shiner (*Notropisatherinoides*)' *J. Agri. Food Chem.*, **32**, 1347–1351.

59. JOSEPHSON, D.B., LINDSAY, R.C. and STUIBER, D.A (1987) 'Influence of processing on the volatile compounds characterising the flavor of pickled fish' *J. Food Sci. 52*, 596–602.

60. GERMAN, J.B. and CREVELING R.K. (1985) 'Lipid oxidation in fish tissue enzymatic initiation via lipoxygenase' *J. Agric. Chem.* **33**, 680–83.

61. HSIEH, R.J., GERMAN, J.B. and KINSELLA, J.E. (1988) 'Lipoxygenase in fish tissue – some properties of the 12 lipoxygenase from trout gill' *J.Agri. Food Chemi.* **36**, 680–5.

62. HSIEH, R.J. and KINSELLA, J.E. (1989) 'Lipoxygenase generation of specific volatile flavour carbonyl compounds in fish tissue' *J. Agri. Food Chem.* **37**, 279–86.

63. HSIEH, R.J., GERMAN, J.B. and KINSELLA, J.E. (1990) 'Relative inhibitory potencies of flavonoids on 12 lipoxygenase of fish gill' *Lipids* **23**, 322–326.

64. GERMAN, J.B. and KINSELLA, J.E. (1986) 'Hydroperoxide metabolism in trout gill tissue – Effect of glutathione on lipoxygenase products generated from arachidonic acid and dococahexaenoic acid' *Biochem. Biophys. Acta* **879**, 378–87.

65. GRUN, I.U. and BARBEAU, W.E. (1995) 'Lipoxygenase activity in menhaden gill tissue and its effect on odor of *n-3* fatty acid ester concentrates' *J. Food Biochem.*, **18**, 199–212.

66. GERMAN, J.B. and CREVELING R.K. (1990) 'Identification and characterization of a 15 lipoxygenase from fish gill' *J. Agri., Food Chem.* **38**, 2144–7.

67. KNIGHT, J., HOLLAND, J.W., BOWDEN, L.A., HALLIDAY, K and ROWLEY, A.F. 'Eicosanoid generating capacities of different tissues from the rainbow trout *Onorhynchus mykiss*' (1995) *Lipids* **70**, 451–8.

68. MOHRI, S., CHO, S.Y., ENDO, Y. and FUJIMOTO, K. (1990) 'Lipoxygenase activity in sardine skin' *J. Agri. Food Chem.*, **54**, 1889–91.

69. MOHRI, S., CHO, S.Y., ENDO, Y. and FUJIMOTO, K. (1992) 'Linoleate 13(S)-Lipoxygeanse in sardine skin' *J. Agri. Food Chem.*, **40**, 573–6.

70. SAEED S. and HOWELL N.K. (2001) '12-lipoxygenase activity in the muscle tissue of Atlantic mackerel (*Scomber scombrus*) and its prevention by antioxidants.' *J. Sci, Food Agri.* **81**, 745–50.

71. SHARP, G.J.E., PETTITT, T.R., ROWLEY, A.F. and SECOMBES, C.J. (1992) 'Lipoxin-induced migration of fish leukocytes' *J. Leuk. Biol.* **51**, 140–5.

72. EVANS R.J. 'Optimizing lipid stability with matural inhibitors' in *Natural Antioxidants Chemistry, Health effects and Application* (1997) (F. Shahidi ed.) chapter 13, p 224–44, AOCS Press, Champaign, Illinois.

73. OHSHIMA T., WADA S. and KOIZUMI C. (1988) 'Influence of heme pigment, non haem pigment, non haem iron and nitrite on lipid oxidation in cooked mackerel meat' *Nippon Suisan Gakkaishi* **54**, 2165–71.

74. EUN J.B., HEARNSBERGER J.O. and KIM J.M. (1993) 'Antioxidants, activators, and inhibitors affect the enzymatic lipid-peroxidation system of catfish muscle microsomes' *J. Food Sci.* **58**, 71–4.

75. DECKER E.A., CRUM A.D., SHANTHA N.C. and MORRISSEY P.A. (1993) 'Catalysis of lipid oxidation by iron from insoluble fraction of beef diaphragm muscle' *J. Food Sci.* **58**, 233.

76. MCCORD J.M. and DAY E.D. (1978) 'Superoxide-dependant production of hydroxy radicals catalysed by iron complex' *FEBS Letters* **86**, 139–42.

77. NOLAN, K.B., DUFFIN, P.A. and MCWEENY, D.J. (1987) 'Effect of phytate on mineral bioavailability *in vitro* studies on Mg^{2+}, Ca^{2+}, Fe^{3+}, Cu^{2+} and Zn^{2+} (also Cd^{2+}) solubilities in the presence of phytate' *J. Sci. Food Agric.* **40**, 79–85.

78. GRAF, E., EMPSON, K.L. and EATON, J.W. (1987) 'Phytic-Acid A natural antioxidant' *J. Biol. Chem* **262**, 11647.

79. GRAF, E., MAHONEY, J.R., BRYANT, R.G. and EATON, J.W. (1984) 'Iron-catalysed hydroxyl radical formation – stringent requirement for free iron coordination site' *J. Biol. Chem.* **259**, 3620.

80. EMPSON, K.L., THEODORE, P.L. and GRAF, E. (1991) 'Phytic acid as a food antioxidant' *J. Food Sci.* **56**, 560–3.

81. BOYD, L.C., GREEN, D.P., GIESBRECHT, F.B. and KING, M.F. (1993) 'Inhibition of oxidative rancidity in frozen cooked fish flakes by tert-butylhydroquinone and rosemary extract' *J. Sci. Food Agric.* **61**, 87–93.

82. LINGNERT, H. and ERIJSSON, C.E. (1981) 'Antioxidant effect of Maillard reaction products' *Prog. Food Nutr. Sci.* **5**, 453.

83. BECKEL, R., LINGNERT, H., LUNDGREN, B., HALL, G. and WALLER, G.R. (1985) 'Effect of Maillard Reaction on the stability of Minced herring in frozen storage' *J. Food Sci.* **50**, 501–2.

84. RAMANATHAN, L. and DAS, N.P. (1993) 'Natural products inhibit rancidity in salted cooked ground fish' *J. Food Sci.* **58**, 318–20.

85. FANG X. and WADA, S. (1993) Enhancing the antioxidant effect of a-tocopherol with rosemary in inhibiting catalysed oxidation caused by Fe^{2+} and hemoprotein, *Food Res. Int.* **26**, 405–11.

86. RAMANATHAN, L. and DAS N.P. (1992) 'Studies on the control of lipid oxidation in ground fish by some polyphenolic natural products' *J. Agri. Food Chem.* **40**, 17–21.

87. JORGENSEN, L.V. and SKIBSTEAD, I.H. (1998) 'Flavonoid deactivation of ferrylmyoglobin in relation to ease of oxidation as determined by cyclic voltammetry' *Free Radical Research* **28**, 335–51.

88. DECKER, E.A. and HULTIN, H.O. (1992) 'Some factors influencing the

catalysis of lipid oxidation in mackerel ordinary muscle' *J. Food Sci.* **55**, 947–50.

89. JIA T.D., KELLEHER S.D., HULTIN H.O., PETILLO D., MANEY R. and KRZYNOWEK J. (1996) 'Comparison of quality loss and changes in the glutathione antioxidant system in stored mackerel and blue fish muscle' *J. Agri. Food Chem.* **44**, 1195–201.

90. DECKER E.A., LIVISAY S.A. and ZHOU S. (2000) 'Re-evaluation of the antioxidant activity of purified carnosine' *Bioch.-Moscow* **65**, 766–70.

91. BRANNAN, R.G. and ERICKSON, M.C. (1996) 'Quantification of antioxidants in channel catfish during frozen storage' *J. Agri. Food Chem.* **44**, 1361–6.

92. NAKANO, T., SATO, M. and TAKEUCHI, M. (1992) 'Glutathione Peroxidase in fish' *J. Food Sci.* **57**, 1116–9.

93. BELL, J.G., COWEY, C.B. and YOUNGSON, A. (1984) 'Rainbow trout liver microsomal lipid peroxidation, the effect of purified glutathione peroxidase, glutathione S-transferase and other factors' *Biochim.. Biophys. Acta* **795**, 91–9.

94. KANSCI, G., GENOT, C., MEYNIER, A. and GANDEMER, G. (1997) 'The antioxidant activity of carnosine and its consequences on the volatile profiles of liposomes during iron/ascorbate induced phospholipid oxidation' *Food Chem.* **60**, 165–75.

95. MITSUMOTO, M. (2000) in *'Antioxidants in Muscle Foods; Nutritional Strategies to Improve Quality'* (E.A. Decker, C. Faustman and C.J. Lopez-Bote eds.) Wiley Interscience, Chichester, Ch. 12, pp. 315–66.

96. MORRISSEY, P.A., SHEEHY, P.J.A., GALVIN, K., KERRY, J.P. and BUCKLEY, D.J. (1998) 'Lipid stability in meat and meat products' *Meat Sci.* **49**, S73–S86.

97. O'NEILL, L.M., GALVIN, K., MORRISSEY, P.A. and BUCKLEY, D.J. (1999) 'Effect of carnosine, salt and dietary vitamin E on the oxidative stability of chicken meat' *Meat Sci.* **52**, 89–94.

98. RASKIN, P., CLINQUART, A., MARCHE, C. and ISTASSE, L. (1997) 'Vitamin E and meat quality' *Annales de Medecine Veterinaire* **141**, 113–26.

99. RILEY, P.A., ENSER, M., NUTE, G.R. and WOOD, J.D. (2000) 'Shelf life and eating quality of beef from cattle of different breeds given diets differing in n-3 polyunsaturated fatty acid composition' *Animal Science* **71**, 483–500.

100. SHEARD, P.R., ENSER, M., WOOD, J.D., NUTE, G.R, GILL, B.P. and RICHARDSON, R.I. (2000) 'Shelf life and quality of pork and pork products with raised n-3 PUFA' *Meat Sci.* **55**, 213–21.

101. BOGGIO, S.M. HARDY, R.W. BABBITT, J.K. and BRANNON, E.L. (1985) 'The influence of dietary lipid source and α-tocopherol level on product quality of Rainbow trout' *Aquaculture* **51**, 13–24.

102. FRIGG, M., PRABUCKI, A.L. and RUHDEL, E.U. (1990) 'Effect of dietary Vitamin E level on oxidative stability of trout fillet' *Aquaculture* **84**, 145–58.

103. ACKMAN R.G., PARAZO, M.P.M. and LALL, S.P. (1997) 'Impact of dietary peroxides and tocopherols on fillet flavor of farmed Atlantic salmon' *ACS*

Symposium series **674**, 148–65.

104. POZO, R., LAVETY, J. and LOVE, M. (1988) 'The role of dietary α-tocopherol in stabilising the canthaxanthin and lipids of rainbow trout' *Aquaculture* **73**, 165–75.

105. HAMRE, K., BERGE, R.K. and LIE, O. (1998) 'Oxidative stability of Atlantic salmon fillet enriched in α, γ and δ tocopherol through dietary supplementation.' *Food Chem.* **62**, 173–8.

106. JENSEN, C., BIRK, E., JOKUMSEN, A., SKIBSTEAD, L.H. and BERTELSSEN, G. (1998) 'Effect of dietary levels of fat, α-tocopherol and astaxanthin on the colour and lipid oxidation during storage of frozen rainbow trout and chilled storage of smoked trout' *Z. Lebensm. Unters. Forsch.* **207**, 189–96.

107. BRUUNJENSEN, L, SKOVGAARD, M., SKIBSTEAD, L.H. and BERTELSEN, G. (1994) 'Antioxidant synergism between tocopherols and ascorbyl palmitate in cooked, minced turkey' *Z. Lebensm. Unters. Forsch.* **199**, 210–13.

108. BJERKENG, B. and JOHNSEN, G. J. (1995) 'Frozen Storage Quality of Rainbow trout as affected by oxygen, illumination, and fillet pigment.' *Food Sci* **60**, 284–8.

109. ANELICH, L.E., HOFFMAN, L.C. and SWANEPOEL, M.J.J. (2001) 'The quality of frozen African sharptooth catfish (*Clarias gariepinus*) fillets under long-term storage conditions'. *Food and Agri.* **81**, 632–9.

110. SANTOS, E.E.M. and REGENSTEIN, J.M. (1990) 'Effects of Vacuum Packing, Glazing and Erythorbic Acid on the shelf-life of Frozen White Hake and Mackerel' *J. Food Sci.* **55**, 64–70.

111. CHISTOPHERSEN A.G., BERTELSEN, G., ANDERSEN, H.J., KNUTHSEN, P. and SKIBSTEAD, L.H. (1992) 'Storage Life of Frozen Salmonoids – Effect of light and packaging conditions on carotenoid and lipid oxidation' *Z. Lebensm. Unters. Forsch.* **194**, 115–19.

112. CHOW, C.J., OCHIAI, Y., WATABES, S. and HASHIMOTO, K. (1989) 'Reduced stability and accelerated autoxidation of tuna myoglobin in association with freezing and thawing' *J. Agri. Food Chem.* **37**, 1391–5.

113. CHOW, C.J., HSIEH, P.P., TSAI, M.L. and CHU Y.J. (1998) 'Quality changes during iced and frozen storage of tuna flesh treated with carbon monoxide gas' *J. Food Drug Analysis* **6**, 615–23.

114. BENJAKUL, S. and BAUER, F. (2001) 'Biochemical and physicochemical changes in catfish (*Silurus glanis Linne*) muscle as influenced by different freeze-thaw cycles.' *Food Chem.* **72**, 207–17.

115. UNDELAND, I. and LINGNERT, H J. (1999) 'Lipid oxidation in fillets of herring (*Clupea harengus*) during frozen storage. Influence of prefreezing storage.' *Agri Food Chem.* **47**, 2075–81.

116. TOAMA, M.C. (1990) 'Study on the influence of freezing rate on lipid oxidation in fish (salmon) and chichen breast muscles' *Int. J. Food Sci. Tech.* **25**, 718–21.

14.12   Acknowledgements

The author would like to acknowledge the support for the EU-FAIR project (No. 97-3228) discussed in this chapter, and thank the project coordinator Christine Davies, Unilever R&D, Colworth Laboratory, U.K. and the collaborators involved in the project, namely Dr Robert Verger and Dr Karim Abousalham, CNRS, Marseilles, France; Jan Willem Borst and Prof. Antonie J.W.G. Visser, Wageningen Agricultural University, Netherlands; and Prof. Bo Ekstrand at SIK, Gothenburg, Sweden.

Part III

Improving quality within the supply chain

15

Quality chain management in fish processing

M. Frederiksen, Danish Institute of Fisheries Research, Lyngby

15.1   Introduction: the fish supply chain

The fish supply chain differs in relation to the fish species and product type. This chapter deals mainly with wild caught fresh fish products (fish in ice at about 0°C) because frozen fish products and farmed fish have a different supply chain structure, which is beyond the scope of this chapter. The current chain structure dealing with fresh fish can be described in general with the following steps in Fig. 15.1. Most processes are only done once in the chain but are listed in Table 15.1 for all the steps where they can be done.

The fresh fish market is in general characterised by small and medium enterprise companies (SMEs) in all steps of the chains. In many countries the collector/auction steps are bypassed and direct sales are made to the wholesaler. In some countries the fishing vessels are owned by the wholesaler/processor, in others every step in the chain is independent of the other steps. All kinds of different chain ownership constellations are present. The French supermarket

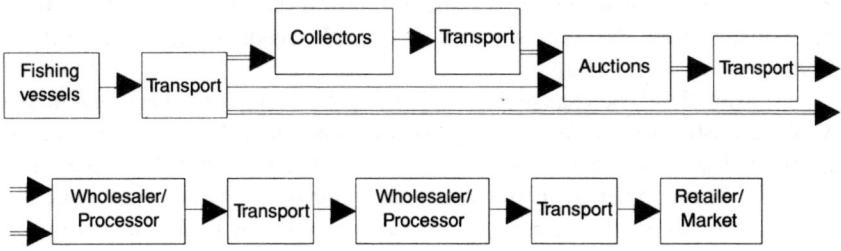

Fig. 15.1   Types of fresh fish chains in the fish industry today. Arrows show the product flow.

Table 15.1   Processes in fresh fish chains today

Step	Processes in the chain
Fishing vessels for fresh fish	Catch, gut, bleed, wash, sort in species, size grade, weigh, icepack, store and unload
Collectors	Size grade, weigh, icepack, store and bring to auction
Auctions	Store and auction (sell)
Wholesalers/ processors	Size grade, process, weigh, icepack, store and sell. There can be one or several steps of wholesalers/processors in a fish supply chain
Transport companies	Load, store and unload
Retailers/markets	Process, weigh, icepack, store and sell

chain 'Intermarche' is one of the few companies in Europe owning all steps in a fish chain and it actively uses its vertically integrated structure to bring the fish direct to the supermarkets without any time-consuming or costly extra steps (Anon., 1998).

Traditionally, every single step competes with other steps in the chain in order to survive. For that reason every step in the chain after the vessel step considers it must have its own buffer store to ensure it is possible to provide product for its customers. In addition each store has to be large enough to ensure that it is possible to provide fish product in case a percentage of the stored fish products prove to be defective (Olsen, 2001). This way of doing business is maybe affordable for one step for a short time, but in the long run all steps in a chain lose profit. In addition it is unlikely that the end customer expectations will be met or surpassed and therefore the market will probably decrease over time. The less competitive step will lose in the long run (Porter, 1988). In the future it is foreseen that the real competition will not be between steps but between supply chains on the market (Christopher, 1998). One example is that the fruit and vegetables chains in the north of Europe are under great pressure to change to achieve closer cooperation. The most sophisticated chains, according to well-established supply chain management (SCM) systems, are in England where direct supply has become common. In each country in northern Europe there are companies organised by well-established SCM systems. The margins in the fresh produce sector are highest in England and it is argued that this is because they have the most sophisticated supply chain organisation (Wilson, 1996).

One example of the changed structure towards chains is the declining inland wholesale fish market in the UK. The inland wholesale market had in 1989 25% of the total domestic retail and catering trade of fresh fish but they have much less now. The number of inland wholesale companies has been declining for several years, as has the volume of fish that is handled by them. The reason for this is the changed pattern of consumption and purchasing behaviour and the intervention of alternative distribution channels (Symes and Maddock, 1989).

As fish has a very short shelf-life compared with other products there are very high requirements placed on the chain for it to be effective. The high-value fresh fish market in Europe in general demands fresh fish at a maximum of 8–9 days from catch and the demand on performance of the logistics system is very dependent on the distance between the harbour for unloading the fish and the retailer/end user in the chain. The transport system for fresh fish in Europe has become more and more effective. Today it is possible to bring fresh fish from the north of Europe to the southern parts within two days. Transport is by refrigerated trucks so the fish products are in good condition when they arrive. The whole operation has been organised in large units. For instance, all fresh fish for export are collected in one location in Denmark before they are transported south. In that way it is possible to ensure full truckloads. Many trucks also have GPS (global positioning system) with satellite connection to the home shipping agent office. Therefore it is possible to enhance the management of the transport company, track each truck and give the end market or retailer precise information about arrival or delays. Thus the current solution to export fresh fish from the northern to the southern part of Europe has been to develop very effective transport systems making it possible to distribute fresh fish all over Europe within two days.

Today there is only minor cooperation between steps in fresh fish chains, and information about the first step in the chain and the basic product information is often lost between fishing vessel and first-hand buyer (Frederiksen and Bremner, 2001). Systems to transfer information from vessels to the next steps in the chain have been developed. Systems available today include sea-scales for onboard weighing, but these are mostly restricted to larger vessels because of the costs and space needed. A simple Internet-based test system 'Info-fisk' has been developed at the Danish Institute for Fisheries Research (DIFRES) (Frederiksen et al., 2002), which will be discussed later in this chapter.

15.2   Definitions

15.2.1   What is a chain/step?
The various features of traceability in food manufacture have been recently outlined (Moe, 1998). The terminology from that paper is also employed here in that

- a step refers to a discrete operation or location at which some task or process is performed on the product
- a chain is composed of the sequence of these steps
- a product can be any material at any stage of processing, e.g., a live fish, a whole fish, or a gutted fish or fillet.

15.2.2   Traceability

A definition of traceability is included in ISO 9000 (ISO, 2000). Traceability is the ability to trace the history, application or location of that which is under consideration. When considering a product, traceability can relate to:

- the origin of material and parts
- the processing history
- the distribution and location of the product after delivery.

In general the term 'trace' is used when the history of product origin is searched and the term 'track' is used for searching its history after delivery.

15.2.3   Supply chain management

Supply chain management (SCM) is the management of upstream and downstream relationships with suppliers and customers to deliver superior customer value at less cost to the supply chain as a whole (Christopher, 1998).

15.2.4   What is fresh fish quality?

The most important factors deciding the properties of fish are the time temperature tolerance (Frederiksen *et al.*, 1997). In general white fish species have 'high quality' in the first 6–8 days after catch (kept at 0°C, caught and treated gently). But in the first 6–8 days fish properties, and hence perceived quality, changes too. The rigor period starts shortly after death dependent on the temperature, stress and species. If the fish are kept at 0°C rigor can last up to about 2–4 days (for cold-water species like cod and plaice at 0°C) (Huss, 1995). The yield can be low if a fish is filleted pre rigor, as the fillets will contract and give drip loss at a later stage. Machine filleting fish in rigor condition also gives a lower yield. The taste also changes through the rigor period. Fishermen and people close to the sea can have preferences for eating pre-rigor fish, but most consumers eat post-rigor fish and prefer that texture and taste. Ideally the pre-rigor and rigor period can be used to transport the fish to the retailer and remaining 'high quality' shelf-life is available for processing and sale of the fish to the customer.

In this chapter, quality is considered to be the time temperature tolerance of the fish/product defined as the number of days it is kept at 0°C, after having been caught and treated gently by approved methods. If the quality of a fish is stated as the catch date in all steps, the customer has all the critical information to make a decision on what to buy. The chosen unit for quality is equivalent days in ice at 0°C (sometimes called 'icedays'). Hours at 0°C is impractical to use and the natural variation in quality attributes of raw material means that greater precision cannot be obtained (Hyldig, 2001). The variation in quality properties is defined as the mixture of fish with different icedays (two different or several icedays). Quality variation is minimised through avoiding mixing fish with different icedays.

15.2.5 What is a quality chain?

If quality assurance is used in all steps in a chain and the quality is defined and co-ordinated in the whole chain it is a quality chain. A typical way to design quality assurance systems in the fish industry is to base them on the HACCP (hazard analysis critical control point) principles as described elsewhere in this book and GMP (good manufacturing practice). With an effective quality assurance system it is default if a product defect occurs that there is a system to ensure it is corrected and the quality assurance system is updated to prevent the defect happening again. Today it is mandatory for fish producers and retailers in the European Union to have so-called 'own check' systems, which are mainly based on HACCP principles (EEC, 1994). The minimum requirement for a quality assurance system is to prevent any hazard to the consumer. That is only a basic requirement; another is to choose a quality level for each customer segment in each step and to determine the acceptable variation in each quality level for each step. That level can be different from step to step in a chain. This coordination of quality level and quality variation for each customer segment is what makes a chain a quality chain.

Another factor is the ability to react to a crisis in a chain. The crises in the meat sector made the focus fall on traceability systems that were capable of making an effective recall from the market. An effective recall system is able to react fast, identify and locate suspect material and reduce to a minimum the amount of goods needed to recall. To handle crises effectively in fresh fish chains traceability is a must. A quality chain has this ability.

15.3 Organising quality chains

A quality fresh fish chain delivers the right product at the right place at the right time. The physical organisation and the management of the chain must support this condition, and to be competitive the number of steps in the chain from fishing vessel to end customer can be reduced to a minimum. A consequence is reduced time and costs in the chain. Ideally the only 'warehouse' in a quality fresh fish chain is onboard the vessels until they reach the quay and in the trucks on the way to the retailers. Each fish should only be physically handled in two steps of the chain after catch.

1. Onboard the vessel: gut, bleed, wash, sort in species, size grade, weigh, icepack, label each box with catch date.
2. At the retailer: process if any, display, icepack and sell to the consumer.

Size grading and weighing the catch is only possible onboard a number of big vessels in Europe today. In some countries containers are used instead of boxes onboard vessels. Then it is very important to control the mixture of ice and water that is used to keep temperature at about 0°C in the containers. The vessels should know the amount of fish sold to the consumer each day to plan the fisheries and the length of journey and the fish should be sold before the vessel enters the harbour.

Planning on behalf of day-to-day consumer demand is not possible in practice because the lengths of most vessel journeys are of up to 10 days depending on the actual catch. Registering the fish by date when caught onboard the vessels, in combination with an efficient quality assurance system and a fixed end date of journey, gives the retailer the possibility to buy the fish remotely. Buying from several vessels will give a retailer the above opportunity, but the vessels must then be able to send catch information to shore online, which today is possible only on big vessels with a satellite connection. Another condition is that there is a supportive computer system that minimises the time the retailer uses to trade. A supportive computer system for retailers is not developed today and satellite equipment will not be on the majority of vessels in the near future.

There are some examples of big business units in fresh fish chains, but most are SMEs, which are independent. Generally, the staff has a relatively low level of education so that implementing management philosophy is often difficult. The job of implementing change is not so much technical as sociological. All the techniques to chill, cool, process, and transport fish products have been known for years, so lack of suitable equipment and knowledge does not represent a major obstacle for development. The main problem is that the steps do not work together to exchange information, forming a chain that is manageable. To manage a quality chain in fish processing a distributed chain management approach must be implemented if there is not a company in the chain large enough to take on the responsibility for management of the whole chain. It is also not expected that a vessel can always sell all its catch to one chain, unless that is a part of the deal to be connected in the chain cooperation.

One way to organise quality chains is to make a chain system so it is possible for each step to trust the other in a higher degree. Ideally, the chain members should act as if the whole chain was one single company and optimise costs and share all the profit. The ideal situation is possible only if the same company owns all steps. What can be done then? Well, first of all the steps must play with open cards. Disclosure of projected margins and requirements must be shared, securely, within the chain to make the whole operation profitable. The author's suggestion for ideal distributed chain management of a quality fresh fish chain is given Fig. 15.2. The number of steps in this quality fresh fish chain has been reduced (the collector and auction steps are bypassed and only one wholesaler/processor is used). Common goals for the chain are reached through discussion and agreement within the chain management team. The representation from the catching sector can be from each vessel or one common representative for all vessels. The transport steps are not represented in the chain management team. It may be better to make precise descriptions to the transport companies of what is expected and then use the same transport companies all the time. Even if the collector and auction steps were present in the chain they should not be represented in the chain management team as they only size grade and price the product and that can be handled by precise descriptions of what is expected of them in the transport steps. Chain cooperation from all participants must be

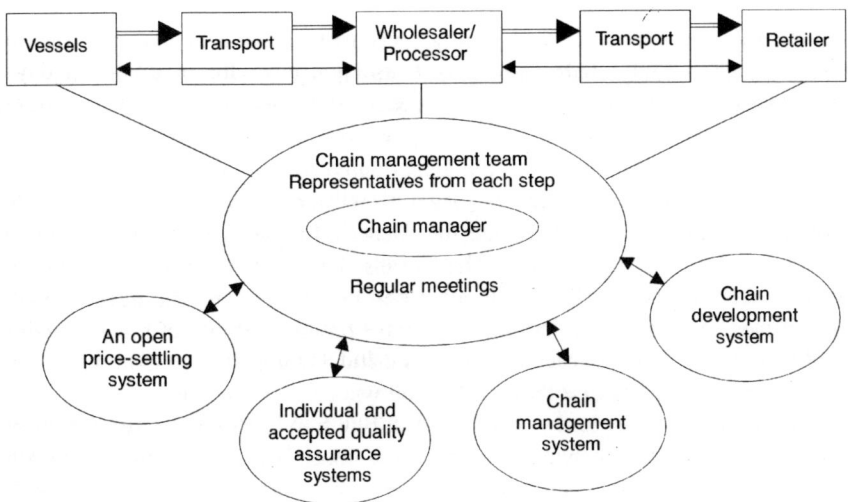

Fig. 15.2 Ideal distributed chain management in a quality fresh fish chain. Double arrows show product flow. Single arrows show information flow.

experienced as an improvement compared to the current or prior situation when considered from a long-term point of view. It must be simple, user friendly and not administrative. Information should be keyed in only once in the chain and simple forms of communication must be used in the whole chain.

All chain members must have a common goal. The major common goal is:

All steps achieve long-term economic benefit from participating in chain co-operation

It is also described as 'win-win' thinking and it does not necessary mean to share 50/50 but all partners should benefit and be better off as a result of cooperation (Christopher, 1998). The four systems shown in Fig. 15.2 are explained in the following sections.

- an open price settling system
- individual and accepted transparent quality assurance systems
- a chain management system
- a chain development system.

15.4 An open price settling system

- A transparent price settling system in the chain (trade by commission sales or to share the total benefit of the chain between the steps).
- It must pay to be honest.
- It must be clear for all chain members what it costs to deliver unmarketable products, but it must also be clear that quality failures are not acceptable.

Ideally the earnings/profit on each product must be common knowledge for the chain members (only chain members). If this is not possible a system must be established to make the transactions transparent so that everyone in the chain must know the pricing in each step based on a certain amount of money on each kg, each box or a percentage of the actual value of the product.

One of the main problems is to get fishermen involved in the chain. In countries where auctions dominate the first-hand trade, fishermen can have a custom to 'play' at the auction. That means it is hard for them to change to direct/contract landings for wholesalers because, by playing the auction, they sometimes get very high prices for an average product because of low deliveries and high demand on a particular day. It is difficult for a chain to deal with open trade at an auction because their competing wholesalers are able to destroy their cooperation when they know the chain will buy from a certain vessel. One way to deal with this is to make a flowing price contract according to the auction price the same day. An example could be based on a certain percentage on top of the average auction price the same day. The benefit is that the chain can get fish of known quality attributes as soon as the size grading and weighing have been completed onboard the vessel or on shore. The vessel skippers and crew then know they will never get a low price compared to the market price. Fish that is not sold to the chain can be sold at the auction as usual. Today the raw material for the fishmeal industry is already paid on the basis of the current world fishmeal and oil price and in some herring industries there is a similar connected raw material price to the export price (Olsen, 2001).

15.5   Quality assurance systems

The main requirements for such a system are that it should be based on the HACCP principles to ensure consumer safety and GMP and it must be an active system that is continuously corrected, updated and developed according to input from customers and results of inspections. There must be

- an intact cold chain in and in between all steps
- intact information flow in and in between all steps (traceability of the products)
- an agreed standard method in the whole chain to measure fish quality on an objective basis (inspection method).

HACCP and GMP that must be implemented in each step will not be treated further here, but the basic requirements are mentioned below.

15.6   Maintaining the cold chain

As mentioned at the start, controlled temperature of the product at 0°C is the key to trusting product quality without actual inspection of it. Inspection only once in the chain, onboard the vessel when gutting the fish will reduce the chain costs significantly. If the cold chain is broken the simple calculation indicating product quality of days elapsed (0°C) since catch will not be true and the product cannot be automatically used in the quality chain. Equipment and methods to ensure an intact cold chain are readily available. It is just a matter of implementing the right equipment and using robust procedures in each step in the chain to make it function. Humans are the problem not the equipment.

15.7   Product traceability

Many food scandals in Europe have resulted in a demand for food traceability. It is now mandatory in the meat industry and it will be mandatory for the fish industry in a few years. The crisis in the meat industry caused by BSE (bovine spongiform encephalitis) has had enormous consequences that will resonate for many years. Although the animal itself formed a highly visible traceable unit (tru) the fact that feed ingredients were implicated as the agents spreading the infection introduced enormous complexity into the traceability systems. At the outset the traceability systems were primitive and not electronically based. Data gathering and retrieval was slow and laborious. When an animal was sent for slaughter its information was sent by postcard from the farm to the central registry. This meant that the meat was often consumed before the data had been entered in a retrievable state. Furthermore the mandatory systems in Europe do not prescribe the speed and efficiency with which they must operate.

For perishable foods, such as fresh fish, the need for an effective management of the whole chain is high as there are only 6–8 days for cold-water species from catch to end-user on the high quality market in Europe. Studies have shown that traceability does not exist in the whole domestic Danish fresh fish chains today and that cooperation between steps is limited (Frederiksen and Bremner, 2001). As well as being a mandatory requirement traceability is also the key to managing and operating a quality chain.

The systems available today that can include full traceability are costly and a business IT package is too dear for most SMEs. At DIFRES a small traceability system has been developed and tested. It has been shown that it is possible to make a ready-to-use simple traceability system with the latest Internet technology (Frederiksen et al., 2002). The advantages of such systems can best be utilised if they are implemented in a production management system in each step. When reliable information on product quality from the first step is speedily and securely made available throughout the whole chain it is possible to generate a high degree of trust between the steps.

The greatest problem occurs if the quality and uniformity of the raw material cannot be trusted and a business then has to buy more than it actually needs to ensure sufficient material to meet its orders. As a consequence, lower prices are offered to take into account possible defects in the raw material. Furthermore, more resources to check raw material, extract defective products or to reclassify material to another quality level will be required, thus decreasing efficiency and adding greater costs to the processing steps. Ideally, quality inspection should be done only once in the chain and that is when the fisherman guts the fish right after catch. That quality inspection is only to discard defective raw material from the catch as storage temperature and handling procedures are well controlled onboard.

If the fisherman gets credit (money and respect) for his work it should be possible to ensure a stable supply of known quality fish from the fishing vessel. With information technology it is also possible to transfer catch information before the vessel enters the harbour. Trust in the quality would be the perfect situation that would make it possible to sell the product to the customer even before the vessel had unloaded the fish on the quay.

15.8   Inspection

Objectivity in measurement and in agreement on quality levels is necessary to forge close relations between the single steps in the chain. The inspection method used in the chain must be as objective as possible. Close relations and mutual trust are necessary also to reconcile differences and to avert and settle any conflicts that may arise. The quality inspection method used in Europe today has four nominal grades E (excellent), A, B, and C (condemned) (EEC, 1996). The problem is that a fishing vessel has a trip of for instance five days and the catch on different days can be mixed to an unknown extent when unloading, auctioning, and repacking. There is no date labelling of fish boxes onboard the vessels and the manual quality evaluation and size grading of the fish on shore takes about one second per fish. At this speed it is obvious that the evaluation cannot be very precise. The collectors claim that they can quality grade to an accuracy within three days of storage at 0°C, but this is not precise enough for a quality chain.

A tool to estimate the quality attributes in a more objective way is the Quality Index Method (QIM) (Bremner *et al.*, 1987; Warm *et al.*, 1998). The QIM is the most objective and precise method existing to decide the actual quality level of a fish, and it will probably be the future standard method for Europe. The method is based on a human sensory analysis and schemes have been developed especially for each species to be evaluated. There are no technical methods available that have the necessary precision to estimate fresh fish quality. All steps must use the same method to estimate the quality attributes. When the methodology in a chain is uniform, then if a failure occurs discussion about the problem can be rational, and talk about quality problems between steps in the chain will be much more constructive. The QIM has a precision of about ±1 day (Jónsdóttir *et al.*, 1991) which is as precise as it is ever likely to be. It takes about one minute per fish to make an estimate with the QIM (Hyldig, 2001). It is now possible to buy handheld terminals with a program that will speed up the QIM evaluation and registration of the data in standard format. It is still not so fast that it can be used on each individual fish, or even on each box as a real production control tool. But in a properly functioning quality chain it should only be necessary as a backup technique, a spot check on systems, or as a troubleshooting tool. To ensure less variation on the market the temperature must be controlled at 0°C and the fish labelled by date of catch and kept separate in the whole chain. The QIM is the best tool possible to ensure an objective quality inspection when the quality is questioned between steps in the chain or quality failure occurs.

15.9   Organising a chain management system

Organisation of the management in a quality fresh fish chain can be done by a management team of chain step representatives. In that group a chain manager either employed by the chain or a part of one of the steps is responsible for management of the meetings, reports and the daily management if that is necessary. The management from each step must participate and receive full support from each organisation to ensure that any change results in closer relations between the steps in the chain. The basic requirements are mentioned below.

- A common accepted chain management philosophy must be found, described and implemented.
- An established standard for communication:
  - means of communication
  - which pieces of information are transferred in between steps in the chain and what information is available on request.
- Cooperation described in the chain:
  - a written rule of what each should do in the chain and the common goals
  - an established system to handle conflicts
  - description of how participants must treat each other as customers on all levels

15.10   A common chain management philosophy

One of many management philosophies is Total Quality Management (TQM). TQM is a philosophy of management whereby everyone contributes to giving the customer what they expect all the time. It is a system for gauging a firm's dedication to constant improvement to serve the customer with direct emphasis on communication between management, employees, and the customer (Gould, 1992). 'Total' means everyone must be involved both inside and outside a company, 'Quality' means uniform quality of the products that are constantly being worked on to effect improvement, whilst remaining competitive.

'Management' means the way to act and work with all inside and outside the company. No fresh fish chains in Denmark appear to have a TQM system in function. In general the fish chains are at a very low level of development compared to TQM development in other industries. Quality programs such as TQM work within the framework of a company's existing processes and seek to enhance them by means of what the Japanese call *kaizen*, or continuous incremental improvement (Hammer and Champy, 1993).

More drastic is another philosophy of management, Business Process Reengineering (BPR) which is more a one-time project to achieve dramatic improvements. The basic definition is: fundamental rethinking and radical redesign of business processes to achieve dramatic improvements in critical contemporary measures of performance, such as cost, quality, service and speed (Hammer and Champy, 1993). Business processes are a set of activities that, taken together, produce a result of value to the customer. BPR shares a lot of common themes with TQM but generally BPR focuses on the processes without taking the present organisation into account and uses information technology to improve performance dramatically.

Supply chain management (SCM) is the management of upstream and downstream relationships with suppliers and customers to deliver superior customer value at less cost to the supply chain as a whole (Christopher, 1998). For a fresh fish chain of SMEs, SCM is a very useful philosophy because it is a basic tenet to focus on the core business and 'out-source' the rest (the rest is procured outside the company). That means that it is not vertical integration that 'normally' implies ownership of upstream and downstream steps in the chain, which is not possible in the existing structures. For an existing fresh fish chain with chain cooperation experience a BPR project can be initiated in case of a crisis where either the whole operation is changed or the companies will die.

Logistics is essentially a planning orientation and framework that seeks to create a single plan for the flow of product and information through a business. SCM builds upon this framework and seeks to achieve linkage and coordination between processes of other entities in the pipeline and the organisation itself (Christopher, 1998).

The latest management philosophy is the virtual supply chain or extended enterprise. Like SCM it is based on out-sourcing and IT as a core means, but the difference is that it is more of a temporary network around a certain production, problem or development.

The characteristic of the virtual supply chain is that the final product or offer is not created until the last possible moment. The idea is that maximum flexibility is achieved if there can be postponement in the creation of what has been termed 'time, place and for utility' (Christopher, 1998). Transferring this approach to fresh fish chains implies that the fresh fish chain or temporary chain member constellation is first created when initiated from the retailer/consumer step. When the promotion campaign for a week is decided and the species of fish is chosen, the chain is created from the best actors in each level of the market and the chain, in that form only exists, until it has done the one week's job. In a mature industry where quality is very well defined and easily measurable and controllable this is a good approach, but in the present fresh fish sector it is not possible without building mutual trust through steady cooperation between steps over a longer period of time. There is also a principal problem in dealing with variable lengths of vessel journeys of up to ten days and the ever-changing catch possibilities from day to day. However, the postpone principle can be used to postpone all chain operations until the last possible moment. The processing of whole fish to fillets in most traditional fish retailer shops already occurs just before the consumer gets the product. But by controlling the quality (and traceability), transport can also be postponed until the day before use (depending on distance) and the flexibility to use the whole fish for another purpose is still available if needed.

15.11   Communication and cooperation

Analysis has shown that primary telephone and fax are used as communication means today. Even in fish chains in Australia and Japan dealing with very high value products, such as bluefin tuna and kuruma prawns, the means of communication in the chains are the same, phone and fax (Frederiksen and Bremner, 2001). The development of the Internet has made it possible to speed up communications and the amount of data to be transferred is not limited once the connection has been made. The old methods are too expensive to use to establish traceability for a relative low-value product such as white fish species. Clearly the Internet will form the prime means for communication in the future. The standard for communication should be to use the Internet for daily use and the telephone to keep relations good and solve chain conflicts. Which pieces of information that should be transferred in between steps in the chain and what information is available on request must be decided by meetings in the chain management group. The basic pieces of information that could be transferred together with fresh fish are: date of catch, vessel name, species, size and a unique box number representing the batch of fish (Frederiksen et al., 2002). This information should be enough for the chain to handle the product without the need for quality inspection of the product. When any failures occur it should be agreed what kind of information can be requested to find and correct the mistake behind the failure or what data is not available for others. Additional information could for instance be: storage room temperatures, different quality assurance procedures in different areas of production but the main thing is that the information chain is kept intact to be able to find and correct any mistakes.

Requests for data should not be in the form of a paper system, as these are too costly and time consuming. A database system is the only future way to deal with traceability data in a low-value fresh fish chain. There is also a fortunate side-effect to transmitting the identity of previous steps to the next step in the chain. In the fishery industry today there is not the same sense of craftsmanship, as in other branches (e.g., pride in the job). The work of seapacking on board vessels developed a sense of pride in the crew of the vessels because their name was put on the product.

The developments in electronics and information transfer systems have made it possible to sell products/information on the Internet and several systems are available with the potential to register and transfer information together with the product. Seapacking systems have been developed since the mid-1990s. Seapacking enables the fishermen to weigh each fish box and register and label it on board the vessel. Today at least four systems for seapacking from different companies that make on-board weighing equipment are available on the market in Europe.

All these systems make it possible to sell the fish before the vessel enters the quay but it is only a few operators that actually use this feature. The reason is that there is no system on shore that is able to 'talk' and transfer information further in the chain and the auctions do not sell the fish before it is physically in the auction hall. If for instance the auction is bypassed and the fish are sold direct from the boat, the truck bringing the fish further on in the chain has several hours more in which to plan the transport. This means the product can reach the end market more quickly and that additional resources are saved in handling the product paperwork.

The newest development in transferring quality information in fresh fish chains is a system called 'Info-fisk' developed at DIFRES (Frederiksen et al., 2002). This system deals with quality information on each fish box and is able to keep the information flow intact in the whole chain from fisherman to retailer. The system is not a commercially available system but it is a test system. It has been proved that with the newest Internet technology it is possible to make information transfer in a secure and effective way without very high investments in time and equipment. Future developments will definitely be in traceability systems integrated in the traditional finance and production control systems used in the fish processing industry today. The written rule of what each should do in the chain and the common goals can only be settled through meetings where the differences between steps can be put on the table and discussed. The same should be done with a system to handle conflicts when they arise. Description of how each participant must treat the other as customers on all levels is a natural part of this work.

15.12 Developing quality chains

The objectives of a chain development system are:

- continuous development of the chain to:
    - improve quality and/or decrease variation in quality for each customer segment
    - optimise the operation/reduce costs
    - identify future strategic goals (e.g. ISO certification, change in organisation/chain members)

- measure the chain performance. The European Foundation for Quality Management (EFQM) excellence model is a suitable tool for this purpose.

It is clear that for a few SMEs forming a chain it is too much to implement all these points from the start. However, it is necessary to implement or at least deal with them and to have them in mind from the start to establish a healthy quality chain that will stay competitive and survive on the market in the long term. All of these above points cannot be considered in detail here. In the next sections some of the points will be studied further focusing on the overall functions to manage the quality chain.

15.12.1 Continuous development of the chain
The continuous development of the chain to improve quality and/or decrease variation in quality for each customer segment depends on the actual market situation, as the chain must be competitive. From an internal chain point of view it will always be an advantage to get as good a quality as possible and to minimise variation in quality, as this provides the capacity to be able to deliver higher quality than competitors or to gain extra time to process and sell the product.

Future strategic goals
In the chain development system, future strategic goals must be settled. An ISO certification could for instance be one of the future strategic goals for a fresh fish chain. Other examples are major changes in organisation, number of chain members and change in management and management philosophy.

15.12.2 Measure the chain performance
To be able to deal with future goals it is necessary to measure the performance of the present chain management. The European Foundation for Quality Management (EFQM) excellence model seems to be a suitable model to measure this. Figure 15.3 shows the EFQM model. This model is a general framework for self-evaluation that can be used by any industry to measure the performance of the management system. It is not a steady framework but is updated continuously by the organisation (EFQM, 2001).

A total (100%) of 1,000 Points is allotted with 500 points for the 'Enablers' side and 500 points for the 'Result' side. The model is divided into nine parts, five parts under Enablers and four parts under Results and the total points are distributed to each part, which also provides weighting and balance to the different functions. Heavy weightings are put on processes, consumer satisfaction and business result. TQM is often used in connection with this model and it can also be seen from the weights that TQM philosophy underlies this model. If a BPR project has been initiated and implemented from one year to another the results of the model may not be comparable to the year before, because of the fundamental redesign of the whole organisation by BPR.

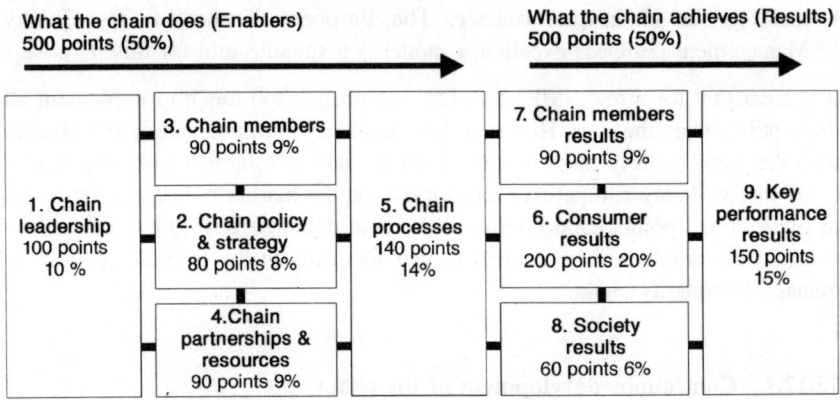

Fig. 15.3 The EFQM model modified to a fresh fish chain (from EFQM, 2001).

Table 15.2 Explanation of the EFQM model's nine parts

Part of the EFQM model	Explanation of part
1. Chain leadership	How leaders develop and facilitate the achievement of the mission and vision, develop values required for long term success and implement these via appropriate actions and behaviours, and are personally involved in ensuring that the organisation's management system is developed and implemented
2. Chain policy and strategy	How the chain organisation implements its mission and vision via a clear stakeholder-focused strategy, supported by relevant policies, plans, objectives, targets and processes
3. Chain members	How the chain organisation manages, develops and releases the knowledge and full potential of its people at an individual, team-based and organisation-wide level, and plans these activities in order to support its policy and strategy and the effective operation of its processes
4. Chain partnerships and resources	How the organisation plans and manages its external partnerships and internal resources in order to support its policy and strategy and the effective operation of its processes
5. Chain processes	How the organisation designs, manages and improves its processes in order to support its policy and strategy and fully satisfy, and generate increasing value for, its customers and other stakeholders
6. Consumer results	What the organisation is achieving in relation to its external customers
7. Chain members results	What the organisation is achieving in relation to its people
8. Chain society results	What the organisation is achieving in relation to local, national and international society as appropriate
9. Key performance results	What the organisation is achieving in relation to its planned performance

Each part of the EFQM model is listed and explained in Table 15.2. By adopting this model it should be possible to measure the performance of the chain management of a fresh fish chain. It is impossible for small SMEs to implement a system like this right away. But it is a very good list to build a system to measure the performance of the cooperation and then develop it in time. More details of the model can be found at (EFQM, 2001).

15.13   Future trends

As developments in information technology have been initiators for many management philosophies such as BPR and virtual chain, this IT-development will probably mature the fish processing industry to be able to make chain cooperation and quality chain management in the future. The marketplaces for fish and fish products on the Internet have been growing rapidly over the last 2–3 years and it is foreseen that they will expand further in future. The different kinds of e-commerce have recently been outlined as have most of the e-commerce websites concerning fish and fish products (Kyprianou, 2001). B2C (business to customers) concepts on the Danish domestic fresh fish market include concepts where consumers buy fresh fish direct from wholesalers on the Internet. Steps in the chain are bypassed and a special packing material including ice ensures cold chain in distribution. A variety of different concepts are available. One provides the consumers with traceability information such as vessel name, catch area, and catch date on each consumer package. Strong competition in this area will possibly bring more concepts on the market and that includes traceability. Many parcel service companies offer a 'track and trace' service that could readily be adopted for fresh fish thus providing consumers with actual transit status for their fish package.

B2B (business to business) approaches are used at 'Pefa.com' towards wholesalers. This company provides an electronic auction system available and operational in most countries of Europe (www.pefa.com). It has made it possible to buy fish on the Internet without actually inspecting the product before sale. The quality is ensured by manual quality evaluation of 'Pefa.Com' employees at the actual harbour of unloading before auctioning the fish. In the rest of the industry B2B is seldom used. When B2B ideas are further developed in the fish industry and when they include traceability, a new area in research will arise for analysis and simulation based on variations in data and chains. Profit will increase as chains are provided as early as possible with the product information necessary to choose the best end-user or end product. When a lot of data becomes available in the chain other decisions about what to transmit and what to make available on request, and who has access to it will need to be decided. User friendly techniques of storage, access and transfer will need to be developed. The present standard, that will probably also be retained into the near future, for transferring data on the Internet is extensible mark-up language (XML). Standardising the variable names and format will enhance easy and cheap information transfer with XML and an EU Concerted Action is now taking place to establish common views as to what data should follow a fish product and how the data should be coded and transmitted (www.tracefish.org).

It is important to stress that focus on ensuring quality and on the cold chain must be continuously maintained and developed together with the work on enhancing cooperation and information transfer in fish chains. If quality assurance is not maintained, the basis for chain cooperation and e-commerce will disappear, as the quality information has no value when the cold chain is lost.

15.14   References

ANONYMOUS (1998), 'French trawlers truck catches back home', *Fishing News*, May 1998.

BREMNER H A, OLLEY J and VAIL A M A (1987), 'Estimating time-temperature effects by rapid systematic sensory method', in: Kramer D E and Liston J, *Seafood Quality Determination*, Amsterdam, Elsevier, 413–35.

CHRISTOPHER M (1998), *Logistics and supply chain management*, London, Financial Times.

EEC (1994), 'Commission Decision 94/356/EC of 20 May 1994 laying down detailed rules for the application of Council Directive 91/493/EEC, as regards own health checks on fishery products', *Official Journal of the European Communities* No. L 156, 23.06.1994, 50–57.

EEC (1996), 'Council Regulation 2406/96/EEC of 26 November 1996 laying down common marketing standards for certain fishery products', *Official Journal of the European Communities* No. L 334, 23.12.1996, 1–15.

EFQM (2001), *The EFQM Excellence Model*. The European Foundation for Quality management, Brussels, www.efqm.org.

FREDERIKSEN M and BREMNER HA (2001), Fresh fish distribution chains, *Food Australia*, 54, 117–23.

FREDERIKSEN M, POPESCU V and OLSEN, K B (1997), 'Integrated Quality Assurance of Chilled Food Fish at Sea', in Luten J B, Børresen T and Oehlenschläger J, *Seafood from producer to consumer, integrated approach to quality*, Amsterdam, Elsevier, 87–96.

FREDERIKSEN M, ØSTERBERG C, SILBERG S, LARSEN E and BREMNER H A (2002), 'Info-fisk. Development and validation of an Internet based traceability system in a Danish domestic fresh fish chain', *Journal of Aquatic Food Product Technology*, 11 (2), 13–34.

GOULD W A (1992), *Total quality management for the food industries*, Maryland, USA, CTI publications.

HAMMER M and CHAMPY J (1993), *Reengineering the cooperation*, New York, USA, HarperCollins Publishers.

HUSS H H (1995), *Quality and quality changes in fresh fish*, Rome, Italy, FAO fisheries technical paper 348.

HYLDIG G (2001), *Personal communication*, Danish Institute for Fisheries Research, Department of Seafood Research, Lyngby, Denmark.

ISO (2000), *Quality management systems – Fundamentals and vocabulary*. Brussels, Belgium, European Standard [EN ISO 9000:2000, Point 3.5.4.], European Committee for Standardisation.

JONSDOTTIR S, LARSEN E, MARTINSDÓTTIR E, BRATTÅR R and GUDJÓNSSON A (1991), '*Kvalitetsnormer på fisk*', A report and manual (sensory evaluation of fish) to the Nordic Industry Foundation.

KYPRIANOU M (2001), 'Fisheries and e-commerce', *INFOFISH International*, 2, 21–4.

MOE T (1998), 'Perspectives on traceability in food manufacture', *Trends in Food Science & Technology*, 9: 211–4.

OLSEN K B (2001), *Personal communication*, Danish Institute for Fisheries Research, Department of Seafood Research, Lyngby, Denmark.

PORTER M E (1998), *Competitive Strategy*, 2nd edn, New York, The Free Press.

SYMES D and MADDOCK S (1989), 'The role of the inland wholesale markets in the distribution of fresh fish in the UK', *British Food Journal*, (91/51989), 7–12.

WARM K, BØKNES N and NIELSEN J (1998), 'Development of Quality Index Method for evaluation of frozen cod (*Gadus morhua*) and cod fillets', *Journal of Aquatic Food Product Technology*. 7, 45–59.

WILSON N (1996), 'The supply chain of perishable products in northern Europe', *British Food Journal*, 98/6, 9–15.

16

New non-thermal techniques for processing seafood

M. Gudmundsson and H. Hafsteinsson, Technological Institute of Iceland, Reykjavik

16.1 Introduction

The idea of trying to develop new methods for preservation of foods that cause the minimum of disruption to the appearance and properties of the foods has stimulated researchers to investigate a variety of novel techniques. Food irradiation is one such method that has been known for a long time but has not gained general acceptance. However, in the last few decades some other novel non-thermal technologies have been emerging that look promising. These are methods like high-pressure processing, high electric field pulse treatments and other less-investigated methods like oscillating magnetic field treatment and use of pulses of light for sterilisation. The main emphasis in this chapter will be on the effect of high pressure and high electric field pulses on seafood.

16.2 The potential application of high pressure

Food processing with high pressure is one of the latest methods for food preservation though it is still in its development phase. Foods preserved with high pressure look promising as they keep their natural appearance, taste and flavour. Even though high-pressure processing has been commercialised for products like juices and jams (Farr 1990) it has still not found commercial application for marine products. Effects of high pressure on seafood have only been tried in a limited number of researches. They have focused on different kind of fish, fish mince, surimi and effects on fish proteins (Shoji and Saeki 1989, Shoji *et al.* 1990, Okamoto *et al.* 1990, Ohshima *et al.* 1992, Yoshioka *et al.* 1992, Murakami *et al.* 1992, Goto *et al.* 1993, Iso *et al.*

1993, Ohshima *et al.* 1993, Yukizaki *et al.* 1993, Yukizaki *et al.* 1994, Ledward 1998).

16.3   Effect on microbial growth

There have been many studies on the effect of high pressure on reduction of microbial growth and sterilisation of food and beverages (Johnson and ZoBell 1949, Jaenicke 1981, Hoover *et al.* 1989, Cheftel 1995). They show that inactivation of microorganisms by high pressure depends on the pressure level and the duration of treatment in order to reduce bacterial growth in many kinds of food like milk, meats, fruits and juices. The inactivation involves denaturation of proteins, i.e., protein unfolding, aggregation and gelation that can lead to inhibition of enzymatic activities (like ATPases) and destruction of vital intracellular organelles for the microorganism (Johnston *et al.* 1992, Suzuki *et al.* 1992, Ogawa *et al.* 1992, Cheftel 1995). High pressure can also affect the membrane structure, chemical reaction and release of intracellular constituents that can contribute to microbial inactivation (Macdonald 1992, Shimada *et al.* 1993, Mozhaev *et al.* 1994, Cheftel 1995). The extent of inactivation depends on type of microorganism, the state of the microbes (i.e. growth phase, stationary phase or spores), the pressure level, the process time and temperature, and the composition of the dispersion medium (Carlez *et al.* 1994, Cheftel 1995).

The pH of the medium apparently has little influence on protection of microbes but salt, sugar and low water content seem to have strong baroprotective effect (Cheftel 1995). In most cases, at ambient temperatures it is necessary to apply pressures above 200MPa in order to induce inactivation of vegetative microorganisms. Some pathogenic microorganisms, like Yersinia enterocolitica and others, need 275MPa for 15–30 minutes at 20°C to be inactivated need 700MPa pressurisation like Salmonella enteritis, Escherichia coli and Staphylococcus aureus (Patterson *et al.* 1995).

A few studies have been done specifically on the effect of high pressure on inactivation of microbes in seafood products. The effect of high pressure on total count of bacteria in tuna and squid samples treated with 450MPa for 15 minutes at 25°C reduced the plate count between one and two log cycles. This treatment provides insufficient reduction to prevent spoilage or to sterilise the sample (Shoji and Saeki 1989).

Perhaps it is necessary to use a combination of moderate pressure and heat treatment in order to inhibit or inactivate vegetative microbes. The combination of heat treatment from zero to 60°C and high pressure up to 400MPA, tested on Lactobacillus casei and E. coli, showed that low temperature treatment (0°C) and high pressure was more effective than other treatments in inactivating microbes (Sonoike *et al.* 1992).

High pressure treatment (500MPa/10 min.) has been shown effectively to kill bacteria like Vibrio parahaemolyticus, Vibrio cholerae and Vibrio mimicus in sea urchin eggs but the eggs still retained their original flavour and taste

(Yukizaki *et al*. 1993). However, only 200MPa pressure for five minutes at 0°C is needed to inactivate Vibrio parahaemalyticus in a buffer solution (Yukizaki *et al*. 1994), indicating that some compounds in sea urchin eggs have a baroprotective effect. A study on oyster preservation by high pressure (López-Caballero *et al*. 2000a) showed that pressurisation of 400MPa for five minutes reduced the total quantity of microorganisms up to five log cycles. The pressurised oysters were stable for 41 days at 2°C but 13 days for the control sample. Another study on preservation of chilled and vacuum packaged prawns by high pressure at 200 and 400MPa showed that the shelf-life was extended by one and two weeks respectively compared to prawns that were only chilled and vacuum packaged (López-Caballero *et al*. 2000b).

16.4   Effect on seafood quality

Flavour, taste and texture are important quality parameters regarding consumer acceptance for seafood products. These quality parameters are affected during storage by many factors that can reduce the perceived freshness or spoil the product. These factors may involve protein denaturation, enzyme activities that produce off-flavour compounds and lipid oxidation. The consequences can be dripping of the fish muscle leading to dry and tough texture, rancidity and off-flavours. High pressure will affect these factors in a way that some of the quality parameters will be improved compared to other preservation methods and others will decrease compared to fresh products.

16.4.1   Effects on microstructure, fish proteins and enzymes

High pressure is said to induce breakdown of ionic bonds due to electrostriction of the water molecules because of the electric field effect of the ion on the dipoles of the solvent and partial disruption of hydrophobic interactions (Heremans 1995) In contrast, hydrogen bonds seem to be strengthened somewhat under pressure and covalent bonds have low sensitivity towards pressure. It depends on the pressure level, whether denaturation of proteins and inactivation of enzymes occurs. Denaturation can involve dissociation of oligomeric structures, unfolding of monomeric structure, protein aggregation and protein gelation (Balny and Masson 1993, Gross and Jaenicke 1994, Funtenberger *et al*. 1995, Cheftel 1995). Denaturation of proteins is the main underlying factor in changes in the microstructure of muscle products and in the inactivation of enzymes. Seafood treated with high pressure is therefore prone to all these changes, which depend on the severity of the pressurisation. The effects of high pressure can be either reversible or irreversible depending on the pressure and temperature. In general, reversible effects of high pressures are observed below 100–200MPa (e.g. dissociation of proteins into subunits).

Above 200MPa, non-reversible effects occur and they may include complete inactivation of enzymes and denaturation of proteins (Balny and Masson 1993).

Denaturation depends also on external parameters, e.g., temperature, pH and solvent composition (sugar, salts and other additives). Hydrophobic interactions are first affected by pressure below 150MPa and therefore the quaternary structure of proteins are first to change. Tertiary structure changes occur above 200MPa and changes of secondary structure take place above 700MPa (Balny and Masson 1993, Cheftel 1995). Thus protein denaturation caused by high pressure involves rearrangement and or destruction of non-covalent bonds such as hydrogen, hydrophobic interaction and ionic bonds of quaternary and tertiary structure of proteins while covalent bonds are not affected (Okamoto et al. 1990, Balny and Masson 1993, Cheftel 1995).

The main proteins of fish muscle are myofibrillar and sarcoplasmic proteins. Myofibrillar proteins are the proteins that determine the structure of the muscle while sarcoplasmic proteins are water soluble non-structural proteins. The myofibrillar proteins constitute between 65–80% of the total proteins in the fish muscle. They are mainly composed of the contractile proteins actin and myosin, regulatory proteins, elastic proteins and some other minor proteins. Myosin denatures at 100–200MPa and actin at 300MPa. Only a few soluble proteins survive a pressure of 800MPa (Balny and Masson 1993).

High-pressure treatment of carp myofibrils at 150MPa for 30 minutes, destroyed the arrangement of myofibrils and the striation pattern was lost (Ohshima et al. 1993). By contrast, myofibrils treated at 38°C for two hours still exhibited a striped appearance, although some unique structural changes had occurred. The mobilities on electrophoretic gels of myosin heavy chain and actin are not changed by high pressure of 150MPa or heating at 38°C (Shoji and Saeki 1989). However, when normal muscle from cod and mackerel is treated with high pressure, certain sarcoplasmic proteins become covalently linked together and are thus resistant to extraction with SDS (Ohshima et al. 1992).

16.4.2 Effects on enzymatic activity

Rigor mortis starts when the ATP level decreases post mortem in fish muscle. ATP is degraded into several compounds by dephosphorylases inherent in the fish muscle. Some of these compounds are intermediate compounds but other can accumulate in the fish during storage. The amount of these compounds is used to evaluate the freshness of the fish by the ratio of two of these compounds to ATP, which is called the k value (Saito et al. 1959, Sakaguchi and Koike 1992). When carp muscle was treated with high pressures of 200, 350 and 500MPa and subsequently stored at 5°C, further suppression of the decrease in inosine 5'monophosphate level (intermediate breakdown compound) was observed at 350 and 500MPa (Shoji and Saeki 1989). These results strongly suggest that the enzymes involved in degradation of ATP undergo protein denaturation and are deactivated during high-pressure treatment. On the other hand, heat promotes breakdown of ATP in fish, even in a very fresh one.

There is a difference between the inactivation of ATPase activity by heat and high pressure. The inactivation of Ca^{2+}ATPase activity by heat follows first-order kinetics (Arai 1977). On the other hand, Ca^{2+}ATPase activity in carp myofibrils that were pressurised at 125 and 150MPa showed a shift in linear relationship with time. After a certain time has lapsed there is an apparent breakpoint in the pressure-time relation of enzyme activity and the actvity decreases at a slower rate. This suggests that the mechanism of denaturation by heat is somewhat different from denaturation of proteins by high-pressure treatment (Ohshima *et al.* 1993).

This is supported to some extent by research on seven different fish species (Iso *et al.* 1994). They measured the effect of heat and pressurisation (200MPa for 13h) on denaturation of myofibrillar proteins by differential scanning calorimetry (DSC). The thermograms of proteins heated in DSC showed three peaks, indicating the denaturation of the two different myosin chains at 44°C and 51°C and actin at 71°C. However, the pressurised fish proteins when heated showed only two peaks, one small peak for the myosin and a broad indistinct peak for actin. The pressurised fish proteins seem to have been partially denatured, as total enthalpy change was noticeably smaller for pressurised fish proteins than untreated proteins. The myosin and actin were both partially denatured by the pressure at the same time.

Studies have been done on the effect of high pressure on enzymes that contribute to the deterioration of seafood (Ashie and Simpson 1995). These enzymes were trypsin, chymotrypsin, cathepsin and collagenase. They found that all the enzymes studied were susceptible to pressure between 100–400MPa and proportional to duration of pressure application. Trypsin was more susceptible to inactivation than chymotrypsin. Lipases are still active in fish muscle during storage at low temperatures and they will eventually release free fatty acids from glycolipids that accumulate in the fish (de Koning and Mol 1990). High-pressure treatment of fish above 405MPa before storage is needed to stop the increase of fatty acids and the decrease of phospholipids (Ohshima *et al.* 1993).

16.4.3 Effects on texture and microstructure

Researches on red meat show that pressurisation does not markedly affect ageing or conditioning of post-rigor meat at pressures below 200MPa and at ambient temperature (Cheftel and Culioli 1997). Also collagen, which is mostly stabilised with hydrogen bonds, is little affected by pressure at ambient temperatures. There are though significant effects of pressure on the organisation and subsequent gelation of myofibrillar proteins in both meat and fish (Ledward 1998). In Fig. 16.1, one can see a comparison of untreated and pressure-treated salmon muscle samples. The effect of 400MPa pressure for 30 minutes on the microstructure is clearly seen as the muscle has disintegrated and the cells have decreased in size compared to the untreated sample (data from the authors).

The stability of different myosins to pressure depends, as with their thermal stabilities, on the environment of the species. Thus, myosins from both turkey

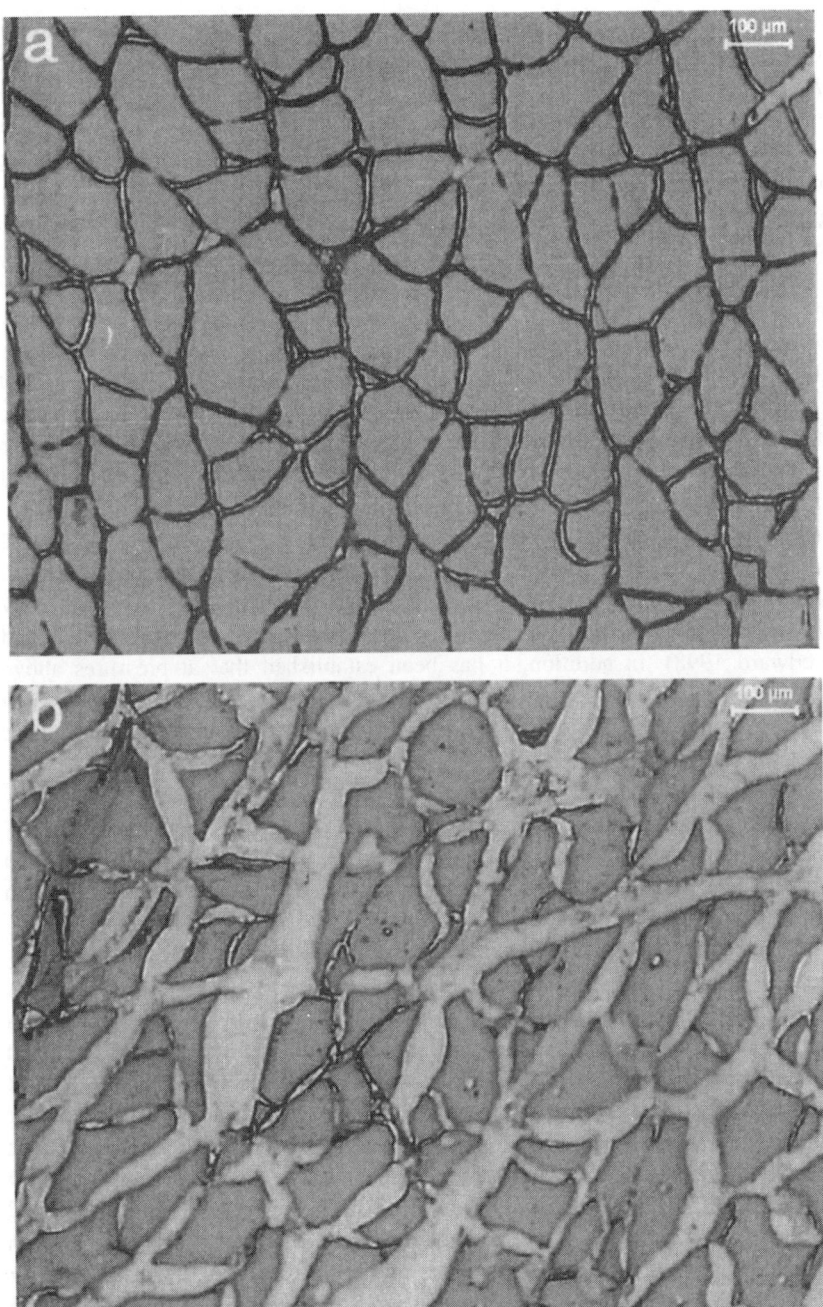

Fig. 16.1

and pork are significantly more stable to pressure than myosin from cold-water fish like cod (Cheah and Ledward 1996, Angsupanich and Ledward 1998 and Angsupanich *et al.* 1999). At relatively low pressures, 100–200MPa, myosin initially aggregates by two heads fusing together to form one headed structure, and they further aggregate and form a clump of heads with the tails extending radially outwards (Cheftel and Culioli 1997). A similar mechanism has been proposed for the initial stages in the thermal gelation of myosin (Yamamoto *et al.* 1990).

Although the initial stages of the aggregation of myosin, either through heat or pressure may be similar, the subsequent gelation mechanisms are very different which is not surprising considering the relative stabilities of hydrogen bonds and hydrophobic interactions to pressure and temperature. Thus, on heat treatment the myosin tails readily unfold and form a gel network or aggregate depending on the conditions and are primarily stabilised by disulphide linkages and hydrophobic interactions (Yamamoto *et al.* 1990). On pressure treatment there is a formation of new or modified hydrogen bonded structures which, when subjected to differential scanning calorimetry, melt a few degrees below the temperature at which native myosin denatures (Angsupanich and Ledward 1998). The hydrogen bonds melt at relatively low temperatures but are relatively insensitive towards pressure in the range 200–800MPa (Angsuspanich and Ledward 1998). In addition, it has been established that at pressures above 400MPa, the myosin heavy chain can form disulphide linkages (Angsupanich and Ledward 1998). Disulphide bonds are also formed on heat gelation in these systems (Lee and Lanier 1995).

In short, the thermally produced myosin gels are stabilised primarily by disulphide linkages and hydrophobic interactions. However, the pressure-induced myosin gel results in a gel network, which is stabilised by both disulphide linkages and a significant number of hydrogen bonds that can subsequently be broken on heat treatment.

The texture of pressure-treated fish is therefore markedly different from heat-treated fish (Angsupanich and Ledward 1998, Angsupanich *et al.* 1999). Pressure-treated fresh cod muscle at 400MPa showed much greater hardness determined by texture profile analysis than heated cod muscle at 50°C. Cod treated with pressures below or above 400MPa showed less hardness than at 400MPa. If, however, the pressure-treated cod muscle is heated it will be similar in hardness to the heat-treated one. The texture of pressure-treated fish is relatively heat sensitive and will soften up on heat treatment at low temperatures. Largest changes were seen in adhesiveness, chewiness and gumminess of cod muscle at pressure treatment below 400MPa compared to untreated samples. However, the results on high-pressure treatment on bluefish showed that a pressure of 101MPa increased the firmness of the fish muscle during storage at 4–7°C but pressurisation of 202 and 303MPa had the opposite effect (Ashie *et al.* 1997). The choice of the evaluation method is important as Ashie and Simpson (1996) obtained different results with sensoric analysis and an Instron compression probe on pressurised bluefish at 300MPa. The pressurised blue fish was judged harder in sensoric analysis than the control sample but the reverse was the case with the compression method.

Oysters also showed increased shear strength after pressurisation compared to untreated oysters (López-Caballero et al. 2000a). Shrimps treated with high pressure of 200 or 400MPa were somewhat harder than control samples (López-Caballero et al. 2000b). It has also been found that high-pressure treatment improved gel forming ability of poor performing fish minces (Pérez-Mateos and Montero 2000). In surimi and other fish meat products, high-pressure treatment readily induces gelation at low temperatures (Ohshima et al. 1993, Shoji et al. 1994).

Hydrostatic pressure of 200MPa for 30 minutes at 25°C was needed to induce a gel from carp crude actomyosin that could support its own weight and maintain its shape. The pressure-induced gel kept its original colour and flavour and was glossy and soft in comparison with heat-induced gels (Ohshima et al. 1993). The gels tended to increase in hardness and to decrease in adhesiveness as the applied pressure was increased. However, they were still soft and had large extensibility and were not fractured by high stress. There are significant differences in appearance and textural properties between pressure- and heat-induced gels. Heat-induced carp gels swell a little and are relatively hard but lack adhesiveness. These results are further indications that the gelation mechanism is different between pressure and heat-induced gels.

A study of the difference between heat- and pressure-treated fish gels has shown that a blue whiting gel formed with high pressure of 200MPa (10°C and ten minutes) had greater breakforce and more cohesiveness than heat-induced gels (Borderias et al. 1997). Pressurisation at 4°C prior to incubation at 25 or 40°C increased the gel strength two- to threefold in uncooked surimi gels that contained transglutaminase (TGase). High pressure rendered protein substrates are more accessible to TGase, thereby enhancing intermolecular cross-link formation and gel strength. The TGase enzyme was not affected by high pressure up to 300MPa at 4°C (Ashie and Lanier 1999).

The effects of high pressure on sarcoplasmic proteins of sardine, walleye pollack, marble sole and horse mackerel has also been investigated (Okamoto et al. 1990). The sarcoplasmic proteins become insoluble and precipitate at pressures above 140MPa and when the concentration of sarcoplasmic was above 50mg/ml, the proteins formed gels. The properties of the gels were affected by many factors such as fish species, pH, protein concentration, pressure and treatment time. The hardness of pressure-induced gels was highest for the sole but the breaking strength was highest for horse mackerel. The strength of the gels increased with increased applied pressure, where the sole gel showed the greatest increase. Water-holding capacity of the gels generally decreased with increased pressure for the sole and pollack but was relatively constant for the sardine and horse mackerel at applied pressures up to 370MPa. Breaking strength showed maximum at pH between 5 and 6 but water-holding capacity was then at minimum (Okamoto et al. 1990). An observation by electronic microscope showed that the gels were porous and rheological measurements showed that they were elastic and quite different from heat-induced gels. The breaking strength of pressure-induced gels made at 470MPa was much greater than for heat-induced gels of the same proteins (Okazaki and Nakamura 1992). The results for sarcoplasmic fish proteins show similar trends when pressurised as for the myofibrillar proteins.

High pressure has also been applied to surimi of both Pacific whiting and Alaska pollack (Shoji et al. 1990, and Chung et al. 1994). Surimi analogues are traditionally made from heat-induced gels at temperatures near 90°C (Lee 1984, Lanier and Lee 1992). A number of fish species may undergo a weakening of gel structure during normal heating regimes because of endogenous proteases in the muscle tissue (Niwa 1992). As the temperature increases during cooking, it will cause gelation of the surimi product. However, the product is heated through an interval 50–60°C, where these proteases are most active. At present this potential weakening of the gel is prevented by the use of protein inhibitors like beef plasma proteins or egg whites (Matsumoto and Noguchi 1992). High-pressure-treated Pacific whiting and Alaska pollack surimi gels with added protease inhibitors showed greatly increased elasticity at all pressure/temperature combinations (100–280MPa and temperatures between 28–50°C) when compared to heat-induced surimi gels. However, the gel strength varied. The gel strength of Alaska pollack surimi gel with added inhibitor was higher than control except gels formed at 50°C and highest pressure used. A pressure-treated Pacific whiting surimi gel without inhibitor had threefold increase in strain and stress values compared to heat-induced gels except at 50°C where the pressurised surimi did not form a gel. This indicates that protease activity is increased under pressure at that temperature (Chung et al. 1994).

Another study (Shoji et al. 1990) on surimi showed that surimi with 2.5% salt formed strong gels when treated with pressure between 200–400MPa at 0°C for ten minutes. The gel formed at 300MPa formed the strongest gel. The pressure-induced surimi gels, as in the study above, formed gels with greater gel strength than heat-induced gels and were more transparent. Fractionation and electrophoresis of the surimi gel proteins suggested that gel formation by high pressure depends largely on the cross-linking of myosin heavy chains (Shoji et al. 1990).

16.4.4   Effects on oxidation of lipids

The marine lipids are characterized by high levels of polyunsaturated fatty acids (PUFA) (Ackman 1990). The PUFAs are generally susceptible to autoxidation and oxidative degradation of lipids in foods and foodstuffs during processing and subsequent storage directly affects the quality of products, including flavour, colour, texture and nutritional value (Eriksson 1982). Highly purified fats and oils are believed to be relatively stable to oxidation when subjected to high pressure. However, with commercial fats the relationship between high pressure and sensitivity of the fat to lipid oxidation is a complex function of water activity (Cheah and Ledward 1995).

There are few studies on the effect of high pressure on fish oils. When extracted sardine oil was treated with hydrostatic pressure of 506MPa for 60 minutes, the oxidation indicators, peroxide value (POV) and thiobarbituric acid (TBA) did not change (Tanaka *et al.* 1991). On the other hand, when cod muscles were exposed to high hydrostatic pressure of 202, 404 and 608MPa for 15 and 30 minutes the POV of the extracted oils increased with increased hydrostatic pressure and processing time. Even more pronounced effects were observed for mackerel muscle lipids (Ohshima *et al.* 1992). These results indicate that pure oils are stable after high-pressure treatment but lipids in the fish muscle are not, probably due to release of metal ions that act as catalysts. This is supported by the work of other researchers (Cheah and Ledward 1995 and Angsupanich and Ledward 1998). They found also that applying high pressure above 400MPa decreased the oxidative stability of the lipids in cod.

Lipid oxidation appears to be catalysed in the range of water activities that are most common in meat and fish products, when subjected to pressures over 400MPa (Cheah and Ledward 1995). The pressure induces changes in fat and tissues that probably release metal ions from specific complexes, which are then able to catalyse the oxidation. This has been supported by a study on incorporation of appropriate anti-oxidants and specific metal chelators that effectively inhibit oxidation (Cheah and Ledward 1997). It is not clear from what compounds the ions are released. Haem compounds are considered unlikely as a catalytic effect is also seen in cod muscle but a complex like haemosiderin and other insoluble complexes are likely candidates as oxidation still occurred despite removal of soluble metal complexes (Cheah and Ledward 1996, Ledward 1998).

16.4.5 Effects on appearance and colour

High pressure can be used in food processing even though it causes denaturation of proteins as it inactivates microorganisms without changing the flavour, colour, vitamins and tastes of foods (Hayashi 1993). The only noticeable colour change in white fish such as cod and mackerel is the loss of translucency; the fish becomes opaque due to denaturation of proteins and the fish looks similar to cooked fish (Ohshima *et al.* 1992, Shoji *et al.* 1990, Cheah and Ledward 1996, Angsupanich *et al.* 1999). These changes take place at pressures between 100–200MPa for cod (Angsupanich and Ledward 1998).

16.5 Other uses of high pressure and future trends

It has been mentioned that high pressure can be used to produce surimi gels and other fish products of better quality (Okamoto *et al.* 1990). High pressure below subzero temperature can also be used to produce rapidly small ice crystals (microcrystallisation) in a product, which would be less detrimental to microstructure and the texture of the product than traditional freezing (Karino *et al.* 1994, Cheftel 1995). This could be of great advantage in frozen fish products. Another potential application is thawing of product under a pressure between zero and 20°C because water is not frozen in this range at a pressure of 210MPa. Thus it is possible to thaw product at subzero temperatures with the help of pressurisation (Kalichevsky *et al.* 1995, Cheftel 1995).

As it is difficult to destroy both microorganisms and microbial spores even at as a high pressure as 450MPa (Miyao *et al.* 1993), it seems necessary to use high-pressure treatment in combination with temperatures either below −20°C or at moderately high heating temperature to obtain acceptable preservation and stability of quality attributes of the fish product. This effect could also be obtained by combination with another minimal processing method. Further studies on the use of high pressure in combination with other minimal processing methods are therefore necessary. High pressures, above 405MPA, have also been shown successfully to inhibit some inherent enzymatic activities which cause undesirable changes in seafood quality attributes. Thus use of high pressures may result in a product which is more stable during storage and has a longer available shelf-life.

High-pressure treatment has been shown to produce kamaboko with a very fine surface and to induce fish gels with very interesting properties. The pressure-induced fish gels give an indication that a range of novel products could be produced from fish or other marine products that have an appearance and texture that is different from traditional products.

16.6   The potential application of high-intensity pulsed electric fields (PEFs)

The use of high-intensity pulsed electric fields (PEFs) is a non-thermal preservation method like high-pressure treatment that has the potential to be used in fish processing. The possible uses of PEF as a food preservation method have been investigated for a number of years (Knorr 1995, Knorr *et al.* 1998, Barbosa-Cánovas *et al.* 1998). The main emphasis has been on inactivation of different types of microorganisms in different phases, i.e., growth and stationary phases or as spores (Sale and Hamilton 1967, Castro *et al.* 1993, Hülsheger and Nieman 1980, Wouters *et al.* 1999). A few studies have been done on the use of PEF to improve the yield of juices (Flaumenbaum 1968, Knorr *et al.* 1994). For the consumer of fish products the sensoric experience is important besides safety issues and the texture is a large part of that experience. Traditional processes for fish products such as frozen storage, drying, salting and canning have from moderate to severe effects on the microstructure of the product compared to fresh product (Duerr and Dyer 1952, Connell 1964, Chu and Sterling 1970, Dunajski 1979, Bello *et al.* 1982, Fennema 1990, Mackie 1993, Sikorski and Kotakowska 1994, Greaser and Pearson 1999). There are only limited researches on the effect of PEF treatment on microstructure of food (Barsotti *et al.* 1999, Fernandez-Diaz *et al.* 2000) and only one is available on fish products as far as we are aware (Gudmundsson and Hafsteinsson 2001).

16.7   Effect on microbial growth

The lethal action of electric fields on living cells has been explained by dielectric breakdown of the cell membrane (Zimmermann *et al.* 1976, Zimmermann 1986, Sale and Hamilton 1967). The applied external electric field induces transmembrane potential, which above a certain critical value of 1V causes pore formation that can be lethal to microorganisms. The irreversible changes occur to a cell when an external electric field between one and 10kV/cm is used for more than 10–15m (Zimmermann *et al.* 1976).

It has been shown that the relative rate of killing bacteria is related to the field strength, duration of application of the field and also the number of pulses and the pulse width. It is also clear that PEF inactivation is a function of type of microorganism and the microbial growth stage, the initial number of microbes, the ionic concentration and conductivity of the suspension (Hülsheger *et al.* 1981, Wouters and Smelt 1997). Bacteria in the growth phase are more sensitive to an electric field than stationary bacteria and spores are the most resistant (Sale and Hamilton 1967, Wouters and Smelt 1997). Some factors in the suspension media or in the sample seem to have a protective effect, for example, cations, proteins and lipids (Hülsheger *et al.* 1981, Zhang *et al.* 1994, Grahl and Märkl 1996, Martín *et al.* 1997). The bactericidal effect of PEF decreases with increased ionic strength (Hülsheger 1981). Therefore it is more difficult to inactivate microorganisms in semi-solid or solid food materials than in dilute buffer solutions as they are rich in ions and other protective substances (Hülsheger *et al.* 1981, Zhang *et al.* 1994). It has also been pointed out that many foods are heterogeneous with areas of different electrical resistivity, which can alter the effects of PEF treatment as some areas will be untreated and others over-treated in such material (Barsotti *et al.* 1999).

To our knowledge no specific study has been published on the effect of PEF treatment on different types of bacteria in seafood. The effect of PEF treatment on total bacteria count has though been reported on lumpfish roes, where treatment of 11kV/cm and seven pulses ($2\mu s$ in width) reduced the total bacteria count by one log cycle (Gudmundsson and Hafsteinsson 2001). In seafood the most potential bacterial pathogens are those of the Vibrionaceae family and the most important of these are Vibrio cholerae, Vibrio parhaemolyticus and Vibrio vulnificus and one can also mention Aeromonadas hydrophila (Wekell *et al.* 1994). Other pathogenic bacteria that can be present in seafood for various reasons are Salmonella species, *E. coli, Shigella, Campylobacter, Yersinia enterocolitica, Clostridium botulinum, Listeria monocytogenes, Staphylococcus aureus* and *Bacillus cereus* (Liston 1990).

Listeria moncytogenes in a stationary phase was reduced by two log cycles and between two and three log cycles in the growth phase at 20kV/cm and 30 pulses respectively (Hülsheger *et al.* 1983). Stapylococcus aureus and E. coli in the stationary phase treated in the same way were reduced between three and four log cycles. Other researches show reduction from two to nine log cycles for E. coli using fields from 20kV/cm up to 70kV/cm (Dunn and Pearlman 1987, Zhang *et al.* 1994). The reduction for Staphylococcus aureus was two log cycles at 27.5kV/cm (Hamilton and Sale 1967) and reduction of four log cycles for Salmonella dublin at 18kV/cm (Dunn and Pearlman 1987). No studies are available for Vibro species, Shigella or Campylobacter. However, it is clear that external electric fields above 20kV/cm are needed to inactivate most types of microorganisms for minimum of two to three log cycles.

16.8   Effect on seafood quality

16.8.1   Effects on proteins and enzyme activity

It has been shown that ovalbumin and other egg white proteins do not denature when treated with PEF with an electric field of 27–33kV/cm and using from 50 to 400 pulses (Jeantet *et al.* 1999, Fernadez-Diaz *et al.* 2000). There are no studies on PEF dealing with denaturation of proteins in seafood, except that it has been reported that treatment of cod with PEF treatment of up to 18.6kV/cm and seven pulses ($2\mu s$ width) did not affect the proteins from cod (Gudmundsson and Hafsteinsson 2001). According to their results, obtained from SDS-electrophoresis, no changes were seen in molecular bands in PEF-treated cod proteins compared to untreated samples.

Studies on the effect of PEF treatment on enzyme activity show that many enzymes are unaffected even at electric fields above 30kV/cm. These are enzymes like amylases, lipase, NADH dehydrogenase, succinic hydrogenase and hexogenase (Hamilton and Sale 1967). However PEF treatment inactivates some enzymes like proteases from Pseudomonas fluorescens at 15kV/cm and 98 pulses and plasmin at 30kV/cm and 50 pulses (Vega-Mercado *et al.* 1995a,b). The enzymes α-amylase, lipase and glucose oxidase were markedly inactivated at very high electric fields of 64–87kV/cm, whereas peroxidase and polyphenoloxidase were more resistant (Ho *et al.* 1997). No studies are available that deal specifically with enzymes from marine sources.

16.8.2   Effects on texture and microstructure

There are only a few studies on the effect of PEF treatment on the properties of foods, mainly on juices and other pumpable foods, which show that sensoric properties are not affected (Knorr *et al.* 1994, Qin *et al.* 1995, Barbosa-Cánovas *et al.* 1996). Research on meat and seafood is very limited. Only one publication on the effect of texture and microstructure of muscle foods has been published to our knowledge (Gudmundsson and Hafsteinsson 2001). Changes in microstructure and texture can be expected as a consequence of the increase in permeability caused by PEF treatment that can induce changes in water-holding properties of the muscle.

Salmon treated with PEF of 1.36kV/cm and 40 pulses caused gaping in the fish muscle and collagen leakage into the extra-cellular gap between the muscle cells as seen under microscope (Gudmundsson and Hafsteinsson 2001). Treatment with high pressure at 300MPa also caused gaping. Teleostic fish including salmon contain low amounts of connective tissue in the muscle (0.66%) (Dunajski 1979, Eckoff et al. 1998). On the other hand, for example, chicken meat contains about 2% of connective tissue (Baily and Light 1989). It is also known that the increased size of cells make them more vulnerable to PEF treatment (Sale and Hamilton 1967, Hülsheger et al. 1993). The muscle cells of salmon are considerably larger than any bacteria as muscle cells are usually between 50–100μm in diameter but bacteria between 0.3–2.0μm (Nester et al. 1983, Wong 1989). These two facts could explain why salmon and fish in general do not tolerate even a mild PEF treatment without damage to the microstructure. The impact of PEF treatment on the microstructure of fish muscle cannot be the result of protein denaturation as far too low an intensity of electric field was used. A probable explanation is punctuation of the cell membranes that causes leakage of cell fluids into extra-cellular space.

Fresh lumpfish roes treated with 12kV/cm and 12 pulses (2μs) were intact after the treatment except for a very low percentage of the roes as can be seen in Fig. 16.2. Firmness of PEF-treated roes measured with a compression test showed also that the PEF treatment only marginally affected the firmness of the roes (Gudmundsson and Hafsteinsson 2001). Another study (Craig and Powrie 1988) on frozen and then thawed salmon roes showed that 46% less energy is needed to rupture such roes than fresh roes. The three-layer membrane of the roes probably gives them the strength to tolerate PEF treatment. The roes can probably tolerate an even stronger PEF treatment, which could then make it more plausible to use PEF treatment on roes for preservation.

16.9   Future trends in PEF

Preservation of fish products with PEF treatment does not seem plausible as a relatively low intensity of electric field pulses have a detrimental effect on fish microstructure and at the same time the low field voltage does not effectively reduce the growth of bacteria. Roes on the other hand seem to tolerate PEF treatment without a visible effect on the microstructure or texture. A PEF treatment could therefore be valuable as a pre-treatment for roes but that needs to be further investigated. Other possible uses of PEF treatment in the fish industry have not been explored but it could be possible to use PEF treatment in a similar way to that used in juice production where PEF treatment ruptures the tissue cells, making it easier to extract the valuable substances. This could be done on waste material and by-products from the fish industry and possible products could be, e.g., enzymes or fish oil.

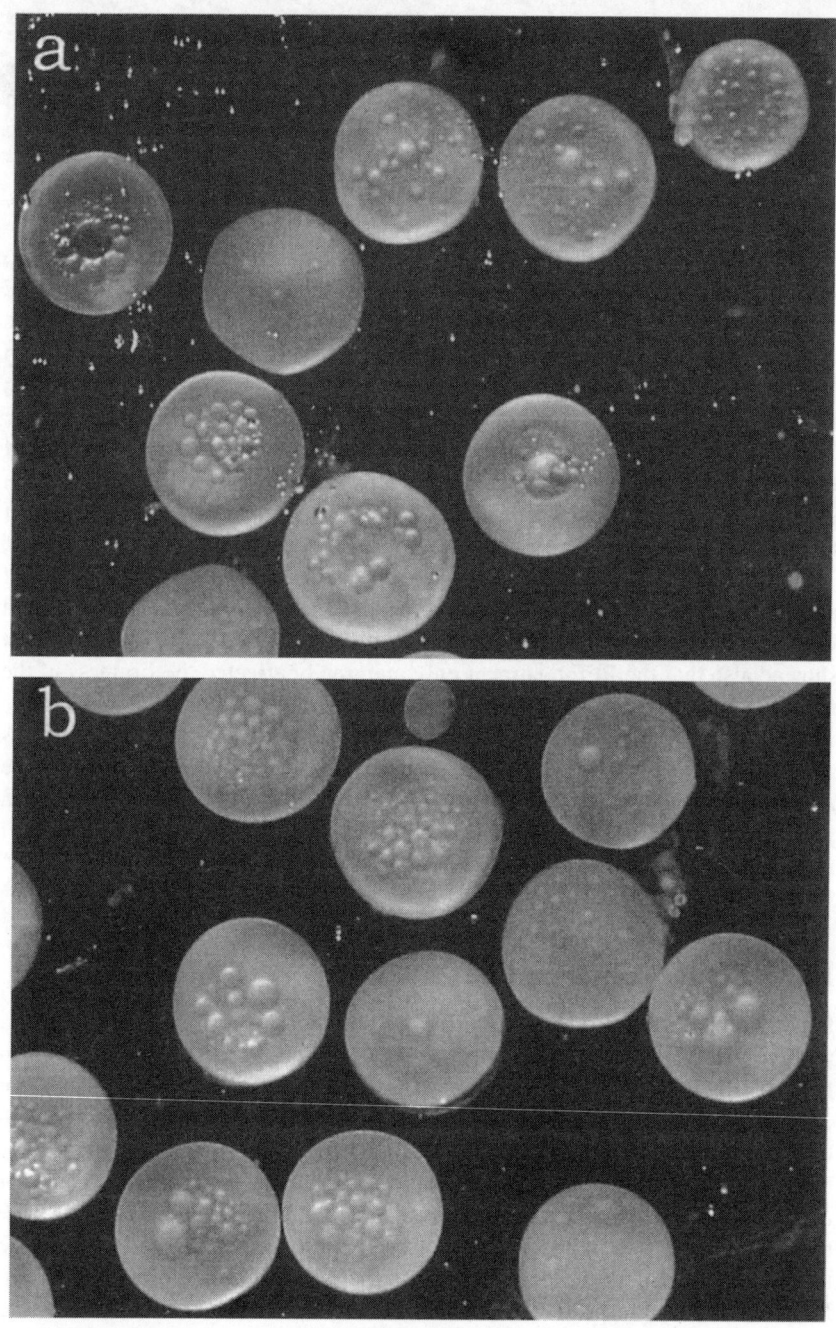

Fig. 16.2

16.10 References

ACKMAN R G (1990), 'Seafood lipids and fatty acids', *Food Review International*, 6 (4) 617–46.

ANGSUPANICH K and LEDWARD D A (1998), 'High pressure treatment effects on cod (Gadus morhua) muscle', *Food chemistry*, 63 (1), 39–50.

ANGSUPANICH K, EDDE M and LEDWARD D A (1999) 'The effects of high pressure on the myofibrillar proteins of cod and turkey', *Journal of Agricultural and Food Chemistry*, 47 (1), 92–9.

ARAI K (1977), 'Fish muscle proteins' Tokyo, Japanese Society of Scientific Fisheries. Kouseisha-kouseikaku, 75–90.

ASHIE I N A and SIMPSON B K (1995), 'High pressure effects on some seafood enzymes', IFT Annual Meeting, Session 71D-9.

ASHIE I N A and SIMPSON B K (1996), 'Application of hydrostatic pressure control enzyme-related seafood texture deterioration', *Food Res Int*, 29, 564–75.

ASHIE I N A, SIMPSON B K and RAMASWAMY H S (1997), 'Changes in texture and microstructure of pressure-treated fish muscle tissue during chilled storage', *Journal of Muscle Foods*, 8, 13–32.

ASHIE I N A and LANIER T C (1999), 'High pressure effects on gelation of surimi and turkey breast muscle enhanced by microbial transglutaminase', *Journal of Food Science*, 64 (4), 704–8.

BAILY A J and LIGHT N D (1989), *Connective Tissue in Meat and Meat Products*, London, Elsevier Applied Science.

BALNY C and MASSON P (1993), 'Effects of high pressure on proteins', *Foods Review International*, 9 (4), 611–28.

BARBOSA-CÁNOVAS G V, QIN B L and SWANSON B G (1996), 'Preservation of foods by pulsed electric fields: system design and key components', in Rodrigo M, Martinez A, Fiszman S M, Rodrigo C, and Mateu A, *Proceedings of the international symposium on Advanced Technologies in Sterilization and Safety of Foods and Non Food Products*, Valencia, Instituto de Agroquímica y Technologica de Alimentos, 273–87.

BARBOSA-CÁNOVAS G V, POTHAKAMURY U R, PALOU E and SWANSON B G (1998), *Non-Thermal Preservation of Foods*, New York, Marcel Dekker Inc, pp 1–276.

BARSOTTI L, MERLE P and CHEFTEL J C (1999), 'Food processing by pulsed electric fields. I. Physical aspects', *Food Review International*, 15 (2), 163–80.

BELLO R A, LUFT J H and PIGOTT G M (1982) 'Ultrastructural Study of Skeletal Fish Muscle after Freezing at Different Rates', *Journal of Food Science*, 47, 1389–94.

BORDERIAS A J, PÉREZ-MATEOS M, SOLAS M and MONTERO P (1997), 'Frozen storage of high-pressure and heat induced gels of blue whiting (Micromesistius poutassou) muscle: rheological, chemical and ultrastructure studies', *Z Lebensm Unters Forch A*, 205, 335–42.

CARLEZ A, ROSEC J P, RICHARD N and CHEFTEL J C (1994), 'Bacterial growth during chilled storage of high pressure-treated minced meat' *Lebensm Wiss Technol*, 27, 48–54.

CASTRO A J, BARBOSA-CÁNOVAS G V and SWANSON B G (1993), 'Microbial inactivation of foods by pulsed electric fields', *J Food Proc Pres*, 17, 47–73.

CHEAH P B and LEDWARD D A (1995), 'High pressure effects on lipid oxidations', *Journal of American Oil Chemist Society*, 72, 1059–63.

CHEAH P B and LEDWARD D A (1996), 'High pressure effects on lipid oxidation in minced pork', *Meat Science*, 43, 123–34.

CHEAH P B and LEDWARD D A (1997), 'Catalytic mechanism of lipid oxidation following high pressure treatment of pork fat and meat' *Journal of Food Science*, 62, 1135–8, 1141.

CHEFTEL J C (1995), 'Review: High pressure, microbial inactivation and food preservation' *Food Science and Technology International*, 1, 75–90.

CHEFTEL J C and CULIOLI J (1997), 'Effects of high pressure on meat: a review', *Meat Science* 46, 211–36.

CHU G H and STERLING C (1970), 'Parameters of Texture Change in Processed Fish: Myosin Denaturation', *Journal of Texture Studies*, 2, 214–22.

CHUNG Y C, GEBREHIWOT A, FARKAS D F, and MORRISSEY M T (1994),'Gelation of surimi by high hydrostatic pressure' *Journal of Food Science*, 59 (3), 523–34.

CONNELL J J (1964), 'Fish Muscle Proteins and some Effect on them of Processing', in Schultz H W and Anglemier A F, *Proteins and their Reactions*, Connecticut, Avi Westport, pp. 255–94.

CRAIG C L and POWRIE W D (1988), 'Rheological Properties of Fresh and Frozen Chum Salmon Eggs with and without Treatment by Cryoprotectants', *Journal of Food Science*, 53 (3), 684–7.

DE KONING A J and MOL T H (1990), 'Rates of free fatty acid formation from phospholipids and neutral lipids in frozen cape hake (Merluccius spp) mince at various temperatures', *Journal of Science of Food and Agriculture*, 50, 391–8.

DUERR J D and DYER W J (1952), 'Protein in fish muscle. IV. Denaturation by salt', *J. Fish Res Bd Can*, 325–31.

DUNAJSKI E (1979), 'Texture of fish muscle' *Journal of Texture Studies*, 10, 301–9.

DUNN J E and PEARLMAN J S (1987), 'Methods and apparatus for extending the shelf-life of fluid food products', US Patent 4,695,472.

ECKHOFF K M, AIDOS I, HEMRE G I and LIE Ø (1998), 'Collagen content in farmed Atlantic salmon (Salmo salar, L.) and subsequent changes in solubility during storage on ice', *Food Chemistry*, 62, 197–200.

ERIKSSON C E (1982),'Lipid oxidation catalysis and inhibitions in raw materials and processed foods', *Food Chemistry*, 9, 3–19.

FARR D (1990), 'High pressure technology in the food industry', *Trends in Food Science and Technology*, 1 (1), 14–16.

FENNEMA O R (1990), 'Comparative water holding properties of various muscle foods', *Journal of Muscle Foods*, 1, 363–81.

FERNANDES-DIAZ MD, BARSOTTI L, DUMAY E and CHEFTEL J C (2000), 'Effects of pulsed electric fields on ovalbumin solutions and dialyzed egg white', *J Agric Food Chem* 48, 2332–9.

FLAUMENBAUM B L (1968), 'Anwendung der Elektroplasmolyse bei der Herstellung von Fruchtschäften', *Flüssiges Obst* 35, 19–22.

FUNTENBERGER S, DUMAY E and CHEFTEL J C (1995), 'Pressure aggregation of β-lactoglobulin isolate in different pH 7 buffers', *Lebensm Wiss Technol*, 28, 410–18.

GOTO H, KAJIYAMA N and NOGUCHI A (1993), 'Changes of soy and fish proteins under high pressure', in Hayashi R, *High pressure Bioscience and Food Science*, Kyoto, San-Ei Publications, 315–21.

GRAHL T and MÄRKL H (1996), 'Killing of microorganisms by pulsed electric fields', *Appl Microbiol Biotechnol*, 45, 148–57.

GREASER M L and PEARSON A M (1999), 'Flesh foods and their analogues', in Rosenthal A J, *Food Texture*, Gaithersburg, Maryland, Aspen Publishers Inc, 228–58.

GROSS M and JAENICKE R (1994), 'Protein under pressure. The influence of high hydrostatic pressure on structure, function and assembly of proteins and protein complexes', *Eur J Biochem*, 221, 617–30.

GUDMUNSSON M and HAFSTEINSSON H (2001), 'Effect of electric field pulses on microstructure of muscle foods and roes', *Trends in Food Science and Technology*, 12, 122–8.

HAMILTON W A and SALE A J H (1967), 'Effects of high electric fields on microorganisms. II. Mechanism of action of the lethal effect', *Biochim Biophys Acta*, 148, 789–800.

HAYASHI R (1993), *High pressure bioscience and food science*, Kyoto, San-Ei Suppan.

HEREMANS K (1995), 'High pressure effects on biomolecules, in Ledward D A, Johnston D E, Earnshaw R G and Hasting A P M, *High pressure processing of foods*, Nottingham, Nottingham University Press, 81–98.

HO S Y, MITTAL G S and CROSS J D (1997), 'Effects of high field electric pulses on activity of selected enzymes', *Journal of Food Engineering*, 31, 69–84.

HOOVER D G, METRICK C, PAPINEAU A M, FARKAS D F and KNORR D (1989), 'Biological effects of high hydrostatic pressure on food microorganisms', *Food Technology*, 43, 99–107.

HÜLSHEGER H and NIEMANN E G (1980), 'Lethal effects of high voltage pulses on E. coli K12', *Radiat Environ Biophys*, 18, 281–8.

HÜLSHEGER H, POTEL J and NIEMANN EG (1981), 'Killing of bacteria with electric pulses of high field strength', *Raditiat Environ Biophys* 20, 53–65.

HÜLSHEGER H, POTEL J and NIEMANN E G (1983), 'Electric field effects on bacteria and yeast cells' Raditiat Environ Biophys, 22, 149–62.

ISO S, MIZUNO H, OGAWA H, MOCHIZUKI Y and ISO N (1994), 'Differential scanning calorimetry of pressurized fish meat', *Fisheries Science*, 60 (1), 127–8.

JAENICKE R (1981), 'Enzymes under extreme conditions', *Ann Rev Biophys Bioeng*, 10, 1–67.

JEANTET R, BARON F, NAU F, ROIGNANT M and BRUTÉ G (1999), 'High intensity pulsed electric fields applied to egg white: Effect on Salmonella enteritis inactivation and protein denaturation', *J Food Protect*, 62, 1381–6.

JOHNSON F H and ZOBELL C E (1949), 'The retardation of thermal disinfection of Bacillus subtilis spores by hydrostatic pressure' *J Bacteriol*, 57, 353–8.

JOHNSTON D E, AUSTIN B A and MURPHY R J (1992), 'The effects of high pressure treatment on skim milk, in Balny C, Hayashi R, Heremans K and Masson P, *High pressure and Biotechnology*, Colloque INSERM/ John Libbey Ltd, 243–7.

KALICHEVSKI M T, KNORR D and LILLFORD P J (1995), 'The effects of high pressure on water and potential food applications', *Trends in Food Science and Technology*, 6, 253–9.

KARINO S, HANE H and MAKITA T (1994), 'Behavior of water and ice at low temperature and high pressure' in Hayashi R, *High pressure bioscience*, Kyoto, San-Ei Suppan Co, 54–65.

KNORR D, GEULEN M, GRAHL T and SITZMANN W (1994), 'Food application of high electric field pulses', *Trends in Food Science and Technology*, 5, 71–5.

KNORR D (1995), 'Advances and limitations of non-thermal food preservation methods', in Ahvenainen R, Mattila-Sandholm T and Ohlsson T, *New Self-life Technologies and Safety Assessments*, Helsinki, VTT Symposium 148, 7–18.

KNORR D, HEINZ V, UN-LEE D, SCHLÜTER O and ZENKER M (1998), 'High pressure processing of foods: Introduction', in Autio K, *Fresh Novel Foods by High Pressure*, Espoo, VTT Symposium 186, Finland.

LANIER T C and LEE C M (1992), *Surimi Technology*, New York, Marcel Dekker INC,

LEDWARD D A (1998), 'High pressure processing of meat and fish' in Autio K, *Fresh Novel Foods by High Pressure*, Espoo, VTT Symposium 186, 165–176.

LEE C M (1984), 'Surimi process technology', *Food Technology*, 38 (11), 69–80.

LEE H and LANIER T C (1995), 'The role of crosslinking in the texturizing of muscle protein sols', *Journal of Muscle Foods*, 6, 125–38.

LISTON J (1990), 'Microbial hazards of seafoods consumption', *Food Technol*, 44 (2), 56–62.

LÓPEZ-CABELLERO M E, PEREZ-MATEOS M, MONTERO P and BORDERÍS JA (2000a), 'Oyster preservation by high-pressure treatment', Journal of Food Protection, 63 (2), 196–201.

LÓPEZ-CABELLERO M E, PÉREZ-MATEOS M, BORDERÍS J A and MONTERO P (2000b), 'Extension of the shelf-life of prawns (Penaeus japonicus) by vacuum packaging and high-pressure treatment', *Journal of Food Protection*, 63 (10), 1381–8.

MACDONALD A G (1992), 'Effects of high hydrostatic pressure on natural and artificial membranes', in Balny C, Hayashi R, Heremans K and Masson P, *High Pressure and Biotechnology*, Colloque INSERM/ John Libbey Ltd, 67–74.

MACKIE I M (1993), 'The Effects of freezing on flesh proteins', *Food Reviews International*, 9, 575–610.

MARTÍN O, QIN B L, CHANG F J, BARBOSA-CÁNOVAS G V and SWANSON B G (1997), 'Inactivation of Echerichia coli in skim milk by high intensity pulsed electric fields', *J Food Process Eng*, 20, 317–36.

MATSUMOTO J J and NOGUCHI S F (1992), 'Cryostabilization of protein in surimi', in Lanier T C and Lee C M, *Surimi Technology*, New York, Marcel Dekker INC, 357–88.

MIYAO S, SHINDOH T, MIYAMORI K and ARITA T (1993), 'Effects of high pressurization on the growth of bacteria derived from surimi (fish paste)', *Nippon Shokuhin Kogyo Gakkaishi*, 40 (7), 478–84.

MOZHAEV V V, HEREMANS K, FRANK J, MASSON P and BALNY C (1994), 'Exploiting the effects of high hydrostatic pressure in biotechnological applications', *Trends in Biotechnol*, 12, 493–501.

MURAKAMI T, KIMURA I, YAMAGISHI T, YAMASHITA M and SATAKE M (1992), 'Thawing of frozen fish by hydrostatic pressure' in Balny C, Hayashi R, Heremans K and Masson P, *High pressure and Biotechnology*, Colloque INSERM/ John Libbey Ltd, 329–31.

NESTER E, ROBERTS C E, LIDSTROM M E, PEARSALL N N and NESTER M T (1983), *Microbiology*, Philadelphia, Holt-Saunders International Edition, p 16.

NIWA E (1992), 'Chemistry of surimi gelation', in Lanier T C and Lee C M, *Surimi Technology*, New York, Marcel Dekker INC, 389–428.

PATTERSON M F, QUINN M, SIMPSON R and GILMOUR M (1995), 'The sensitivity of vegetative pathogens to high hydrostatic pressure treatment in phosphate-buffer saline and foods', *J Food Protect*, 58, 524–9.

PÉREZ-MATEOS M and MONTERO P (2000), 'Response surface methodology multivariate analysis of properties of high-pressure induced fish mince gel', *Eur Food Res Technol* 211, 79–85.

OGAWA H, FUKUHISA K and FUKUMOTO H (1992), 'Effect of high hydrostatic pressure on sterilization and preservation of citrus juice', in Balny C, Hayashi R, Heremans K and Masson P, *High pressure and Biotechnology*, Colloque INSERM/ John Libbey Ltd, 269–78.

OHSHIMA T, NAKAGAWA T and KOIZUMI C (1992), 'Effect of high hydrostatic pressure on the enzymatic degradation of phospholipid in fish muscle during storage', in Blight, E G, *Seafood Science and Technology*, Oxford, Fishing News Books, 64–75.

OHSHIMA T, USHIO H and KOIZUMI C (1993), 'High-pressure processing of fish and fish products', *Trends in Food Science and Technology*, 4, 370–75.

OKAMOTO M, KAWAMURA Y and HAYASHI R (1990), 'Application of high pressure to food processing: Textural comparison of pressure- and heat-induced gels of food proteins', *Agric Biol Chem*, 54 (1), 183–9.

OKAZAKI E and NAKAMURA K (1992), 'Factors influencing texturization of sarcoplasmic protein of fish by high pressure treatment', *Nippon Suisan Gakkaishi*, 58 (11), 2197–2206.

QIN B L, POTHAKAMURY U R, VEGA H, MARTÍN O, BARBOSA-CÁNOVAS G V and

SWANSON B G (1995), 'Food pasteurization using high-intensity pulsed electric fields', *Food Technology*, 49 (12), 55–60.

SAITO T, ARAI K and MATSUYOSHI M (1959), 'A new method for estimating the freshness of fish', *Bull Jpn Soc Sci Fish*, 24, 749–50.

SAKAGUCHI M and KOIKE A (1992), 'Freshness assessment of fish fillets using the Torrymeter and k-value', in Huss H H, Jakobsen M and Liston J, *Quality Assurance in the Fish Industry*, Amsterdam, Elsevier, 321–32.

SALE A J H. and HAMILTON W A (1967), 'Effects of high electric fields on microorganisms I. Killing of bacteria and yeasts', *Biochim Biophys Acta*, 148, 781–8.

SHIMADA S, ANDOU M, NAITO N, YAMADA N, OSUMI M and HAYASHI R (1993), 'Effects of hydrostatic pressure on the ultrastructure and leakage of internal substances in the yeast Saccharomyces cerevisiae', *Appl Microbiol Biotechnol*, 40, 123–31.

SHOJI T and SAEKI H (1989), 'Processing and preservation of fish meat by pressurization' in Hayashi R, *Use of High Pressure in Food*, Kyoto, San-Ei Publications, 75–87.

SHOJI T, SAEKI H, WAKAMEDA A, NAKAMURA M and NONAKA M (1990), 'Gelation of salted paste of Alaska pollack by high hydrostatic pressure and change in myofibrillar protein in it', *Nippon Suisan Gakkaishi*, 56 (12), 2069–76.

SHOJI T, SAKEI H, WAKAMEDA A and NONAKA M (1994), 'Influence of ammonium sulfate on the formation of pressure-induced gel formation walleye pollack surimi', *Nippon Suisan Gakkaaishi*, 60, 101–9.

SIKORSKI A E and KOTAKOWSKA A (1994), 'Changes in protein in frozen stored fish', in Sikorski Z E, Pan B S and Shahidi F, *Seafood Proteins*, New York, Chapman and Hall, 99–112.

SONOIKE K, SETAYMA T, KUMA Y and KOBAYASHI S (1992), 'Effects of pressure and temperature on the death rate of Lactobacillus casei and E. coli', in Balny C, Hayashi R, Heremans K and Masson P, *High pressure and Biotechnology*, Colloque INSERM/ John Libbey Ltd, 297–301.

SUZUKI A, KIM K, HONMA N, IKEUCHI Y and SAITO M (1992), 'Acceleration of meat conditioning by high pressure treatment', in Balny C, Hayashi R, Heremans K and Masson P, *High pressure and Biotechnology*, Colloque INSERM/ John Libbey Ltd, 217–27.

TANAKA M, XUEYI Z, NAGASHIMA Y and TAGUCHI T (1991), 'Effect of high pressure on lipid oxidation in sardine meat', *Nippon Suisan Gakkaishi*, 57 (5), 957–63.

VEGA-MERCADO H, POWERS J, BARBOSA-CÁNOVAS G V and SWANSON B G (1995a), 'Plasmin inactivation with pulsed electric fields', *Journal of Food Science*, 60, 1143–6.

VEGA-MERCADO H, POWERS J, BARBOSA-CÁNOVAS G V, SWANSON B G and LÜDECKE L (1995b), 'Inactivation of protease from Pseudomonas fluorescens M3/6 using high voltage pulsed electric fields', Anaheim CA, Annual IFT 1995 Meeting, June 3–7, Book of abstracts, paper no. 89-3, p. 267.

WEKELL M M, MANGER R, COLBURN K, ADAMS A and HILL W (1994), 'Microbiological quality of seafoods: bacteria and parasites', in Shahidi F and Botta J D, *Seafoods: chemistry, processing technology and quality*, London, Blackie Academic and Professional, 196–219.

WONG D W S (1989), *Mechnism and theory in food chemistry*, New York, AVI Book, 48–109.

WOUTERS P C and SMELT J P P M (1997), 'Inactivation of microorganisms with pulsed electric fields: Potential for food preservation', *Food Biotechnology*, 11 (3), 193–229.

WOUTERS P C, DUTREUX W, SMALT J P P M and LELIEVELD H L M (1999), 'Effects of pulsed electric fields on inactivation kinetics of Listeria innocua', *Applied and Environmental Microbiology*, 65 (12), 5364–71

YAMAMOTO K, MIURA T and YASUI T (1990), 'Gelation of myosin filament under high hydrostatic pressure', *Food structure*, 9, 269–77.

YOSHIOKA K, KAGE Y and OMURA H (1992), 'Effect of high pressure on texture and ultrastructure of fish and chicken muscles and their gels', in Balny C, Hayashi R, Heremans K and Masson P, *High pressure and biotechnology*, Colloque INSERM/ John Libbey Ltd, 325–7.

YUKIZAKI C, KANO M and TSUMAGARI H (1993), 'The sterilization of sea urchin eggs by hydrostatic pressure', in Hayashi R, *High Pressure Bioscience and Food Science*, Kyoto, San-Ei Publications, 225–8.

YUKIZAKI C, KAWANO M, JO N and KAWANO K (1994), 'High pressure sterilization and processing of sea urchin eggs', in Hayashi R, Kunagi S, Shimada S and Suzuki A, *High Pressure Bioscience*, Kyoto, San-Ei Suppan Co, 248–58.

ZHANG Q, CHANG F J, BARBOSA-CÁNOVAS G V and SWANSON B G (1994), 'Inactivation of microorganisms in a semisolid model food using high electric voltage pulsed electric fields', *Lebensm Wiss u Technol*, 27, 538–43.

ZIMMERMANN U, PILWAT G, BECKERS F and RIEMANN F (1976), 'Effects of external electrical fields on cell membranes', *Bioelectrochem Bioenergetics*, 3, 58–83.

ZIMMERMANN U (1986), 'Electrical breakdown, electropermeabilization and electrofusion', *Rev Physiol Biochem Pharmacol*, 105, 176–256.

16.11   Acknowledgement

Experimental studies were carried out at Matra (at the Technological Institute of Iceland) and were supported by the Icelandic Research Council, the Nordic Industry Fund and The European Commission under the FAIR programme.

17

Lactic acid bacteria in fish preservation

G. M. Hall, Loughborough University

17.1 Introduction

This chapter will describe the nature of lactic acid bacteria (LAB) as a group and their role in food fermentations generally. The commodities briefly treated here will include dairy, vegetable, cereal and meat fermentations. The various mechanisms of the protective effects of lactic fermentation will be described and also the concept of the probiotic effect. These general principles will then be applied in detail to the LAB fermentations of fish and fish products. In this context the word, 'fish', includes other aquatic species, particularly crustacea, which feature prominently as fermented products. Fermentations of note here include the fish sauces and pastes of South East Asia and the ensilation of waste fish and fishery by-products to generate animal feeds and other added-value products. The use of LAB fermentation as a preservative for fish fillets will also be described. Finally, some thoughts on the future of LAB fermentation will be put forward and sources of information to complement the academic references included.

17.2 The lactic acid bacteria (LAB)

17.2.1 General definition

The LAB are a group of bacteria composed of several genera with a number of morphological, physiological and metabolic characteristics in common. These characteristics include: Gram-positive staining; anaerobic, micro-aerophilic or aero-tolerant; catalase negative; rods or cocci. Also, by definition, they produce lactic acid as a sole or main product of their metabolism. They are generally perceived as benign organisms with the ability to preserve various food types usually associated with favourable changes in texture and flavour and the probiotic effect. Recent developments in bacterial taxonomy have led to a widening of the description of LAB bringing into the fold organisms which are sporeformers, pathogenic, motile and pseudo-catalase positive. However, for the purposes of this chapter only the beneficial aspects of the LAB will be considered. In terms of environments the LAB are also found on plants, sewage and in the genital, intestinal and respiratory tracts of humans and animals.

17.2.2   Metabolism

The production of lactic acid by LAB can be described as homolactic (only lactic acid produced) or heterolactic (produces carbon dioxide, lactate, acetate and sometimes ethanol). The starting point is a hexose sugar, usually glucose, and lactic acid is the end product of the well-known Embden-Meyerhof-Parnas (EMP) glycolytic pathway. The heterolactic route will be taken under certain conditions and it might be necessary to promote the homolactic route in food fermentations if lactic acid is to be the only end-product. Lactose (a disaccharide) and pentose sugars can also be metabolised yielding varying amounts of lactic acid. The provision of fermentable sugars in a food to be preserved by the action of LAB is an essential prerequisite for a successful fermentation.

Figure 17.1 describes the homolactic and heterolactic routes from glucose. Some LAB can catabolise amino acids by deamination or decarboxylation with the production of carbon dioxide, ammonia and volatile fatty acids (VFA). These compounds can contribute to the flavour of fish products although the presence of ammonia can prevent low pH. Arginine is common in fish tissues and is metabolised, but any amino acid can be utilised. Decarboxylation can give rise to toxic biogenic amines and sulphur-containing amino acids can give rise to hydrogen sulphide (Han-Ching et al., 1992).

17.2.3   The genera

The recent developments in bacterial taxonomy mentioned above have come about through advances in metabolic classification methods and molecular biology such that new genera have been created and movements of species groups between genera have occurred (Wood and Holzapfel, 1995). The classical LAB genera include: Lactobacillus, Streptococcus, Pediococcus, Lactococcus, Leuconostoc, Bifidobacterium, Carnobacterium, Enterococcus and Sporolactobacillus. Table 17.1 gives examples of species from these genera which are important in fermentation. Of these genera, the Lactobacillus with 56 species, Streptococcus (39) and Bifidobacterium (29) are the most numerous. Several species from these genera have been found associated with fermented fish products (see later) although some workers have suggested that the presence of terrestrial LAB in sea and fresh water is a sign of contamination. The Carnobacterium genus consists of six species which were originally

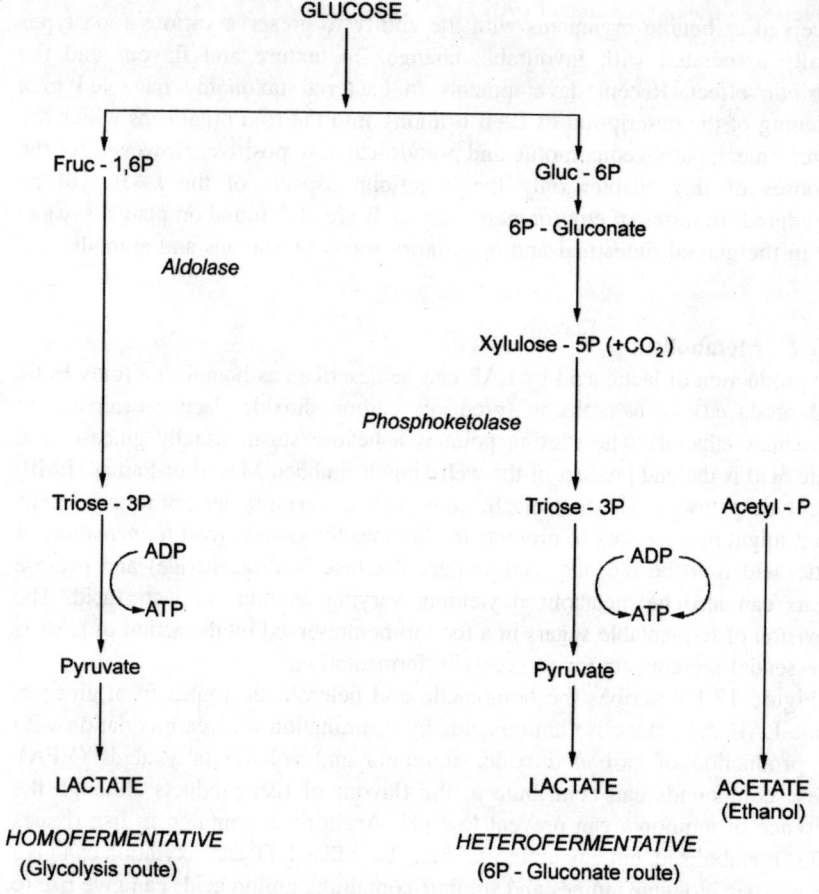

Fig. 17.1 Main metabolic pathways for LAB production of lactic acid.

described as unusual members of the *Lactobacillus* genus isolated from meat, poultry, fish and seawater. *C. piscicola* has been isolated from several fish species and *C. funditum* and *C. alterfunditum* have been isolated from seawater.

17.3 Inhibitory effects

The preservation of foods through LAB fermentation is due to the inhibition of a wide range of spoilage and pathogenic organisms by the various end-products of the fermentation.

17.3.1 Acid production

The accumulation of lactic and other organic acids generated by LAB reduces the pH of the environment with an inhibitory effect on Gram-positive and Gram-

Table 17.1 Classical genera of LAB for fermentation

Genus	Typical species
Lactobacillus	Lb acidophilus
	Lb brevis
	Lb casei subsp casei
	Lb delbrueckii
	Lb paracasei
	Lb plantarum
Pediococcus	P acidilactici
	P halophilus
	P parvulus
Lactococus	Lc lactis subsp lactis
	Lc lactis subsp cremoris
Leuconostoc	L mesenteroides (various subsp)
	L lactis
Bifidobacterium	B bifidum
	B longum
	B breve
Carnobacterium	C piscicola
	C funditum
	C alterfunditum

negative bacteria. In addition, the undissociated form of the organic acid can penetrate the microbial cell wall where the higher pH of the cell contents promotes dissociation with the release of hydrogen ions and the acid anion and both of these species interfere with cell metabolism. The pK value of the organic acid is important because the undissociated species will predominate at a pH below the pK for that acid. Hence, acetic acid (pK 4.76) will have greater antimicrobial activity than lactic acid (pK 3.86) at a given pH. It is often useful for food preservation to be achieved at a relatively high pH for palatability reasons, as will be seen later when considering ensilation.

17.3.2 Hydrogen peroxide and carbon dioxide

When oxygen is present LAB can generate hydrogen peroxide which, in turn, can generate hydroxy radicals which cause the peroxidation of membrane lipids and microbial cell susceptibility. These effects are well known for many organisms. Carbon dioxide is an end-product of heterolactic fermentation and, on occasions, by decarboxylation of amino acids by LAB. Carbon dioxide promotes an anaerobic environment, reduces pH and can help to destroy the integrity of the microbial cell wall.

17.3.3 Bacteriocins

Bacteriocins have been described in many ways over the years. They are, 'antibiotic-like', molecules, usually protein in nature, of varying molecular

Table 17.2　Bacteriocins produced by LAB

Bacteriocin	Produced by	Active against
Nisin(s)	*Lc lactis* subspp *lactis*	*Bacillus cereus*
		Clostridium botulinum
		Staphylococcus aureus
Pediocin(s)	*P acidilactis*	*Listeria monocytogenes*
		Clostridium sporogenes
Reuterin	*Lb reuteri*	*Listeria monocytogenes*
		Salmonella typhimurium
		Shigella spp

weight, mode of action and specificity of action (see Table 17.2). They are produced by many LAB, either naturally or induced, with some LAB producing a single form whilst others produce two or three forms. Recent classification has divided bacteriocins into four groups: lantibiotics (contain lanthionine); small heat-stable peptides; large heat-stable proteins and complex proteins which require a carbohydrate or lipid prosthetic group. The importance of bacteriocins in food preservation has been recognised by much research into their properties and production as bio-preservatives in their own right. Nisin is a bacteriocin produced by *Lactococcus lactis* with activity against Gram-positive bacteria but not the important pathogenic Gram-negative bacteria such as *Escherichia coli* and *Salmonella*. Recent research has looked for synergistic effects between nisin and other compounds and processing techniques to improve its efficacy against the Gram-positive organisms.

17.4　Probiotic effect

The word probiotic is derived from the Greek 'for life' and first used to describe the positive effects of a substance secreted by one microorganism on another (the opposite of antibiotic). Further refinements of this approach have led to the recognition of the positive benefits of microorganisms on higher animals including humans. Metchnikoff (1907) saw the benefits of regular consumption of fermented milk products on the health and longevity of the Russian peoples. The importance of LAB in this phenomenon was also appreciated from the start. A modern definition of a probiotic (Naidu *et al.*, 1999) is 'a microbial dietary adjuvant that beneficially affects the host physiology by modulating mucosal and systemic immunity, as well as improving nutritional and microbial balance in the intestinal tract'. Now there are several commercial products available claimed to alter the intestinal flora towards a beneficial population with prophylactic effects. The effect of probiotic LAB spp against *Escherichia coli* 0157:H7 has been reported (Erkkila *et al.*, 2000). The influence of probiotics has been extended to positive immunomodulatory and anti-tumour effects.

Following from the interest in probiotic effects in humans there has been much research into effects in poultry and pigs and the control of plant pathogens, but relatively little on the effects in aquaculture (Wood, 1992). The use of probiotics in aquaculture has been suggested to prevent diseases, and the use of antibiotics to treat them, giving rise to a reduced environmental impact by aquaculture (Kumar and Sharma, 2001).

17.5   LAB fermentation of foods

As this chapter is devoted to LAB fermentation of fish we will not dwell on other LAB fermentations. However, there are a number of lessons to be learned from the study of these ferments which can be applied in the fish fermentation area.

17.5.1 Dairy fermentations

The LAB fermentation of milk is a classic process for the preservation of a perishable food. Although known from ancient times it is still an important process in areas of the world where refrigeration is not easily available. The milk of cows, sheep, goats, camels, water buffalo, reindeer and mares are all used for LAB fermentation. The range of products is wide and they can be produced by naturally occurring LAB or, on an industrial scale, by the addition of carefully controlled inocula. Lactose is the main carbon source in milk and can be enzymically hydrolysed into D-glucose and D-galactose which can both be converted into lactic acid which is the major end-product. The conversion rate varies for the milk from the different species mentioned above but the overall effect is the development of an acidic flavour due to lactic acid and the precipitation of casein at its iso-electric pH.

Other physical changes take place such as gelling and syneresis of a thin, 'whey-like' solution. Kosikowski (1977) classifed LAB fermented milks into four groups: acid/alcohol types such as kefir; high-acid types such as Bulgarian sour milk; medium-acid types such as yoghurt and low-acid types such as cultured buttermilk and cream. Table 17.3 lists some representative types of LAB fermented milks. Dairy LAB fermentations in general indicate the need for a fermentable carbon source and the possibility for physical changes brought about by pH in addition to the development of flavours.

17.5.2 Vegetable fermentations

The LAB fermentation of vegetables is a process equally as old as that for milk and to this day certain fermented vegetables remain as the staple diet of some countries. Classical examples of fermented vegetables include sauerkraut and pickled cucumber in Europe and kimchi (pickled cabbage, cucumber or radish) in Korea. Sauerkraut production is a closely regulated commercial process whilst kimchi

Table 17.3   LAB fermented milk products

Fermentation type	Product	Milk	Responsible organisms
Acid/alcohol	Kefir	Goat, sheep, cow	*Lb brevis*
	Koumiss	Mare	*Lb delbrueckii* subsp *bulgaricus* (and yeasts)
High acid	Bulgarian sour milk	Cow	*Lb delbrueckii* subsp *bulgaricus*
Medium acid	Yoghurt	Cow, buffalo, sheep, goat	*Lb delbrueckii* subsp *bulgaricus*
Low acid	Cultured buttermilk	Skim milk	*Streptococcus lactis* subsp *diacetylactis*

production is home based. Although there are differences in these two products the LAB fermentation of vegetables generally has some common attributes. Firstly, the addition of salt to draw out water-soluble nutrients for the LAB to utilise (and to inhibit competing flora) and the development of flavours due to sulphurous compounds and lipid-based compounds modified by the presence of lactic acid. Table 17.4 gives a list of some common fermented vegetables.

17.5.3   Cereal fermentations

The LAB fermented cereal products come in two forms: acid-leavened breads and porridge-like foods. The leavened breads are found in many areas such as the Indian sub-continent, the Philippines and North East Africa (Ethiopia and Sudan) although the demand for such breads does exist in Europe and the USA. The porridge-like products are found in many areas of Africa. Table 17.5 lists some examples of these food types. The cereals used include rice, rye, wheat, maize, tef, and has been extended to include legumes and cassava. A characteristic of these ferments is that the lactic acid production occurs alongside a yeast-mediated ferment. Such an arrangement relies on the provision of fermentable sugars suitable for both organisms. For sourdough breads *Torulopsis holmii* is a yeast that can utilise glucose which has been produced by

Table 17.4   LAB fermented vegetable products

Product	Raw material	Responsible organisms
Sauerkraut	White cabbage	*Lb brevis*, *Lb plantarum*, *Lb mesenteriodes*, *P cercivisiae*
Kimchi	Korean cabbage, radish, cucumber	*L mesenteroides*, *L, brevis*, *P cerevisiae*, *L plantarum* (and aerobes, yeasts)
Hum choy (Chinese sauerkraut)	Gai-choy	As above

Table 17.5   LAB fermented cereal products

Product	Raw material	Responsible organisms
Breads		
Sourdough	Rye, wheat	*Lactobacillus* spp
Enjera	Tef (Ethiopia)	Moulds, yeasts, such as *Candida guilliermondii* various LAB
Porridges		
Ogi (Nigeria)	Maize (and sorghum millet)	Various moulds, yeasts, bacteria, *Lb plantarum*
Uji (Kenya)	As for ogi	As for ogi and *P acidilactici, Lb fermentum*
Gari (Nigeria)	Cassava	*Lb plantarum, Streptococcus* spp, *Leuconostoc* spp, *Candida*
Kenkey (Ghana)	As for gari	Fungi, yeasts, *P cerevisiae, L mesenteroides, Lb fermentum*

Lactobacillus spp from maltose which itself has been produced by the action of amylases on starch (Steinkraus, 1996).

17.6   LAB fermentation of fish

As mentioned earlier, in this section the term 'fish' will include other aquatic species, particularly crustacea, and although the emphasis will be on marine resources the same principles could be applied to freshwater fish.

17.6.1   European products

In Europe the fermentation of fish has a long history which now is mainly associated with the Scandinavian countries. A product called, 'garum' was much prized by the ancient Romans and consisted of a fermented and autolysed fish paste. Gaffelbitar, Tidbits and Surstromming are fermented products made from the Atlantic herring (*Clupea harengus*) whilst Rakefisk is made from trout (*Salmo trutta*). Gaffelbitar and Tidbits are fermented for 12–18 months with a mixture of salt, sugar and spices present. The fermented fish are filleted and packed for sale. Flavour development is due mainly to endogenous enzyme activity but the contribution of LAB is significant. The organisms involved have included: *Pediococcus, Lactobacillus* and *Leuconostoc* spp. Surstromming and Rakefisk are fermented in salt for a shorter time, up to two weeks, when a very strong aroma is generated due to methylmercaptan production (Han-Ching *et al.*, 1992)

Table 17.6   LAB fermented fish products of S.E. Asia

Product sources	Country	Fish species
Nuoc-mam	Vietnam	*Stelophorus, Engraulis, Rastrelliger*
Patis	Philippines	*Stelophorus, Leiognathus, Sardinella*
Nam-pla	Thailand	*Stelophorus, Rastrelliger*
Ketjap-ikan	Indonesia	*Stelophorus, Leiognathus, Clupeidae*
Budu	Malaysia	Ikan-bilis (*Anchoviella*)
Pastes		
Bagoong	Phillipines	*Engraulis, Clupeidae* (and shrimps)
Trassi	Indonesia	Shrimps
Belachan	Malaysia	Shrimps (*Acetes* spp)
Prahoc	Cambodia	Freshwater fish (*Cyprinidae*)

Note: The addition of rice and salt in various proportions occurs, giving further variety in products.

17.6.2   South East Asian products

In contrast to the few LAB fermented fish products mentioned above the fermented products of South East Asia are almost bewildering in number, name and distribution. Although fish and crustacea are the main raw materials many other marine creatures are utilised. Some products are species specific whilst others use any available species. The naming of similar products in different countries can differ or, where linguistic connections cross national borders, the same name is used in more than one country. Table 17.6 lists some of the better known products as exemplars. Common characteristics of the fermentations include the use of small, shoaling or seasonally, available fish that are not prime eating species; the use of varying amounts of salt and the use of carbohydrate sources, particularly rice, in the fermentation. The importance of these products in the diet should not be underestimated. For poor people a diet of boiled rice and dried fish can be enriched by the use of fish sauces and pastes with interesting, 'meaty', flavours and partially hydrolysed protein with a good amino acid content. Their production can be very simple and with salt and rice as the main added ingredients they are relatively cheap to buy. The presence of salt and a lactic ferment give a good storage life.

17.6.3   Classification of LAB fermented fish products

Given the apparent complexity of LAB fermented fish products in South East Asia several workers have tried to classify them according to various rules or characteristics of the ferments. Thus, Subba Rao (1967) recognised three groups according to the final appearance of the product: those in which the fish obviously retain their original form or large parts are kept; products which are reduced to a paste and products which are in a liquid form. Amano (1962) described three groups defined by the mechanism of protein breakdown, as follows: traditional salted products where endogenous enzymes from fish viscera and flesh are responsible; traditional salted products where enzymes are added as an inoculum and non-traditional products where accelerated protein breakdown is promoted by commercial enzymes or chemical hydrolysis. In yet another approach, Orejana (1983), described four groups: high salt (15–20%) products; true LAB fermented products, sometimes with added carbohydrates; products with added mineral acids and, finally, products with added organic acids. Another, more recent, classification is due to Adams *et al.* (1985) based on the substrate used, either fish/salt mixtures or fish/salt/carbohydrate mixtures. Saisithi (1987) proposed a classification based on fermentation characteristics: fermentation by fish enzymes and LAB enzymes; fermentation by LAB enzymes present in the fish/salt carbohydrate mixture and, finally, a LAB ferment with carbohydrate fermented by added yeasts and moulds.

These classifications might seem of academic interest only but they do point to one area of controversy. Although the production of fish sauces and pastes is described as a 'fermentation', some workers argue that they are really the result of endogenous fish enzymes with little or no impact from microbial (LAB) activity (Adnam and Owens, 1984). The enzymes involved are active at acid-pH and are thought to include pepsins (from viscera) and cathepsins (released from cells); these enzymes being most active at the start of the fermentation and bacterial enzymes becoming involved at a later time. The influence of the LAB would appear to vary from product to product and may involve high protease activity and flavour development in some cases. As LAB can be detected in many of these products (Ohhira *et al.*, 1988; Pederson, 1979), even if in small numbers, to ignore their contribution to the development of the product would seem to be unrealistic.

It has been noted that similar products made in different locations are considered to be of different quality (in terms of flavour, colour, clarity, etc.) much as wines vary from region to region. The evolution of a specific LAB flora in these regions over generations of production could well occur giving rise to these quality differences. The recent developments in the naming and movement of LAB species between genera mentioned earlier (Section 17.2.3) suggest that the whole issue of their presence in fermented fish products should be re-investigated with better identification of those involved. The classifications also indicate the important criteria for successful fermentations.

Fermentable carbohydrates

The presence of fermentable carbohydrates (mono- and disaccharides) is essential and is usually provided by the addition of rice (or cassava). However, starchy substrates must be broken down and enzymes with amylolytic activity have been demonstrated. Controlled experiments have been conducted to demonstrate these effects with glucose and sucrose (Adams *et al.*, 1987) and rice and cassava (Twiddy *et al.*, 1987).

Presence of salt
The presence of salt inhibits spoilage bacteria and promotes LAB and halophiles generally by lowering the water activity. The effect varies depending on the phase in which the salt is found, being highest in the solid state and less in solution allowing the LAB to proliferate (Han-Ching *et al.*, 1992). As the LAB will be inhibited by a lowered water activity the amount of salt added strongly influences the length of the fermentation (Adams *et al.*, 1987).

Initial pH
The initial pH of the raw material will vary and hence the pH drop due to any specific LAB activity will depend on this value. The general requirement is for a rapid drop in pH and the presence of easily fermentable carbohydrates, as indicated above, is essential for this to occur.

Temperature
Temperature has an enormous effect on fermentation but in most traditional products ambient temperatures alone prevail but as this is usually in the range 25–35°C a vigorous ferment is assured with a tendency to promote tissue breakdown and favour sauce production.

17.6.4   Specific products
A description of the production processes for some specific products will be useful in illustrating the points mentioned above. As examples, a fish paste and a fish sauce from the Philippines will be used and it should be noted that similar fish paste/sauce combinations can be found in every country of South East Asia.

Balao balao
This is a fermented product made from a mixture of shrimps, salt and boiled rice. The shrimp, usually *Peneaus indicus* and known as suahe, is a small saltwater species and should be fresh, preferably live, at the start of the process. Solar salt is used at various levels depending on the process adopted and the rice is boiled to a pasty consistency (see Fig. 17.2). The main variable in the process is the amount of salt added, which can be from 3–20% (w/w), and which determines the fermentation time to give an acceptable product – the higher salt content products take longer to ferment. At a salt content of 3%, as a minimum requirement, a pH of less than 4.0 is desirable for a good product and this can be achieved in four days. Physical characteristics of the process are liberation of gas, liquefaction, development of a red colour and softening of the shrimp shell. The microbial characteristics of the process show peaks at intervals with different species predominating at each peak. For a ten-day ferment, Solidum (1979) described a peak at three days due to *Leuconostoc mesenteroides*, a second at five days due to *Pediococcus cerevisiae* and at seven days peak numbers of *Lactobacillus plantarum* occurred. This sequential process where different organisms predominate at different times (and pH) is a common feature

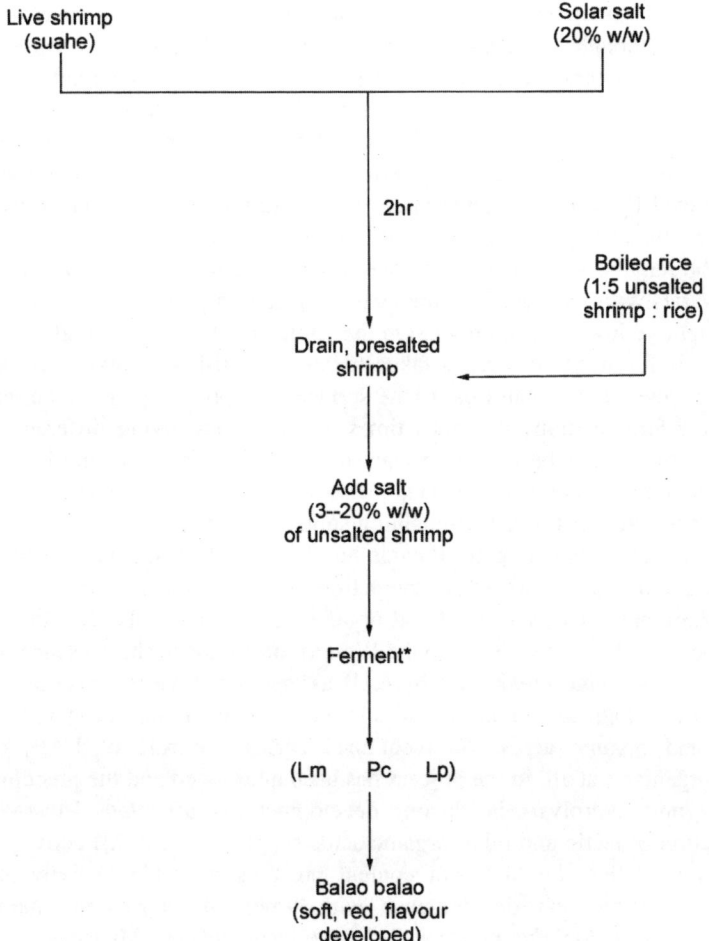

Fig. 17.2   Production stages for the Philippino shrimp paste balao balao.

of LAB fermentations. As no inoculum is added these organisms must be present in the components of the system. The product is used whole with the softened shell being edible and it can form a major or minor component of the diet.

Patis
This is a fish sauce produced from fish-solar salt mixtures and often produced alongside a fish paste called bagoong (which differs from balao-balao in not

having added rice). The raw material can be any fish or shrimp species readily available although some are preferred, such as *Stolephorus, Dacapterus* and *Sardinella* and production ranges from home to large commercial scale. The amount of salt added varies but is commonly 3:1 salt:fish with the proportion being the controlling mechanism of fermentation. On the home scale the fish and salt are mixed and left for a long fermentation (six to twelve months) which is encouraged by mixing at intervals. Once the fermentation is considered to be complete the clear patis is decanted off, more sauce can be produced by pressing the solid remains with the final residue being bagoong. The colour and clarity of patis varies with fermentation time and fermentation practice; the colour varies from light yellow to dark brown. On the commercial scale essentially the same process is followed but care is taken to press the fish-salt mixture because as liquor is released fish can float to the surface and spoil. Liquor can be removed from the fermentation at various times and positions giving different quality patis. Removal can be by scoop from the top or by tap from the bottom and lower-quality patis can be produced by washing the solids with salt-water at the end of the fermentation (the remaining solid is bagoong).

The development of patis flavour has been studied and is mainly due to peptides and amino acids (from the action of endogenous enzymes), ammonia and additional nucleotide, lipid and organic acid components. The flavour has been described as cheese-like (to the Western nose) due to the presence of VFA generated by amino breakdown by LAB as mentioned earlier (Section 17.2.2). Flavour development can be followed by increases in amino-nitrogen, free fatty acids and organic acids. As mentioned earlier the role of LAB, or any microorganisms at all, in the process has been questioned and the predominance of enzymic hydrolysis in flavour development is accepted. However, the production of lactic and other organic acids suggests some LAB activity. It has been argued that the high salt content prevents microbial activity and the numbers of bacteria do decrease with time, although some halophiles (*Pediococcus halophilus* in particular) have been isolated. Microbial proteases might contribute to protein breakdown even after the death of the bacteria themselves. Spoilage bacteria give rise to ammoniacal odours which are considered to be undesirable.

The characteristics of balao-balao, patis and bagoong production are repeated in most other fermented fish products in the region and individuality comes from the fish species used, localised bacterial flora and fermentation practice.

Quality and standards

The vast majority of fermented fish products are home made or, if produced on a commercial scale, they are consumed locally or at most within the country of origin. Different quality attributes have been described usually related to the length of fermentation and the means of extraction of sauces. The export market has been limited with some trade within the region but trade with the North America, Europe and Australia has grown along with an interest in, 'ethnic', foods and immigration from South East Asia. Thus, the application of legal food standards relating to quality, composition and toxicology have developed only in recent years. As an example, standards for nam-pla have been described in Thailand (Virulhakul, 2000) including parameters such as: sodium chloride content, % amino acid nitrogen, glutamic acid/total nitrogen, pH and relative density.

Nutritional information concentrates on amino acid content (where the provision of essential amino acids is good). Toxicological interest centres on the biogenic amines with histamine as an indicator compound; values have been set at 20mg/100g (max) in the USA and Canada. High values indicate poor handling of the raw material prior to fermentation. It would seem inevitable that similar regulation will be applied to the wider range of fermented fish products in the future. It should be kept in mind that these traditional products have been produced under what appear to be insanitary conditions for many years with few problems probably due to the presence of salt and an effective LAB flora. Early attempts to reproduce these fermentations under clean conditions in Western factories have produced poor-quality products with pathogenic bacteria present suggesting that local know-how has been lacking

17.7   LAB in ensilation

Ensilation is the process of preservation of raw materials by acid conditions and it has been applied to grasses, poultry offals and various fish wastes. The acid can be added in the form of inorganic or organic acids such as hydrochloric, sulphuric, formic, propionic and acetic acids, either alone or in combination. The problem with inorganic acids is that the undissociated species predominates at low pH (see Section 17.3.1) so the feed must be neutralised before use. Both inorganic and organic acids are corrosive and difficult to handle in bulk. The alternative means of acidification is to generate acid in situ and in this case the LAB and the presence of a fermentable sugar are prerequisites although salt is never added as an aid to fermentation. The characteristics of a silage are similar to those of the fish sauces mentioned above: considerable liquefaction, autolysis of proteins by endogenous enzymes and release of lipid (for fatty species). Unlike fish meal the water content of the raw material is retained in the silage, unless removed by drying, which can be a limitation for transport to farms.

Fish silage was seen as a means of utilising trash fish, by-catch and processing wastes whilst supplying high-quality protein for animals such as poultry, pigs, calves and other species such as mink. Early feeding trials of added-acid silages with poultry proved to give erratic results although this might have been related to the use of poor-quality raw material, different ensilation conditions and variable feeding trial regimes. Recent trials have been more favourable (Vizcarra-Magana et al., 1999). Feeding trials of added-acid silages with pigs have been generally good. Fewer trials have been done in aquaculture over the years but they have been favourable (Jackson et al., 1984; Stone et al., 1998; Heras et al., 1994).

Many of the issues surrounding added-acid silages have been addressed in an early review (Raa and Gildberg, 1982). In contrast, feeding trials with LAB silages have been more successful, suggesting that the presence of the LAB might be beneficial over and above the nutritional value of the protein content of the raw material. As mentioned earlier (Section 17.4) probiotic effects have been demonstrated for poultry and pigs.

17.7.1   LAB of crustacean waste

The ensilation of crustacean processing waste by added acid has been described previously where it has been used to prepare a salmonid feed containing natural pigments (Raa *et al.*, 1983). The use of a LAB fermentation of crustacean wastes was proposed for the recovery of chitin, proteins and pigments from tropical prawns at a later date (Hall and de Silva, 1992). This approach was one of several attempts to apply microbial technology to the production of chitin and its conversion to commercially viable derivatives (Hall, 1997). Chitin is a polysaccharide which forms a major part of the exoskeleton of crustacea in conjunction with proteins and calcium carbonate; other proteins are present from the organic parts of the crustacean. The conventional method of chitin production involves demineralisation by inorganic acid and deproteination by alkali to remove the calcium and shell/organic proteins respectively. This process uses large volumes of acid and alkali at elevated temperatures and large volumes of wash water with attendant pollution. In addition, the use of acid/alkali treatments can lead to random depolymerisation of the chitin molecule which in turn leads to a variable product.

Ensilation involves the addition of an inoculum of LAB and a fermentable sugar leading to a drop in pH (from 7.5–8.0 to 4.0–4.5 depending on the nature of the waste), liquefaction, protein hydrolysis and release of calcium from the prawn shell. This results in the recovery of a partially purified chitin (90% and 67% removal of protein and calcium respectively), a usable protein hydrolysate and possible pigment recovery (Zakaria, 1997; Shirai-Matsumoto, 1999). A crustacean waste silage has been produced by the LAB fermentation of cassava and used as a poultry feed (de Silva, 1998). An aquaculture feed was made from the protein hydrolysate produced by LAB fermentation of prawn processing waste (Plascencia-Jatomea, 2000). The hydrolysate was used as a substitute for fish meal in diets for tilapia with favourable results including possible probiotic and nutritional benefits. This is a potentially profitable outlet for crustacean processing wastes as problems are foreseen with the provision of fish meal in an expanding aquaculture industry (Tacon, 1995; 1998).

Aquaculture competes with terrestrial animal feeds for fish meal and the supply might reach a maximum sustainable level soon as capture fisheries reach about 100 million tonnes per year. The supply might even diminish as 'industrial fish' only suitable for conversion to fish meal are recognised as a vital part of the food chain for edible species leading to moves to protect them. Fish meal substitution in aquaculture by other protein sources, particularly of plant origin, is being investigated but proves difficult for carnivorous species. The crustacean protein hydrolysates could play a part in improving their acceptability by acting as 'attractants' to increase the feed intake. The use of the large volumes of crustacean processing waste generated in the many countries involved in the industry from South East Asia, China and Latin America would also lead to pollution abatement (Subasinghe, 1994).

17.8   LAB fermentation of food fish

The products described above are alike in utilising fish which are unattractive for direct consumption such that they are transformed into sauces and pastes (or animal feeds) and used as accompaniments to the staple diet. Attempts have been made to apply LAB to whole fish or fillets to be eaten as such. The objectives have been to preserve the fish between capture and processing and to generate novel products in their own right. In Europe traditional means of fish preservation such as salting/smoking/pickling/drying have fallen in popularity as consumers have turned their backs on 'chemical', preservatives such as salt, vinegar and smoke components whilst embracing lighter flavours. The use of LAB could replace some of the preservative effect of these components with those due to LAB fermentation.

A question to be asked is whether the conditions of a LAB ferment (low pH and presence of lactic acid) should have a neutral effect on flavour and texture or whether changes should be accepted or, indeed, encouraged. An EU-funded project applied LAB isolated from the marine environment to the preservation of whiting (minced and filleted) and smoked salmon with emphasis on spoilage control, sensory modification and functional properties of the fish muscle proteins (Hall et al., 1995). Carnobacterium spp were found to lower pH at chill temperatures with no proteolytic or lipolytic activity. With minced whiting the effect of LAB on spoilage was variable but they had no effect on chemical or organoleptic characteristics. Smoked salmon inoculated with LAB, vacuum-packed and kept at chill temperatures had good sensory properties (judged by taste panels) although chemical and microbial indices were difficult to interpret. Morzel et al, (2000) applied LAB fermentation with Lactobacillus sake to Atlantic salmon fillets and texture changes were studied by electron microscopy and texture profile analysis (TPA). Texture changes were found due to pH, moisture loss and ultrastructural changes with pH being the dominant factor. Glatman et al. (2000) performed similar experiments with minced yellowfin tuna and Leuconostoc mesenteroides, Pediococcus pentosaceus and Lactobacillus plantarum. Under chill conditions the presence of L mesenteroides gave changes towards a 'meaty' flavour and juicy texture, low values for chemical indices of spoilage and extended shelf-life (over four weeks).

The use of LAB fermentation in combination with other preservative techniques can contribute to the control of microbial safety and quality (hurdle technology). In particular, a combination of LAB and modified-atmosphere-packaging (MAP) or vacuum packaging at chill temperatures is effective. In MAP the air in a pack is replaced by oxygen, nitrogen, carbon dioxide or mixtures of these gases (Davies, 1997); the use of carbon dioxide being selective for *Lactobacillus* spp (Han-Ching *et al.*, 1992).

17.9   Future trends

It should be remembered that research on LAB in every respect is intense with conferences, papers and books appearing constantly and the reader interested in applications for fish should be aware of this wider picture. Research into LAB/ fish combinations was strong in the 1970–80s but since then has continued sporadically with two areas needing particular attention. Firstly, the use of LAB for the industrial-scale production of animal feeds by ensilation from fish and crustacean wastes. Research into ensilation in general has been unpopular for many years probably due to the poor performance of the added-acid silages as mentioned earlier. However, the technology deserves a revisit with attention to the selection of appropriate LAB inocula and characterisation of the protein hydrolysis products with a view to their nutritional and probiotic properties. The demand for fish meal substitutes mentioned earlier could be the driving force for such research. This should be allied to plant design and the development of process control parameters. It is remarkable that the ensilation, which appears to be a random process, when applied consistently to a uniform raw material is capable of yielding a consistent product which is essential for application in the animal feed industry. These positive characteristics should be exploited and developed to an industrial scale.

The application of LAB to food fish has been given attention recently but is deserving of more given that the value of such products is high and the technology could be applied in countries wishing to export to the markets of the USA and Europe in particular.

17.10   Sources of further information and advice

The reference section which follows gives details of those sources used directly in writing this chapter, of which, some are essential reading for specific aspects of LAB microbiology and technology. For the greatest historical perspective on LAB fermentations across a wide range of foods the *Handbook of Indigenous Fermented Foods*, edited by Keith H Steinkraus is unsurpassed, whilst the books on, *The Lactic Acid Bacteria*, edited by Brian Wood are equally valuable. The reader who wishes to contribute to the literature on all aspects of fish products should visit the web site developed by the UN FAO headquarters in Rome called 'onefish' (http://www.onefish.org) which is also linked to an online fish database at: http://www.nelfish.com.

17.11   References

ADAMS M R, COOKE R D, PONGPEN R (1985), 'Fermented fish products of South-East Asia', *Trop Sci*, 25, 61–73.

ADAMS M R, COOKE R D, TWIDDY D R (1987), 'Fermentation parameters involved in the production of lactic acid preserved fish-glucose substrates', *Int J Food Sci Technol*, 22, 105–114.l

ADNAM N A M, OWENS J D (1984), 'Technical note: Microbiology of oriental shrimp paste', *J Food Tech*, 19, 499–502.

AMANO K (1962), 'The influence of fermentation on the nutritive value of fish with special reference to fermented fish products of South-East Asia', in Heen E and Kreuser R, *Fish and Nutrition*, London, Fishing News, 180–200.

DAVIES A R (1997), 'Modified-atmosphere packaging of fish and fish products', in Hall G M, *Fish Processing Technology*, 2nd edn, Glasgow, Blackie Academic and Professional, 200–23.

DE SILVA, L L S S K (1998), 'Poultry feeds prepared from fermented prawn waste, PhD thesis, Loughborough University.

ERKKILA S, VENELAINEN M, HIELM S, PETAJA E, PUOLANNE E, MATTILA-SANDHOLM T (2000), 'Survival of *Escherichia coli* 0157:H7 in dry sausage fermented by probiotic lactic acid bacteria', *J Sci Food Agric*, 80, 2101–104.

GLATMAN L, DRABKIN V, GELMAN A (2000), 'Using lactic acid bacteria for developing novel fish food products', *J Sci Food Agric*, 80 (3), 375–80.

HALL G M, DE SILVA L L S S K (1992), 'Lactic acid fermentation of shrimp (*Peneaus monodon*) waste for chitin recovery, in Brine C J, Sandford P A and Zikakis J P, *Advances in Chitin and Chitosan*, London, Elsevier Applied Science, 633–8.

HALL G M, MESCLE J-P, HAN-CHING L (1995), 'Application of a lactic acid fermentation to the preservation of fish and fish products', Final report for publication, FAR Project UP-2-514.

HALL G M (1997), 'Biotechnology approaches to chitin recovery', in Stevens W F, Rao M S and Chandrkrachang, *Chitin and Chitosan, Environmentally friendly and Versatile Biomaterials*, Proceedings of the 2nd Asia Pacific Symposium on Chitin and Chitosan, Bangkok, 26–33.

HAN-CHING L, IN T, MAUGUIN S, MESCLE J-P (1992), 'Application of lactic acid fermentations', in Hall G M, *Fish Processing Technology*, 1st edn, Glasgow, Blackie Academic and Professional, 193–211.

HERAS H, MCLOED C A, ACKMAN R G (1994), 'Atlantic dogfish silage vs. herring silage for Atlantic salmon (*Salmo salar*): growth and sensory evaluation of fillets', *Aquaculture*, 125, 93–106.

JACKSON A J, KERR A K, BULLOCK A M (1984), 'Fish silage as a dietary ingredient for salmon. II. Preliminary growth findings and nutritional pathology', *Aquaculture*, 40, 283–91.

KOSIKOWSKI E (1977), *Cheese and fermented milk foods*, Ann Arbor, Michigan, Edwards Bros Inc Printers and Distributors.

KUMAR G, SHARMA R (2001), 'Probiotics – the mainstay in aquaculture health management', *Infofish international*, 5/2001, 42–7.

METCHNIKOFF E (1907), *The Prolongation of Life*, London, Heinemann.

MORZEL M, HEAPES M M, REVILLE W J, ARENDT E K (2000), 'Textural and ultrastructural changes during the processing and storage of lightly preserved salmon (*Salmo salar*)', *J Sci Food Agric*, 80 (11), 1691–7.

NAIDU A S, BIDLACK W R, CLEMENS R A (1999), 'Probiotic spectra of lactic acid bacteria (LAB), *Crit Rev Food Sci Nutrit*, 38 (1), 13–126.

OHHIRA L, CAHN MIN C, MIYAMOTO T, KATAOKA K, NAKAE T (1988), 'Identification of lactic acid bacteria isolated from traditional side-dish fermented foods of southeast asia', *Jap J Dairy Food Sci*, 37, 185–93.

OREJANA F M (1983), 'Fermented fish products', in Chan H T, *Handbook of tropical foods*, New York, Marcel Dekker, 255–95.

PEDERSON C S (1979), 'Nutritious Fermented Foods of the Orient', in *Microbiology of Food Fermentations*, Westport Connecticut, AVI, 310–33.

PLASCENCIA-JATOMEA M (2000), 'Recuperacion de proteinas a partir de desechos de camaron y su aplicacion de dietas para acuicultura', Maestro de Biotecnologia thesis, Universidad Autonoma Metropolitana-Iztapalapa, Mexico City.

RAA J, GILDBERG A (1982), 'Fish silage', *Crit Rev Food Sci Nutrit*, 16 (4), 383–419.

RAA J, GILDBERG A, STROM T (1983), 'Silage production – theory and practice', in Ledward D A, Taylor A J, Lawrie R A, *Upgrading waste for feeds and food*, London, Butterworths, 117–132.

SAISITHI P (1987), 'Traditional fermented fish products with special reference to Thai products', *Asian Food J*, 3 (1), 3–10.

SHIRAI-MATSUMOTO C K (1999), 'Utilizacion de desperdicios de camaron para recuperacion de quitina, proteinas y pigmentos por via microbiana', Doctor en Ciencias Biologicas thesis, Universidad Autonoma Metropolitana-Iztapalapa, Mexico.

SOLIDUM H (1979), 'Chemical and microbiological changes during the fermentation of balao-balao', *J Food Sci Tech* (Manila, Philippines), 3, 1–16.

STEINKRAUS K H (1996), *Handbook of indigenous fermented foods*, 2nd edn, New York, Marcel Dekker.

STONE F E, HARDY R W, SHEARER K D, SCOTT T M (1989), 'Utilisation of fish silage for rainbow trout (*Salmo gairdneri*)', *Aquaculture*, 76, 109–18.

SUBASINGHE S (1994), 'Greening the seafood processing industry', *Infofish International*, 1/94, 27–34.

SUBBA RAO G N (1967), '*Fish processing in the Indo-Pacific area*', Bangkok, Indo-Pacific Fisheries Council Regional Studies, 4, 75–76, 81.

TACON A G J (1995), 'The potential for fish meal substitution in aquafeeds', *Infofish International*, 3/95, 29–34.

TACON A G J (1998), 'Issue: Dependence on agricultural and fishery resources',

Infofish International, 2/98, 19–25.

TWIDDY D R, CROSS S J and COOKE RD (1987) 'Parameters involved in the production of lactic acid preserved fish-starchy substrate combinations', *Int J Food Sci Technol*, 22, 115–21.

VIRUHAKUL P (2000), 'The processing of Thai fish sauce', *Infofish International*, 2/2000, 49–53.

VIZCARRA-MAGANA L A, AVILA E, SOTELO A (1999), 'Silage preparation from tuna fish wastes and its nutritional evaluation in broilers', *J Sci Food Agric*, 79, 1915–22.

WOOD B J B (1992), *The lactic acid bacteria: Volume 1: The lactic acid bacteria in health and disease*, London, Elsevier Applied Science.

WOOD B J B, HOLZAPFEL W H (1995), *The lactic acid bacteria: Volume 2: The genera of lactic acid bacteria*, Glasgow, Blackie Academic and Professional.

ZAKARIA Z (1997), 'Lactic acid purification of chitin from prawn processing waste using a horizontal rotating bioreactor', PhD thesis, Loughborough University, UK.

18

Fish drying

Peter E. Doe, University of Tasmania, Hobart

18.1   Introduction

In many parts of the world there is no access to refrigeration or ice. In the absence of cold, placing a stress on the physical, chemical and biological processes that lead to spoilage can slow deterioration of freshly caught fish. Reducing the moisture content through drying, smoking or curing results in a stable source of protein that can be transported to communities with limited access to fresh fish. Smoking, curing and drying of fish either as a means of prolonging shelf-life, or to produce desired flavours and texture has been practised by many societies for centuries. Dried products range from the wood-hard *bonito* of Japan to the lightly smoked salmon products that sell at a premium in developed countries.

Paralleling the range of smoked and dried products are the methods employed to produce those products. Traditional methods in tropical countries include direct sun drying with the fish placed either directly on the ground or on mats or racks. Some of these processes involve brining or dry-salting. In Northern Europe and elsewhere, where refrigeration is generally available, fish are smoke-dried in mechanical smokers to meet a demand for particular products. The quality of smoked, cured and dried fish can be assessed using a range of physical, chemical and organoleptic methods. Modern quality control procedures such as HACCP (Hazard Analysis Critical Control Point) are increasingly used in the smoked, cured and dried fish industries.

18.2   The drying process

The drying of fish is a well-understood physical process. When a fish is exposed to air its surface begins to dry as moisture evaporates from the surface into the air by the process known as convective mass transfer. The rate of this evaporation is primarily governed by the humidity of the air stream, but is also affected by air speed. While there is free water at the surface, the evaporation will proceed at a constant rate. This process is referred to as 'convection governed' or 'constant rate' drying. At some stage, however, the surface will begin to dry and this establishes a moisture gradient near the surface that will cause more water to move from the interior of the fish towards the surface. Water moves within the fish through a combination of liquid diffusion, vapour diffusion, molecular motion, and osmosis. At the same time, salt may be moving into the fish if the salt concentration at the surface of the fish is greater than that within.

As the fish dries through the movement of the interior water to the surface, the moisture concentration gradient gradually decreases. Thus the driving force for the water transport decreases, and the drying rate slows. This part of the drying process is referred to as 'diffusion governed' or 'falling rate' drying. Drying will proceed at a progressively decreasing rate until equilibrium is reached, at which stage the fish is said to have reached 'equilibrium moisture content'.

18.3   Spoilage of smoked, cured and dried fish

Fish is said to have spoiled when it is unfit for human or animal consumption. This may be due to physical, chemical or microbiological effects either separately or in combination. Physical effects include loss due to birds, dogs, etc., and the effects of beetle attack or fly larvae infestation. Preventative measures include packaging; the use of mechanical or enclosed dryers; salting; insecticides; and good house-keeping practice such as the proper disposal of offal to minimise fly breeding.

Chemical spoilage includes chemical contamination from fuel oil, kerosene, and insecticides. This can be minimised by proper food-handling practices, hygiene and effective packaging. The other common form of chemical spoilage is due to oxidation of lipids leading to rancidity. These reactions are accelerated by the high temperatures associated with hot smoking, or direct sun-drying. However in some traditional products the brown colouring produced by the Maillard type reductions of endogenous sugars may be desirable.

A mixture of anti-oxidants (butylated hydroxyl anisole (BHA) 19g, butylated hydroxyl toluene (BHT) 1g, citric acid 5g, and prolylene glycol 100ml) has been found to be effective for preventing discolouration and reducing the incidence of moulds and red halophiles when added at a rate of 1ml per litre of brine in wet-curing process and 1ml per kg of salt for dry curing (Doe, 1998).

Enzymic degradation begins as soon as the fish is killed. Some fish have naturally high concentrations of very active hydrolytic enzymes that cause the fish to autolyse rapidly resulting in belly burst. Endogenous bacteria present in the gut or introduced in the brine, ice or handling equipment will reproduce in the favourable conditions of temperature and nutrient availability encountered in on-board storage, or during transportation to the market, particularly if there is insufficient ice as is often the case. Every species of bacteria has a particular range of temperatures and water activity (see below) for optimum growth, and some have particular nutritional requirements, salt for example. Bacterial growth can be controlled through icing or drying. A particular species, for example the mesophilic bacterium *E.coli* which has an optimum temperature for growth of 41°C, would not grow if the temperature is kept below 3°C; however, at that temperature a psychrophile such as *Psudomonas* sp. would be encouraged.

Microbial spoilage can be reduced by good handling and processing practices. Keeping freshly caught fish separate from partially spoiled fish, keeping surfaces and boxes clean, using clean ice and having good hygiene practices all contribute to a lowering of microbial load. Smoking fish has the effect of partially drying the fish, and thus reducing microbial growth, but can also provide a deposit of bacteriostatic compounds on the surface of the fish.

18.4   Water activity and its significance

Water activity is a measure of the water available to sustain the reproduction and growth of micro-organisms within a food. The value of water activity, A_w, expressed as a fraction, is numerically equal to the humidity of air in equilibrium with the food. This can best be explained through one of the methods used to measure water activity. A sample of the food is enclosed in a container and the humidity of the head-space is measured. If a humidity of 94%, say, is measured, the water activity is equal to 0.94.

It was established early last century that lowering water activity places a stress on micro-organisms (Scott, 1936). Moreover it has been established that certain micro-organisms cannot reproduce and grow unless the water activity is greater than a particular value. Table 18.1 shows the minimum values of water activity for the growth of various types of micro-organisms found on fresh and dried fish. It can be seen from Table 18.1 that if the water activity of dried fish is reduced below 0.6 there can be no microbial activity. However, that does not prevent the fish being eaten by beetles!

Further stress can be placed on a micro-organism by increasing the acidity. A simple 4-parameter model for the growth rate, k, of bacteria under sub-optimal conditions subject to the combined effects of temperature, water activity and pH was developed by McMeekin *et al.* (1992):

$$\sqrt{k} = b(T - T_{min})\sqrt{(A_w - A_{w_{min}})}\sqrt{(pH - pH_{min})}$$

Table 18.1 Minimum values of water activity for microbial growth (Burt, 1988)

Micro-organisms	Water activity
Bacteria	0.91
Yeasts	0.85
Moulds	0.80
Halophilic bacteria	0.75
Xerophilic moulds	0.65
Osmophilic yeasts	0.6

where T, A_w and pH are respectively the temperature, water activity and pH, and T_{min}, $A_{w_{min}}$ and pH_{min} are the intercepts of the extrapolated response of the model to their respective axes for the value of growth equal to zero. For example, the growth of the bacterium *Stapylococcus xylosis* can be described by the equation:

$$\sqrt{k} = 0.0205(T - 275.9)\sqrt{(A_w - 0.835)} \qquad 18.2$$

This method of applying a combination of stresses on the micro-organisms in order to slow down spoilage is known as 'hurdle technology'.

Lowering the water activity of fish can be accomplished by drying, salting or a combination of both. Many traditional drying/curing processes use salt as a means of lowering water activity. The salt acts to bind water molecules. A saturated solution of common salt has a water activity of close to 0.75. Thus salting is an effective way of avoiding toxins associated with the growth of bacteria such as *Clostridium* sp. for example. The measurement of water activity is complicated by the methodology, and by variations within the substance being tested. Also the water activity will change as the substance dries. The simplest method of measuring water activity is to seal a sample in an airtight container, allow the air in the headspace to come to equilibrium with the sample, then measure the humidity of the air. The problem with this method is that many hygrometers (instruments for measuring humidity) suffer from hysteresis at high humidity. Commercial A_w meters such as the LLUFT, which relies on a mechanical hygrometer with a hair sensor, require constant calibration against saturated salt solution standards.

Many of the problems associated with absorption hygrometers are overcome by using a dew point instrument. The Aqualab CX2 Water Activity Meter (Decagon Devices Inc, Pullman, Washington, USA) is particularly suited to measurement of high levels of water activities in foods. Dew point instruments rely on the detection of condensation on a mirror that is repeatedly cooled and heated. The accuracy of this instrument is within +0.003 units. Readings typically take five minutes (Doe, 1998). The water activity for dried salted fish can also be estimated from its fat, moisture and salt content (Doe *et al.*, 1982).

18.5   Drying methods

The simplest form of drying is placing the fish, either whole or split, directly on the ground, or on mats placed on the ground, in the sun, for a period of one to three days. Fish dried by this method are likely to be contaminated by sand or dust, infested with fly larvae, and partly rancid. This method is preferred only for very small fish species (e.g. anchovies) which can be dried within hours.

Salting the fish by dipping in brine helps to reduce the incidence of fly larvae infestation, as does raising the fish off the ground onto racks. With the fish placed on the ground the fly larvae can move easily into the fish and return to the ground when the fish are either too hot or too dry. Having the fish on racks also allows air to circulate below the fish. It is also more convenient to gather up the fish for storage under cover overnight, or when it rains.

Solar dryers have been developed and tested in many parts of the world. Most designs have a glass or plastic cover that increases the temperature of the air around the fish, and hence accelerates the drying. A test of three types of solar dryers and direct sun drying with fish placed on black volcanic rocks was conducted in the Galapagos Islands (Trim and Curran, 1983). All three solar dryers produced fish that was judged to be of a higher quality than that from the sun dryers. A solar tent dryer made from clear and black polythene sheeting stretched over a wooden 'A' frame (Doe *et al.*, 1977) was slightly cheaper than the solar cabinet dryers of Excell and Kornsakoo (1978) and much cheaper than the Brace Research Institute (1973) cabinet dryer. All three solar dryers were more efficient (as indicated by mass of fish dried/m^2/day) than direct sun drying. A large (1 tonne capacity) solar dryer based on the design of a horticultural greenhouse has been successfully tested in Yeman and Gambia (Sachithananthan *et al.*, 1986). The principles of design, operation and economics of solar driers are described by Brennendorfer *et al.* (1985).

Many types of agro-waste dryers have been designed and tested. (Clucas and Ward, 1996). A fundamental problem with agro-waste dryers that rely on natural convection for inducing a draft is that natural convection is enhanced by increasing the air temperature, however, if the air temperature is too high (above about 40°C) the fish will begin to cook resulting in the denaturation of the protein.

18.6   Dried and cured fish products

Dried fish products can be roughly categorised into 'fully dried' and 'partly dried' products. The former have been dried until their moisture content is close to uniform and water activity is close to, or below 0.75. Examples of 'fully dried' fish products are (Burt, 1988):

- Bonito (Japan) – fillets without skin are boiled for 60–90 minutes, inoculated with moulds (*Aspergillus glaucus, A.melleus* and *Penicillium glaucum*), broiled in a smoking furnace, then sun dried for 1 month (A_w = 0.76).

- Anchovy (Malaysia) – whole fish are boiled in 10% brine in bamboo baskets for one minute with agitation on board boat. Baskets are removed from brine and allowed to drain and cool. Fish are spread to dry in the sun for one day (A_w = 0.79).
- Stockfish (Norway) – cod (*Gadus morrhua*) is split lengthwise but remains joined at the tail, rinsed in fresh water then hung to dry for 1½–2 months in a relative humidity of 70–80%; post-drying in well-aired storerooms (A_w = 0.74).
- Skipjack (Sri Lanka) – dry salted for ½ hour or cooked in brine for a few minutes. Smoked and dried on coals for 10–12h or sun dried for 6–7 days (A_w = 0.64–0.67).

'Fully dried' fish have a shelf-life of between one week and several months provided that they are properly packaged and stored.

Examples of 'partly dried' fish are (Burt, 1988):

- Mackerel (Indonesia) – 24h in 30% salt (by weight) and saturated brine, dried with solar drier for 48h. (A_w = 0.89).
- Sand lance (Japan) – whole fish boiled in 10–12% brine for 10 minutes followed by 2–3 days sun-drying (A_w = 0.98).
- Herring kippers (Norway) – fish are split, washed, soaked in 80% brine for 15 minutes, hung to drain for 1h, smoked at max 29°C in mechanical kiln for 4h. (A_w = 0.98).
- Catfish (Malaysia) – fish are split, gutted and snout cut off. Put in a concrete vat with pickle from previous salting batches for 1–2 days. Sun dried for 2–3 days (A_w = 0.82).

These 'partly dried' fish typically have a shelf-life of up to one week and are usually kept refrigerated before consumption.

18.7   Recent developments

Recent research has focused on developing new dried products from trawl by-catch and other non-traditionally used fish, and methods of improving traditional products.

The Indian product 'masmin' is heavily smoked fully-dried tuna meat. The traditional method involves boiling in sea-water which results in a loss of proteins and nutrients. Muraleedharan and Valsan (1980) have developed an improved method in which strips of tuna are washed, dipped in saturated brine for 30 minutes, then steamed on open-mesh screens for one hour before drying and smoking. The improved product has a protein content of 76.6% (compared with 66.3% for the traditional product) and has a salt content of 1.6% (7.6%).

Experiments with solar driers in Thailand have proven effective in drying squid. Preservation with gamma radiation has also been tried with satisfactory results in preventing mould growth. A combination of sorbate dip and irradiation resulted in a six-month shelf-life for bulk packages of dried squid. This was more effective and cheaper than vacuum packaging which had a shelf-life of four months (Doe, 1998).

In Japan a new product – seasoned dry squid – is gaining popularity. This product involves dressing, skinning, boiling, seasoning, smoking, cutting, seasoning a second time, drying and packaging (Doe, 1998).

18.8   Quality assurance and control

The distinction between quality assurance and quality control is that the former specifies processes and procedures, the adherence to which should guarantee or assure a level of quality appropriate to the end use of the product. Whereas quality control involves sampling of the product and the measurement of quality indices which, if within tolerance and given the adequacy of the sampling protocol, will confirm that the product meets the required standard.

18.8.1   Quality control

Examples of quality control measures for cured and dried fish are:

- proximate analysis
- Total Viable Count, TVC (a measure of bacterial contamination)
- Total Volatile Base, TVB (a measure amino acid degradation)
- Peroxide Value (a measure of oxidative rancidity)
- Thiobarbituric Acid Test (TBA) (a measure of oxidative rancidity)
- protein digestibility.

All of these tests require skilled technicians with, in some cases, complex and expensive laboratory equipment. Besides, some tests developed for fresh or frozen food can give misleading results when used for dried fish. For example, a TVC using standard media .will fail to grow one of the predominant bacteria found on dried fish, *Vibrio parahaemolyticus* which has a specific need for salt, and will not grow in the absence of salt. Another example is the use of TVB for elasmobranches (sharks, rays) which contain high amounts of urea. During curing and subsequent storage the urea changes to ammonia giving high amounts of TVB in good-quality fish.

Another problem with laboratory tests is the lack of standard methodologies and reporting. For example with the most commonly used *in vitro* method for fish meal digestibility, pepsin digestibility enzyme concentrations ranging from 0.02g/l to 2.5g/l and reaction times from 0.5h to 48h are used by different laboratories (Doe *et al.*, 1998).

A different approach to quality control is the so-called 'demerit point' system (Doe and Olley, 1990). In this system a number of physical characteristics are rated on a 0–1 or 0–2 scale. For example, physical damage would be rated as absent (0), moderate (1) or extreme (2); mould would rate 0 for absent, 1 for present, and 2 for excessive. The system for dried fish had a maximum demerit score of 12 which would be considered 'poor', while a score of 5 would be considered 'fair', and a score of 0, 'good'.

18.8.2 Quality assurance

The system of quality assurance based on Hazard Analysis Critical Control Points (HACCP) has been used for cured and dried fish. A critical control point (CCP) can be defined as a location, procedure or processing step at which a hazard can be controlled. When a CCP fully controls or eliminates a hazard it is designated CCP-1. When it minimises but does not completely control the hazard, CCP-2. There is a complication when applying HACCP to cured and dried fish production in that in most cases the fresh fish supplier and the dried fish processor are not the same, and meet only at the fresh fish market. For a HACCP system to be effective it must apply to the whole process from capture to plate. Hazards associated with the capture and landing of fresh fish include physical and chemical contamination from external sources such as fuel oil, microbiological contamination from belly burst and flies, the production of toxins, autolysis, protein denaturing, oxidation and other biochemical reactions, all of which can be minimised, but not eliminated, hence CCP-2. During curing and drying there are further opportunities for physical, microbiological and biochemical degradation all of which are CCP-2 (Kow et al., 1998). Table 18.2 shows a Hazard Audit Table for the processing of dried salted fish.

Table 18.2 Hazard Audit Table for the processing of dried salted fish

Hazard analysis Critical Operation	Potential Risk	Critical control point	Preventative, control, monitoring measures
Fish landing	Microbial growth	Time ~temperature control (CCP2)	Use clean ice. Ice as soon as possible. Temp. 1–5°C
Washing	Bacterial and chemical contamination	Hygiene (CCP2)	Use clean water
Salting	Microbial contamination	Hygiene (CCP2)	Use clean salt. Check for halophilic bacteria. Use clean salting tank
Drying	Microbial growth Maggot infestation	Time~temperature control (CCP2)	Reduce A_w to 0.91 within 48 hours. Insect-free environment
Storage	Microbial growth Insect infestation	Humidity, hygiene, temperature, and time (CCP2)	Humidity < 65%. Insect-free environment. Temperature < 10°C

Thus with the most rigorously monitored HACCP, applied to the whole process from catch to plate, the best that can be achieved is a minimisation of the hazard.

An example of the application of HACCP to the production of seasoned dry squid is given in Doe (1998). Also in this reference is a government-recommended code of practice for fresh and cured fish in Indonesia, and an Indian code of practice for processing and storage of dried fish and fishery products. The extent to which traditional dried fish producers adopt these and other national guidelines is a measure of their commitment to quality assurance. Adoption of quality assurance measures is a necessary step in developing international markets for dried fish products.

Time-temperature monitors may be used to monitor the shelf-life of chilled or frozen fish. A TTM senses the temperature of the product and integrates the effects of the times the product experienced different temperatures. The cumulative effect is indicated by a colour change in some types of TTM (Olley and Lisac, 1985). Another TTM measures temperature and calculates growth rates of likely spoilage bacteria from the growth model which relates the square root of the bacterial growth rate to temperature (Owen and Nesbitt, 1984).

The model for bacterial growth in equation 18.1 could be used as the basis for an electronic device to monitor the rate of growth of spoilage bacteria in cured and dried fish. Measured values of temperature, water activity, and pH could be input to calculate the growth rate of a particular bacterium and, together with the time, calculate the number of bacteria given an assumed initial count. The challenge would be to make this device small enough, cheap enough, and sufficiently robust to 'live' with fish through the entire curing and drying process. An alternative could be a biochemical or biological device. The 'colour-change' TTMs rely on a temperature sensitive chemical. Is there a chemical that is sensitive to temperature, water activity and pH? If so, could it be tailored to give a colour change which corresponds to the combined effect of all three?

Another possibility could be called the 'miner's canary'. Is there a benign bacterium which could be inoculated into a growth medium which, while not contaminating the fish, will 'spoil' at the required rate and give a visible indication? This seems to be the most likely candidate. Keeping the bacterium viable until it is activated at the start of the process might be a problem.

18.9   References

BRACE RESEARCH INSTITUTE. 1973. 'How to make a solar cabinet dryer for agricultural produce.' Ste Anne deBekkevue, Canada: McGill University. 8p.

BRENNENDORFER, B., KENNEDY, L., OSWIM BATEMAN, C.O., TRIM, D.S., MEMMA, G.C. and WEREKO-BROBBY, C. 1985. Solar Dryers- their role in post-harvest processing. Commonwealth Scientific Council, Commonwealth

Secretariat, Marlborough House, Pall Mall, London SW1Y 5HX, 337p.

BURT, J.R. (ed.) 1988. *Fish smoking and drying: the effect of smoking and drying on the nutritional properties of fish.* Elsevier Science Publishers Ltd. Barking, England. 166p.

CLUCAS, I.J.. and WARD, A.R. 1966. *Post-harvest Fisheries Development: A Guide to Handling, Preservation, Processing and Quality.* Chatham Maritime, Kent ME44TB, United Kingdom, 443p.

DOE, P.E. (ed.) 1998. *Fish drying and smoking production and quality.* Technomic Publishing Co Inc. Lancaster. 250p.

DOE, P.E., AHMED, M., MUSSELMUDDIN, M. and SACHITHANANTHAN, K. 1977. 'A polythene tent dryer for improved sun drying of fish.' *J. Food Technol.*, 17:124–134.

DOE, P.E. and OLLEY, J.N. 1990 'Drying and dried fish products'. In *Seafood: Resources, Nutritional Composition and Preservation.* (ZE Sikorski ed.) CRC Press, Boca Raton, Florida, Chap 8 125–45.

DOE, P.E, OLLEY, J.N, HAARD, N.F. and GOPAKUMAR, K. 1998' Methodology for quality measurements' in *Fish Smoking and Drying Production and Quality* (PE Doe ed.) Technomic Publishing Company Inc, Lancaster, USA Chap 5, 117–35.

DOE P.E., RAHILA HASHMI, POULTER R.G. and JUNE OLLEY 1982.' Isohalic sorption isotherms I. Determination for dried salted cod (*Gadus morrhua*)'. *J.Fd Technol.* 17:125–34.

EXCELL, R.H.B. and KORNSAKOO, S. 1978. 'A low cost solar rice dryer.' *Appropriate Technology*, 5(1):23–4.

KOW, F., MOTOHIRO, T., DOE, P.E., HERUWATI E.S. 1998 'Quality assurance'. in *Fish Smoking and Drying Production and Quality* (P.E. Doe ed.) Technomic Publishing Company Inc, Lancaster, USA Chap 6, 137–55.

MCKEEKIN, T.A., ROSS, T and OLLEY, J. 1992. 'Application of predictive microbiology to assure the quality and safety of fish and fish products.' *Int.J.Food Microbiol.* 15:13–32.

MURALEEDHARAN, V. and VALSAN, A.P. 1980. 'Preparation of masmin – an improved method' *Fish Technol.*, 17(2):99–101.

OLLEY, J.N. and LISAC, H. 1985 'Time-temperature monitors'. *Infofish Marketing Digest* **3**, 45–7.

OWEN, D. and NESBITT, M.A. 1984 'A versatile time-temperature function integrator'. *Lab. Pract.* **33**, 30–75.

SACHITHANANTHAN, K., TRIM, D.S. and SPEIRS, C.I. 1986. *Proc FAO Expert Consultation on Fish Technology in Africa*, Lusaka, Zambia, FAO Fish. Rep.No 329, Food and Agricultural Organisation, Rome. 161p.

SCOTT, W.J. 1936. 'The growth rate of micro-organisms on ox muscle. 1. The influence of water content of substrate on growth at 1°C' *J. Counc. Sci. Ind. Res. Aust.*, 9:177–90.

TRIM, D.S. and CURRAN, C.A. 1983. *Comparative study of solar and sun drying in Ecuador.* London: Report of the Tropical Products Institute, L60. 44p.

19

Quality management of stored fish

E. Martinsdóttir, Icelandic Fisheries Laboratories, Reykjavik

19.1 Introduction: quality indices for fish

Quality and production management of fish as raw material on storage is of great importance in the fish sector. Fish is a very perishable product. Supplies of fish are unstable and fresh fish can be stored only for a short time. Freshness is one of the most important aspects of fish and fish products. For all kinds of fish and fishery products freshness makes a major contribution to the quality of fish and fishery products (Olafsdóttir, 1997). From the moment the fish is caught the deterioration process starts and the quality of the fish for use as a food product is affected. Changes occur in composition and structure because of (bio) chemical, physical, enzymatic and bacterial influence.

Sensory changes are perceived with the human senses, i.e. appearance, odour, texture and taste. Sensory methods have great advantages. They can be very fast, reliable, non-destructive on raw fish and no expensive instruments are needed. However the panellists need training and retraining under supervision of experienced panel leaders using fish samples of known freshness stage.

Sensory evaluation of raw whole fish in fish auctions and landing sites and during storage of fish in ice before processing is done by assessing appearance, texture and odour. Most of the scoring systems are based upon changes taking places during storage in melting ice. Characteristic changes vary however depending on the storage method. Knowledge of time/temperature history and handling of the fish gives important information about the quality. A characteristic pattern of the deterioration of fish stored in ice can be detected by sensory evaluation and divided into four phases (Huss, 1988). In the first phase, the fish is described as fresh with delicate taste, but in phase two, it loses its characteristic odour and taste. In phase three signs of spoilage occur but in phase four, the fish can be characterised as spoiled and putrid. Textural changes are very evident after catch at onset of *rigor*. Muscles become hard and stiff, rigor resolves and the muscle relaxes and becomes limp and not so elastic as before.

Utilisation and interest in sensory analysis in the fish sector is growing (York and Sereda, 1993). Sensory analysis will continue to be essential, even if better, cost-effective instrumental methods are developed. Sensory evaluation is used as a tool for grading according to product standards and for studying specific properties of fish species in connection with evaluation of quality, shelf life, storage conditions and product development (Nielsen, 1997). Sensory methods performed in a proper way are rapid, accurate and give unique information about food. They give direct measurement of perceived attributes and provide information assisting in better understanding of consumer responses.

Methods to verify freshness are needed at different transaction points in the fishery chain from catch to consumer, at landing site and the auction and, at all levels of trade from auctioning via wholesale to retail (Oehlenshläger, 1997). In-house evaluation of raw materials, during production and of products is carried out regularly in fish processing plants; in quality control for utilisation of incoming raw-material, and during processing and of products for compliance with a given specification. Detection of bruising, bones, scales, parasites, blood-spots etc. is a part of the inspection. The sensory evaluation is mostly performed by a single expert or to a less extent by a small group. In fish processing plants small sized panels frequently test cooked samples according to specification of buyers and retailers.

Bremner and Sakaguchi (2000) put forward an approach to the overall idea of freshness – the totality of characteristics of a recently harvested product that bear on its ability to meet stated or implied requirements. The product is undamaged and shows no signs of spoilage. Freshness evaluated by senses not only describes the freshness sensory properties of the fish but also includes factors such as bleeding and storage that may normally be considered as workmanship. In this chapter quality indices for fish will be dealt with, which are mainly freshness and handling or storage indices, criteria related to the specific definitions and sensory methods for determining the criteria.

19.2   Guidelines for sensory evaluation of fish

The sampling system, methods and procedures for sensory evaluation must be very well defined for sensory evaluation to serve its purpose in quality management. The sampling plan for pre-packaged food as described in the Codex sampling plans for pre-packaged foods (Codex XPT 13-1969) might be used as a basis for sampling plans. Lots and batches have to be defined before deciding the number of samples taken from each batch.

Sensory evaluation can be practised at different levels in fish processing. Sensory evaluation of whole fish is generally carried out by trained assessors in the reception or processing halls of fish factories or at the auction site. However sensory evaluation of cooked fillets must be carried out in rooms with special facilities.

The Codex guidelines for the sensory evaluation of fish and shellfish in laboratories (Codex CAC-GL31-1999) describe facilities, procedures and training of assessors and can be used as a basis for practising sensory evaluation in the fish sector. The guidelines are written with the Codex requirements in mind but can be used where sensory evaluation is used in testing fishery products for conformity requirements. The guidelines also cover a list of reference documents. In quality assurance systems it is most common to use a few types of sensory test and a small number of highly trained inspectors.

19.2.1 Facilities for sensory evaluation

In sensory evaluation within the fish sector a sensory panel or trained inspectors perform sensory analysis on the daily production. To avoid errors in the sensory evaluation in daily quality control, it is necessary to follow well-defined grading systems or guidelines and standards. To get good results with sensory evaluation, assessors must be trained and have clear and descriptive guidelines. Textbooks on sensory evaluation of foods often describe the facilities required for sensory evaluation. There are international and national standards and guidelines for the design and construction of sensory assessment rooms (ISO 8589, 1988; Meilgaard et al., 1999). The recommendations in these publications are intended for establishments or situations where sensory evaluation is a major activity, e.g. R & D laboratories of food companies and research institutes. Sensory evaluation for quality control purposes must be carried out no less accurately and conscientiously than in R & D laboratories, but the requirements need not be as elaborate. They are intended to be used by specialists who need to apply sensory methods when using criteria based on sensory attributes on the products (Botta, 1995).

The effects of the surroundings should be minimised as much as possible; lighting and ventilation is very important, noise level, foreign odours and distracting elements should be minimised, etc.

19.2.2 Training of panellists

Screening, selection and training of panellists is extensively described in international guidelines and standards (Codex CAC-GL31-1999; ISO-8586-1, 1993). Panellists should be able to perceive odours and basic tastes and describe these in a consistent manner. They should have normal colour vision and be able to learn terminology etc. The participants are screened for perception of odour, colour and texture if needed. The Codex guidelines also suggest a syllabus for training courses for assessors in the sensory assessment of fish and fishery products.

For general guidelines for panels the ISO-8586-2 (1994) is recommended for the panels of R&D work. Companies using sensory evaluation in their quality control use in-house panels. Personal characteristics are very important when selecting people for the sensory work such as conscientiousness and accuracy. People should be interested in sensory evaluation and food in general. Depending on the regular duties of the individual they should be readily available and healthy. Training is an expensive job and the panellists are needed on a regular basis.

Highly specialised evaluators and larger panels using attribute assessments are also described by Botta (1995).

19.3   Sensory evaluation of fish

During the last 50 years many schemes have been developed for sensory evaluation of raw fish. The Torry Research Station (Shewan et al., 1953) developed the first modern and detailed method. In the article score sheets for sensory evaluation of white raw fish are described and the sensory factors are classified. The general appearance and odour, texture of the fish and the flesh, including belly flaps is described. The odour, flavour and texture of cooked fish is also described.

Descriptive testing can be used for quality determination and shelf life studies. In structured scaling the panellists are presented an actual scale showing degrees of intensity. A few detailed attributes are chosen often based on work from trained panels. Descriptive words are selected and panellists trained so that they agree on the terms and objective terms used instead of subjective (Nielsen, 1995, ISO 11035, 1994).

19.3.1   Evaluation of whole fish: EU scheme

In Europe today, the method most used and recommended for quality assessment of raw fish in the industry and the inspection service is the EU scheme, according to the Council Regulation (EC) No 2406/96 of November 26, 1996 (Anon., 1996). In this scheme, three grades of freshness are established: E, A and B, corresponding to various stages of spoilage. E (Extra) is the highest possible quality, while below B is the level where fish is considered unfit for human consumption. This method gives rather limited information about the condition of the fish, as it is not species-related and does not therefore take into account the differences between species. The EU-scheme is commonly accepted at auction levels, however its use has been disputed. The use of the EU scheme is not giving results that can provide information to predict shelf life or remaining shelf life. The degradation is described in steps that are not continuous. A grading system like the EU-scheme has several serious drawbacks. Before the fish can be categorised into grades, different quality criteria are evaluated. Whenever the sensory characteristics do not agree with all of the subscriptions of any specific grade, the grader becomes confused. This increases the time to decide the grade and decreases the objectivity of the evaluation (Botta, 1995).

Statistical analysis of data from grading scale has been disputed by Land and Sheperding (1984). Treating scales as interval scales is to be discouraged. Data treatment of quality scores can be very problematic. Unidimensional scales should be used when possible. However quality scales are often multidimensional even for single defects like oxidation (Lawless, 1994).

19.3.2   Evaluation of whole fish: Quality Index Method

It is critical for a sensory system used in quality management that it reflects the different quality levels in a simple and documented way. Therefore, new and improved seafood freshness and quality grading systems that are both rapid and objective have been under development for various species. The Quality Index Method (QIM) is one such system and has several unique characteristics. QIM is based upon a scheme originally developed by the Tasmanian Food Research Unit (Bremner, 1985). The QIM- method has to be adapted to each fish species.

To date the QIM-system incorporates fresh herring (*Clupea harengus*), cod (*Gadus morhua*) (Jonsdóttir, 1992; Larsen *et al.*, 1992), Atlantic mackerel (*Scomber scombrus*), horse mackerel (*Trachurus trachurus*) and European sardine (*Sardina pilchardus*) (Andrade *et al.*, 1997), red fish (*Sebastes mentella/marinus*), brill (*Rhombus laevis*), dab (*Limanda limanda*), haddock (*Melanogrammus aeglefinus*), pollock (*Pollachius virens*), sole (*Solea vulgaris*), turbot (*Scophtalmus maximus*) and shrimp (*Pandalus borealis*) (Luten 2000a; Martinsdóttir *et al.*, 2001), gilthead seabream (*Sparus aurata*) (Huidobro *et al.*, 2000) and farmed salmon (*Salmo salar*) (Sveinsdóttir *et al.*, 2001; 2002). Warm *et al.* (1998) described development of QIM for frozen cod. Jensen and Jörgensen (1997) used QIM on thawed whole cod and concluded that the method was well suited for industrial quality grading of frozen raw material. Huidobro *et al.* (2001) investigated the effect of washing gilthead seabream on the results of sensory evaluation by QIM. QIM has several advantages, including estimation of past and remaining storage time in ice (Botta, 1995; Hyldig and Nielsen, 1997; Luten and Martinsdóttir, 1997). The method is based on characteristic changes that occur in raw fish. QIM is based on significant, well-defined characteristic changes of outer appearance attributes (eyes, skin, gills, smell) for raw fish and a score system from 0 to 3 demerit (index) points. The scores for all of the characteristics are summarised to give an overall sensory score, the so-called Quality Index. In Table 19.1 an example of a QIM-scheme is given. The scientific development of QIM for various species aims at having the Quality Index increase linearly with the storage time in ice (Nielsen, 1995).

The descriptions of each score for each parameter are listed in the QIM scheme. The assessor must evaluate all the parameters involved in the scheme. As the Quality Index increases linearly with storage time in ice, the information is well suited to use in production management. QIM is well suited to teach inexperienced people to evaluate fish, train panellists and monitor performance of panellists.

Table 19.1 The QIM scheme for farmed salmon (Sveinsdóttir *et al.*, 2002)

Quality parameters		Description	Points
Skin:	Colour/	Pearly-shiny all over the skin	0
	appearance	The skin is less pearl-shiny	1
		The skin is yellowish, mainlt near the abdomen	2
	Mucus	Clear, not clotted	0
		Milky, clotted	1
		Yellow and clotted	2
	Door	Fresh seaweed, neutral	0
		Cucumber, metal, hay	1
		Sour, dish cloth	2
		Rotten	3
	Texture	In rigor	0
		Finger mark disappears rapidly	1
		Finger leaves mark over 3 s	2
Eyes:	Pupils	Clear and black, metal shiny	0
		Dark gray	1
		Matt, gray	2
	Form	Convex	0
		Flat	1
		Sunken	2
Gills:*	Colour/	Red/dark brown	0
	appearance	Pale red, pink/light browm	1
		Grey-brown, brown, gray, green	2
		Tramsparent	0
	Mucus	Milky, clotted	1
		Brown, clotted	2
		Fresh, seaweed	0
	Odour	Metal, cucumber	1
		Sour, mouldy	2
		Rotten	3
Abdomen	Blood in	Blood red/not present	0
	abdomenn	Blood more brown, yellowish	1
	Odour	Neutral	0
		Cucumber, melon	1
		Sour, fermenting	2
		Rotten/rotten cabbage	3
Maximum sum (Quality Index):			24

* Examine the side where the gills have not been cut through.

When applying the QIM schemes, the outer appearance of the fish, eyes, gills and texture is evaluated. The odour of gills is evaluated, and for some species the odour and mucus of the skin is also evaluated. The colour of blood and fillets (or the cut surface at the flaps) is evaluated in gutted fish. For some fish species that are not gutted, such as redfish, dissolution of viscera is evaluated as well.

The QIM method has been reported as a rapid method (Larsen *et al.*, 1992). For use of QIM and other measurements of quality in a quality assurance system a sample has to be taken from a lot. A homogeneous lot has to be defined. It could be a catching day but a catch from one catching day could consist of different batches if the boat is fishing from different fishing grounds within a day. Individual fish spoil at different rates. Some conclusions can be drawn on sample sizes from controlled storage experiments. In such experiments wild fish might be taken from the same haul of a trawler or farmed fish slaughtered from different days from the same processor for the whole set-up. Sveinsdóttir *et al.* (2002) reported that assessing three fish per lot storage time of salmon might be predicted within 2.0 days at the 95% significance level, but examining greater number of salmon per lot might increase the precision. Larson *et al.* (1997) reported that when using an average of the assessor's score it is possible to predict the remaining storage life of the fish to ± one day. The method was very fast (approximately 5 minutes to assess ten fish resulting in an average demerit point). In a reference manual (Martinsdóttir *et al.*, 2001) guidelines are given for freshness assessment of whole fish by the QIM-method. A minimum of three fish to a maximum of ten fish should be sampled randomly from different places in fish boxes from a homogenous lot to cover some of the biological difference in spoilage rate of fish.

19.3.3   Evaluation of raw fillets
Grading schemes for fillets for quality need to describe different sensory attributes like appearance, odour and textures. The main sensory attributes are odour, gaping, blood veins in belly flaps, bloodstains, bruises, colour of muscle, number of areas damaged by guts and visible cod-worms. Some are related to freshness some are due to handling. The defects like bruises and blood-spots can be measured and counted and worms can be counted and compared to standards. A grading scale based for fillets was reported by Learson and Ronisvalli (1969). Scores were given from 5 to 0 and for each score odour and appearance are described. A freshness quality grading scheme for cod fillets was suggested by Martinsdóttir and Stefansson (1984). The results of the QIM system for evaluating fillets have been reported to correlate with the length of chilled storage (Bremner *et al.*, 1987). A quality index method scheme for fillets from thawed cod has been developed (Warm *et al.*, 1998) and a similar scheme could be developed for fillets from unfrozen fish of different species.

19.3.4   Evaluation of cooked fillets: Torry-scale, QDA-method
For sensory evaluation of fish fillets, it is common to cook the fillets and evaluate their odour and flavour. The Torry-scale is the most-used scale for evaluating the freshness of cooked fish (Martinsdóttir, 1997), but sensory profiling is also used in research laboratories in Europe (Hyldig and Nielsen, 1997).

The Torry-scale is also used in the fish industries of some countries and by buyers of fish products. It is a descriptive 10-point scale that was developed at the Torry Research Station and has been developed for lean, medium fat and fat fish species. Scores are given from 10 (very fresh in taste and odour) to 3 (spoiled). It is considered unnecessary to have descriptions below 3, as the fish is then not fit for human consumption. The maximum storage time of fish can be determined by sensory evaluation of cooked samples. The average score of 5.5 has been used as the limit for consumption (Martinsdóttir et al., 2001). At that stage members of the sensory panel detect evident spoilage characteristics, such as sour taste and hints of 'off' flavours.

In storage studies in the EU-project 'Development and Implementation of a Computerised Sensory System (QIM) for Fish Freshness' a linear relationship was found between QI of raw material and Torry-score of cooked fillets (haddock and cod from two seasons) (Martinsdóttir et al., 2001; Luten et al., 2000a). This indicates that using QIM on whole raw fish could replace sensory evaluation of cooked samples. QIM is more rapid and is performed earlier in the production chain.

The Torry-scale provides limited information about how the individual characteristics of the cooked fish change through the storage time, but by using Quantitative Descriptive Analysis (QDA), much more detailed information can be gained. The QDA (Stone and Sidel, 1992; Stone and Sidel, 1998) is a sensory method, which may be used for the determination of maximum shelf life in addition to a detailed description of the sensory profile for a product. With the QDA, all detectable aspects of a product are described and listed by a trained panel under guidance of a panel leader. The panellists make a list of concepts/words describing the product. The list is then used to evaluate the product and the panellists quantify the sensory aspects of the product using an unstructured scale. The panellists are trained in using an unstructured scale for each of the concepts, before participating in the sensory analysis of the product. The words used to describe the odour and flavour of the fish can be grouped into 'positive sensory parameters' and 'negative sensory parameters', depending on whether they described fresh fish or fish at the end of the storage period (Sveinsdóttir et al., 2001; 2002). The fish had reached marginal acceptability when the negative attributes dominated.

19.4   Developing a quality index

Information or detailed guidelines on how to develop a quality index are scarce in the literature. Hyldig and Nielsen (1997) described test procedures for development of QIM-schemes. Whenever QIM is used for new species, storage studies must be conducted to ensure that appropriate criteria and their corresponding defined characteristics are included in the QIM-schemes. There are mainly three steps in the development of a new QIM scheme (Sveinsdóttir et al., 2001a). Firstly a pre-observation is conducted. Two or three experts in sensory evaluation of fish observe fish that have been stored for different periods in ice. All changes occurring in appearance, odour and texture during storage are listed in a preliminary scheme (ISO 11035, 1994). The next step is development of the QIM scheme and training of a QIM panel. Prior to a shelf life study several sensory sessions should be done for the development of the scheme and training of the QIM panel. In each session, 3–4 different groups of fish that have been stored different periods of time in ice are observed. During the first sessions the preliminary scheme is explained to the panellists while they evaluate fish identified as being stored for different periods. Special attention is given if the evaluation of some quality parameters is destructive to the fish. Suggestions of changes should be taken into consideration. The panellists are trained during the next sessions by evaluating the samples without knowledge of the storage period prior to the sessions. During the development of the scheme some parameters might be removed if they are destructive to the sample when evaluated. Changes in the selection of words to describe the changes more precisely might be made.

In the third step QIM assessment is used in a full-scale shelf-life study where throughout the storage trial the fish must not be handled, so each time a new sample is taken. In parallel with this a trained sensory panel working in recommended facilities (ISO 8589, 1988) should conduct a sensory evaluation of cooked samples to estimate the reasonable maximum shelf life that can be obtained in this way. The QIM scheme should be applied to evaluate the fish at least every third day during the shelf life of the fish. Preferably five fish should be evaluated each of the storage days, but for fish of small sizes more may be included. During storage experiments chemical and microbiological indices might also be measured to follow the spoilage pattern and to use for comparison (Sveinsdóttir et al., 2001; 2002). The shelf life study should be repeated to observe if the same slope was found between the Quality Index and storage time in ice.

Data analysis is an important part of the development. O'Mahony (1986) describes statistical methods and procedures for handling data from sensory analysis. The results from the shelf life studies should be fitted into a linear relationship and studied. The linearity of the Quality Index should be checked. The changes of all attributes during the storage should be studied and weight of scores might be changed to obtain a QI with higher correlation to storage time. To obtain a better understanding of how the different quality parameters of fish change with storage time the results could be analysed with multivariate statistical methods (i.e. principal component analysis, PCA) (Martens and Jørgensen, 1997). PCA might also be used to decrease the numbers of attributes needed. However, it must be taken into consideration that the attributes have to be of a sufficient number to give a possible score of reasonable magnitude (Bremner et al., 1987).

19.5   Using quality indices in storage management and production planning

19.5.1   Shelf life

QIM for whole fish can be used to predict storage time in ice, remaining shelf life and Torry-scores of cooked fillets in storage management and production planning. The end of shelf life is defined as the number of days that whole, fresh (gutted) fish can be stored in ice until it becomes unfit for human consumption. Shelf life of fish is thus the whole period of time in which it is regarded as being fit for human consumption. Spoilage due to microbial activity is the main limitation of the shelf life. Another cause of spoilage may be rancidity, especially in fat fish species. Estimated storage time in ice is defined as the number of days that the fish has been stored in ice. From these results, a prediction can be calculated for the remaining shelf life (= shelf life − estimated storage time).

It is emphasised that remaining shelf life should be used with some caution due to the uncertainty in the estimation. Various factors can affect the remaining shelf life. It depends on the handling of the fish. Rapid cooling after the catch and an uninterrupted cold storage, different fishing gear, bleeding and gutting methods are important and the season and catching ground can also have an effect.

In a reference manual (Martinsdóttir et al., 2001) an estimated shelf life of 12 fish species is given, assuming optimal storage conditions, i.e. storage in ice without fluctuations in temperature. The shelf life and the estimated storage time in ice are based upon the outcome of very well-controlled storage experiments with whole, fresh (gutted) fish stored in ice under good manufacturing conditions on board the vessel, which implies proper gutting, washing and use of fish/ice ratio. The end of storage time is defined when a trained sensory panel detects spoilage flavour in cooked samples of the fish. Estimated shelf life for most species is from 13–18 days in ice but may be shorter (8 days for herring and 6 days for shrimp) but longer for salmon (20 days).

A linear relationship between the Quality Index and storage time in ice has been found and the best fit of the regression lines calculated for each species is also shown in the manual. The regression lines are used to predict storage time in ice after evaluation of the Quality Index and find remaining shelf life. This information can be used in quality and product management of stored fish in ice.

19.5.2   QC-charts and statistical analysis

Whenever the results of freshness grading are conducted as a part of a quality control programme, analysis of the data might consist of comparing the results to a lower limit and upper limit or to both lower and upper limits, which have previously been established by the management and the buyer. When QIM results are used these limits refer to the total of demerit points or the Quality Index. When control charts are used to monitor seafood freshness of a specific seafood over time they permit immediate detection of trends and out of control conditions, allowing appropriate handling and processing procedures to be corrected and the variability to be reduced. The preparation of bar graphs of the QIM-scores of a particular group of samples, such as a specific batch of samples or the samples from an entire production lot reveal the variability and central tendency of the QI-scores. Control charts are used for monitoring the output of a process to determine if the process is in control. Users of control charts have statistical criteria to distinguish between random variability and assignable causes. When sensory panel data are used in a statistical process control programme it is important to distinguish between panel mean for an individual sample of product and the mean of a small group of production samples. If only a single sample of product is collected during each of the periodic QC samplings, the sensory panel yields only one piece of raw data (panel mean) about the state of the process at that point of time regardless of the number of individuals on the panel. A third source of variability of particular importance to sensory data is measurement variability. The instrument is the panel, which can be sensitive to a variety of factors that can influence the evaluations. Quality standards can be used for data from a sensory panel the same as for any other analytical data. There are two components of variability, within session variability and between session variability. Within session variability can be reduced by increasing the number of panellists who participate in the session; assuming they are well trained and calibrated. Between sessions variability might arise from changes in testing environment, general shifts in the calibration of the panellists etc. Good analytical test controls must be exercised to keep those sources of variability to a minimum. The maintenance of the panel is very important (Munoz et al., 1992).

Using analysis of variance to analyse some types of result from grading systems can be problematic. The results may not meet the underlying assumptions necessary to conduct analysis of variance (O'Mahony, 1982). Analysis of QI-scores is based on the total demerit points not on individual assigned grades for each attribute. These ranges are much larger than the range normally used for grading and some assumptions when statistically analysing results of QI-scores may be used (Botta, 1995). The sum of scores may be normally distributed even though the individual scores are not.

19.6   Keeping fish under different storage conditions

Bremner et al. (1987) described estimation of time-temperature effects by the QIM sensory method. Two major determinants of deterioration in chilled seafood, namely microbial spoilage and nucleotide degradation, follow similar temperature functions. There is also evidence that the sum total of all effects of all the degradative processes also follow a similar pattern up to 15°C when other factors may intrude. Whatever method of calculating change in the stored seafood is used it must respond to temperature in the same way as the seafood itself. If the method does not have this property it can only be useful at or close to particular temperature. The use of sensory methods as integrators of time and temperature has rarely been considered. Most experiments are done at one temperature – generally in melting ice. Furthermore some scoring systems do not allow for continuous assessment or sufficient differentiation between samples. In the Torry-scale the fish undergoes a series of discrete changes during storage and the fish is given an appropriate score according to its description. Descriptive scales are usually constructed to give a linear progression. Bremner (1985) outlined an alternative approach to have one common measurement for the quality of fish with different temperature histories. The system should be used with fish in any form such a whole, gutted or in modified forms for fillets. Once the pattern of spoilage for a particular species has been established the demerit point score can be related directly to a suitable reference temperature such as days in ice. The remaining shelf life or time elapsed post-mortem can be readily calculated (Branch and Vail, 1985). The fact that the demerit points of different categories appear sequentially does not affect the system because the relative rate spoilage curve is based on the relative time at different temperatures to reach specified spoilage levels. Most degradation phenomena follow similar patterns. Research has shown that relative rates of increase in K value and the demerit points were similar for two fish species kept in ice and at 24–26°C. Most emphasis has been on chilled storage but the system should be applicable to frozen seafood. The system is suited for in plant control and any commodity that alters with time and temperature.

Further research is needed to evaluate the applicability of the QIM for fish stored under different conditions such as frozen – thawed fish, storage in ice slurry, temperature abuse during storage, etc. This is very important because in reality fish is not always kept under best storage conditions in ice and furthermore new packaging techniques extend the shelf life of fish alter the spoilage pattern and this has to be taken into account when using the QIM evaluation. The ideas of Bremner should be studied further about time-temperature integration and building models using the sensory method based on QIM for prediction of shelf life at different storage time and temperature.

19.7   Future trends

Research effort during the past years funded by the EU has focused on developing more objective sensory methods for each fish species. The need for quality measurements, monitoring and labelling of fish has also been discussed. One of the outcomes of the EU-funded project 'Evaluation Fish Freshness' (EU AIR3 CT942283) at the end of 1997 was that the Quality Index Method (QIM) was considered by the European fish research institutes as a scientifically valid objective sensory method for determining fish freshness (Olafsdóttir et al., 1997). Results from a recently finalised EU-funded project: 'Development and implementation of a computerised sensory system (QimIT) for evaluating fish freshness' (CRAFT FAIR FA-S2-9063), has shown that QIM is a rapid and reliable method to assess freshness in practical circumstances of auctions and processing sites (Luten, 2000a). A computerised system and appropriate QIM software for 12 important European fish species is now available (Martinsdóttir et al., 2001) and the number of QIM schemes developed at research level is growing steadily. The results from the on-going EU Concerted Action 'Fish Quality Labelling and Monitoring' (EU Concerted Action, 1998) shows that the level of acceptance of QIM by the partners within the fishery chain is growing (Luten, 1999; Luten, 2000b). However, a critical phase of a further and broader implementation of QIM in the fishery chain has now been reached. The establishment of QIM-Eurofish in April 2001 is therefore a step forward to implement the QIM method in the fishery chain (www.qim-eurofish.com). The mission of QIM-Eurofish is to stimulate the use of the QIM quality assessment method as a versatile tool within the fishery distribution and production chain in Europe (QIM-Eurofish, 2000).

The last 25 years has seen an increasing interest in improved quality assurance procedures in the fish processing industry (Howgate, 1987; Luten and Martinsdóttir, 1997). It is very important that all partners in the fishery chain: fishermen, fish auctions (ports of landing), fish processors and retailers agree on using the same standardised method for evaluating the quality (freshness) of fish.

To speed up and shorten the time between catch and processing, it is now common to sell fish via computerised fish auctions and in some cases fish is even sold before landing. However, buying fish unseen can be difficult, as it requires a reliable information on fish quality. Using the same method for quality/freshness evaluation of whole fish will facilitate trade in fish. Commerce via the Internet will increase and computerised information on the freshness of fish will be a necessity. Fish auctions and the fish processing industry are interested in having valid methods to facilitate grading of the raw material and to ensure its quality.

Information systems are used in the fish industry to ensure traceability in the inner control system of companies as required by the EC directives. Modern quality assurance systems require monitoring, controlling and recording of important quality parameters and parameters that might be critical throughout the production chain. Information about temperature and time from catch are of course of major importance. To verify this information sensory methods like QIM are very useful tools. It is foreseen that the QI will be useful to give feedback to crew members of fishing vessels about the quality of their catch, which may influence better handling onboard. Fish processing plants using raw material from their own vessels have records of time/temperature but they also have to rely on supplies from auctions or other sources and would like to have information on how fresh these catches are. Buyers of fish often use different sensory methods and sensory schemes and often demand that the processors use the same method. It would minimise cost and effort if the buyers and sellers used the same method.

For tracing and tracking of fish throughout the whole fishery chain it is also recommended to apply accepted standardised methodologies for the determination of quality (freshness). The quality attributes of a batch can be defined at each stage and labelled. To ensure traceability and labelling, methods are needed to measure and verify the quality at certain stages. An EU-funded project 'Traceability of Fish products' (EU Concerted Action, 2000) is now dealing with traceability of fish and fish products. Information is needed about properties of fish products and the inevitable changes that occur in the production processes (http://www.tracefish.org). This facilitation of full-chain traceability for fish products will in the end aid the consumer in guaranteeing safe and healthy products with well-documented characteristics.

Consumers are demanding more information about the quality of fish and fish products and consumer acceptance of fish is related to its freshness. Consumers might not be able to detect all the freshness/spoilage stages of fish when preparing their meal, but the sulphuric and nitrogenous compounds formed during storage are usually not appealing to the consumers. In some countries the consumers more often have their meals of fish in restaurants than in their everyday meal at home. The demand for freshness or knowledge of the freshness stage will be from retailers. Retailers play a significant role in influencing consumer perception of quality (Bisogny et al., 1987). Most of the fish is now sold at supermarkets instead in the old fishmonger's shops. The supermarkets are already demanding information on freshness of fish and fish-products from their suppliers even though this information is not stated on the packages in the supermarket. How consumers perceive edible fish products, what sensory attributes they use to describe and discriminate between species and what sensory factors contribute most to consumption have been studied by several authors. Several authors have also studied the consumer awareness of quality factors related to fish. High correlation was found between descriptions of trained panels and the opinions of consumers even though the trained panel used a wider range of intensity scale (Sawyer et al., 1988) Bech et al. (1997) showed it is possible to establish a link between consumers' demand for taste quality and attributes from sensory profiling. Food choice is under a wide range of influences beyond sensory considerations (Marshall, 1988). However, perception of taste determines a major part of overall attitudes of buying fish (Bredahl and Grunert, 1997). Perceived taste has been found to be significant in determining the frequently of consumption (Nauman et al., 1995; Myrland et al., 2000). Consumer criteria regarding freshness have not yet been established. Currently the interest in quality labels for fish products has brought attention to defining what should be included in quality labels and what kind of measurements are needed to monitor quality described by labels (Luten, 1999; Luten, 2000b).

Objective, standardised sensory methods would make information on fish quality more reliable and readily accessible and would facilitate and enhance quality- and process management in the fish industry. Moreover, a standardised sensory method would facilitate communication between buyers and sellers of fish and fulfil the demands of inspection authorities and regulations for tracking and tracing information about the quality of fish.

19.8   References

ANDRADE A, NUNES M L and BATISTA I (1997), 'Freshness quality grading of small pelagic species by sensory analysis', in Olafsdóttir G *et al.*, *Methods to determine the freshness of fish in research and industry. Proceedings of the Final Meeting of the concerted Action 'Evaluation of Fish Freshness' AIR3CT942283, Nantes Conference, Nov 12–14*, Paris, International Institute of Refrigeration, 333–8.

ANON. (1996), 'Council regulation (EC) No. 2406/96 of 26. November 1996 laying down common marketing standards for certain fishery products', *Official Journal of the European Communities*, No. L334., 1–14.

BECH A C, KRISTENSEN K, JUHLA H J and POULSEN C S (1997), 'Development of farmed smoked eel in accordance with consumer demands', in Luten J, Børresen T and Oehlenschläger J, *Seafood from producer to consumer, integrated approch to quality, Proceedings of the International Seafood Conference on the occasion of the 25th anniversary of the WEFTA, held in Noordwijkerhout, The Netherlands, 13–16 November, 1995*, Amsterdam, Elsevier Science B.V., 21–30.

BISOGNY C A, RYAN J and REGENSTEIN J M (1987), 'What is fish quality? Can we incorporate consumer perceptions?', in Kramer D E and Liston J, *Seafood Quality Determination, Proceedings of the International Symposium on Seafood Quality Determination, Coordinated by the University of Alaska Sea Grant College Program, Anchorage, Alaska, U.S.A., 10–14 November 1986*, New York, Elsevier Science B.V., 547–73.

BOTTA, J R (1995), *Evaluation of Seafood Freshness Quality*, New York, VCH Publishers Inc.

BRANCH A C and VAIL A M A (1985), 'Bringing fish into the computer age', *Food Technol. Aust.*, 37, 352–5.

BREDAHL L and GRUNERT K G (1997), 'Determinants of the consumption of fish and shellfish in Denmark: an application of the theory of planned behaviour', in Luten J, Børresen T and Oehlenschläger J, *Seafood from producer to consumer, integrated approch to quality, Proceedings of the International Seafood Conference on the occasion of the 25th anniversary of the WEFTA, held in Noordwijkerhout, The Netherlands, 13–16 November, 1995*, Amsterdam, Elsevier Science B.V., 3–19

BREMNER H A (1985), 'A convenient easy to use system for estimating the quality of chilled seafood' in Scott D N and Summers C, *Proceedings of the fish processing conference, Nelson, New Zealand, 23–25 April 1985. Fish Processing Bulletin* , 7, 59–703.

BREMNER H A, OLLEY A and VAIL, A M V (1987). 'Estimating time-temperature effect by a rapid sensory method' in Kramer D E and Liston J, *Seafood Quality Determination, Proceedings of the International Symposium on Seafood Quality Determination, Coordinated by the University of Alaska Sea Grant College Program, Anchorage, Alaska, U.S.A., 10–14 November 1986, New York*, Elsevier Science B.V., 413–36.

BREMNER A and SAKAGUCHI M (2000), 'A Critical Look at Whether 'Freshness' Can be Determined'. *J Aquatic Fd Prod Techn*, 9, 5–24.

CODEX STANDARDS FOR METHODS OF ANALYSIS AND SAMPLING, 'Sampling Plans for Prepackaged Foods (AQL 6.5)', XPT 13-1969, Rome, FAO/WHO Codex Alimentarius.

CODEX STANDARDS FOR FISH AND FISHERY PROJECT, 'Guidelines for the sensory evaluation of fish and shellfish in laboratories' CAC-GL 31-1999 Rome, FAO/WHO Codex Alimentarius.

EU CONCERTED ACTION (1998) FAIR PL 98–4174, 'Fish quality labelling and monitoring', http://www.fqlm.nl

EU CONCERTED ACTION (2000) QLK1-2000-00164 'Traceability of fish products', http://www.tracefish.org

HOWGATE P (1987), 'Fish inspection and quality control in Europe' in Kramer D E and Liston J, *Seafood Quality Determination*, Amsterdam, Elsevier Science B.V., 605–13.

HUIDOBRO A, PASTOR A and TEJADA M (2000), 'Quality Index Method Developed for Raw Gilthead Seabream (*Sparus aurata*)', *J. Fd Science*, 65 (7), 1202–5.

HUIDOBRO A, PASTOR A, LOPEZ-CABALLERO M E and TEJADA M (2001), 'Washing effect on the quality index method (QIM) developed for raw gilthead seabream (*Sparus aurata*)', *European Fd Research and Technol*, 212 (4), 408–12.

HUSS, H H (1988), *Fresh fish quality and quality changes*, FAO Fisheries Series 29, Italy, 27–61.

HYLDIG, G and NIELSEN J (1997), 'A rapid sensory method for quality management', in Olafsdóttir G and others, *Methods to determine the freshness of fish in research and industry. Proceedings of the Final Meeting of the concerted Action 'Evaluation of Fish Freshness' AIR3CT942283, Nantes Conference, Nov 12–14*, Paris, International Institute of Refrigeration, 297–306.

ISO 8586-1 (1993), 'Sensory Analysis – general guidance for the selection, training and monitoring of assessors. Part 1: Selected Assessors', The International Organization for Standardization, Geneva, Switzerland.

ISO 8586-2 (1994), 'Sensory Analysis – general guidance for the selection, training and monitoring of assessors. Part 2: Experts', The International Organization for Standardization, Geneva Switzerland.

ISO 8589 (1988), 'Sensory Analysis – general guidance for the design of test rooms', The International Organization for Standardization, Geneva, Switzerland.

ISO 11035 (1994), 'Sensory Analysis – Identification and selection of descriptors for establishing a sensory profile by a multidimensional approach', The International Organization for Standardization, Geneva, Switzerland.

JENSEN H S and JØRGENSEN B M (1997), 'A sensometric approach to cod-quality measurement', *Food Quality and Preferences*, 8 (5/6), 404–7.

JONSDOTTIR S (1992), 'Quality index method and TQM system', in Ólafsson R and Ingthorsson A H, *Quality Issues in the Fish Industry*, Reykjavik, Iceland, the Research Liaison Office, University of Iceland, 81–94.

LAND D and SHEPERDING R (1984), 'Scaling and Ranking methods', in Piggott J R, *Sensory analysis of Foods*, New York, Elsevier, 141–9.

LAWLESS H T (1994), 'Getting results you can trust from sensory evaluation', *Cereal Foods World*, 39 (11), 809–14.

LARSEN E, HELDBO J, JESPERSEN C M and NIELSEN J (1997), 'Development of a method for quality assessment of fish for human consumption based on sensory evaluation', in Huss H H M and Liston J. *Quality Assurance in the Fish Industry, Proceedings of an International Conference, Copenhagen, Denmark, 26-30 August 1991*, Amsterdam, Elsevier Science Publishers B. V., 351–8.

LEARSON R J ánd RONISVALLI L J (1969), 'A new approach for evaluating the quality of fish products' *Fishery Industry Research* 4(7), 249–59.

LUTEN J B and MARTINSDÓTTIR E (1997), 'QIM – a European tool for fish freshness evaluation in the fishery chain', in Olafsdóttir G *et al.*, *Methods to determine the the freshness of fish in research and industry ustry. Proceedings of the Final Meeting of the concerted Action 'Evaluation of Fish Freshness' AIR3CT942283, Nantes Conference, Nov 12–14*, Paris, International Institute of Refrigeration, 287–96.

LUTEN J B (1999), 'Annual Reports EU Concerted Action PL98-4174 'Fish Quality Labelling and Monitoring', November 1999 (ISBN 90-74549-03-9)', Wageningen, The Netherlands, RIVO The Netherlands Institute for Fisheries Research, p 22.

LUTEN J B (2000a), 'Development and implementation of a computerised sensory system (QIM) for evaluating fish freshness. CRAFT FAIR CT97 9063. Final Report for the period from 01-01-98 to 31-03-00', Wageningen, The Netherlands, RIVO The Netherlands Institute for Fisheries Research, p 18.

LUTEN J B (2000b), Annual Reports EU Concerted Action PL98-4174 'Fish Quality Labelling and Monitoring', December 2000, (ISBN 90-74549-05-5), p 23.

MARSHALL, P W (1988), 'Behavioural variables influencing the consumption of fish and fish products', in Thompson D M H, *Food Acceptability*, London, Elsevier Applied Science, 219–31.

MARTENS M and JØRGENSEN B M (1997), 'Multivariate data analysis used for investigation of the sensory quality of fish', in *Methods to determine the the freshness of fish in research and industry. Proceedings of the Final Meeting of the concerted Action 'Evaluation of Fish Freshness' AIR3CT942283, Nantes Conference, Nov 12–14*, Paris, International Institute of Refrigeration, 325–32.

MARTINSDÓTTIR E (1997), 'Sensory evaluation in research of fish freshness', in Olafsdóttir G *et al.*, *Methods to determine the freshness of fish in research and industry. Proceedings of the Final Meeting of the concerted Action 'Evaluation of Fish Freshness' AIR3CT942283, Nantes Conference, Nov 12–14*, Paris, International Institute of Refrigeration, 306–12.

MARTINSDÓTTIR E and STEFANSSON G (1984), 'Development of a new grading system for fresh fish' in Möller A, *Fifty Years of Fisheries Research in Iceland*, Reykjavik, Iceland, Icelandic Fisheries Laboratories, 23–31.

MARTINSDÓTTIR E, SVEINSDÓTTIR K, LUTEN J, SCHELVIS-SMIT R and HYLDIG G (2001) *Sensory Evaluation of Fish Freshness. Reference manual for the Fish Sector*, IJmuiden, The Netherlands, QIM-Eurofish.

MEILGAARD G, CIVILLE V and CARR B T (1999), *Sensory Evaluation Techniques*, 3rd. ed., New York, CRC Press, 23–36.

MUNOZ A M, CIVILLE G V and CARR B T (1992), *Sensory Evaluation in Quality Control*, New York, Van Nostrand Reunhold.

MYRLAND O, TRONDSEN T, JOHNSTON R S and LUND E (2000), '*Determinants of seafood consumption in Norway: lifestyle, revealed preferences and barriers to consumption*', Food Quality and Preferences, 11(3), 169–88.

NAUMAN F A, GEMPESAW C M and BACON J R (1995), 'Consumer Choice for Fresh Fish: Factors Affecting Purchase Decisions', *Marine Resource Economics*, 10, 117–42

NIELSEN, J (1995), 'Sensory methods', in Huss H H, *Quality and quality changes in fresh fish*. Rome, FAO Fisheries Technical Paper, No 348, 130–9.

NIELSEN, J (1997), 'Sensory analysis of fish', in Olafsdóttir G and others, *Methods to determine the freshness of fish in in research and industry. Proceedings of the Final Meeting of the concerted Action ' 'Evaluation of Fish Freshness' AIR3CT942283, Nantes Conference, Nov 12–14*, Paris, International Institute of Refrigeration, 279–86.

OEHLENSCHLÄGER, J (1997), 'Sensory evaluation in inspection', in Olafsdóttir G and others, *Methods to determine the freshness of fish in research and industry. Proceedings of the Final Meeting of the concerted Action ' 'Evaluation of Fish Freshness' AIR3CT942283, Nantes Conference, Nov 12–14*, Paris, International Institute of Refrigeration, 339–44.

ÓLAFSDÓTTIR G, MARTINSDÓTTIR E, OEHLENSCHLÄGER J, DALGAARD P, JENSEN B, UNDELAND I, MACKIE I M, HENEHAN G, NIELSEN J and NILSEN H (1997), 'Methods to evaluate fish freshness in research and industry'. *Trends Food Sci. Technol.*, 258–65.

O'MAHONY, M (1982), 'Some assumptions and difficulties with common statistics for sensory analysis', *Food Technol.* 36 (11), 75–82.

O'MAHONY, M (1986), *Sensory evaluation of food. Statistical methods and procedures*, New York, Marcel Dekker INC.

QIM-Eurofish (2000) web page, http:www//qim-eurofish.com.

SAWYER F M, CARDELLO A V and PRELL P A (1988), 'Consumer evaluation of sensory properties of fish'. *J.Food Sci.* 53 (1), 12-24.

SHEWAN J M, MACKINTOSH R G, TUCKER C G and EHRENBERG A S C (1953), 'The development of a numerical scoring system for the sensory assessment of the spoilage of wet white fish stored in ice', *J. Sci. Food Agric.*, 4, 283–98.

STONE H and SIDEL J L 1992. *Sensory evaluation practices*. Academic Press, Inc., Orlando, Florida

STONE H and SIDEL J L (1998), 'Quantitative Descriptive Analysis:

Developments, Applications and the Future', *Fd. Technol.*, 52 (8), 48–52.

SVEINSDÓTTIR K, HYLDIG G, MARTINSDÓTTIR E, JØRGENSEN, B and KRISTBERGSSON, K. (2001), 'Quality Index Method (QIM) scheme developed for Farmed Atlantic Salmon (Salmo salar) *(in press)*.

SVIENSDÓTTIR K, MARTINSDÓTTIR E, HYLDIG G, JØRGENSTEIN B and KRISTBERGSSON, K (2002), Application of Quality Index Method (QIM) Scheme in Shelf Life Study of Farmed Atlantic Salmon (Salmo salar). *J. Fd. Science*, 67 (4).

WARM K, BØKNÆS N and NIELSEN J (1998), 'Development of Quality Index Methods for Evaluation of Frozen Cod (*Gadus morhua*) and Cod Fillets', J. Aquat. Food Prod. Tech., 7(1), 45–59.

YORK, R K and SEREDA, L M (1993), 'Sensory assessment of quality on fish and seafood', in Shahidi F and Botta J R, *Seafood: Chemistry, Processing, Technology and Quality*, New York, Blackie Academic and Professional, 233–62.

19.9   Acknowledgements

Special thanks to my colleagues and co-authors of *Sensory Evaluation of Fish Freshness. Reference manual for the Fish Sector:* Kolbrún Sveinsdóttir, IFL, Iceland, Joop Luten and Rian Schelvis-Smit, RIVO, The Netherlands and Grethe Hyldig, DIFRES, Denmark. I also wish to acknowledge other authors of published sources from which I have used information.

20

Maintaining the quality of frozen fish

N. Hedges, Unilever R&D, Sharnbrook

20.1 Introduction

The process of freezing as a method of preserving fish quality has been long established. It was back in the early 20th century that Clarence Birdseye discovered that seafood and meats frozen in the bitter arctic winter tasted better than those frozen in the milder spring and autumn. It was from this observation that he developed, and refined what was described as 'the quick freeze machine'.[1] The process of quick-freezing is still a good, and arguably the best, way of preserving fish in a natural and safe manner for periods of many months or even years. For other produce, such as blanched vegetables, it has been established that freezing can maintain the nutritional content on subsequent frozen storage for at least six months.[2,3] Therefore, for vegetables freezing offers distinct advantages over other preservation methods, and may even offer benefits over ostensibly fresh produce that has been stored for a number of days at chill or ambient temperatures. The challenge for producers of frozen fish and frozen fish products is to optimise the natural preserving ability of the freezing process to provide products of consistently high quality from raw materials that are mostly from a wild origin. In addition, with the increasing pressure on the stocks of traditionally processed fish, another key challenge is to make better use of fish that are currently under-utilised for human consumption and that are also from sustainable sources. It is the aim of this chapter to describe some of the issues relating to the frozen fish supply chain, to highlight some of the key factors that need to be controlled in order to maintain quality and to describe, where possible, how this control might be exercised.

20.2   Frozen supply chains

The following section will describe the supply chains relevant to producing fish fillet products, but will not describe processes relating to minced products or to surimi. Lanier and Lee give a detailed review of surimi and surimi manufacture in their book *Surimi Technology*.[4] This chapter will focus on the supply chains for marine fish, as fish fillets from marine origin represent the largest tonnages of fish found in products.

Fish are currently either frozen into regular blocks or individually as fillets or portions. For frozen blocks the fillets are placed into cardboard liners within rectangular stainless steel moulds. The liners are sealed around the fish and lids placed on the moulds. The sealed moulds are slotted into plate-freezers, whereby the moulds are clamped between two extremely cold plates (between $-30°C$ and $-40°C$). Ice is less dense than water and on freezing expansion will occur. However, because the moulds are sealed the expansion of the frozen fish fillets is restricted. This causes any trapped air to be expelled and gives a regular shaped block. Regularity of shape and elimination of voids is an important part of block manufacture. This is because blocks need to be subsequently cut into portions of consistent shape, and irregularities in the block will lead to excessive losses during this process. An alternative to block manufacture is to freeze the fillets either whole or as portions cut from the fillets prior to freezing. These types of product are referred to as individually quick-frozen (IQF) fillets or portions. One of the issues with processing IQF fish portions is in controlling the weight of the final product(s) in the product box or bag. The following sections will describe the types of supply chain that produce the frozen fish.

The first part of the supply chain for fish fillet products involves catching the fish, removing their heads and guts, skinning and filleting followed by freezing. These stages are referred to as the primary process, whereby a frozen raw material is produced. This frozen raw material may either be sold to consumers 'as is', for example as bags of IQF fillets, or converted by secondary processing into, for example, coated fish products such as fish fingers and coated fish steaks. The types of supply chain for frozen fish may be distinguished from each other by where the latter stages of the primary processing operation take place, in particular where the filleting, skinning and final freezing are carried out. Thus, there are either land-based or sea-based operations. A sea-based operation requires a factory ship that has the capability to catch, process and freeze the fish. All of the primary processing for 'frozen at sea' fish may be carried out only a short time after catching. This eliminates the requirement for storage of the catch in boxes containing ice. One of the issues for sea-frozen raw materials is whether or not the fish are frozen before the fillets have gone into rigor mortis, i.e., whether or not the fillets are frozen with sufficient energy stores to be able to undergo muscle contraction. The rigor process will be described briefly in Section 20.5.3. The benefits and problems associated with freezing pre-rigor fish will also be described in the same section.

For land-based processing other factors become more significant. In a typical land-based process the fish are caught and the guts and sometimes the heads are removed. The fish are then stored on ice until the vessels return to shore. The time that fish are stored on ice, therefore, is dependent upon the time that the vessels are at sea. A trawler may be at sea for several days before sufficient fish are caught. Therefore, it is possible that fish may have been stored for several days on ice prior to being landed. In addition, fish caught early on in the catching cycle will be stored for longer on ice than those caught towards the end of the time at sea. Clearly fish which have experienced different lengths of time in ice storage may ultimately be processed together. This may have implications for the consistency of the raw material. The potential problems associated with ice storage will be discussed in more detail in Section 20.5.4. How the chill/ice storage process can be improved to minimise quality changes is discussed in Sections 20.5.5.

A third, hybrid, supply chain also exists. In this supply chain only part of the primary processing is carried out at sea. The fish have the guts and in some cases the heads removed. The headed and gutted fish are then frozen, often with a water glaze, and typically in a vertical plate freezer. In this way many fish can be frozen together, surrounded by ice. The fish 'block' can then be easily wrapped in plastic and packed into cardboard cartons. The frozen fish are then transported to land where they are thawed, filleted, skinned and re-frozen. Because two freezing stages are involved in this supply chain the resulting raw material is often referred to as double or twice frozen. This supply chain is most commonly used for Alaska pollack.

Within each of the supply chains described there are stages that are critical to the final quality of the product. Before discussing these critical stages the consequences of freezing fish fillets will be described.

20.3   Freezing of fish tissue

Figure 20.1 shows a photograph of a cod fillet with an individual muscle fibre partially excised from it. The muscle fibre, whose diameter is in the region of 50 to 200 μm (about the same thickness as a human hair), is a muscle cell and will therefore possess a cell membrane and cellular contents. The muscle cell is unusual in that it is long and thin and is filled with many myofibrils. The myofibrils are the contractile elements responsible for muscle movement in the live animal. Figure 20.2 shows a simplified structural hierarchy within fish muscle from the fillet down to the organisation of myofilaments within an individual myofibril. Most of the water that is present in a fish fillet (some 80% of the total weight) resides within the muscle fibres, and most of the water within the fibre resides within the myofibrils. Figure 20.3 shows a transmission electron micrograph of myofibrils within a cod muscle fibre. The circular structures that may be observed between the myofibrils are the membranes of the sarcoplasmic reticulum. Under most commercial freezing processes ice

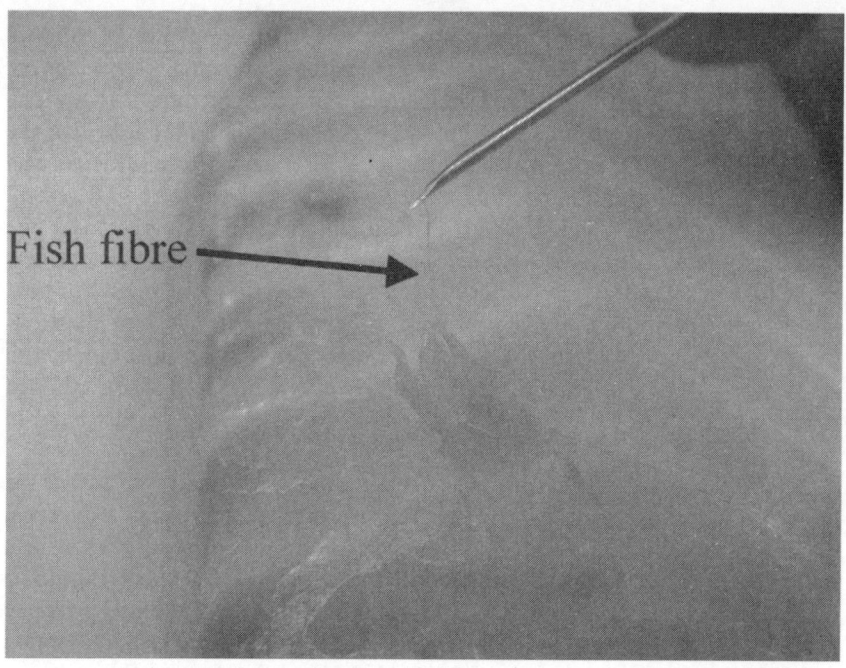

Fig. 20.1   Excised muscle fibre (circa 100μm in diameter) from a cod fillet.

forms between the muscle fibres, and the fibres shrink transversely as water is abstracted from them to form extra-cellular ice. Figure 20.4a shows a frozen section taken from a frozen cod fillet which had been air-blast frozen at −30°C. Areas of ice and areas of concentrated muscle fibres may be observed. Surprisingly, even at quite low temperatures there will be a fraction of water, the non-frozen fraction, which will not be converted to ice. This fraction exits because as ice forms the solute concentration within the fibres increases, this in turn depresses the freezing point within the fibre. A point is reached whereby the freezing point of the non-frozen fraction equals the temperature at which the fish are being frozen. Once this point is reached, equilibrium is established between the ice phase and the non-frozen fraction. Table 20.1 shows the amount of unfrozen water in cod fillet as a function of temperature.[5] Clearly, the amount of non-frozen fraction will vary as a function of temperature, and so will its viscosity. In this viscous phase the diffusion rates of reactants and products become significant. Consequently, the kinetics of many time-dependent processes may not be predicted by a simple Arrhenius relationship, but by a non-linear form of kinetics, with extraordinarily large temperature dependence. Reaction kinetics in viscous media has been described by an empirical equation proposed by Williams, Landel and Ferry.[6,7]

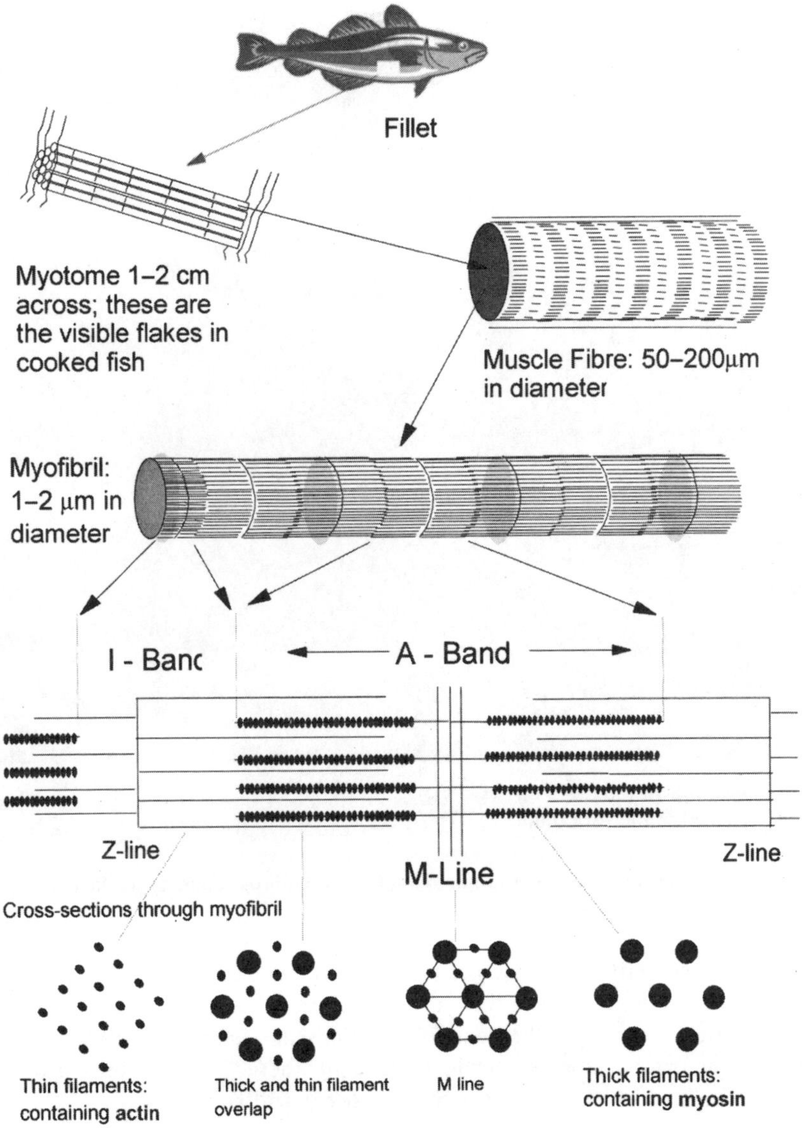

Fillet

Myotome 1–2 cm
across; these are
the visible flakes in
cooked fish

Muscle Fibre: 50–200μm
in diameter

Myofibril:
1–2 μm in
diameter

I - Band ←———— A - Band ————→

Z-line

M-Line

Z-line

Cross-sections through myofibril

Thin filaments:
containing **actin**

Thick and thin filament
overlap

M line

Thick filaments:
containing **myosin**

Fig. 20.2 Structural hierarchy within fish fillets.

20.4   Texture and flavour changes on frozen storage

One of the problems associated with freezing and frozen storage of fish and fish-based products is that changes in the flavour and texture of the final cooked product may occur. By forming a highly concentrated phase, through freezing, potential reactants (for example, enzymes and substrates) will be brought

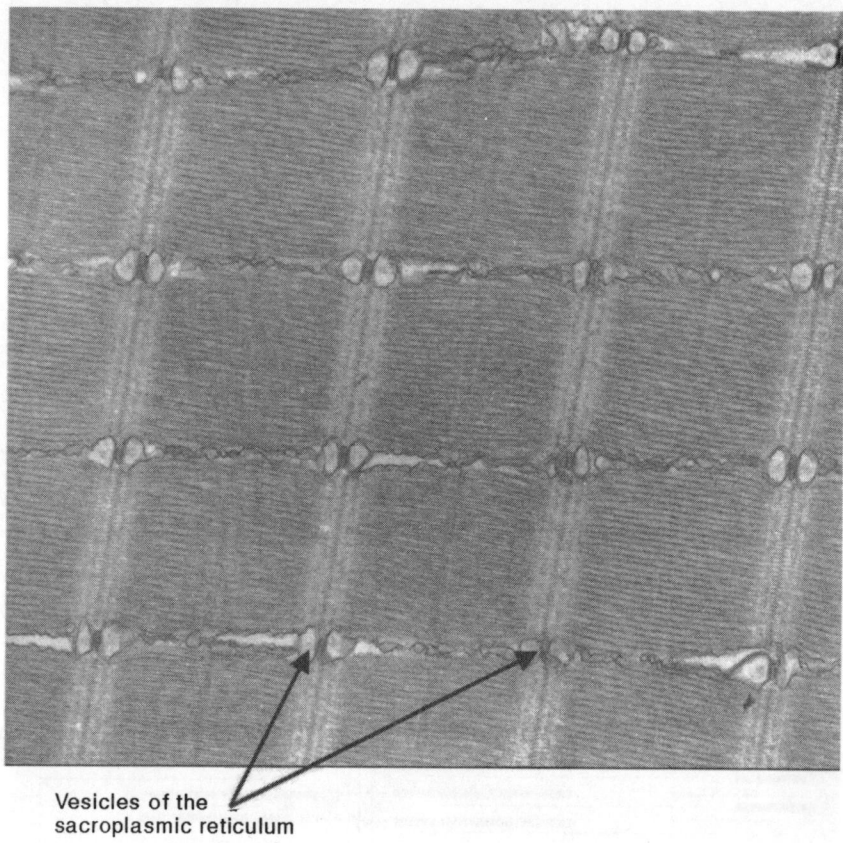

Vesicles of the
sacroplasmic reticulum

Fig. 20.3   Transmission electron micrograph of myofibrils (diameter circa 1–2 μm) of
very fresh cod.

together and diffusion distances shortened. Indeed, reactants that were
inaccessible to each other in the unfrozen cell may become far more intimately
mixed after freezing. In the next four sections the nature of the changes that
occur on frozen storage, measurement of these changes and proposed
mechanisms which describe how changes occur will be discussed.

20.5   Texture changes on frozen storage

It should be remembered that if good-quality fish are frozen reasonably rapidly,
and stored for only a short time in the frozen state the changes in sensory
attributes on subsequent thawing and cooking are quite small. The major losses
in quality attributes tend to occur on frozen storage. During frozen storage of
certain species of fish changes occur which result in deterioration of the textural
attributes of the cooked fillet. The textural change has been described as a

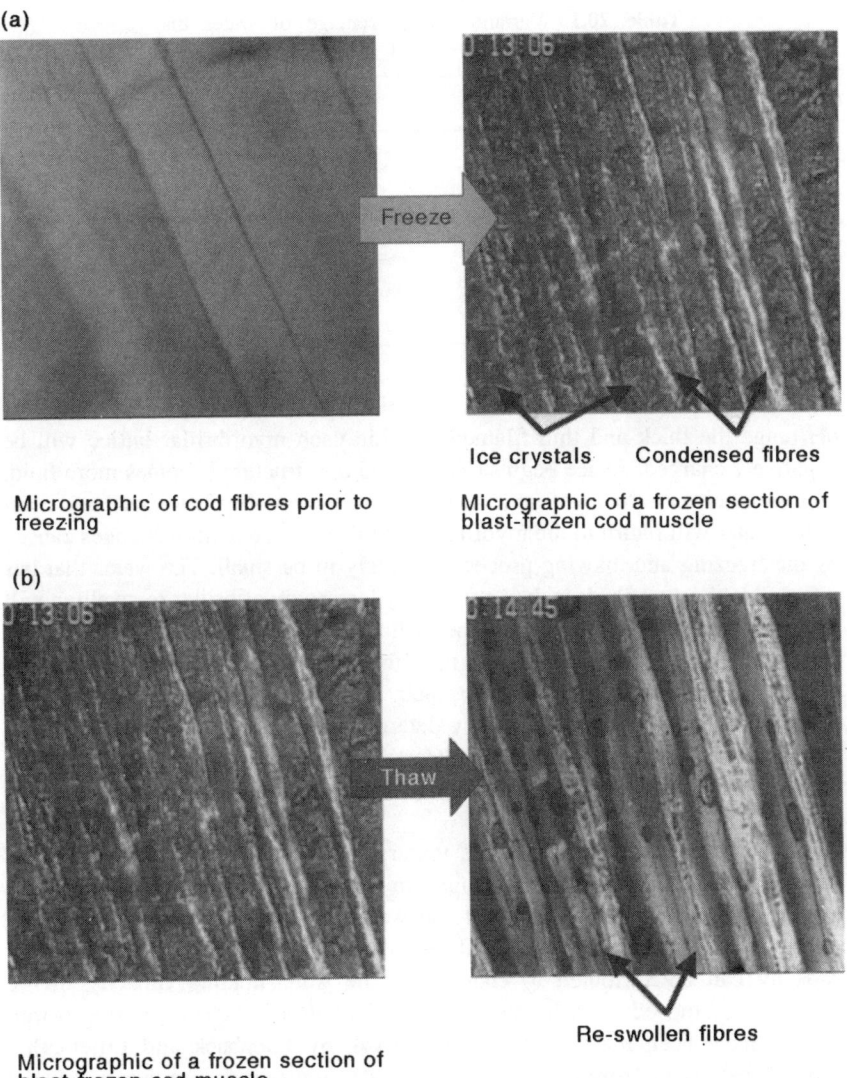

Fig. 20.4 Micrographs of frozen and thawed cod muscle.

tendency to express liquid on initial compression in the mouth, and for the remaining material to be hard, dry and fibrous. The problem is most acute for the white flaky fish fillets of cod, haddock, hake, etc.[8,9,10,11]

Figure 20.4b shows a light micrograph of the frozen section shown in Fig. 20.4a after thawing on the microscope slide. As the ice has turned back to water, much of the liquid abstracted by the formation of ice has returned to the muscle fibres. Consequently the fibre diameters increase. The driving force for this re-swelling is most likely to be one of charge repulsion. The pH of fish fillets of,

Table 20.1   Variation of percentage of water that remains unfrozen for cod fillet

Temperature (°C)	% Unfrozen water
−1	92
−2	48
−3	33
−4	27
−5	21
−10	16
−20	11

for example, cod and Alaska pollack are typically in the range pH6.8–7.0. In this pH range the thick and thin filaments within each myofibrillar lattice will be negatively charged. As ice begins to melt, and the structure becomes more fluid, repulsion between the myofilaments will cause the myofibrils to swell and most of the water will return to the myofibrils. In this case the texture changes caused by the freezing and thawing process are likely to be small. The water that has returned to the myofibrils is then not readily lost, so myofibrillar re-swelling will influence directly the ability of the fish to retain water after thawing and cooking. If less water returns to the fibres they will have higher protein concentrations and that may affect their mechanical properties. It has been proposed that fibre diameter is a key determinant of firmness when comparing fish species[12,13] (smaller fibre diameters correlating with increased textural firmness) and that the rheological properties of fish fibres may change on frozen storage.[14]

Generally it is reported that the textural changes occurring during frozen storage are due ultimately to changes in the myofibrils, although the exact mechanism is not clear. Studies of the water-holding capacity of other meat systems have shown that most of the observed changes in water-holding capacity can be attributed to changes in the water-holding capacity of the myofibrils.[15] Indeed for fish, observations of myofibrillar change during frozen storage have been made. For example, work by Jarenback and Liljemark[16] showed that during frozen deterioration of cod (*Gadus morhua*) a decrease in the dimensions of the thick filament lattice was observed, coupled with disturbances to the regular hexagonal lattice spacing. It was also noted that the myofibrils within a thawed, frozen deteriorated muscle fibre were pushed closer together, with a commensurate disappearance of the sarcoplasmic reticulum vesicles lying between the myofibrils.

Results reported by Garcia *et al.* for frozen cod fillets also indicated that increased deformation of the hexagonal array of thick filaments occurred during frozen storage.[17] Similar effects have also been reported for Alaska pollack muscle that had been subjected to −20°C frozen storage for two months.[18] On thawing, the myofibrils within each muscle cell did not recover their original diameters and appeared closer together. Further frozen storage (i.e., 12 months at $-20°C$) rendered the myofibrils even less able to effect a recovery of their original diameters on thawing, and a substantial loss of the order within individual myofibrils was also observed. Thus, the changes in water-holding capacity of fish muscle on frozen storage may be due to two events. The first is the pushing or collecting together of the myofibrils within the fibres as ice is formed externally to the myofibrils. The second involves changes within the myofibrils that render them increasingly unable to swell back up once the fish is thawed and/or cooked. It is likely that it is the changes that occur within the myofibrils that give rise to the texture changes observed on frozen storage.

A number of hypotheses have been proposed to explain the observed changes. One of the most regularly reported mechanisms is the formation of the protein cross-linking agent, formaldehyde, from the degradation of trimethylamine oxide (TMAO). TMAO is present in gadoid fish and tends to be found only in marine fish where it may serve as an osmo-regulator.[19] It is soluble in water and is broken down by an enzyme (TMAO-ase) to produce formaldehyde and dimethylamine during frozen storage.[20,21,22,23] Generation of formaldehyde and dimethylamine has been correlated with textural change in the frozen state.[10,24,25,26] The presence of TMAO, and the rate at which it is degraded during frozen storage have also been used to explain the observed differences in rates of textural loss between species.[8,27] However, a number of inconsistencies exist that suggest that TMAO degradation is not a complete explanation of the reported observations.

Firstly, it has been shown that fillets from haddock, another member of the gadoid family, deteriorate on frozen storage. Haddock also uses TMAO as an osmoregulator but TMAO is reported not to be degraded to formaldehyde and dimethylamine during frozen storage of whole fillets.[10,28] Thus, it appears that haddock fillets may change in texture on frozen storage without any concomitant development of formaldehyde or dimethylamine. Mincing of haddock fillets, however, is reported to stimulate the break-down of TMAO,[29] suggesting that TMAO-ase activity may be present in the fillets.

Secondly, it has been reported that minces are not stabilised by washing.[30,31,32] For example, it has been reported that minces from red hake (also a Gadoid) are more prone to structural changes within the myofibrils after freezing and thawing than minced muscle that has not been washed.[32] The same authors also reported that washing produced an increased textural 'hardness' after three freeze thaw cycles, or six months frozen storage at $-20°C$, compared to unwashed mince (hardness being measured as a shear force, i.e., the force required to shear through a cooked fish mince patty with a blade). These data bring into question an exclusive role of TMAO in frozen deterioration because if TMAO degradation products are the root cause of frozen deterioration in species such as hake then its removal by washing ought to slow the process of textural change. In addition, other authors have reported alternative mechanisms for myofibrillar aggregation. Other hypotheses include interaction of myosin with lipids and with lipid oxidation products.[33,34,35] Also, it has been observed that myosin from species that inhabit warmer waters have higher thermal stability[36,37] and that species more resistant to frozen deterioration of texture have muscle proteins, particularly myosin, with increased thermal stability.[38] These observations suggest a link between thermal stability of myosin, which reflects the environmental temperature of the fish and reduced myofibrillar aggregation during frozen storage.

20.5.1   Measurement of texture changes

Electron microscopic techniques have been used to probe changes in the ultra-structure of fish muscle during frozen deterioration of texture. However, it must be remembered that these techniques probe only a very small volume of the sample. Thus, other techniques may be employed to give measurements of change within larger amounts of sample. One such probe of myofibrillar change during frozen deterioration has been the response of myofibrils to high concentrations (0.6mol/dm^3 NaCl) of salt, referred to as salt solubility. At a high enough salt concentration the myosin molecules within the thick myofilaments dissolve. At the same time the actin filaments also go into solution, usually intact, as they become detached from the Z-line, although some uncertainty exists as to how this dissolution is effected. The consequence of these events is that actomyosin (myosin bound to actin) goes into solution, with each actin filament regularly 'decorated' with myosin molecules. If fresh cod myofibrils are prepared substantial dissolution results when they are treated with 0.6mol/dm^3 concentrations of salt,[39] but as frozen deterioration progresses it has been observed that the myofibrils become increasingly resistant to dissolution.[40,41]

Another consequence of the inability of the fish fibres to recover their original volumes upon thawing is that water that was originally within the myofibril prior to freezing/frozen deterioration is now found externally to the fibres. A technique that may be employed to report upon the water distribution changes in muscle is pulsed proton nuclear magnetic resonance (NMR).[42,43] The spin-spin or T_2 relaxation time of water protons is dependent upon the motional correlation time of the water molecule. The motional correlation time reflects the overall isotropic reorientation, and can be defined as the average time required for the water molecule to rotate by one radian. Thus, a water molecule experiencing restricted motion at, for example, the surface of rigid protein structures within a myofibril, will have a shorter spin-spin relaxation time than water molecules that are unable to contact the surface. Thus, a distribution of relaxation times may be observed, which reflects the extent to which water molecules interact with the myofibrillar structure.[42]

Indeed, it has been shown that the spin-spin relaxation times of water in whole muscle are complex,[44] but reflect the spacial heterogeneities that exist within the structure. Also, is has been shown that changes in the distribution of spin-spin relaxation times reflect changes in the ultra-structure, water-holding capacity and sensory properties of meat and fish.[45,46,47,48] Thus, the NMR relaxation technique is a useful tool in probing the water distribution changes associated with frozen deterioration of fish texture, and may also report on changes in sensory attributes relating to water distribution changes.

In order to measure textural changes some studies have employed mechanical tests to monitor changes in the firmness of fish minces and fillets.[49,50] These tests allow the changes in textural toughening to be measured. However, in evaluating the results from some mechanical tests it must be remembered that few models exist to relate mechanical/failure properties in whole fillets to sensory attributes. For example, the relative contribution of connective tissue and myofibrillar elements to the perceived texture is not clear therefore how changes in these structural elements relate to changes in sensory perception may be difficult to judge.

The site of ice formation and the size of ice crystals may have important effects on the structure of the muscle cells and their organelles. This in turn can impact upon the storage stability of the fish in the frozen state. One group of measurements that is often ignored in many studies of texture (and flavour) change is the measurement of changes in the ice crystal structure. For example, in some cases the effects of different freezing regimes are reported but no measurements of ice crystal size and location are made. Thus, it is sometimes unclear whether different freezing or storage regimes have produced differences in the ice crystal structure or location. Therefore, although many authors report biochemical changes on frozen storage, how much of the observed change is driven by changes in the ice crystals is unclear. The impact of freezing rate on ice crystal size and location will be discussed in more detail in Section 20.6.

20.6   Flavour changes on frozen storage

Most fish that are used in the manufacture of fish products belong to the superorder Teleostei. These are the boney fish with vertebrae. This superorder includes all of the freshwater fish and many of the marine fish. The remainder belong to the Elasmobrachii, or cartilaginous fish, which include rays, skates and sharks. Teleost fish may be divided into two groups: the demersal (or bottom dwelling) and the pelagic (surface dwelling) fish. Demersal fish tend to drift with ocean currents, therefore do not swim actively for long periods. Consequently, they do not possess much of the red, or dark muscle required for prolonged aerobic metabolism of energy reserves. The second group of teleost fish are the pelagic, or oily fish. These fish swim actively for long periods and consequently their muscles are equipped for prolonged aerobic muscle activity. Therefore, these fish have more dark muscle, required for aerobic metabolism, and they have fat deposits throughout the muscle tissue and just beneath the skin.

The development of rancid off-flavours, essentially the oxidation of unsaturated fats, is problematic in many fish types and seriously limits their shelf life.[51,52] Because of this problem much of the work studying flavour changes in frozen fish has focused on preventing the development of rancid off-flavours in pelagic species of fish which have a relatively high oil content. Low-fat species of fish such as cod (*Gadus morhua*) and haddock (*Gadus aeglefinus*), predominantly store lipid in the liver. The muscle tissues of such fish contain 0.5–1.5% lipid in the form of functional lipid such as lipoproteins or phospholipids of the cellular membranes. High-fat-content fish have more than 5% lipid present in the tissues, usually in the form of triglycerides.[53] Examples of high-fat species are Atlantic mackerel (*Scomber scombrus*) and salmon (*Salmo salar*). Total lipid content will depend upon additional factors such as availability of food and, in mature fish, the time in the breeding cycle. In some species the seasonal variation in fat content can be very marked, for example, in Atlantic herring (*Clupea harengus*), it can vary from 1–25%. Susceptibility to rancidity depends not only on the amount of lipid present, but also the lipid composition and its location in the fish tissue.[54] Fish contain high levels of highly unsaturated fatty acids and uniquely high levels of n-3 fatty acids, for example, eicosapentaneoic acid (EPA) and docosahexanoic (DPA). It is for this reason that they are susceptible to oxidative rancidity.

Although the process of lipid oxidation is highly favourable thermo-dynamically, the direct reaction between oxygen and even highly unsaturated lipids is kinetically hindered.[54] Hence, an activating reaction is necessary to initiate free radical chain reactions. It has been proposed that lipid oxidation in fish may be initiated and/or promoted by a number of mechanisms. These include the production of singlet oxygen; enzymatic and non-enzymatic generation of partially reduced or free radical oxygen species (i.e., hydrogen peroxide, hydroxy radicals); active oxygen iron complexes and thermal or iron-mediated homolytic cleavage of hydroperoxides. The exact details of these mechanisms will not be covered here but have been reviewed extensively in other articles.[55,56] From a practical point of view some of the key factors influencing the rate and extent of rancidity development are oxygen concentration in the local environment, surface area exposed to oxygen, fatty acid composition of lipids, levels of endogenous antioxidants and/or endogenous oxidative catalysts and the storage temperature. It has also been suggested that lipid oxidation products can affect the colour and nutritional value of food.[57] Nutritionally important unsaturated fats, such as n-3 fatty acids, may become degraded in purified fish oils.[58] However, the degree to which these molecules are degraded in whole fish fillets is still an area of some uncertainty.[59]

A topic that has received much less attention is the development of off-flavours in many of the white flaky fish. These species are demersal fish, and have low lipid contents in their muscles, however flavour changes may still occur on frozen storage. The flavour changes have been described as the development of 'cardboard' or 'cold store' flavours.[60,61] Cold store flavours have been attributed to the development of oxidation products of the phospholipids present in cell membranes. Little work has been done in reducing the development of cold store flavour in fish, although it has been suggested that starvation prior to slaughter may offer a benefit.[62] A second topic that has received only limited attention is the loss of positive flavour notes associated with fresh fish. This area will be discussed in more detail in section 20.7.4 on ice storage.

20.6.1   Measurement of rancidity changes leading to off-flavour

Many of the most frequently used methods to monitor rancidity development in fish and fish products may be divided into colorimetric tests, other spectroscopic tests and chromatographic tests. Of the colorimetric tests the more often used are peroxide values (PV) and thiobarbituric acid (TBA) or thiobarbituric reactive substance (TBARS) tests.

The PV is determined by titration of iodine released from the reaction between a lipid extract from the fish and potassium iodide solution. Problems occur due to low sensitivity, interference from any coloured lipid soluble components and in the extraction of the lipids from the fish. Modifications to this method have been made to improve sensitivity when assaying small amounts of lipid extracted from food. The modified method uses the detection of a red thiocyante complex formed by the reaction between sodium thiocyanate and ferric iron. Ferric iron is produced by the oxidation of ferrous iron by peroxides present in the extracted oils.[55]

The TBA test is one of the most commonly used methods to study the degree of lipid oxidation in meat products. The test relies upon the reaction between TBA and aldehydes to give a coloured alkanal whose absorbance can be measured spectrophotometrically at 450nm. Despite its widespread use there are a number of criticisms of the technique. These include uncertainty about the nature of the colour generation reaction, lack of specificity of the TBA reaction (TBA may react with compounds that are not generated by lipid oxidation to give red colours) and TBA only reacts with a minor component of lipid oxidation (malonaldehyde). The term TBA value is now often replaced by the term TBARS tests to reflect the lack of specificity. However, the test can be improved by using high-performance liquid chromatography (HPLC) to separate the TBA reaction products. Despite these improvements it is generally recommended to use TBARS tests in conjunction with other assay methods. Although PV and TBA determinations have limitations, they can be useful indicators of rancidity and may help to distinguish between the effectiveness of different treatments and process regimes.[63] For example, Hwang and Regenstein[64] used PV to measure rancidity development in minced menhaden treated with various antioxidants and molecules that have a synergistic effect on antioxidant action.

Direct spectroscopic tests (i.e., tests that do not require an additional 'reaction' to produce a colour change) have also been developed to give an estimation of the extent of rancidity in fats. When fatty acids are oxidised to hydroperoxides, double bonds within the molecule may become conjugated (i.e., the double bonds are separated by a single bond). Conjugation is characterised by absorption of ultraviolet light at 232nm. Similarly conjugated dienes, that may result from decomposition of hydroperoxides, also absorb at this wavelength. Other secondary oxidative products absorb elsewhere in the UV range. For example, diketones absorb at 268nm.

Techniques based on fluorimetry have been developed to measure lipid oxidation, primarily because of potentially much higher sensitivity than is possible with TBA tests. Although not used widely, there are instances of it being used to monitor oxidative changes in fish tissues. Smith et al.[65] used fluorescence to measure lipid oxidation in Indonesian salted-dried marine catfish (*Arius thalassinus*).

Gas chromatography (GC) is becoming increasingly more widely used to measure oxidative rancidity in oils and foods. Volatile oxidation products can be measured directly by, for example, headspace GC methods or as more stable derivative products. Headspace methods can be used to determine the volatile components in the gas (or headspace) above the food. This method will give an accurate profile of the volatile components. This is of value since some of these volatile molecules contribute to the rancid odour and/or flavour. Sampling of the headspace can be an issue due to limitations in sampling volume. Also, although the peak areas are proportional to the concentration of each component the palate response is dependent on the chemical structure of each molecule. Thus, the relationship of peak area to organoleptic response must also be investigated in order to establish which volatiles are crucial to off-flavour development, and what threshold level needs to be exceeded before tasters can detect off-flavours.

By using a purge and trap method problems with sensitivity of headspace GC can be reduced. In this technique an inert gas, usually helium or nitrogen, is passed over the sample and the volatiles are collected (trapped) in a cartridge containing, for example, activated charcoal or Chromosorb. The volatiles are subsequently desorbed from the cartridge by rapid heating and analysed by GC. Limitations of this technique include alteration or decomposition of volatiles during desorbtion from the cartridge, low recovery of volatiles with a high boiling temperature and transfer of water from the cartridge to the GC column.

As with measurements for textural deterioration, it should be remembered that there is no industrial-standard measure for rancidity or off-flavour development. Thus, it is reliant upon the individual researcher to establish the significance of the changes in volatile profile that are observed. Flavour changes must be related to sensory changes, as assessed by a trained panelist, and ultimately changes must be related to consumer reaction to (or rejection of) the product(s).

20.7   Pre-freezing factors influencing storage stability

The aim of this section is to discuss some of the factors that may affect the quality of frozen fish products. There are a number of factors and treatments that may have a positive or a negative impact upon frozen storage stability and final quality of fish products as experienced by the consumer. These have been set out in approximately chronological order from initial catch to the freezing process.

20.7.1   Species type and texture change

It has been reported by a number of authors that different species of fish have different susceptibilities to deterioration of texture attributes on frozen storage.[8,9,10,11] The reason for this is not clear, but a correlation between the body temperature of the fish and increased stability on frozen storage has been suggested.[36,37,38] In addition, it is also reported that flat fish, such as sole and plaice tend to be lest prone to textural change.[27,66,67] Some authors have cited the presence of TMAO and subsequent rates of degradation to formaldehyde and dimethylamine as the reason for different rates of textural decline on frozen storage. However, this hypothesis does not offer a complete explanation of all the observed results. Consequently, there is no simple explanation for why different species of fish deteriorate on frozen storage at different rates or why some pre-freezing treatments are beneficial or detrimental.

20.7.2   Time of catching and handling procedures

It has been well documented that the season in which fish are caught can have a profound influence on properties. The most significant seasonal effect is that of spawning. The main consequence of spawning is that much of the lipids and proteins in the fillets are converted to either eggs or sperm. As a result of this, fillets tend to become much softer and more watery. In addition, once spawning is complete fish need to replenish their depleted protein and energy stores. This in turn may lead to build-up of muscle glycogen once spawning is over. The result of this is that the final post-mortem pH of the fillets may be lower than usual. The consequences of high glycogen levels, how this affects the final pH post-mortem and the consequences of low pH are explained in more detail in the next section.

Rough handling may also cause loss of desirable properties. For example bruising of fish may induce an increased rate of rancidity development on subsequent frozen storage. Therefore, it has been recommended to use chilled sea water (CSW) rather than ice for reducing bruising and to improve flavour stability of snapper (*Chrysophrys auratus*)[68] and sardine (*Sardina pilchardus Walbaum*).[69] It has also been reported that whole fish may deteriorate less rapidly than fillets do,[70] suggesting that the filleting process could induce damage, likely to increase rates of deteriorative change when the fish are frozen and stored. Certainly, mincing of fillets is often cited as resulting in increased texture and flavour change on frozen storage, when compared to the intact fillets.

Contamination of the fillets with blood or viscera, especially material from the kidneys, is also likely to increase the rate of textural change on frozen storage. This is particularly problematic when preparing minces. It has been reported that addition of cod kidney extracts to minced cod may reduce stability on frozen storage.[71,72] It has been suggested that this could be due to a high concentration of the TMAO-ase enzyme in the kidney extract generating formaldehyde. Similarly it has been reported that addition of cod frame mince (which may also be contaminated with kidney and blood material) may induce instability in otherwise relatively stable lemon sole mince.[73]

20.7.3 Effect of pH

After death the blood supply to the muscle ceases, and oxygen and nutrients no longer reach the muscles. The compound that directly supplies energy for the metabolic processes within each muscle cell is adenosine triphosphate (ATP). In order to maintain ATP levels the energy stores within the muscle cells (particularly glycogen) are metabolised anaerobically. The consequence of this is that lactic acid will be generated in the muscle and this cannot be removed. As the amount of lactic acid increases the pH will fall. The amount by which the pH falls depends upon the level of energy reserves present in the fish tissue prior to death. The amounts of such reserves will depend upon the nutritional status of the fish prior to catching and the amount of energy the fish expends during the catching process. Despite differences in the stores of energy immediately post mortem a point is reached whereby regeneration of ATP ceases and the residual ATP is expended. It is at this point the fish enters rigor mortis and becomes stiff. Rigor mortis is the binding of the thick filaments to the thin filaments within the myofibrils.

The ultimate pH of the fish tissue may have a direct impact on texture of fresh fish. For example, it has been shown that cod fillets of lower pH tend to have firmer texture.[74,75] Also pH may influence the rate of change of texture and flavour on frozen storage. Lower pH cod tend to become tougher and drier more rapidly, whilst higher pH cod tend to develop an unacceptable flavour before textural toughening becomes problematic.[76] Changing the pH of the fish muscle will have a direct impact on the water-holding capacity of the myofibrils.[15] As the pH of the fillet decreases the charge on the filaments of the myofibril lattice also decreases. This decrease in net negative charge reduces charge repulsion between the filaments and this causes the myofibrillar lattice to contract. This contraction expels liquid from the myofibrils and increases their protein density. Thus, if fish have higher energy reserves when they are caught the final pH will tend to be lower and this may impact directly on the texture of the fish and on storage stability.

20.7.4 Ice storage

During storage in ice a number of changes may occur. These changes may be caused by enzymes present in the fish muscle or by the increase in microbial load on the fish surface. The changes may affect the initial flavour and texture of the fish and also the resistance of the fillets to changes during frozen storage.

The flavours of species of gadoid fish, for example cod, haddock and whiting are typically described as sweet and meaty. The sweet notes develop as the fish goes into and through rigor mortis, and the greatest intensity of flavour is around 1–2 days after death: the fish having been held in ice.[77] The intensity of the sweet notes decreases on further ice storage until, in cod for example, after about 7–8 days the taste is perceived as being bland. After that, spoilage flavours caused by microbial growth can be detected. In haddock, the loss of sweetness is slower than in cod and there is not a completely bland stage during iced storage before spoilage flavours are apparent.[78] The loss of the fresh, sweet flavours is reported to be a result of enzyme reactions.[79] Samples of fillets prepared and stored under sterile conditions can be held for much longer at 0°C without the development of spoilage flavours.[79] In this case the change in the observed flavour tends to be a loss of sweet flavours over a few days followed by a prolonged period where the flavour is bland.

One consequence of microbial growth on the surface of fillets from marine fish is the development of 'fishy' aromas. Although the development of spoilage flavours is complex, a molecule that has received a lot of attention in trying to understand fish flavour change is trimethylamine (TMA). This volatile amine is generated from microbial action on TMAO. Because freshwater fish do not have TMAO present in their muscle they tend not to develop 'fishy' odours on chill storage, and consequently, may remain acceptable for longer. TMA usually stays at a low level for approximately the first ten days of ice storage. Its concentration then increases dramatically, and because of this, it has potential to be used as a marker for the onset of microbiological spoilage. Simple methods by which TMA can be measured have been reported.[80,81,82] Once the fish are frozen the TMA will remain at its pre-frozen level, but DMA, generated from TMAO on frozen storage will increase. Therefore combined DMA and TMA measurements have the potential to be used to assess the frozen and chilled history of some gadoid species of fish (with the notable exception of haddock).

The time that fish spend on ice, prior to processing can have a significant effect on texture. Post mortem storage leads to softening of the muscle tissue. In mammalian muscle a period of chill storage is often necessary to tenderise muscle. In fish tissue the post mortem conditioning process can lead to excessive softening and the rate may depend upon the temperature of the water in which the fish live.[83] Although the main myofibrillar proteins, actin and myosin are believed to remain largely intact during conditioning, many of the cytoskeletal proteins (the proteins which join the myofibrillar elements together), for example connectin, nebulin and desmin, are believed to be degraded.[84] There have been many recent studies of post mortem conditioning in fish and the consensus seems to be that the process may not be the same in all fish species. Certainly the rate at which the degradative changes and the associated softening occurs varies considerably between fish species.

20.7.5 Pre-chilling

Traditionally, ice is used to chill fish prior to being frozen. This has several advantages. One is that good contact between the fish and the ice can be achieved, allowing good heat transfer from the fish to the ice. A second is that to melt ice requires a large amount of heat energy to be removed from the fish. The disadvantages of using ice are that it can be labour intensive and, for fish in boxes, the contact between the fish and ice may be poor. An alternative to chilling on ice is to partially freeze the fish. By reducing the temperature of the fish to −3°C or −4°C fish can be partially frozen. It has been claimed that partially freezing fish in this manner will extend the chill storage shelf-life by several weeks.[85,86] In addition, it is suggested that freezing fish that are already partially frozen will offer benefits in terms of freezing efficiency.

20.7.6   Freezing at sea

Clearly, the effects of ice storage on texture and flavour of fish can be alleviated by freezing fish soon after catch, but freezing at sea can cause some problems. If the fillets are frozen before the energy stores are depleted, the fillets may 'come alive' again during the thawing and cooking process. This can lead to an extensive shortening of the fillet, which is referred to as thaw rigor, and results in excessive liquid loss from the fish and twisting of the product's shape. In addition, filleting pre-rigor fish may also lead to uncontrolled muscle contraction that can cause tearing of the fillets.

20.7.7   Addition of cryoprotectants to preserve texture

Cryoprotectants are compounds that improve the quality and extend the shelf-life of frozen foods. A wide variety of cryoprotective compounds are available. These include sugars, amino acids, polyols, methyl amines, carbohydrates, some proteins and even inorganic salts, such as potassium phosphate and ammonium sulphate. The selection of cyroprotectants will depend on whether the application is for a comminuted product or for whole fillets. In minces the cryoprotectant can be intimately mixed with the muscle proteins. For whole fillets the ability to diffuse the protectant into the correct location within the structure must also be considered. For example, use of high molecular weight protective agents may not be possible in fillets simply because they will not penetrate the fish structure sufficiently well. Many cryoprotective agents have been added to fish fillets and minces and have been reported to lead to improvements. These additives include polyphosphates, sugars, carboxyllic acids and milk proteins.[72,87,88,89,90] MacDonald and Lanier have given a good overview of the use of cryoprotective agents in maintaining the texture of frozen fish.[91] In choosing a cryoprotective agent, however, thought must be given to regulatory approval for its use in fish products. In fact, for whole fillet products very few additives have regulatory approval for use in the European Community.

A newer area of cryoprotection is the use of antifreeze proteins. It has been reported that the addition of AFPs to meat will control the ice crystal size. In this way drip loss can be reduced and textural quality maintained.[92,93] However, the method of addition involved either soaking the muscle in an AFP solution, or injection of an AFP solution into the bloodstream pre-slaughter. For whole fish fillets and poultry muscle, the difficulty is in finding an expedient route of perfusing AFP molecules into the muscle structure without causing disruption/ damage. However, antifreeze proteins do occur naturally in some fish species and this may offer some benefits[94] by controlling the ice crystal structure.

20.7.8   Control of rancid off-flavour and the addition of antioxidants

It should be remembered that addition of antioxidants is only one way to control oxidative changes in frozen fish products. Other options may also include vacuum[95,96] or controlled atmosphere packaging.[97] Both these packaging options effectively reduce the exposure of the product to atmospheric oxygen. Other ways of reducing direct exposure to the atmosphere are to use barriers such as ice or saturated lipid glazes applied to the surface of a product. Also, reducing the temperature at which products are stored whilst frozen will offer benefits.[98,99]

Since many of the reactions leading to rancidity are oxidative, the addition of antioxidants can improve storage stability. Many herb and spice seasonings contain substances which help control rancidity in meats. For example, oleoresin extract from rosemary can provide inhibition of oxidative rancidity and retard the development of 'warmed-over' flavour in some meat products.[100] It has also been reported that rosemary extracts may offer benefits in controlling oxidative changes in fish.[101] However, in selecting the use of natural antioxidants in herbs and spices the impact of these molecules on product flavour and appearance must also be determined. Also in selecting antioxidants it is important to check their regulatory approval.

20.8   The effect of freezing rate

In Section 20.3 the effect of freezing on the fibrous structure of fish fillet was described. The freezing rate employed was such as to produce ice crystals between the muscle fibres leading to them becoming condensed. The effect of freezing rate has been, by and large, poorly studied. Often, different freezing regimes have been imposed upon fish samples, but no record of the ice structure has been made.

Slow freezing has been reported to cause increased drip loss[102] and, conversely, initial advantages of using rapid/cryogenic freezing are reported.[103,104] However, in other studies of rapid freezing no advantage was observed.[105] Under conditions of ostensibly 'rapid' freezing the possibility exists whereby outer parts of the 'rapidly' frozen sample do experience very rapid freezing, but the bulk of the sample may not.[106] The different results observed may, therefore, reflect whether the different freezing regimes (rates) actually produce significant differences in the ice crystal size and location in the bulk of the sample. In addition, some species may be more affected by 'poor' freezing than others. Furthermore, advantages gained initially through rapid freezing may be lost due to ice crystal growth of subsequent frozen storage.

20.8.1   Effect of storage temperature and temperature cycling

It has been well documented that the overall quality of frozen foods is dependent on the storage temperature, and that lowering the storage temperature can slow down detrimental physical, chemical and biochemical changes. It is therefore, invaluable for food manufacturers, distributors and retailers to be able to predict the changes in food quality that are likely to occur in the frozen state. One of the most common assumptions is that 'quality' loss can be predicted from a simple relationship between storage time and temperature. However, many frozen food systems are not in a stable equilibrium, and their storage stability may depend upon complex kinetic and diffusion processes. The effect of storage temperature has been investigated, and generally the rate of quality loss, as measured by some change in properties, will decrease with decreasing temperature. Predictive models, however, that relate time-temperature history to quality of the final product have been much harder to establish.

There are papers that suggest that temperature fluctuations will increase the rate at which undesirable changes in fish occur.[107,108] Indeed, Love suggested that, due to a logarithmic relationship between quality loss and temperature, short periods at higher temperature may cause rapid changes and should, therefore, be avoided.[109] Due to the complex nature of frozen systems, however, Scudder[110] observed that the degradation of TMAO to formaldehyde and dimethylamine was at a maximum rate around at $-18°C$. This observation reflects the complex influence of temperature and freeze-concentration on 'quality loss' reactions in frozen fish.

20.8.2   Requirements of a low-temperature supply chain

It has been suggested that the storage temperature on lean demersal fish should be between $-24°C$ to $-30°C$.[110] For oily fish, storage temperatures of $-30°C$ to $-70°C$ have been recommended[98,99] in order to reduce the rate of oxidative changes. However, low-temperature storage is not currently employed as the cost to the consumer for the final product is too high. In some frozen foods, reaction rates may become diffusion controlled[6,7] at temperatures above that of the glass transition (due to the high viscosity of the non-frozen fraction). Indeed, rates of diffusion-controlled reaction are likely to be extremely slow below the glass transition temperature, due to the massive increase in viscosity as the non-frozen fraction forms a glass. This will lead to a significant decrease in reaction rates. By increasing the glass transition temperature of a food the temperature at which that food can be stored whilst maintaining quality may be elevated. A review of glassy states in foods and how they may be manipulated is described in the book *Glassy States in Foods*.[111]

20.9   Summary

Firstly, in any fish product supply chain there are many points at which improvements may be made. However, to obtain real benefits the supply chain must be viewed in its entirety. Thus, there may be little to be gained by controlling the quality attributes of raw materials if inappropriate freezing and storage conditions are employed during processing and distribution. Generally, better raw materials tend to survive better during freezing and frozen storage.

The processes occurring during the freezing and frozen storage of fish products are complex. A number of measurement techniques may be employed in order to monitor these changes. However, the significance of the measured changes must always be related to observable sensory changes and ultimately to consumer response.

It is hoped that this chapter gives the reader sufficient background information about factors which are likely to impact upon quality, however, to ascribe relative importance to each step in determining the final fish quality is difficult as each fish supply chain will be different.

20.10   Future trends

The frozen foods market has grown throughout the twentieth century and since the time of Clarence Birdseye a much wider selection of frozen foods is available. Products tend not to be just the individual components of a meal such as fish fillets or joints of meat. Now whole meals are available and the increased availability of such products is consistent with the changes in Western lifestyle. The trends of more people living alone and for more households with both partners working are likely to increase. Thus, the requirement for good-quality, nutritious, safe foods without compromising convenience is also likely to increase. The frozen food format is ideally suited to be able to deliver all of these consumer requirements.

Generally fish is viewed as a safe and nutritious source of protein, but the most important issue for the fishing industry is the depletion of the current fishing stocks. It will be increasingly necessary to impose greater restrictions on the catches of 'traditional species': For example, strict quotas have been imposed on the catching of cod in the North Sea and it is likely that these restrictions will be maintained or even increased in future years. Thus, it will become increasingly important to maximise the use of current fish raw materials and also to improve the utilisation of newer species. Also, farming of fish may become more common. Salmon farming has already developed to an extent where fresh, farmed salmon is now cheaper than cod in supermarkets in the UK.

With an increasing amount of foreign travel, consumers are becoming more familiar with a wider range of fish species and fish dishes. It is likely therefore that consumers will look for a greater range of good-quality fish products than is currently available. Also, the product formats are likely to change, with fish pieces rather that whole fillets becoming the key ingredients in meals. The use of smaller pieces of fish will bring its own problems in preventing oxidative damage due to increased exposure to air.

Finally, the trend towards natural foods seems likely to impact upon the type of molecules the food scientist can use to maintain the texture and flavour quality of fish. Thus, the challenge will be to select the most natural options to offer products with the quality expected by increasingly discerning consumers.

20.11   Further reading

ERICKSON, M.C. and HUNG, Y.C. *Quality in Frozen Foods* New York International Thomsom Publishing (1997).

LOVE R.M. *The Food Fishes: their intrinsic variation and practical implications.* London Farrand Press (1988).

ALLEN, J.C. and HAMILTON, R.J. *Rancidity in Foods (3rd edn)* Gaithersburg Maryland Blackie Academic and Professional (1994).

BLANCHARD J.M.V. and LILLFORD P.J. *The Glassy State in Foods* Nottingham University Press (1993).

20.12   References

1. FUCINI, J.J. and FUCINI, S.F. *Entrepreneurs* Boston. 1985.
2. TRIFIRO, A., SACCANI, G. FRANZONI, A. FRULLANTI, B., ZANOTTI, A. and GHERARDI, S. 'Vitamin content of frozen vegetables' *Industria Conserv.* (2001) **76**(2) 151–66.
3. FAVELL D. 'A comparison of the vitamin content of fresh and frozen vegetables' *Food Chem.* (1998) **62**(1) 59–64
4. LANIER T.C. and LEE C.M. *Surimi Technology* New York Marcel Dekker Inc 1992.
5. NOIKOV V. *Handbook of Fishery Technology Vol. I* New Dehli Amerind Publishing Co Pvt. Ltd. (1981).
6. SLADE L., LEVINE H., IEVOLELLA J. and WANG M. 'The glassy state phenomenon in applications for the food industry – application of the food polymer science approach to structure-function relationships of sucrose in cookie and cracker systems' *J. Sci. Food Agric.* (1993) **63**(2) 133–76.
7. REID D.S., HSU J. and KERR W. *The Glassy State in Foods* Nottingham University Press (1993).
8. HSIEH Y.L. and REGENSTEIN J.M. 'Texture changes of frozen stored cod and ocean perch minces' *J. Food Sci.* (1989) **54**(4) 824–6.
9. KIM Y.J. and HELDMAN D.R. 'Quantitative analysis of texture change in cod muscle during frozen storage' *J. Food Proc. Eng.* (1985) **7**(4) 265–72.
10. GILL T.A., KEITH R.A. and SMITH LALL B. 'Textural deterioration of red hake and haddock muscle in frozen storage as related to chemical parameters and changes in the myofibrillar proteins' *J. Food Sci.* (1979) **44**(3) 661–7.
11. HILTZ D.F., LALL B.S., LEMON D.W. and DYER W.J. 'Deteriorative changes during frozen storage of fillets and minced flesh of silver hake (*Meluccius*

bilinearis) processed from round fish held in ice and refrigerated sea water' *J. Fisheries Res. Board Can.* (1976) **33**(11) 2560–7.

12. HURLING R., RODELL J.B. and HUNT H.D. 'Fibre diameter and fish texture' *J. Texture Studies* (1996) **27**(6) 679–85.

13. HATAE K., YOSHIMATSU F. and MATSUMOTO J.J. 'Role of muscle fibres in contributing firmness of cooked fish' *J. Food Sci.* (1990) **55**(3) 693–6.

14. PARTMANN W. 'Contractability of muscle fibres as a possible aid to judging textural changes in frozen meats and fish' *J. Texture Studies* (1971) **2**(3) 328–38.

15. OFFER G. and TRINICK J. 'On the mechanism of water holding in meat: the swelling and shrinking of myofibrils' *Meat Sci.* (1983) **8**(4) 245–81.

16. JARENBACK, L. and LILJEMARK, A. 'Ultrastructure changes during frozen storage of cod. I: Structure of myofibrils as revealed by freeze etching preparation' *J. Food. Technol.* (1975) **10**(2) 229–39.

17. GARCIA M.L., MARTIN-BENITO J., SOLAS M.T. and FERNANDEZ B 'Ultrastructure of the myofibrillar component of cod *(Gadus morhua)* and hake (*Merluccius merluccius L.*), stored at 20°C as a function of time.' *J. Agric. Food Chem.* **47**(9) (1999) 3809–15.

18. TANAKA, T. 'Electron microscopic studies on toughness in frozen fish' *Proc. Int. Symp.* (1964) 121–6.

19. LOVE R.M. *The Food Fishes: their intrinsic variation and practical implications* London Farrand Press (1988) 156–7.

20. REY-MANILLA M.D., SOTELO C.G., AUBOURG S.P., REHBEIN H., HAVERMEISTER W., JORGENSEN B. and NIELSEN M.K. 'Localisation of formaldehyde production during frozen storage of European hake (*Merluccius merluccius*)' *European Food Res. Tech.* (2001) **213**(1) 43–7.

21. SOTELO C.G., PINIERO C. and PEREZMARTIN R.I. 'Denaturation of fish proteins during frozen storage – role of formaldehyde' *Zeitschrift Lebensmittel Untersuch. Fors.* (1995) **200**(1) 14–23.

22. CASTELL, C.H., NEAL, W. and SMITH, B. 'Formation of dimethylamine in stored frozen sea fish' *J. Fish. Res. Bd. Can.* (1970) **10** 1685–90.

23. SOKORSKI, Z. and KOSTUCH, S. 'Trimethylamine N-oxide demethylase: its occurrence, properties and role in technological changes in frozen fish' *Food.Chem.* (1982) **9**(3) 213–22.

24. REHBEIN H. 'Relevance of trimethylamine oxide demethylase activity and haemoglobin content to formaldehyde production and texture deterioration in frozen stored minced fish muscle' *J Sci. Food Agric.* (1988) **43**(3) 261–76.

25. RODRIGUEZ, C., MASOUD, T. and HEURTA, M.D., 'Degradation of trimethylamine oxide for evaluation of quality of frozen fish' *Alimentaria* (1997) No. 288 125–9.

26. TOKUNAGA T. 'The effect of decomposition products of trimethlyamine oxide on the quality of frozen Alaska pollack fillets' (1974) *Bul. Jap. Soc. Sci. Fish.* **40**(2) 167–71.

27. DINGLE J.R., KEITH R.A. and LALL B. 'Protein instability in frozen storage included in mince muscle of flatfshes by mixing with muscle of red hake'

Can. Inst. Food Sci. Tech. J. (1977) **10**(3) 143–6.

28. AUBOURG S.P. and MEDINA I. 'Influence of storage time and temperature on lipid deterioration during cod (*Gadus morhua*) and haddock (*Melanogrammus aeglefinus*) frozen storage' *J. Sci. Food Agric.* (1999) **79**(13) 1943–8.

29. LAIRD W.M., MACKIE I.M. and REGENSTEIN J.M. 'Deterioration of frozen cod and haddock minces' *Refrigeration Sci Tech.* (1981) **1981** –4 395–400.

30. TRAN V.D. and HAN-CHING L. 'Preliminary studies on use of cryoprotective agents in comminuted fish' *Revue Travaux Institut Peches Maritimes* (1981) **45**(3) 215–37.

31. MIYAUCHI D., KUDO G. and PATASHNIK M. 'Effect of processing variables on storage characteristics of frozen minced Alaska pollack' *Marine Fish. Rev. Nat. Oceanic Atmos. Admin.* **39**(5) (1977) 11–14.

32. YOON K.S, LEE C.M. and HUFNAGEL L.A. 'Effect of washing on the texture and microstructure of frozen fish mince' *J. Food Sci.* (1991) **56**(2) 294–8.

33. SHENOUDA, S.Y.K. and PIGGOT, G.M. 'Lipid-protein intraction during aqueous extraction of fish proteins: myosin-lipid interaction' *J. Food Sci.* (1974) **39**(4) 726–34.

34. SIKORSKI Z.E., KOSTUCH S. and OLLEY J. 'Protein changes in frozen fish' *CRC Critical Rev. Food Sci.* (1976) **8**(1) 97–129.

35. SAEED S., FAWTHROP S.A. and HOWELL N.K. 'Electron spin resonance (ESR) study on free radical transfer in fish lipid-protein interaction' *J. Sci. Food Agric.* (1999) **79**(13) 1809–16.

36. HASTINGS R.J., ROGER G.W., PARK R., MATHEWS A.D. and ANDERSON E.M. 'Differential scanning calorimetry of fish muscle: the effect of processing and species variation' *J. Food Sci.* (1985) **50**(2) 503–6, 510.

37. DAVIES J.R., LEDWARD D.A., BARDSLEY R.G. and POULTER R.G. 'Species dependence of fish myosin stability to heat and frozen storage' *Int. J. Food Sci. Tech.* (1994) **29**(3) 287–301.

38. HOWELL, B.K., MATHEWS, A.D. and DONNELY, A.P. 'Thermal stability of fish myofibrils: a differential scanning calorimetry study.' *Int. J. Food Sci. Technol.* (1991) **26**(3) 283–95.

39. JARENBACK L. and LILJEMARK A. 'Ultrastructural changes during frozen storage of cod (*Gadus morhua*) II Structures of extracted myofibrillar proteins and myofibril residues' (1975) *J Food Technol.* **10**(3) 309–25.

40. DEL MAZO M.L., TORREJON P., CARECHE M. and TEJADA M. 'Characteristics of the salt soluble fraction of hake (*Merluccius merluccius*) fillets stored at −20 and −30 degrees C' (1999) *J. Agric. Food Chem.* **47**(4) 1372–7.

41. MACKIE I.M. 'The effect of freezing on flesh proteins' (1993) *Foods Rev. Int.* **9**(4) 575–610.

42. LILLFORD P.J., CLARK A.H. and JONES D.V. 'Distribution of water in foods and model systems' (1980) *A.C.S. Symp. Ser* **127** 177–95.

43. PEARSON R.T., DUFF I.D., DERBYSHIRE W. and BLANCHARD J.M.V. 'NMR investigation of rigor in porcine muscle' (1974) *Biochim. Biochem. Acta* **362** 188–200.

44. HAZELWOOD C.F., CHANG D.C., NICHOLS B.L. and WOESSNER D.E. 'Nuclear magnetic resonance transverse relaxation times of water protons in skeletal muscle' (1974) *Biophysical J.* (1974) **14**(8) 583–602.

45. FJELKNERMODIG S. and TORNBERG E. 'Water distribution in porcine Longissimus dorsi in relation to sensory properties' (1986) *Meat Sci.* **17**(3) 213–31.

46. LAMBELET P., RENEVEY F., KAABI C. and RAEMY A. 'Low-field nuclear magnetic resonance relaxation study of stored or processed cod' (1995) *J. Agric. Food Chem.* **43**(6) 1462–6.

47. STEEN C. and LAMBERLET P. 'Texture changes in frozen cod mince measured by low-field nuclear magnetic resonance spectroscopy' (1997) *J. Sci. Food Agric.* **75**(2) 268–72.

48. SAMSON A.D. 'Textural changes in frozen gadoid minces stored at various temperatures' (1985) *Dis. Abs. Int. – B* **44**(8) 2374.

49. KIM Y.J. and HELDMAN D.R. 'Quantitative analysis of texture change in cod muscle during frozen storage' *J. Food Proc Eng.* **7**(4) 265–72

50. JEPSEN M.J., PEDERSEN H.T. and ENGELSEN S.B. 'Application of chemometrics to low-field ^{1}H NMR relaxation data of intact fish flesh' (1999) *J.Sci. Food Agric.* **79** 1793–1802.

51. RAMANATHAN L. and DAS N.P. 'Studies on the control of lipid oxidation in ground fish by some polyphenolic natural products' (1992) *J. Agric. Food Chem.* **40**(1) 17–21.

52. TILLACK J. 'The storage characteristics of deep-frozen trout and slice salmon' *Archiv Lebensmittelhygiene* (1975) **26**(2) 69–73.

53. CASTELL, C.H. 'Metal-catalyzed lipid oxidation and changes of proteins in fish' (1971). *J. Am. Oil Chem. Soc.* **48**(11) 645–9.

54. LABUZA, T.P. 'Kinetics of lipid oxidation in foods' (1971) *CRC Critical Reviews in Food Technology.* **2**(3) 355–405.

55. ALLEN J.C. and HAMILTON R.J. *Rancidity in Foods* Gaithersburg Maryland Aspen Publishers Inc (1999).

56. ERCKSON M.C. and HUNG Y.C. *Quality in Frozen Foods* New York International Thomsom Publishing (1997) 141–73.

57. TIMMERMAN F. and MEGREMIS C. 'More to it than meets the eye' (1996) *Food Ingred. Anal. Int* **18**(3) 41–3.

58. MENDEZ E., SANHUEZA, J., SPEISKY H. and VALENZUELA A. 'Validation of the Rancimat test for the assessment of the relative stability of fish oils' (1996) *J. Ail Oil Chem. Soc.* **73**(8) 1033–7.

59. POLVI S.M., ACKMAN R.G., LALL S.P. and SAUNDERS R.L. 'Stability of lipids and omega-3 fatty acids during frozen storage of Atlantic salmon' (1991) *J. Food Proc. Pres.* **15**(3) 167–81.

60. MCGILL A.S., HARDY R., BURT J.R. and GUNSTONE F.D. 'Hept-cis-4-enal and its contribution to the off-flavour in cold stored cod' (1974) *J. Food Sci. Agric.* **25**(10) 1477–89.

61. MCGILL A.S., HARDY R. and GUNSTONE F.D. 'Further analysis of volatile components of frozen cold stored cod and the influence of these on

flavour' (1977) *J. Food Sci. Agric.* **28**(2) 200–5.

62. ROSS D.A. and LOVE R.M. 'Decease in the cold store flavour developed by frozen fillets of starved cod (*Gadus morhua* L.)' (1979) *J. Food Tech.* **14**(2) 115–22.

63. SANTOS E.E.M. and REGENSTEIN J.M. 'Effects of vacuum packaging and erythorbic acid on the shelf-life of frozen white hake and mackerel.' (1990) *J. Food Sci.* **55**(1) 64–70.

64. HWANG, K.T. and REGENSTEIN, J.M. 'Protection of Menhaden mince lipids from rancidity during frozen storage.'(1989) *J. Food Sci.* **54**(5) 1120–4.

65. SMITH, G., HOLE, M. and HANSON, W.H. 'Assessment of lipid oxidation in Indonesian salted-dried marine catfish (*Arius thalassinus*).'(1990) *J. Sci. Food Agric.* **51**(2) 193–205.

66. KIM H.K., ROBERTSON I. and LOVE R.M. 'Changes in the muscle of lemon sole (*Pleuonectes microcephalus*) after very long cold storage' (1977) *J. Sci. Food Agric.* **28**(8) 699–700.

67. SAMSON A.D. 'Textural changes in frozen gadoid minces stored at various temperatures' (1984) *Diss. Abs. Int. – B* **44**(8) 2374.

68. HARVIE R.E. 'The importance of chilling in producing top quality snapper (*Chrysophrys auratus*) for the Japanese market' (1982) *Refrigeration Sci. Tech.* **1982-1** 163–8.

69. BARHOUMIE M., FAYE A.A., TEUTSCHER F. LIMA-DOS-SANTOS C.A.M. and VIKE E. 'Storage characteristics of sardine (*Sardina pilchardus Walbaum*) held in ice and chilled seawater' (1981) *Refrigeration Sci. Tech.* **1981–4** 243–9.

70. PEREZ-VILLAREAL B. and HOWGATE P. 'Deterioration of frozen hake (*Merluccius merrlucius*). (1991) *J. Food Sci. Agric.* **55**(3) 455–69.

71. REHBEIN H. and ORLICK B. 'Comparison of the contribution of formaldehyde and lipid oxidation products on protein denaturation and texture deterioration during frozen storage of ice-fish fillet (*Champsocephalus gumnnari* and *Pseudchaenichthys georgianus*)' *Int. J. Refrigeration.* (1990) **13**(5) 336–41.

72. CHANG C.C. and REGENSTEIN J.M. 'Texture changes and functional properties of cod mince proteins as affected by kidney tissue and cryoprotectants' (1997) *J. Food Sci.* **62**(2) 299–304.

73. LAIRD W.M., MAACKIE I.M. and REGENSTEIN J.M. 'Deterioration of frozen cod and haddock minces' *Refigeration Sci. Tech.* **1981–4** 395–400.

74. LOVE R.M., ROBERTSON I., SMITH G.L. and WHITTLE K.J. 'The texture of cod muscle' (1974) *J. Texture Stud.* **5**(2) 201–12.

75. LOVE R.M. 'The post-mortem pH of cod and haddock muscle and its seasonal variation' (1979) *J. Sci. Food Agric.* **30**(4) 433–8.

76. KELLY T.R. 'Quality of frozen cod and limiting factors on its shelf life' (1969) *J. Food Tech.* **4**(2) 95–103.

77. MCGILL, A.S., HOWGATE, P., THOMSON, A.B., SMITH, G.L., RITCHIE, A. and HARDY, R. 'The flavour of white fish and the relationship between the chemical analyses and sensory data.' *Progress in Flavour Research 1984*, Amsterdam Elsevier Science Publishers B.V. (1985) 149–64.

78. LOVE R.M. *The Food Fishes: their intrinsic variation and practical implications* London Farrand Press (1988) 149–60.
79. FLETCHER, G.C. and STATHAM, J.A. 'Shelf-life of sterile yellow-eyed mullet (*Aldrichetta fosteri*)'. (1988) *J. Food Sci.* **53**(4) 1030–5.
80. CHANG G.W., CHANG W.L. and LEW K.B.K. 'Trimethylamine-specific electrode for fish quality control' (1976) **41**(3) 723–4.
81. VALLE M., EB P., TAILLEZ R. and MALLE P. 'New method for evaluating bacterial reduction of trimethylamine N-oxide and its application to bacterial populations in fish muscle' *J. Rapid Methods Aut. Microbiol.* (1999) 7(2) 119–33.
82. LOUGHRAN M. and DIAMOND D. 'Monitoring of volatile bases in fish sample headspace using an acidochromic dye' (2000) *Food Chem.* **69**(1) 97–103.
83. SUMNER J.L., GORCZYCA E., COHEN D. and BRADY P. 'Do fish from tropical waters spoil less rapidly in ice than fish from temperate waters?' (1984) *Food Tech. Aus.* **36**(7) 328–29.
84. SEKI N. and TSUCHIYA H. 'Extensive changes during storage of carp myofibrillar proteins in relation to fragmentation' (1991) *Bul. Jap Soc. Sci. Fish.* **57**(5) 927–33.
85. KE-LIANG B.C., JEJIA C., CHYUAN Y.S. and PAN B.S. 'Biochemical, microbiological and sensory changes of sea bass (*Lateolabrax japonicus*) under partial freezing and refrigerated storage' (1998) *J. Agric. Food Chem.* **46**(2) 682–6.
86. NOWLAN S.S., DYER W.J. and KEITH R.A. 'Superchilling – a new application for preserving freshness of filets during marketing' (1974) *Can. Inst. Food Sci. Tech. J.* **7**(1) A16–A19.
87. PARK J.W. and LANIER T.C. 'Cryoprotective effects of sugars, polyols and/or phosphates on Alaska pollack surimi' (1988) *J. Food Sci.* **53**(1) 1–3.
88. NOGUCHI S. and MATSUMOTO J.J. 'The control of denaturation of fish muscle proteins during frozen storage. IV. Preventive effects of carboxyllic acids' (1975) *Bull. J. Soc. Sci. Fisheries* **41**(3) 329–35.
89. KRUEGER D.J. and FENNEMA O.R. 'Effect of chemical additives on toughening of frozen Alaska pollack *(Theragra chalcogramma)* (1989) *J. Food Sci.* **54**(5) 1101–6.
90. ANESE M. and GORMLEY R. 'Effects of dairy ingredients on some chemical, physico-chemical and functional properties of fish mince during freezing and frozen storage' (1996) *Lebensmittel Wissenschaft Tech.* 29(1/2) 151–7.
91. ERCKSON M.C. and HUNG Y.C. *Quality in Frozen Foods* New York International Thomsom Publishing (1997) 197–232.
92. PAYNE S.R., SANDFORD D., HARRIS A. and YOUNG O.A. 'Effects of antifreeze proteins on chilled and frozen meat' (1994) *Meat Sci.* **37**(3) 429–38.
93. PAYNE S.R. and YOUNG O.A. 'Effects of pre-slaughter administration of antifreeze proteins on frozen meat quality' (1995) *Meat Sci* **41**(2) 147–55.
94. PAYNE S.R. and WILSON P.W. 'Comparison of the freeze/thaw characteristics of Antarctic cod (*Dissostichus mawsoni*) and black cod (*Paranotothenia augusta*)-possible effects of antifreeze glycoproteins.' (1994) *J. Muscle*

Foods **5**(3) 233–55.

95. SANTOS E.E. and REGENSTEIN J.M. 'Effects of vacuum packaging, glazing and erthorbic acid on the shelf life of frozen white hake and mackerel' (1990) *J. Food Sci.* **55**(1) 64.

96. SIROIS M.E, SLABYJ B.M, TRUE R.H. and MARTIN R.E. 'Effect of vacuum packaging on changes associated with frozen cod fillets' (1991) *J. Muscle Foods* **2**(3) 197–208

97. ALLEN, J.C. and HAMILTON, R.J *Rancidity in Foods (3rd edn)* Gaithersburg Maryland Blackie Academic and Professional (1994) 256–72.

98. CHAPMAN K.W., SAGI I., HWANG K.T. and REGENSTEIN J.M. 'Extra-cold storage of hake and mackerel fillets and mince' (1993) *J. Food Sci.* **58**(6) 1208–11.

99. INOUE C. and ISHIKAWA M. 'Glass transition of tuna flesh at low temperature and effects of salt and moisture' (1997) *J. Food Sci.* **62**(3) 496–9.

100. MURPHY A., KERRY J.P., BUCKLEY J. and GREY I. 'The antioxidative properties of rosemary oleoresin and inhibition of off-flavours in pre-cooked roast beef slices' (1998) *J. Sci. Food Agric.* **77**(2) 235–43.

101. AKHTAR P., GRAY J.I., GOMAA E.A. and BOOREN A.M. 'Effect of dietary vitamin E and surface application of oleoresin rosemary on iron/ascorbate stimulated lipid oxidation in muscle and microsomes from rainbow trout (*Oncorhynchus mykiss*)' (1998) *J. Lipid Foods* **5**(1) 73–86.

102. BILINSKI E., JONAS R.E.E., LAU Y.C. and GIBBARD G. 'Treatments before storage affecting thaw drip formation in Pacific salmon' (1977) *J. Fisheries Res. Board Can.* **34**(9) 1431–5.

103. CHEN Y.C. and PAN B.S. 'Morphological changes in tilapia muscle following freezing by air-blast and liquid nitrogen methods' *Int. J. Food Sci. Tech.* (1997) **32**(2) 159–68.

104. PAN B.S. and YEH W.T. 'Biochemical and morphological changes in grass shrimp (*Penaeus monodon*) following freezing by air blast and liquid nitrogen methods' (1993) *J. Food Biochem.* **17**(3) 147–60.

105. CUTTING C.L. 'The influence of freezing practice on the quaility of meat and fish' (1977) *Aus. Refrigeration Air Cond. Heat.* **31**(2) 25–8, 37–41.

106. CREPY J.R. and CORBIC G. 'Liquid nitrogen freezing. Experimental for sea foods. Quality of frozen products' *Rev. Gen. Froid* (1974) **65**(2) 133–9.

107. LEBLANC E.L., LEBLANC R.J. and BLUM I.E. 'Prediction of quality in frozen cod (*Gadus morhua*) fillets' (1988) *J. Food Sci.* **53**(2) 328–40.

108. BILINSKI E., JONAS R.E.E. and PETERS M.D. 'Treatments affecting the degradation of lipids in frozen Pacific herring *Clupea harengus pallase*' (1981) *Can Inst. Food Sci. Tech. J.* **14**(2) 123–7.

109. LOVE R.M. *The Food Fishes: their intrinsic variation and practical implications* London Farrand Press (1988) 121–40.

110. SCUDDER B. 'Icelandic research warning on frozen fish'. (1995) *Seafood Int.* **10**(6) 51.

111. BLANCHARD J.M.V. and LILLFORD P.J *The Glassy State in Foods* Nottingham University Press (1993).

21

Measuring the shelf-life of frozen fish

H. Rehbein, Institute of Fishery Technology and Fish Quality, Hamburg

21.1 Introduction

In the first part of this chapter parameters relevant for quality and spoilage of frozen fish are discussed and physical and chemical reactions leading to deterioration are briefly described. Then traditional and advanced methods for measuring the shelf-life of frozen fish are presented, and finally future trends for quality measurement are envisaged. In this chapter the term 'fish' is used for whole fish, fillet and minced fish, unless differences between these products are discussed.

21.2 Deterioration in frozen fish

European and U.S. surveys have revealed that the image of frozen fish is not very positive. Whereas European consumers emphasised the neutral and insipid taste of these products (Anonymous, 1993), many U.S. consumers believed that frozen fish is less nutritious and bonier than fresh, and has a tough and dry texture, bad smell and inferior taste (Peavey *et al.*, 1994). However, by careful selection of raw material, processing and storage conditions, frozen fishery products of high quality can be produced (Boknaes *et al.*, 2001). The many reasons for quality loss of frozen food, as well as techniques to avoid or retard deterioration, have been described recently (Erickson and Hung, 1997). In the case of frozen seafood any study of quality has to take into consideration (i) the high number (several thousands world wide) of fish species possessing different composition and biochemical properties, and (ii) the fact that most of the raw material is from wild, not from farmed, animals, in contrast to meat production (Love, 1988).

Deterioration of frozen fishery products depends on extrinsic and intrinsic factors. The most important extrinsic factors are the speed of freezing, storage temperature, fluctuation of temperature, penetration of oxygen into the product during storage and, not to forget, the mode of thawing or heating the product. Intrinsic factors are given by the biochemical properties of the fish or shellfish. Enzymatic equipment, type of fatty acids in the lipid fraction and presence of other metabolites, which are precursors of undesirable compounds, are responsible for main deteriorative processes (Love, 1988).

The temperature of frozen fish should be low enough to keep the product in a glassy state with limited diffusion of molecules (Brake and Fennema, 1999b). The temperature at which frozen fish muscle enters the glassy state has been determined to be about $-12°C$ for cod and mackerel (Brake and Fennema, 1999a). Below this temperature fish muscle can be considered to be well protected from deteriorative reactions that are diffusion limited.

Freezing of fish is accompanied by formation of ice crystals resulting in concentration of salt and organic compounds, and pH changes in the liquid phase. These processes are influenced by the freezing rate, the storage temperature (Haard, 1992), and temperature fluctuation. Muscle proteins are dehydrated and denatured, and membranes are destroyed (Haard, 1992; Hultin, 1995; Mackie, 1993). The two most important pathways are lipid hydrolysis and oxidation (Shewfelt, 1981; Hultin et al., 1992) and in gadoids and a few other species, cleavage of trimethylamine oxide into formaldehyde and dimethylamine by the enzyme TMAOase (Sotelo and Rehbein, 2000).

The spoilage pattern of seafood may be influenced by biological parameters (Haard, 2000; Love, 1970, 1980, 1988):

• *Fish species*
  Development of rancidity is different for lean fish and fatty fish, and also for pelagic and bottom-living fish differing in the content of red muscle, i.e., the pro-oxidative proteins myoglobin and hemoglobin.
• *Physiological condition of the fish*
  Sexual maturity of fish is dependent on season. The fat content of herring, mackerel and many other species varies considerably over the year. The content and functional properties of actomoysin of pre- and post-spawning fish are also different. For example, in the case of pre-spawning Argentine hake (*Merluccius hubbsi*) the denaturation of myofibrillar proteins during frozen storage proceeds much faster than in post-spawning fish (Roura et al., 2000).
• *Fishing ground*
  Environmental conditions, like water temperature and food supply, can have a great influence on the activity of muscle enzymes. Adaptation to low temperature often is associated with increased activities of glycolytic enzymes and ATPases (Haard, 2000). Changes in the rate of production and utilisation of high-energy phosphates may have an effect on the course of rigor mortis and the pH value of fish muscle. It has been shown recently that

freezing fish pre-, in- or post-rigor may result in different qualities of frozen products (Schubring, 1999d).

- *Size of the fish*
  The metabolic capacity of fish muscle, and the structure and properties of collagen can vary with the age and size of fish. The texture of frozen fish is influenced by the firmness of collagen.

- *Sex of the fish*
  Metabolism, muscle structure, and enzyme activities during spawning are different for male and female fish (Haard, 2000). The influence of sex, sexual maturity and size on quality attributes of frozen-stored fillets of Argentine hake has been determined by sensory assessment and measurement of chemical and physical parameters (Fuselli *et al.*, 1996).

  It was found that the content of formaldehyde, dimethylamine, salt soluble protein, and drip were affected by sex, size and maturity, but the total volatile bases content was not affected. The average values of fibrousness of cooked samples were significantly higher for males than for females of similar size and equal sexual state.

- *Composition of feed for farmed fish*
  It is well known that the quality attributes of farmed fish can be controlled to a large extent by the composition of the feed (Morris, 2001). For example, the content and composition of muscle lipids, as well as the colour of the flesh of farmed Atlantic salmon can be adjusted according to consumer preferences. Most of the salmon is processed to cold smoked fillets, but freezing of the fresh raw material or the smoked product is commonly done in order to balance changes in demand. Thus, the processing step of freezing has to be taken into account when the composition of the feed is defined.

The conditions of *catching and processing* are of paramount importance for the deterioration of frozen fish. Quality is affected by:

- *Method of catching, length of trawling time*
  The method of catching (gillnet, handline, longline, trap) cod (*Gadus morhua*) significantly affected caloric, moisture and protein contents (Botta *et al.*, 1987a), as well as some sensory attributes like muscle firmness (Botta *et al.*, 1987b). In times when stocks are decreasing, long trawling times of up to 16 hours may be necessary to fill the net, conditions being very stressful to the fish. When the fish is brought onboard, it will be in different states of exhaustion or rigor mortis.

- *Stunning and killing procedure*
  The relationship between killing methods and quality has been discussed recently (Robb, 2001). The more a fish struggles during catch and killing, the faster the pH falls after death. As low pH (~ 6.0) results in denaturation of muscle proteins during frozen storage of fish, texture may become firm and dry if stressed fish is processed.

- *Rigor mortis*

Apart from biological factors the conditions of catching and killing fish have a great influence on the course of rigor. Struggling results in depletion of high-energy phosphates and glycogen stores, and may lead to a rapid rigor onset. Fish can be processed either pre- or post-rigor for high-quality products (Boknaes *et al.*, 2001; Erikson, 2001). When fish is frozen pre-rigor, thawing may be accompanied by thaw-rigor, which is characterised by rapid and strong muscle contraction causing high drip loss, gaping (fillet rupture) and a firm texture (Torrissen, 2001).

* *Bleeding*
  Bleeding the fish carefully prior to processing is beneficial to quality for two reasons. Colour and appearance are improved by bleeding, as the presence of bloodstains leads to devaluation of thawed fillets (Warm *et al.*, 1998). Furthermore, lipid oxidation is accelerated by hemoglobin (Hultin *et al.*, 1992), and the blood of gadoid species contains higher TMAO-ase activity than light muscle. Thus, residual blood may catalyse degradation of TMAO to formaldehyde and DMA (Rehbein, 1988).

* *Single and double freezing*
  Frozen fillets are prepared directly after the catch on board freezer-trawlers or from fish that has been frozen at sea, transported to a land-based factory, thawed and filleted. Depending on fish species and processing conditions, small (Hurling and McArthur 1996; Schubring, 2000b) or large (Thiemig and Oelker, 1999) differences in quality have been found between single and double frozen fillets.

* *Glazing and coating*
  Glazing protects frozen fillets against freezer-burn and lipid oxidation (Josephson *et al.*, 1985). As edible coatings can reduce the rate of moisture and oxygen transfer between the fish flesh core and the surrounding atmosphere (Stuchell and Krochta, 1997), battered and breaded fish sticks or fillet portions have a longer shelf-life than untreated fillet portions.

* *Type of product: whole fish, fillet, minced fish*
  Gutted and frozen fish is used either as an intermediate product, which is processed further into canned, smoked or ripened fish, or it is sold directly in the supermarket, e.g., as in the case of rainbow trout. However, the main final products of frozen fish are fillets, fillet portions, fish cakes and fish sticks. Fish sticks may contain various amounts of minced fish.

  The flesh of whole fish may be protected by its 'natural coating', the skin, against moisture loss and oxygen penetration. On the other hand, the layers of fat and red muscle at the skin side of the fillet may accelerate rancidity of the flesh (Undeland, 2001). These layers may be removed by deep-skinning to reduce lipid oxidation during frozen storage of fillets. Comparisons of lipid oxidation in skinned fillets and minces gave inconsistent results (Undeland, 2001). During mincing oxygen is brought into the fish flesh, but on the other hand mincing may activate anti-oxidative enzymes.

  During filleting, and much more pronounced during mincing of fish, the structure of muscle is destroyed and the compartimentation is abolished. In consequence particle-bound enzymes are released and may come into contact with substrates from which they are completely separated *in vivo*. Lipolytic, proteolytic, and other enzymic reactions (e.g., degradation of nucleotides or trimethylamine oxide), are accelerated in minced fish (Grantham, 1981).

Several types of minces are used by the fish industry: mince prepared from pieces of fillet is of highest quality, whereas mince made from filleting waste or whole fish is inferior in respect of colour, texture and shelf-life. The reason for the lower quality and stability is due to the contamination of minced muscle by blood, kidney and other tissues being rich in enzymes and heme-proteins.

- *Storage conditions*
  Temperature and time of frozen storage have a large influence on shelf-life of frozen stored fish. Temperature should be as low as economically possible, but at least less than −20°C. A weak point in the cold chain is the temperature of deep freezers in supermarkets, which often is too high and fluctuates over the day. Other conditions, like vacuum-packaging or protection from light, may have less influence on deterioration. A trial of storing catfish fillets at −20°C for 11 months did not reveal significant differences in the sensory attributes of fillets, which either had been vacuum packaged or oygen-permeable packaged (Anelich *et al.*, 2001).
- *Freezing and thawing conditions*
  It is well known that the speed of freezing has a great effect on the properties of the product. Slow freezing, which promotes growth of large ice crystals, should be avoided to minimise protein denaturation. In industry it is common practice to thaw fish in air or water. More sophisticated techniques, like vacuum thawing, or thawing by the use of microwave, dielectric or electrical resistance heating, are faster but more expensive. In any case, the thawing system should avoid localised overheating and dehydration, excessive drip loss, and bacterial growth (Garthwaite, 1997).

From this section the conclusion must be drawn that loss of quality attributes during storage of frozen fish depends on many different factors, making it very difficult to describe these processes by the use of one or two chemical or physical methods.

21.3 Indicators of deterioration in frozen fish

For many years attempts have been made to determine the shelf-life of frozen fish by a number of chemical, biochemical and physical methods but still today the gold-standard is sensory assessment by a well-trained panel. Some sensory characteristics of spoilage of frozen fish, together with the reactions they are based on, are summarised in Table 21.1.

Thirty years after the publication of the small book *On Testing the Freshness of Frozen Fish* (Gould and Peters, 1971), the statement in the preface of the

Table 21.1 Sensory description of thawed cooked fish flesh, and underlying chemical or physical reaction

Aspect of quality	Underlying chemical or physical reaction
Appearance	
Dry	Protein denaturation
Gaping	Breakdown of connective tissue
Freezer burn	Sublimation of ice
Yellowish	Lipid oxidation; formation of formaldehyde
Odour	
Cold-store odour (cardboard)	Formation of carbonyls by lipid oxidation
Sour	Formation of carbonyls by lipid oxidation
Rancid	Formation of carbonyls by lipid oxidation
Amine	TMAO degradation into DMA and TMA
Flavour	
Cold-store flavour	Formation of carbonyls by lipid oxidation
Sour	Formation of carbonyls by lipid oxidation
Rancid	Formation of carbonyls by lipid oxidation
Soapy	Lipolysis
Amine	TMAO degradation into DMA and TMA
Texture	
Dry	Protein denaturation, loss of muscle structure
Firm, tough	Reaction between formaldehyde and protein
Soft, mushy	Proteolysis

second edition is still correct saying that there are no commonly accepted, simple and reliable tests for quality measurement of frozen fish. In consequence of this lack of official methods, no threshold values exist for frozen fish, as e.g. the TVB-N values for fresh fish.

21.4   Biochemical indicators

Biochemical indicators can be divided into three categories; those that indicate:

1.   protein denaturation, like extractability, hydrophobicity, viscosity, electrophoretic pattern
2.   a decrease or increase of enzyme activity, release of particle-bound enzymes
3.   changes in metabolite concentrations (e.g. amines, aldehydes, and degradation products of nucleotides).

21.4.1   Extractable protein.
The myofibrillar and sarcoplamic proteins of fresh fish are highly soluble in solutions of 0.5–1M sodium, potassium or lithium chloride at neutral pH. The

loss of extractability of myofibrillar proteins during frozen storage of fish has often been taken as a measure for protein denaturation (Del Mazo *et al.,* 1999; Fuselli *et al.,* 1996; Lou *et al.,* 2000). It was found to be significantly correlated with storage time (Del Mazo *et al.,* 1999; Kelleher and Hultin, 1991), but the influence of temperature was less pronounced. However, determination of extractable protein has not been widely used for calculation of shelf-life of frozen fish, presumably because the technique is time consuming and the correlation between the amount of extractable protein and sensory or instrumental texture scores still has to be established for many commercially important species. The careful comparison of different salts clearly demonstrated that LiCl was a better extractant of fish muscle proteins over a broader range of conditions (concentration of salt, temperature during extraction, blending time, pH) than either NaCl or KCl (Kelleher and Hultin, 1991). The solubilised or homogenised myofibrillar protein can be studied further by more sophisticated techniques like viscosimetry, measurement of surface hydrophobicity or electrophoresis. The first two methods may be useful for getting information on the functionality of the protein, and by means of the latter proteins can be identified, that are more susceptible to denaturation than others, e.g., the myosin heavy chain (Careche *et al.,* 1998).

Apparent viscosimetry has been proposed as a method to indicate protein denaturation and aggregation of frozen fish (Borderias *et al.,* 1985; Barroso *et al.,* 1998). Viscosity was found to be more sensitive than solubility, because soluble protein of frozen fish expressed decreased viscosity values (Cofrades *et al.,* 1996). The analytical parameters affecting viscosity determination as a quality control for frozen fish have been optimised previously (Borderias *et al.,* 1985), and the method was applied for comparative quality determination of several hake species (Barroso *et al.,* 1998). However, viscosity measurement has rarely been used for shelf-life determination. In the initial stages of protein denaturation a greater exposition of hydrophobic groups of proteins occurs. The conformational changes of fish muscle actomyosin can be detected by measuring surface hydrophobicity, which increases during frozen storage of fish (Cofrades *et al.,* 1996).

21.4.2   Changes in enzyme activities

Protein denaturation in frozen fish should be accompanied by a decrease of enzyme activities. Therefore, a number of enzymes (Table 21.2) have been tested for a regular disappearance of activity during frozen storage. The results of earlier studies, which had been summarised and critically discussed (Gould and Peters, 1971) were often disappointing, as correlation with storage conditions (time, temperature) was poor. Furthermore, the problems of seasonal fluctuations and differences due to species had not been addressed.

Recently, activity of cytochrome oxidase, an enzyme located in the inner mitochondrial membrane was measured in muscle of fresh and frozen cod (*Gadus morhua*). Activity was enhanced by freezing, and then declined during frozen storage depending on the temperature (Godiksen and Jessen, 2001). The

Table 21.2   Enzymes tested as an indicator of quality of frozen fish (Gould and Peters, 1971; Yamanaka and Mackie, 1971; Godiksen and Jessen, 2001; Chawla *et al.*, 1988)

Enzyme	Results and conclusions
Myofibrillar ATPase	Drop in activity was not correlated with protein solubility or toughness
Aldolase	Decline in soluble enzyme activity observed for cod and haddock stored at −14°C
Malic enzyme (ME)	Latent form of ME showed a decrease in activity when fish was stored for 5 months at −7°C, but no change at −29°C.
Alpha-glycerophosphate dehydrogenase	The test, determination of activity constant from a double reciprocal plot of reaction velocity, was very time-consuming
Sarcoplasmic ATPase	Rapid decrease during frozen storage
Cytochrome oxidase	Activity decreased during frozen storage of cod and it was possible to distinguish between frozen storage temperatures −9, −20, and −40°C.
Acid phosphatase	Activity decreased during frozen storage of cod at −30°C
5′-nucleotidase	Activity decreased during frozen storage of cod at −30°C
Phospholipase	Activity increased in the first weeks of frozen storage of cod at −30°C, then it declined to reach the original level

authors conclude that cytochrome b may have potential as an indicator of frozen storage conditions. Release of particle-bound enzymes due to freezing and thawing has been used for differentiation between fresh and frozen fish. Mitochondrial (HADH, GOT, ME, fumarase), as well as lysosomal enzymes (acid alpha glucosidase, beta-N-acetylglucosaminidase), were found to be suitable (Rehbein, 1992) depending on the fish species.

21.4.3   Changes in metabolite concentrations

In frozen gadoid fish the TMAOase catalysed production of formaldehyde (FA) and dimethylamine (DMA) from trimethylamine oxide (TMAO) occurs at temperatures down to −30°C. Both metabolites, which are formed in equimolar amounts from TMAO, can be used as parameters for deterioration of frozen fish (LeBlanc *et al.*, 1988; Licciardelleo *et al.*, 1982), as binding of FA to proteins was found to result in texture toughening (Sotelo *et al.*, 1995). Recently a rapid and simple semiquantitative method for the determination of 'free' FA in fishery products has been developed (Rehbein and Schmidt, 1996). The test is based on reflectometrically reading of test strips, and may have potential for industrial quality control of gadoid fish. As the highly reactive FA forms bonds of various degrees of stability with proteins, the total amount of FA is difficult to measure, and usually has to be calculated from the content of DMA (Rehbein, 1987).

Free fatty acids (FFA) are formed extensively from phospholipid hydrolysis in frozen fish (Shewfelt, 1981). Increase in FFA concentration during storage of haddock (*Melanogrammus aeglefinus*), cod and hake (*Merluccius merluccius*) was found to be correlated with storage time (Aubourg and Medina, 1999; Aubourg *et al.*, 1999). Despite these promising results the method is seldom used for shelf-life determination.

Rancid and sour off-odours and off-flavours limit the shelf-life of frozen-stored fatty fish to a few months. The lipid oxidation products that are responsible for these undesired sensory properties are carbonyl compounds formed from the intermediate products of tasteless and odourless hydroperoxides (Undeland, 2001). Lipid oxidation can be followed by measuring the increase in hydroperoxides (peroxide value, POV) (Chapman and MacKay, 1949) and by determining the formation of thiobarbituric acid reactive substances (TBARS), mainly malondialdehyde (MA) (Vyncke, 1970). The concentration of MA has been mostly determined colorimetrically, but recently more specific HPLC methods have been developed (Tsaknis *et al.*, 1999). Both types of metabolites, hydroperoxides and carbonyls, are very reactive against other compounds of fish muscle. By reaction with amino groups, fluorescent compounds are formed, which also have been used for assessment of frozen stored lean fish or fatty fish (Aubourg *et al.*, 1982; Aubourg *et al.*, 1999).

Development of POV and TBARS-values in frozen stored cod and haddock were found to be different at -10 and $-30°C$. During frozen storage at $-10°C$ the change of values was characterised by an increase in the beginning, which was followed by a plateau phase, and ended in a decrease indicating the further reaction of hydroperoxides and carbonlys with other fish muscle compounds (Aubourg and Medina, 1999). In contrast to this observation no final decrease of values was found for fish stored at $-30°C$. Similar observations have been made in many other studies on frozen fish, limiting the usefulness of these parameters for estimation of quality properties or shelf-life (Jia *et al.*, 1996) of fish stored at higher sub-zero tempeatures (-10 to $-20°C$).

21.4.4   Nucleotides

In frozen fish degradation of adenosine triphosphate (ATP) and related nucleotides occurs mainly in the temperature range between -5 and $-15°C$, and was found to be very slow at lower temperatures (Love, 1966). Therefore, determination of adenosine monophosphate (AMP), inosine monophosphate (IMP) or hypoxanthine has not been applied very much for the quality measurement of frozen fish.

21.5   Physical indicators

The denaturation of fish muscle proteins during frozen storage leads to a decreased water-binding capacity, and a dry, firm and tough texture, if the time-temperature profile of storage is unfavourable. The state of water in frozen-thawed fish has been determined as thaw drip (TD), water-holding capacity (WHC), water-binding capacity (WBC), or cooking loss (CL). The amount of water released during thawing (thaw drip) can be used as a simple method to get initial information on the properties of the product.

Water-holding capacity of raw muscle is either measured by centrifugation (Eide *et al.*, 1982) or by collecting the fluid released during texture measurement. Water-binding capacity is determined in a similar way, but here a defined amount of water is added to the sample before measurement. WHC was determined in most experiments to describe protein-water interaction during frozen storage of fish. Some examples are compiled in Table 21.3 showing the suitability for the technique for quality assessment and shelf-life determination of frozen fish.

Texture measurement has been used in numerous studies as an objective method for evaluating muscle structure of frozen thawed fish. A short review of the relation between texture and technological properties of fish has appeared recently (Torrissen *et al.*, 2001). The influence of instrumental parameters (compression, speed) on the results of texture profile analysis (TPA) has been critically discussed (Schubring, 1999a, 2000a). The different phases of the diagram of force over time recorded during a TPA cycle can be correlated with the texture properties of hardness, springiness, cohesiveness, gumminess, chewiness, resilience and adhesiveness. A few applications of texture

Table 21.3   Water holding capacity of frozen and thawed fish (Fuselli *et al.*, 1996; Hurling and McArthur, 1996; Bechmann and Jorgensen, 1998; Henry *et al.*, 1995a; Montero and Gomez-Guillen, 1999)

Fish species and type of product	Storage conditions	Results
Whole cod (*G. morhua*)	Different combinations of time and temperature	WHC was strongly negatively correlated with the chemical quality parameters DMA and FA
Fillets of Argentine hake (*Merluccius hubbsi*)	7 weeks at −7°C	WHC was affected by sex, size and sexual maturity of the fish
Blue crab meat (*Callinectes sapidus*)	32 weeks at −29°C	WHC was improved by cryoprotectants
Single- and double-frozen fillets of cod (*G. morhua*)	9 months at −22°C	WHC decreased considerably during storage of double-frozen fillets, but only slightly for single-frozen fillets
Minced prawn flesh (*Penaeus* spp.)	90 days at −12°C	Initial WHC was very high; the slight decline during freezing was diminished by cryoprotectants

Table 21.4 Texture measurement of frozen and thawed fish (Botta *et al.*, 1987b; Schubring, 2000; Barroso *et al.*, 1998; LeBlanc *et al.*, 1988; Henry *et al.*, 1995a)

Fish species and type of product	Storage conditions	Results
Different species of hake	Different temperatures: -12, -18, $-40°C$, several months	Texture was inversely correlated to viscosity
Fillets of cod	1 week at -15, then 2 weeks at $-40°C$	Firmness of cooked muscle was affected by method of catching and time of year
Blue crab meat (*Callinectes sapidus*)	32 weeks at $-29°C$	Texture was affected by storage time and cryprotectants
Fillets of cod	90 days at -12, -14, -22, $-30°C$	Texture was related to many other chemical and physical indices; peak force increased with time and temperature
Single- and double-frozen fillet and mince from saithe (*Pollachius virens*) and haddock (*M. aeglefinus*)	$-24°C$	Texture measurement was compared to sensory assessment; tensile force was affected by double freezing

measurement of frozen thawed fish are listed in Table 21.4, demonstrating the suitability of this technique for quality determination. Other physical techniques, like electronic noses, colour measurement (Schubring, 1999b) or differential scanning calorimetry (Hurling and McArthur, 1996; Schubring, 1999c; Herrera *et al.*, 2000; Lambelet *et al.*, 1995) have been used only occasionally for characterisation of frozen fish.

21.5.1 Advanced physical methods

A number of spectroscopic methods have been applied quite recently to measure the quality-determining properties of frozen fish. Nuclear magnetic resonance (NMR), infra-red (IR) and Raman spectroscopy may offer the chance for rapid evaluation of the state of water and protein in frozen fish. Some results from the application of these techniques are compiled in Table 21.5. Low resolution NMR and near IR may have the greatest potential in the future.

21.6 Sensory assessment

As mentioned above, the best way of quality determination in food science is of course sensory evaluation, if it is performed properly. The sensory properties mainly decide whether a product will be accepted by the consumer or not. In the case of frozen fish, the descriptors given in Table 21.1 have been used in

Table 21.5 New spectroscopic techniques for quality assessment of frozen-thawed fish (Bechmann and Jorgensen, 1998; Careche *et al.*, 1999; Howell *et al.*, 1996; Lambelet *et al.*, 1995; Pink *et al.*, 1999; Steen and Lambelet, 1997)

Fish species, type of product, storage conditions	Spectroscopic method	Results
Whole cod, various storage conditions	Near infra-red (NIR) reflectance spectra of the skin	NIR reflectance measurement provided determination of WHC
Fillets of hake (*M. merluccius*) were stored at −10, −30 and −80°C	Raman spectroscopy	Changes in secondary protein structure were related to changes in viscosity and texture
Fillets of cod (*G. morhua*) and haddock (*M. aeglefinus*) were stored at −20 and −30°C	High-resolution NMR, magnetic resonance imaging (MRI)	NMR enabled measurement of metabolites (TMAO, TMA, DMA, creatine phosphate); by MRI changes in muscle structure were observed
Fillets of cod were stored at −10, −20 and −40°C	Low-field NMR relaxation	Transverse relaxation time of protons was related to time and temperature dependent deterioration
Minced red hake (*Urophycis chuss*)	Fourier transform NIR	Spectral region of 1530–1866 nm was significantly correlated to DMA concentration
Minced fillet of cod was stored at −10, −20 and −70°C for up to 4 months	Low-field NMR relaxation	Water proton relaxation time was correlated to instrumental and sensory texture scores and DMA concentration

combination with different scoring schemes for quality assessment (Anelich *et al.*, 2001; Barroso *et al.*, 1998; Henry *et al.*, 1995b).

A new approach for rapid assessment of quality characteristics of frozen fish is named QIM, the quality index method, which was originally developed for whole fresh fish (Warm *et al.*, 1998). Three different grading schemes have been used for cod, one for thawed whole fish, one for fillet from thawed cod, and one for cooked fillet. The grading schemes are based on parameters that vary considerably with frozen storage conditions. Each scheme consists of two parts, the quality parameters and the characteristics for each parameter. The characteristics are scored, '0' being the highest score. For example, in the case of cooked fillet from thawed cod, the quality parameter 'colour' is described by the characteristics 'white and opalescent' (score 0), 'loss of whiteness' (score 1), 'greyish, one small blood stain' (score 2), 'slightly yellow, a few more blood stains (score 3), 'light brown, discoloured with blood' (score 4). The scores for all single parameters are added to give the total quality index.

For cod the schemes were compared with physical and chemical measurements (WHC, FA concentration, amount of TVB-N, content of dry matter) by principal component analysis (PCA). QIM scores were correlated with FA concentration, TVB-N value and dry matter, and inversely related to WHC. QIM scores also reflected the frozen storage time of the fish.

21.7   Conclusions

Despite so many years of intensive research only a few methods exist that can be recommended for determination of quality attributes of frozen fish. These include: (i) sensory assessment, (ii) determination of salt soluble protein, (iii) measurement of DMA and FA in case of gadoids, (iv) measurement of TBARS for fish stored at low temperatures or for a short period, (v) texture profile analysis, (vi) determination of water-holding capacity. The usefulness of spectroscopic techniques has to be demonstrated by further studies.

21.8   References

ANELICH L E, HOFFMAN L C, SWANEPOEL M J (2001), 'The quality of frozen African sharptooth catfish (*Claria gariepinus*) fillets under long-term storage conditions', *J Sci Food Agric*, 81, 632–9.

ANONYMOUS (1993), *Parallel Food Testing in the European Union: Fish*, International Consumers Research & Testing Limited, London.

AUBOURG S, SOTELO C, PEREZ-MARTIN R (1982), 'Assessment of quality changes in frozen sardine (*Sardina pilchardus*) by fluorescence detection', *J Am Oil Chem Soc*, 75, 575–80.

AUBOURG S P, MEDINA I (1999), 'Influence of storage time and temperature on lipid deterioration during cod (*Gadus morhua*) and haddock (*Melanogrammus aeglefinus*) frozen storage', *J Sci Food Agric*, 79, 1943–8.

AUBOURG S P, REY-MANSILLA M, SOTELO C (1999), 'Differential lipid damage in various muscle zones of frozen hake (*Merluccius merluccius*)', *Z Lebensm Unters Forsch*, 208, 189–93.

BARROSO M, CARECHE M, BARRIOS L, BORDERIAS A J (1998), 'Frozen hake fillets quality as related to texture and viscosity by mechanical methods', *J Food Sci*, 63, 793–6.

BECHMANN I E, JORGENSEN B M (1998), 'Rapid assessment of quality parameters for frozen cod using near infrared spectroscopy', *Lebensm Wiss u Technol*, 31, 648–2.

BOKNAES N, GULDAGER H S, OSTERBERG C, NIELSEN J (2001), 'Production of high quality frozen cod (*Gadus morhua*) fillets and portions on a freezer trawler', *J Aquatic Food Product Technology*, 10, 33–47.

BORDERIAS A J, JIMENEZ-COLMENERO F, TEJADA M (1985), 'Parameters affecting viscosity as a quality control for frozen fish', *Marine Fish Rev* 47 (4), 43–5.

BOTTA J R, KENNEDY K, SQUIRES B E (1987a), 'Effect of method of catching and time of season on the composition of Atlantic cod (*Gadus morhua*)', *J Food Sci* 52, 922–4, 927.

BOTTA J R, BONNELL G, SQUIRES B E (1987b), 'Effect of method of catching and time of season on sensory quality of fresh raw Atlantic cod (*Gadus morhua*)', *J Food Sci* 52, 928–38.

BRAKE N C, FENNEMA O R (1999a), 'Glass transition values of muscle tissue', *J Food Sci*, 64, 10–14.

BRAKE C N, FENNEMA O R (1999b), 'Lipolysis and lipid oxidation in frozen minced mackerel as related to Tg', molecular diffusion, and presence of gelation', *J Food Sci*, 64, 25–32.

CARECHE M, DEL MAZO M L, TORREJON P, TEJADA M (1998), 'Importance of frozen storage temperature in the type of aggregation of myofibrillar proteins in cod (*Gadus morhua*)', *J Agric Food Chem*, 46, 1539–46.

CARECHE M, HERRERO A M, RODRIGUEZ-CASADO A, DEL MAZO M L, CARMONA P (1999), 'Structural changes of hake (*Merluccius merluccius*) fillets; effects of freezing and frozen storage', *J Agric Food Chem*, 47, 952–9.

CHAPMAN R, MACKAY J (1949). 'The estimation of peroxides in fats and oils by the ferric thiocyanate method', *J Am Oil Chem Soc*, 26, 360–3.

CHAWLA P, MACKEIGAN B, GOULD S P, ABLETT R F (1988), 'Influence of frozen storage on microsomal phospholipase activity in myotomal tissue of Atlantic cod (*Gadus morhua*)', *Can Inst Food Sci Technol*, 21, 399–402.

COFRADES S, CARECHE M, CARBALLO J, COLMENERO F J (1996), 'Freezing and frozen storage of actomysin from different species', *Z Lebensm Unters Forsch*, 203, 316–19.

DEL MAZO M L, TORREJON P, CARECHE M, TEJADA M (1999), 'Characteristics of the salt-soluble fraction of hake (*Merluccius merluccius*) fillets stored at −20 and −30°C', *J Sci Food Agric*, 47, 1372–7.

EIDE O, BORRESEN T, STROM O (1982), 'Minced fish production from capelin (*Mallotus villosus*)', *J Food Sci*, 47, 347–54.

ERICKSON MC, HUNG Y-C (1997), *Quality in Frozen Food*, New York, Chapman and Hall.

ERIKSON U (2001), 'Rigor measurements', in Kestin S C, Warriss P D, *Farmed Fish Quality*, Oxford, Fishing News Books, 283–97.

FUSELLI S R, ALMANDOS M E, CIARLO A S, BOERI R L, GIANNINI D H (1996), 'The influence of sexual maturity, sex and size on quality aspects of frozen Argentine hake (*Merluccius hubbsi*)', *J Aquatic Food Product Technology* 5, 81–94.

GARTHWAITE G A (1997), 'Chilling and freezing of fish', in Hall G M, *Fish Processing Technology*, 2nd edn, London, Blackie Academic & Professional, 93–118.

GODIKSEN H, JESSEN F (2001), 'Cytochrome oxidase as an indicator of ice storage and frozen storage', *J Agric Food Chem*, 49, 4488–93.

GOULD E, PETERS J A (1971), *On Testing the Freshness of Frozen Fish*, London, Fishing News Books.

GRANTHAM G J (1981), *Minced Fish Technology: A Review*, Rome, FAO Fisheries Technical Paper no. 216.

HAARD N F (1992), 'Biochemical reactions in fish muscle during frozen storage', in Bligh E G, *Seafood Science and Technology*, Oxford, Fishing News Books, 176–209.

HAARD N F (2000), 'Seafood enzymes: The role of adaptation and other intraspecific factors', in Haard N F, Simpson B K, *Seafood Enzymes*, New York, Marcel Dekker, 1–36.

HENRY L K, BOYD LC, GREEN D P (1995a), 'Cryoprotectants improve physical and chemical properties of frozen blue crab meat (*Callinectes sapidus*)', *J Sci Food Agric*, 69, 15–20.

HENRY L K, BOYD L C, GREEN D P (1995b), 'The effects of cryoprotectants on the sensory properties of frozen blue crab (*Callinectes sapidus*) meat', *J Sci Food Agric*, 69, 21–26.

HERRERA J J R, PASTORIZA L, SAMPEDRO G (2000), 'Inhibition of formaldehyde production in frozen-stored minced blue whiting (*Mikromesistius poutassou*) muscle by cryostabilizers: an approach from the glassy state theory', *J Sci Food Agric*, 48, 5256—62.

HOWELL N, SHAVILA Y, GROOTVELD M, WILLIAMS S (1996), 'High-resolution NMR studies on fresh and frozen cod (*Gadus morhua*) and haddock (*Melanogrammus aeglefinus*)', *J Sci Food Agric*, 72, 49–56.

HULTIN H O, DECKER E A, KELLEHER´S D, OSINCHAK J E (1992), 'Control of lipid oxidation processes in minced fatty fish', in Bligh E G, *Seafood Science and Technology*, Oxford, Fishing News Books, 93–100.

HULTIN H O (1995), 'Roles of membranes in fish quality', Paper presented at the Nordic Conference on 'Fish Quality-Role of Biological Membranes', Hillrod, Denmark, *TemaNord*, 624, 13–55.

HURLING R, MCARTHUR H (1996), 'Thawing, refreezing and frozen storage effects on muscle functionality and sensory attributes of frozen cod (*Gadus morhua*)', *J Food Sci*, 61, 1289–96.

JIA T, KELLEHER S, HULTIN H, PETILLO D, MANEY R, KRZYNOWEK J (1996), 'Comparison of quality loss and changes in the glutathione antioxidant system in stored mackerel and bluefish muscle', *J Agric Food Chem*, 44, 1195–201.

JOSEPHSON D B, LINDSAY R C, STUIBER D A (1985), 'Effect of handling and packaging on the quality of frozen whitefish', *J Food Sci*, 50, 1–4.

KELLEHER S D, HULTIN H (1991), 'Lithium chloride as a preferred extractant of fish muscle proteins', *J Food Sci*, 56, 315–17.

LAMBELET P, RENEVEY F, KAABI C, RAEMY A (1995), 'Low-field nuclear magnetic resonance study of stored or processed cod', *J Sci Food Agric*, 43, 1462–6.

LEBLANC E, LEBLANC R J, BLUM I E (1988), 'Prediction of quality in frozen cod (*Gadus morhua*) fillets', *J Food Sci*, 53, 328–40.

LICCIARDELLO J J, RAVESI E M, LUNDSTROM R C, WILHELM KA, CORREIA F F, ALLSUP M G (1982), 'Time-temperature tolerance and physical-chemical quality tests for frozen red hake', *J Food Qual*, 5, 215–34.

LOU X, WANG C, XIONG YL, WANG B, LIU G, MIMS S D (2000), 'Physicochemical stability of paddlefish (*Polyodon spathula*) meat under refrigerated and frozen storage', *J Aquatic Food Product Technology*, 9, 27–39.

LOVE R M (1966), 'The freezing of animal tissue', in Meryman H T, *Cryobiology*, New York, Academic Press, 317–405.

LOVE R M (1970), *The Chemical Biology of Fishes*, Vol I, London, Academic Press.

LOVE R M (1980), *The Chemical Biology of Fishes*, Vol. II, London, Academic Press.

LOVE RM (1988), *The Food Fishes, Their Intrinsic Variation and Practical Applications*, London, Farrand Press.

MACKIE I M (1993), 'The effect of freezing on flesh proteins', *Food Rev Intern*, 9, 575–610.

MONTERO P, GOMEZ-GUILLEN M C (1999), 'Frozen storage of minced prawn flesh: effect of sorbitol, egg white and starch as protective ingredients', *Z Lebensm Unters Forsch* A, 208, 349–54.

MORRIS P C (2001), 'The effects of nutrition on the composition of farmed fish', in Kestin S C, Warriss, P D, *Farmed Fish Quality*, Oxford, Fishing News Books, 161–79.

PEAVEY S, WORK T, RILEY J (1994), 'Consumer attitudes toward fresh and frozen fish', *J Aquatic Food Product Technology*, 3, 71–87.

PINK J, NACZK M, PINK D (1999), 'Evaluation of the quality of frozen minced red hake: use of Fourier transform near-infrared spectroscopy', *J Sci Food Agric*, 47, 4280–4.

REHBEIN H (1987), 'Determination of formaldehyde content in fishery products', *Z Lebensm Unters Forsch*, 185, 292–8.

REHBEIN H (1988), 'Relevance of trimethylamine oxide demethylase activity and hemoglobin content to formaldehyde production and texture deterioration in frozen stored minced fish muscle', *J Sci Food Agric*, 43, 261–76.

REHBEIN H (1992), 'Physical and biochemical methods for differentiation between fresh and frozen-thawed fish or fillets', *Ital J Food Sci*, 4, 75–86.

REHBEIN H, SCHMIDT T (1996), 'A rapid and simple method for the determination of formaldehyde in fishery products', *Inf Fischwirtsch*, 43, 37–9.

ROBB D H F (2001), 'The relationship between killing methods and quality', in Kestin S C, Warriss P D, *Farmed Fish Quality*, Oxford, Fishing News Books, 220–33.

ROURA S I, MONTECCHIA C L, ROLDAN H, PEREZ-BORLA O, CRUPKIN M (2000), 'Ultrastructure of actomyosin in pre-and post-spawning hake (*Merluccius hubbsi Martini*) during frozen storage', *J Aquatic Food Product Technology*, 9, 85–94.

SCHUBRING, R (1999a), 'Untersuchung von Einflußfaktoren auf die instrumentelle Texturprofilanalyse (TPA) von Fischerzeugnissen, 1. Einfluß der Kompression', *Deutsche Lebensmittel-Rundschau*, 95, 373–86.

SCHUBRING R (1999b), 'Determination of fish freshness by instrumental colour measurement', *Fleischwirtschaft*, 79, 26, 28, 29.

SCHUBRING R (1999c), 'DSC studies on deep frozen fishery products', *Thermochimica Acta*, 337, 89–95.

SCHUBRING R (1999d), 'Einfluß des Doppelgefrierens auf Qualitätsmerkmale des Filets von Seelachs (*Pollachius virens*) während der TK-Lagerung in Abhängigkeit vom Rigor-Stadium', *Deutsche Lebensmittel-Rundschau*, 95, 161–71.

SCHUBRING R (2000a), 'Untersuchung von Einflußfaktoren auf die instrumentelle Texturprofilanalyse (TPA) von Fischerzeugnissen, 2. Einfluß von Meßgeschwindigkeit und Kompression', *Deutsche Lebensmittel-Rundschau*, 96, 45–50.

SCHUBRING R (2000b), 'Double freezing of fillets and minces prepared from saithe and haddock: influence on selected sensory and physical attributes', *Annales Societatis Scientiarium Faeroensis Supplementum*, XXVIII, 169–79.

SHEWFELT R L (1981), 'Fish muscle lipolysis – a review', *J Food Biochem*, 5, 79–100.

SOTELO C G, PINEIRO C, PEREZ-MARTIN R I (1995), 'Denaturation of fish proteins during frozen storage: role of formaldehyde', *Z Lebensm Unters Forsch*, 200, 14–23.

SOTELO C G, REHBEIN H (2000), 'TMAO-degrading enzymes', in Haard N F, Simpson B K, *Seafood Enzymes*, New York, Marcel Dekker, 167–90.

STEEN C, LAMBELET P (1997), 'Texture changes in frozen cod mince measured by low-field nuclear magnetic resonance spectroscopy', *J Sci Food Agric*, 75, 268–72.

STUCHELL Y M, KROCHTA J M (1997), 'Edible coatings and films', in Erickson M C, Hung Y-C, *Quality in Frozen Food*, New York, Chapman & Hall, 264–74.

THIEMIG F, OELKER P (1999), 'Weiterentwicklung der gaping-Bestimmungsmethode für unzubereitete Gefrierfischerzeugnisse(Improvement of the gaping method for raw frozen fish products)', *Fleischwirtschaft*, 79, 82–5.

TORRISSEN O J, SIGURGISLADOTTIR S, SLINDE E (2001), 'Texture and technological properties of fish', in Kestin S C, Warriss P D, *Farmed Fish Quality*, Oxford, Fishing News Books, 42–57.

TSAKNIS J, LALAS S, EVMORFOPOULOS E (1999), 'Determination of malondialdehyde in traditional fish products by HPLC', *Analyst*, 124, 843–5.

UNDELAND I (2001), 'Lipid oxidation in fatty fish during processing and storage', in Kestin S C, Warriss P D, *Farmed Fish Quality*, Oxford, Fishing News Books, 261–75.

VYNCKE W (1970), 'Direct determination of the thiobarbituric acid value in trichloroacetic acid extracts of fish as a measure of oxidative rancidity', *Fette Seifen Anstrichm*, 72, 1084–7.

WARM K, BOKNAES N, NIELSEN J (1998), 'Development of quality index methods for evaluation of frozen cod (*Gadus morhua*) and cod fillets', *J Aquatic*

Food Product Technology, 7, 45–59.

YAMANAKA H, MACKIE I M (1971), 'Changes in the activity of a sarcoplasmic adenosinetriphosphatase during iced-storage and frozen-storage of cod', *Bull Jap Soc Sci Fish*, 37, 1105–9.

22

Enhancing returns from greater utilization

A. Gildberg, Norwegian Institute of Fisheries and Aquaculture Research, Tromsø

22.1 Introduction: the range of byproducts

Byproducts occur from every kind of food processing. For the fisheries sector the utilization of byproducts is more important for economical viability than in most other sectors. This is so because fishery byproducts normally comprise a substantial fraction of the total catch and, in certain cases, may be even more valuable than the main product if properly treated. In traditional small-scale fisheries almost every part of the fish was utilized for food or feed. Since the industrialized fisheries developed during the twentieth century, the volume of wasted byproducts has increased dramatically. A striking example is the industrial shrimp trawling where the bycatch sometimes may amount to 90% of the total catch volume, most of this being discarded or poorly utilized.[1] Also in white fish fillet production a major part (about 60%) is comprised of byproducts. Some of this is utilized if processing takes place on land, but normally wasted during processing on factory trawlers at sea.

The present chapter deals with various aspects concerning the utilization of byproducts from the fisheries and the fish aquaculture sectors. In this introductory part the term 'byproduct' is defined and a coarse estimate of byproduct quantities and economic potential is given before the various categories of byproducts are described. The possibilities of enhanced returns by more extensive utilization of byproducts for food is discussed, and a special focus is put on molecular aspects related to certain byproducts which may be interesting raw materials for the production of high-value biochemical compounds for pharmaceutical or biotechnological application.

22.1.1   What are byproducts?

Often fish byproducts have been regarded as the new term for fish offal, however, this is rather misleading. Whereas the term 'offal' signals some inedible waste which will be thrown away, 'byproducts' surely is a more positive concept indicating something which can be utilized. Today the most common understanding of the term 'byproducts' are all the raw material, edible or inedible, left over during the preparation of the main product. This can be examplified by the production of white fish fillet, where fillet cuts, backbone, head, liver, gonads and the guts are all byproducts. In certain cases, however, this definition is not valid. One example of this is the sturgeon fisheries where the roe obviously is the main product. In the present chapter also bycatch fish is regarded as a byproduct since this too may be looked upon as a raw material left over from the production of the main product.

22.1.2   Quantities and present and potential value

To estimate quantities and value of byproducts on a world basis is extremely difficult. Hence, most figures presented in this section must be regarded as purely indicative. During the last decade the total world fish catch has stabilized at about 90 million t (Table 22.1), and according to experts it is not likely that it will increase significantly in the future.[3] About 30 million t are processed to fish meal and oil. Such processing gives essentially no byproducts. Accordingly about 60 million t are available for human consumption. In addition the prosperous aquaculture sector provides about 30 million t annually, two-thirds of this being fish, the rest molluscs and crustaceans.[4] This means that the total amount of raw materials available for human consumption at present is about 90 million t. A considerable amount of this material can be recovered as byproducts during processing. In industrialized fisheries large quantities of byproducts are produced. When Atlantic cod is processed to fish fillets, about 60% of the total weight is byproducts and during the processing of Northern shrimp, more than 70%. In many parts of the world, however, fractions like liver, gonads, skin, head and even stomach and swimbladder are utilized for food. Although it is difficult to estimate the available quantity of byproducts on a world basis, it is likely that at least 25 million t could be recovered if every part of the fish had

Table 22.1   World catches of fishes and aquatic invertebrates in million t*

	1995	1996	Year 1997	1998	1999	Average
Larger fishes	57.3	58.6	58.2	56.6	55.9	57.3
Small pelagic fishes	22.0	22.3	21.6	16.7	22.7	21.1
Aquatic invertebrates	12.6	12.6	13.9	13.6	14.2	13.4
Total annual catch	91.9	93.5	93.8	86.9	92.9	91.8

* The values are calculated from data in FAO Fishery statistics.[2]

been brought ashore. In addition it is estimated that more than 20 million t of bycatch fish is wasted at sea.[5] Hence, the total estimate would be about 45 million t.

There are no statistical data available on the value of byproducts on a world basis, and the potential value of such an extremely inhomogeneous material is also hard to estimate. To give an indication, however, I will use some data which is available from the Norwegian fisheries. The Norwegian fisheries sector includes the capture of both large marine species, small pelagics, crustaceans and molluscs as well as salmon aquaculture, and may to a certain extent reflect the situation in world fisheries. In 1998 the amount of fishery byproducts utilized in Norway was 0.47 million t representing about 22% of the total fish production. This provided a value addition of about 125 million US$.[6] Extrapolated to the estimated world byproduct quantity of 45 million t, this corresponds to a value of about 12 billion US$. In Norway, present processing to feed and food gives a similar contribution to the value addition although only 10% of the quantity is processed to food. By processing more of the byproducts to food and high-cost speciality products, it is estimated that the value addition by processing byproducts will increase by a factor of five within ten years.[6]

Since the world fish catch is not expected to grow in the future, any increased potential of fishery byproducts is supposed to come from production growth in aquaculture. According to FAO[3] a growth in aquaculture production of about 2.5% per year is expected in China and about 5% in the rest of the world. This means that the global aquaculture production would reach some 55 million t by the year 2015. Again using Norwegian data showing that about 25% of the aquaculture production are byproducts,[6] the global byproduct quantity from this sector will increase from presently about 8 million t to about 14 million t in 2015.

22.2   Physical products

Fishery byproducts and products can be categorized into four major groups; material used for fertilizer, for feed, for food and for speciality products. In many coastal societies dependent on both fisheries and agriculture, fish waste has been, and is still being used as a nitrogen fertilizer.[7] Although good results have been achieved regarding both crop growth and health, such utilization is regarded as the least profitable of the four categories. Traditionally the major utilization of byproducts has been for feed production. Even if fish byproducts are valuable supplements in many feed products, the profitability in such production is also quite low. Processing byproducts to food or food ingredients normally gives a far better return. Even better profitability can be achieved by extracting and purifying high-value biochemical compounds from special byproduct fractions. Such compounds are bioactive molecules like enzymes, biopolymers or special peptides that may find biotechnological or medical application or may be utilized as ingredients in functional foods or cosmetics.

Preparation of such products, however, is very demanding and often dependent on long-term research and development as well as considerable capital investment in production facilities and advanced equipment. Finally, good market knowledge and a close collaboration with international distributors are necessary to succeed with such products.

22.2.1 Physical and chemical description of the major product categories

Normally, freshness of raw material is of premium importance for the value of the final product. When fish waste is used for fertilizer, however, microbial deterioration may be a part of the processing. Composting of fish waste by mixing with a bulking carbon-rich plant material like sawdust is one way of producing crop fertilizer.[8] Whereas composting yields a fertilizer rich in both carbon and nitrogen, the nitrogen is most important when fertilizer is prepared by ensilaging. By this method the fish waste is processed to a protein hydrolysate during storage of acid preserved material. It has been shown that such hydrolysates, in addition to enhancing crop growth, provide interesting biological effects.[7]

Today fish meal is by far the most important product made from bycatch fish or other fish byproducts. Although some special high-quality fish meals are used for human consumption the major utilization is for animal feed. The quality of fish meal used for feed production is, of course, very dependent on both raw material and processing conditions but normally it is a fine greyish/brown powder containing about 70% protein, 10% minerals, 9% fat and 8% moisture.[9]

Fish silage is the second largest feed product made from fish byproducts.[1,10] This liquefied acid preserved fish raw material can be used directly as a feed ingredient, but both the content of acid and fat sets limits to acceptable quantities being given to domestc animals. In many industrialized countries there are strict rules for the acceptability of both microbial and chemical quality of fish silage to be used for animal feed. In Norway, microbial counts of 100,000 per g and volatile nitrogen content of 100mg/100g dry matter have been indicated as maximal acceptable levels.[11] In large-scale silage production most of the oil is removed by centrifugation at high temperatures (about 95°C). Such processing also provides partial sterilization of the product. After oil removal the approximate chemical composition of fish silage is 80% water, 14% protein, 2% ash, less than 1% oil and 2–3% preservative acid (normally formic acid). Before mixing with various dry feed componets to a final dry product, the silage is normally evaporated to 45% dry matter.[12]

In many developing countries the expenses of buying preservative acids may prohibit the use of this method. In such cases, preservation by adding lactic acid bacteria and a low-price fermentable carbohydrate material, like molasses, cassava starch or dehydrated whey, may be an interesting alternative. If a culture of homofermentative lactic acid bacteria (e.g. *Lactobacillus plantarum*) is added to a mixture of minced fish and carbohydrate (15–20%), a stable fish silage with pH < 4.5 may be achieved within one week of fermentation. Fermented fish silage has proved to be a nutritious feed ingredient both for fish and domestic animals.[1,13,14]

Since ancient times considerable amounts of fish byproducts have been used for animal feed either directly or after boiling. This is certainly still taking place in many rural areas all around the world.

It is difficult to indicate any common chemical and physical properties typical for the two last product categories, food and speciality products. What is evident, however, is that byproducts used for such purposes must be treated with the same care and with the same hygienic conditions as other raw materials to be used for food applications. In certain cases, when raw materials are used for the production of bioactive molecules for medical or biotechnological applications, extremely hygienic precautions are necessary, since only minor microbial contamination may ruin the final product.

Many fishery byproducts have a nutritional value similar to fish muscle.[15] Most fish proteins have a high content of essential amino acids like lysine and methionine which can balance the normally low content of these amino acids in vegetable proteins.[1] Obviously byproducts represent great potential as raw materials for food production. Some of them, like liver and roe from many fish, are already genuine high-quality consumption products if handled properly. Other byproducts, which often are processed to fish meal, could just as well be utilized for food production. Fillet cuts and muscle in head and backbone belong to this category. Both chemical composition and nutritional value of fish mass prepared from such material are quite similar to fish fillet.

Generally food habits are different in different societies. What is considered to be a low-value byproduct only suitable as feed or fertilizer in one part of the world may very well be considered as a delicacy elsewhere. This tells us that strict definitions of product categories within this field are impossible.

22.2.2   Quality at catch and delivery and the potential for improved utilization as consumption products

The quality of the raw material at the processing site will decide the manufacturing possibilities. A key factor for the quality of fishery byproducts is handling onboard. Byproducts, and especially the visceral fraction, are very susceptible to deterioration if not properly preserved. The highly nutritious extractive fluids are excellent microbial growth substrates which during storage are further enhanced by the enzymatic hydrolysis caused by digestive and lysozomal enzymes. Blood contamination also causes deterioration both by stimulating microbial growth and by accelerating lipid oxidation.[16]

One important strategy is rapid fractionation of the raw material and low-temperature storage of the separate fractions. In most cases frozen storage is the best preservation. With certain products, like liver and roe, the freezing/thawing process may reduce the quality because it damages fragile membranes and causes leakage of lysozomal enzymes involving further tissue degradation. Storage on ice is a good alternative, particularly in tropical regions where both microorganisms and endogenous enzymes have low activity at low temperatures.

Space and logistics onboard the fishing vessel are universal problems. The main product always has first priority, and in the worst case this means that every other raw material is thrown overboard. In some industrialized countries even governmental regulations for fishing boat construction have resulted in reduced utilization of byproducts. However, a growing awareness of this problem in most places has improved the situation, and today many boats are constructed with facilities for total recovery of the whole fish catch. If cold storage is possible on board and the storage time before processing is less than three days, the best product quality at the processing site may be achieved when the fish remains ungutted until processing. This has been shown by chemical and microbial analyses of both muscle and visceral fractions of Atlantic cod.[17]

22.3   Products from enzymatic modifications

Since ancient times, and long before the word enzyme was invented, the effects of enzymes have been utilized in fish processing. In all ripening and maturation processes the enzymes are of paramount importance. In most cases both enzymes present in the raw material, endogenous enzymes, and microbial enzymes play a role. Endogenous enzymes normally are most important in rapid processes (fish silage), whereas microbial enzymes are principal during long term fermentation (fermented fish).

Recently, the active use of external enzymes has been introduced as a biotechnological alternative to traditional mechanical processing.[18] Such enzymatic methods have been developed both for manufacturing major fish consumption products and to increase the value of various byproduct fractions. A great variety of enzymes are utilized, particularly in ripening and fermentation. However, in most processes discussed in the following, the protein digesting enzymes – the proteases – play a key role.[19]

22.3.1   Ripening and maturation

Enzymatic ripening and maturation are important processes constantly taking place in every semi-preserved cold stored fish product which has not been heat treated. This means that both textural and organoleptic properties change during storage. Smoked and salted products belong to this category. All biological tissues are rich in lysozomal enzymes which leak into surrounding tissues when the lysozomal membranes are disrupted after death. The lysozomes contain a large number of different enzymes but the cathepsins are most important for the modification of texture and flavour in such products since these are proteases active at neutral and weak acid conditions.[20] Although present in low concentrations, these enzymes partly digest muscle proteins and even connective tissues during long storage, giving the products a softer texture and a more rich flavour. Application of such methods normally concerns major fish products like fillets or whole fish, but salting is also applied for semi-preservation of byproducts like tongue, cheeks and even cod swim bladder which is an attractive consumption product in Southern Europe.[21]

22.3.2   Removal of skins, membranes and parasites

Various preparations of commercial enzymes can be used to make byproducts suitable for human consumption. In most cases the enzymes used are protease preparations of either plant or microbial origin, but the exact specification of the enzymes is normally kept secret. By carefully choosing enzymes with the right specificity and applying the right chemical and physical conditions, it is possible to degrade certain tissue fractions whereas others are left unchanged. In most cases, the aim is to solubilize or modify connective tissues without damaging muscle tissues. This is often challenging since native connective tissues normally are more resistant to enzymic digestion than muscle proteins. Collagen type I is the major connective tissue protein, and only collagenases, which are very expensive enzymes, can digest this protein in its native form. After moderate heating or acid treatment, however, the collagen is denatured and becomes susceptible to digestion by most proteases. Such methods are frequently applied when external membranes or skins are removed or modified.

This technique can be used to remove the skin from skate-wings. After a gentle heat treatment the skate-wings are incubated at low temperature in an enzyme solution containing both proteases and carbohydrate degrading enzymes. After a few hours the degraded skin can be washed off. A similar method has been developed to remove the black membrane surrounding cod swim bladder.[21]

Traditionally, in Northern Europe and North America, squid has been regarded as a low-value product mainly used for feed and bait. One reason for this is that the rubbery membrane covering all surfaces is difficult to remove by technical means. If tubes and tentacles are soaked in a specific protease solution and given a short heat treatment (1 min., 80°C), the rubbery texture dissappears.[22] This makes membrane removal unnecessary, and the membrane can be sold as a part of the product offered for consumption.

Canned cod liver is a highly priced delicacy product in some European countries. At present the canning industry pays about 1 US$/kg for first-quality cod liver, but infestation of liver with large numbers of small worm parasites is a problem in the Northern Atlantic, where seals are intermediate hosts. Most parasites are located just beneath the external membrane and can easily be removed if the membrane is disrupted. An industrial method has been developed where cold intact cod liver inside a perforated rotating drum is rapidly soaked in hot water before it is transferred to a protease solution at moderate temperature (Fig. 22.1). After a short enzyme treatment both membranes and parasites are washed off and the liver is ready for canning.[23] Enzymatic removal of connective tissues provides more gentle treatment and often higher yield than

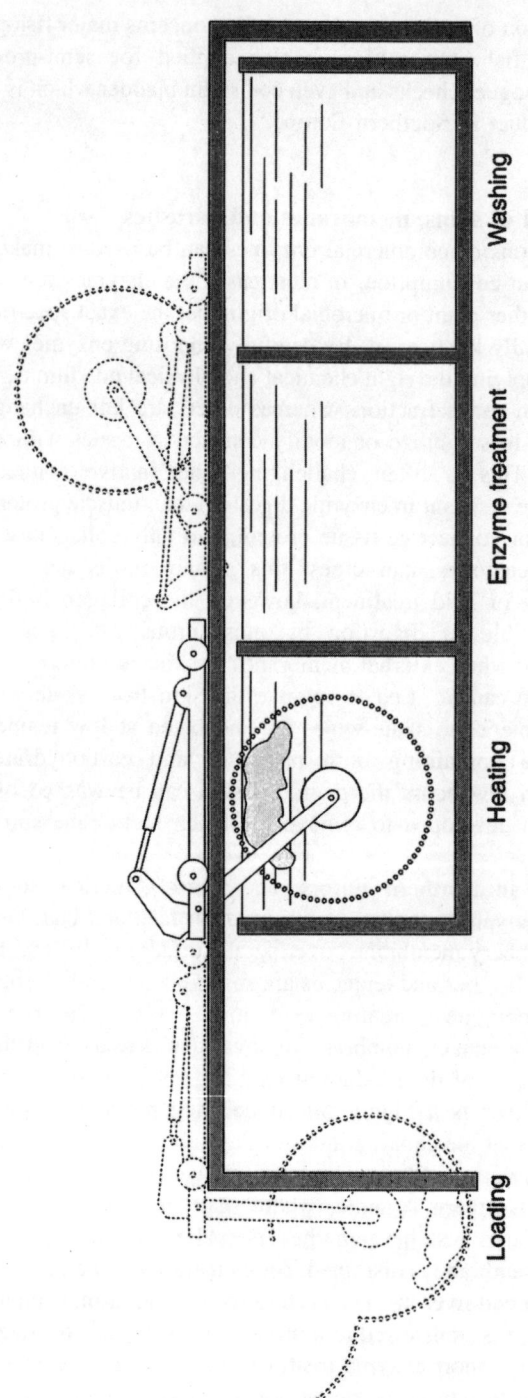

Fig. 22.1 Processing unit for enzymatic removal of parasites from cod liver.[23]

traditional mechanical processing. A problem may be that enzymatic methods are often more time consuming, production is normally by batch and the incubation period may allow bacteria to become established.

22.3.3 Caviar production

Although fish roe is a valuable raw material for the production of delicate caviar products, it is also in many cases a byproduct. The profitability of caviar production is always closely linked to the yield of undamaged fish eggs after processing. In many fish species like sturgeon, trout and salmon, the roe is tightly connected to sheets of connective tissues. Traditionally the eggs are released by rubbing the roe on a metal grid. During such processing a considerable number of eggs are damaged. By a gentle enzyme treatment the eggs can be released without physical damage. Various enzymes, including collagenases from crab hepatopancreas and fish pepsins, have been used succesfully for this purpose.[24] After a short treatment of the roe sacs in a stirred tank, connective tissues are separated from the roe by flotation. By introducing the enzymatic method for production of rainbow trout caviar, product recovery was increased from less than 70 to about 90%.[25] About 50 tons of caviar have been produced annually by this method in Canada and Scandinavia.[26]

22.3.4 Production of hydrolysates, fish sauce and fish silage

Fish protein hydrolysates are made by enzymatic hydrolysis of protein-rich fish raw materials. Most frequently the raw materials are low-price bycatch fish, other byproducts or small pelagic species like anchovies or sardines, and the enzymes used may be either endogenous enzymes, commercial enzymes of plant or microbial origin or enzymes provided by microbial cultures added to the raw materials.

Fish protein hydrolysates is a very complex product category including crude silage used for feed or fertilizer,[1,7] fish sauce for human consumption,[27] fish flavour ingredients[28] as well as more sophisticated bioactive preparations which may be used for growth promotion,[29] as antioxidants[30] or as immuno stimulants.[31,32] The oldest documented use of fish protein hydrolysates is preparation of the fish sauce, liquamen or garum, made from fish intestines and other byproducts in Southern Europe.[33] The ancient Romans and Greeks consumed this product more than two thousand years ago. Fish sauce production in Europe vanished, but more recently fish sauce became an important consumption product in South-East Asia, where the annual production is about 250,000 t.[26,27] Fish sauce is used as a condiment on vegetable dishes, but is also an important supplement of essential amino acids to many people. The extreme storage stability (several years) means that fish sauce may be available in every season, independent of seasonal fisheries.

Fish sauce is prepared by mixing about three parts of fish raw material with one part of sea salt and storing it at ambient tropical temperatures for 6–12 months.[27,34] During this storage period endogenous and microbial enzymes dissolve the tissue proteins, and the fish sauce can be drained off as an amber liquid with pleasant flavour normally containing 8–14% digested protein and about 25% salt. Although fish sauce production is simple and not dependent on sophisticated equipment, there is a need for large storage capacity due to the long production time. Efforts have been made to accelerate the process by adding external enzymes,[35–37] by heating[38] or by adjusting the chemical conditions to increase enzyme activities.[39–41] Although South-East Asia is still by far the most important market for fish sauce, there is a growing interest in such products in the western world.[41,42] In South-East Asia the price of a bottle (0.7 l) high-quality fish sauce is about 1 US$, whereas prices in Europe are about 3 US$. It takes at least 1 kg raw material to produce one bottle of fish sauce.

Fish silage is a more recent development which is almost entirely utilized for feed. All kinds of low-value fish and fish byproducts are used for production of fish silage, which is normally made by mixing 2–3% formic acid into the minced raw material and storing it at ambient temperatures until the tissues are dissolved by endogenous enzymes.[1,10] A well preserved fish silage has pH 3-4, which is optimal for protein digestion by fish pepsins.[43] Provided that the raw material has a sufficient content of pepsins and other proteases active at acid conditions (cathepsins), most tissues are solubilized after a few days of storage. Such solubilized silage may either be used directly as animal feed, or it can be further processed by centrifugation and evaporation to yield oil and a concentrated protein solution suitable as ingredients in standardized feed products.[12] The obvious advantage with fish silage production is the low requirement for sophisticated production equipment and low capital investment. The major disadvantage is high transportation costs of the product due to its high content of water.

Fish silage has been regarded as an alternative to fish meal but has never reached similar production quantities. At present Norway is a major fish silage producer with an annual production of about 140,000 t of raw material – mainly byproducts from salmon aquaculture.[6] Although fish silage is a low-price product, it has a high nutritional value,[44,45] and it is a valuable alternative for utilizing some byproducts which would otherwise be wasted.

There are no data available regarding the quantities of fermented fish silage produced. It is well known, however, that research and development on the production and utilization of fermented fish silage is taking place in tropical regions all around the world, and most reports indicate that the fermentation method is suitable for small-scale transformation of low-value byproducts into high-value feed in rural areas where both fisheries and animal production is taking place.[13,46–48]

Large-scale commercial production of fish protein hydrolysates by the addition of proteolytic enzymes first developed in France in the 1960s.[49] The raw material was bycatch from trawlers, and the end product was fish protein concentrate to be used as food additives, animal feed or milk replacer. This production still exists and the major factory, CTPP in Boulogne, annually processes about 60,000 t of raw material to valuable ingredients for food, feed and cosmetics (Charles DeLannoy, CTPP, Boulogne, personal communication).

The general procedure for the preparation of fish protein hydrolysate is quite simple. Minced raw material is mixed with an equal amount of water in a stirred tank before a preparation of proteolytic enzymes is added. The enzyme incubation may last about 1–2 hrs at 40–60°C before the enzymes are inactivated by heating above 90°C for a few minutes. Normally oil and solids are removed by centrifugation before the hydrolysate is concentrated by evaporation and dried. Gentle methods like spray drying or vacuum drying are often recommended to retain good flavour and functional properties.[18,50] In most cases low-price commercial enzymes of plant or microbial origin are used for the preparation of fish protein hydrolysates. Of the plant enzymes papain, which is prepared from papaya latex, is most frequently used whereas various preparations of Bacillus proteases made by Novo Nordisk are the major microbial enzymes.

The taste of protein hydrolysates is very dependent on the kind of enzyme used and the hydrolysis conditions. A bitter taste due to formation of bitter peptides is a common problem. These are normally small peptides (4–10 amino acids) with a hydrophobic end. The formation of such peptides may be avoided by applying gentle hydrolysis conditions. This may, however, reduce both yield and solubility of the hydrolysate. Hence, the opposite strategy, extensive digestion to non-bitter very small peptides and free amino acids, may be a better alternative.[28,51]

A great variety of fish protein hydrolysates are produced, and the value may vary from a low-price category comparable to fish meal to highly priced peptones used as nitrogen sources in sophisticated microbial growth media.[52] In the US the major market for fish protein hydrolysates is as a milk replacer to early weaned pigs. For this purpose a product is made by gentle hydrolysis and spray drying, and the price is about 3 US$ per kg.[53] Similar products are also made in Japan, where it has been shown that fish protein hydrolysates are very suitable as a protein component in the feed to juvenile animals with a poorly developed digestive system, including calves, piglets and fish fry.[54]

22.4   Functional and pharmaceutical byproducts

Many old traditions handed down in coastal populations reveal great knowledge of the valuable functional properties of various marine raw materials, but the use of this knowledge in the manufacture of commercial products occurred only quite recently. The marketing of marine oil as a vitamin-rich food supplement, is probably the first example of promoting a functional food of fish origin. Chitosan made from crustacean shells is another early development which has been marketed as a product with a great variety of functional properties useful in removal of water pollution, functional foods, cosmetics and in the production of contact lenses and biodegradable shopping bags.[55] More recently several speciality products for pharmaceutical and biotechnological applications have been extracted from marine byproducts, and some of them have initiated highly profitable commercial production.[24,56]

22.4.1 Functional biopolymers: chitosan and gelatine

Among the bioactive molecules recovered from fishery byproducts the biopolymers yield the largest quantity. Chitin is the major structural compound in the exoskeleton of marine invertebrates. Chitin is a nonsoluble polymer of N-acetyl glucosamine units. By deacetylation at strong alkaline conditions chitin is transformed to chitosan which is a 'water-soluble cationic polymer with a great variety of chemical and biotechnological applications. Shellfish processing waste contains 14–35% chitin on a dry matter basis.[55] The total available quantity of chitin from the annual world landings of crustaceans is estimated to be about 40,000 t,[57] whereas the present annual production of chitin/chitosan is about 2,200 tons to a value of about 2 billion US$. The major products are crude preparations used for waste water treatment, to add surface strength to paper or in animal nutrition as dietary fibre and prebiotic to promote growth of bifidobacteria.

Chitosan products with high purity and strictly defined chemical properties are produced for the cosmetic and pharmaceutical market. A high-price chitosan made from shrimp shells has shown excellent properties supporting film-forming capacity for moisturizing creams and adding structural strength and elasticity to hair when it is added to shampoos or other hair-care products.[58,59] Pharmaceutical applications of chitosan are numerous, including as an anticoagulant, artificial kidney membrane, sutures, immunostimulants, anti-tumour agent and for aggregation of cells.[55,60,61] Prices of chitosan vary over a wide range depending on purity and functional properties. Whereas crude chitosan for waste water treatment may have a price about 90 US$/kg, the most expensive pharmaceutical preparations may cost as much as 70 US$/g.[62]

Collagen, particularly type I collagen, is the major structural protein in animal connective tissues like skin, bone and tendon. This protein is arranged in a very defined triple-helical structure. By extracting connective tissues in warm weak acid or alkaline solutions, the collagen solubilizes and denatures, yielding randomized high molecular structures called gelatine. When the temperature decreases below a certain limit, which is dependent on its specific chemical composition, the gelatine forms a soft gel with excellent water-holding capacity.[63] Most gelatine is produced from pig skin or cattle hide splits, and such mammalian gelatines are solid at room temperature. Due to this they are being used as gel-formers in a great number of food products. The total world production of gelatine is estimated to be about 200,000 t.[64]

Although most of the gelatine is used as a food additive, there are also a number of technical applications including preparation of biodegradable microcapsules for pharmaceutical use, coating of photographic paper[65] and in microchip preparation. Gelatine made from fish skin or bone has a melting point below room temperature.[66,67] This is generally not compatible with food technology applications, but very useful during encapsulation of thermolabile pharmaceutical preparations and in coating of photographic paper. Another specific application where fish gelatine has proved to be useful is as a water-soluble-base for the photoresist in television tubes.[68]

22.4.2   Bioactive peptides

Many biological functions are regulated by bioactive peptides. Such peptides may be synthesized specifically, but may also be released randomly during protein hydrolysis. Supplementing hydrolysates of fish byproducts containing bioactive peptides may add value both to functional foods and feeds. It has been shown that medium-size (3–10 kDa) peptides in hydrolysates from cod stomach and head stimulated the growth of mouse fibroblasts in *in vitro* cultures, whereas very small peptides in shrimp and sardine waste hydrolysates had secretagogue activity.[29] The level of these bioactivities, which are valuable both in animal feed and functional foods, is strongly dependent, not only on the raw material, but also on the hydrolytic conditions.[69]

Peptides inhibiting the activity of angiotensin I-converting enzymes also have relevance to functional foods since oral intake may reduce blood pressure. It has been shown that both hydrolysates of bonito viscera and pollack skin contain such peptides.[70,71] Immuno stimulants activate the non-specific immune response system and have relevance both to functional foods and feeds, but may be of particular interest in aquaculture to improve disease resistance in shrimp and juvenile fish.[72] *In vitro* experiments have shown that small peptides from a cod stomach hydrolysate stimulated the activity of salmon leucocytes.[31] Similar activities were also detected in protein hydrolysates from shrimp waste.[73]

At present there is a growing interest in injecting protein hydrolysates from fillet cuts and other byproducts into fish fillets or fish mince products. This is done mainly to improve recovery and water holding capacity. An interesting additional effect is that certain peptides may prevent oxidation. Antioxidative peptides have been detected both in hydrolysates of sardine myofibril protein and of gelatine from pollack skin.[30,74]

22.4.3   Fish and crustacean trypsins

Trypsins are among the most important digestive enzymes in fish particularly at juvenile stages before gastric digestion at acid conditions is fully developed. Application of pancreatic enzymes as a digestive aid in feed to fish larvae has been investigated, however, the results are variable and not very promising.[75,76]

In Iceland, a small commercial production of Atlantic cod trypsin has been established, and a great variety of applications have been investgated.[21] In 1996 a firm manufacturing a very refined enzyme preparation (Penzim) for cosmetic and pharmaceutical application was founded (Jon Bragi Bjarnason, Zymetech,

Reykjavik, personal communication). The preparation is used in a skin oinment to provide moisturizing effects and to reduce skin irritation.

Some tryptic enzymes extracted from crustacean wastes may find medical applications for the removal of damaged dermatic tissues. A chymotrypsin-like enzyme with collagenolytic activity has been isolated from the hepatopancreas of king crab.[77] *In vivo* experiments with rabbits showed that this enzyme was highly efficient for wound debridement. Similar results have been obtained with a mixture of tryptic enzymes extracted from Antarctic krill. This preparation is now being tested for approval as a pharmaceutical product.[78,79]

22.4.4   Marine oils: mono- and polyunsaturated fatty acids

Although the composition of marine oils depends on the raw material, a high content of unsaturated fatty acids apparently is a common property.[9] For several decades particular interest has been concentrated on the high content of the long-chain polyunsaturated n-3 fatty acids EPA and DHA. Medical investigations have shown that intake of oil with high content of these acids may have beneficial effects on human health and, particularly, lower the susceptibility to coronary heart diseases.[80] This was first proposed by Dyerberg and Bang,[81] who studied the diet of Greenland eskimos who have an exceptionally low frequency of such diseases. More resent research indicates that also the high intake of monounsaturated fatty acids from marine mammals may contribute to protection against coronary heart diseases.[82]

Apparently, the intake of certain polyunsaturated fatty acids is of vital importance for the development of the brain and nervous system. Hence, a sufficient dietary supplement of these compounds at juvenile stages seems to be essential both for health and for the ability to learn.[83] Formulated feed for fish larvae must contain a certain amount of polyunsaturated phospholipids to support growth and survival.[84] Fish roe, milt, brain and eye tissues are rich sources of such lipids.

Today a number of companies manufacture encapsulated concentrates of polyunsaturated fatty acids. The products are marketed world wide as vital dietary supplements and they provide high profitability. In Norway refined cod liver oil has been used as a dietary supplement for more than half a century, and presently the price of such products is about 13 US$ per litre, but still several thousand tons of cod liver are wasted at sea every year.

22.5   Useful enzymes

During the last 25 years great efforts have been made to explore the possibilities of commercial utilization of enzymes from fish and shellfish byproducts. Research has revealed a wide spectrum of possibilities ranging from the use of minced pancreatic tissue for fish maturation to biotechnological application of very specific extensively purified enzyme preparations. Although only a few applications have been commercialized, research and development within this sector has contributed significantly to our general knowledge about special properties of fish and shellfish enzymes. The present state of the art within this field was recently updated in the book *Seafood Enzymes*.[85]

Most of the fish and shellfish enzymes studied have similar counterparts of mammalian origin, and it is not possible to indicate general properties which discriminate fish enzymes from mammalian enzymes. Often, however, the low environmental temperature of most marine waters is reflected in the temperature characteristics of the fish and shellfish enzymes. In several of the commercialized concepts it is actually the low temperature properties of these enzymes that are utilized.

22.5.1   Fish pepsins and trypsins

Salmon pepsin was probably the first fish enzyme to be purified and characterized in the late 1930s.[86] Since then a great number of fish pepsins have been studied.[43] Although the molecular structure of fish and mammalian pepsins are quite similar,[87] the chemical properties are different. The major differences are that fish pepsins are more active at low temperature and less acid condition than mammalian pepsins. This is being utilized in at least two commercialized processes; that is caviar production, which has already been mentioned (Section 22.3.3), and enzymatic descaling. There is a commercial market for descaled skin-on fillets from certain fish species like hake and haddock. These species have a soft muscle tissue which is very susceptible to damage during mechanical descaling. After gentle pepsin treatment at weak acid conditions and moderate temperature the scales can easily be removed by water-jet washing. In Norway an industrial scale method has been developed for this purpose, and cod pepsin has proved to be a very suitable processing aid.[88] A special procedure for silage processing of cod stomach is used to produce a crude preparation of cod pepsin for industrial application.[89]

There is a general interest in new kosher enzyme preparations for food processing. A major challenge has been to find good kosher alternatives to calf rennet in cheese production. The possibilities of using various fish pepsins for this purpose have been investigated, and although some promising results have been achieved,[90–92] no commercial cheese production using fish pepsins has been reported.

Trypsins and chymotrypsins are among the most important agents in ripening and maturation of fishery products. Normally the enzymes present in the raw material provide sufficient activity for this purpose, but certain manufacturers prefer to apply more rapid methods where specific amounts of enzymes are added. Such procedures are often kept secret, but it is well known that various preparations of fish trypsins have been used for this purpose. In a patented method,[93] it is claimed that herring fillets can become properly matured within only 1–5 days if homogenized fish intestines are added directly to the brine. In traditional procedures for maturation of de-headed herring, the storage time is 3–7 months. Fish trypsins have also proved to be suitable for maturation of squid mantles.[94] Other reported applications of fish trypsins have been as an efficient enzymic aid during extraction of carotenoprotein from shrimp processing wastes and as a hydrolytic agent during preparation of spermal extracts from skipjack tuna.[95,96]

22.5.2   Shrimp alkaline phosphatase and nuclease

The processing factories for Northern shrimp (Pandalus borealis) receive the raw material as frozen blocks. The thawing water, which is kept at a moderate temperature (about 15°C) is a rich source of several enzymes with commercial interest.[97] Alkaline phosphatase is an enzyme frequently applied in molecular biology, and traditionally an enzyme extracted from calf intestines is used. It is a problem that fairly high temperatures are necessary to inactivate this enzyme. Alkaline phosphatase from shrimp thawing water is very active at low temperatures and can be inactivated at lower temperatures than the calf enzyme.[98] Due to this it has been possible to commercialize this enzyme, and today ultra pure preparations of the enzyme isolated from shrimp waste water are sold on the world market with high profitability. Recently the gene of the enzyme has been isolated, and the viability of commercial production of the recombinant enzyme is being evaluated.[99] A nuclease with high specificity towards double strand DNA may be the second enzyme from shrimp waste water to be commercialized. A method for using this enzyme to degrade carry-over products during DNA-amplification (PCR-reaction) is being patented.[100]

22.5.3   Shellfish lysozymes

Lysozyme is an antibacterial enzyme present in most animals. The most extensively characterized is egg white lysozyme which has specific lytic activity for the cell wall of Gram-positive bacteria.[101] In shellfish lysozyme and lysozyme-like enzymes are essential factors in the non-specific defence against pathogenic bacteria.[102,103] Biopreservation by the use of such enzymes may be an alternative to the application of chemical preservatives. Chlamysin, a lysozyme-like enzyme extracted from the viscera of the marine bivalve Chlamys islandica, has revealed interesting properties which may warrant profitable commercial production.[103] In contrast to egg white lysozyme, chlamysin is an efficient growth inhibitor of both Gram-negative and Gram-positive bacteria. The enzyme is very active at low temperature and, in contrast to most cold adapted enzymes, remarkably stable even at very high temperatures. The chlamysin gene has been isolated and recombinant expression is in progress.[104]

22.6   Future trends

As mentioned in the introductory part of the chapter, there is no expected growth in marine fisheries in the future. Any increase in the total quantity of fish raw materials available is supposed to come as a result of increased aquaculture production. In the byproduct sector, however, there is still a large potential for improved recovery both from fisheries and aquaculture.

Most probably the steady growth in aquaculture production will continue. This will increase the demand for suitable raw materials for fish feed. Although vegetable products may be used for this purpose, it is evident that marine raw materials, and particularly marine oil, may become limiting at least for the marine aquaculture sector. Accordingly, an increased value of oil-rich byproducts for the production of aquaculture feed is expected.

In Norway the value of byproducts from fisheries and aquaculture is three times higher today than it was ten years ago, and optimistic prognoses indicate that the value may be increased a further five times during the coming ten years.[6] Although the major quantity of byproducts is used for feed production, it is evident that manufacturing more of the raw materials into food and speciality products represents the largest potential for value addition. To make this possible, it is essential that we stop treating the byproducts as *by*products. Both onboard and shore facilities must be adapted to meet such demands. Educating all the personnel within the fisheries sector to deal with this new situation is also of primary importance if this is to succeed.

When personnel and material facilities have been organized to recover all fractions of the raw material from the fisheries sector, it is also important that national and international authorities follow up by providing strong restrictions against wasting raw materials at sea or other places. The growing demand for fish as food in the heavily populated Eastern hemisphere will surely be a driving force to push forward this development.

The desire and need to manufacture more of the raw material for high-quality food will provide the impetus to increase the value of byproducts. Some speciality products may also contribute significantly, particularly where a significant fraction of the raw material is a compound with some useful functional property such as the biopolymers chitosan and gelatine. The incorporation of marine oils in functional foods is another important possibility ready for further development.

More spectacular products like enzymes or other bioactive molecules, which normally are present in very low concentrations in specific byproduct fractions, may also contribute to increased profitability, at least for a certain period. It is very likely, however, that most of these products will be produced either synthetically or by recombinant technology within a few years after their first appearance on the market. At this stage the byproduct has contributed by disclosing its secret, but is no longer interesting as a raw material.

22.7   Sources of further information and advice

For obvious reasons statistical data for quantities and value of byproducts and bycatch fish are very limited. However, based on the present accumulated knowledge about chemical and fractional composition of various fish raw materials and available capture and production statistics, it is possible to give reasonable estimates. FAO provides very detailed annual statistics of both world fisheries and aquaculture, and is probably the most valuable source of such information. Regarding the chemical and fractional composition of fish and fishery products, a great number of valuable sources exist, and surely the easiest way of obtaining such information is by searching in various databases for scientific publications. Books and Proceedings are still valuable sources of the more differentiated information which is often important to stimulate curiosity and creativity. Even old books may provide essential information within this field. *The Chemical Biology of Fishes* written by Love in 1970[105] is an excellent example.

Among more recent books related to the utilization and manufacturing of byproducts I will recommend the following:

Advances in Fisheries Technology and Biotechnology for Increased Profitability, editors: Voigt M N and Botta J R, Technomic Publishing, Lancaster, 1990 (566 p).

Making Profits out of Seafood Wastes, editor: Keller S, Alaska Sea Grant College Program, Anchorage, 1990 (239 p).

Fish and Fishery Products, editor: Ruiter A, CAB International, Wallingford, 1995 (387 p).

Seafood Enzymes, editors: Haard N F and Simpson B K, Marcel Dekker, New York, 2000 (681 p).

22.8 References

1. RAA J and GILDBERG A, 'Fish silage: a review', *CRC Crit Rev Food Sci Nutr*, 1982, **16**, 383–420.
2. ANON, *FAO Yearbook Fishery Statistics – Capture Production, Vol. 88/1*, Rome, FAO, 1999.
3. ANON, 'Issues of sustainable resource use and international fish trade', 7th session, *Committee on Fisheries, Sub-Committee on Fish Trade*, Bremen, FAO, 2000.
4. ANON, *FAO Yearbook Fishery Statistics – Aquaculture Production, Vol. 86/2*, Rome, FAO, 1998.
5. BRINK K and MOONEY H A, *Sustaining marine fisheries*, Committee on Ecosystem Management for Sustainable Marine Fisheries Ocean Studies Board. Commission on Geosciences, Environment, and Resources, National Research Council, Washington D.C., National Academy Press, 1999.
6. OLSEN S L, *From Waste to Income* (in Norwegian), Trondheim, RUBIN, 2000.
7. WYATT B and MCGOURTY G, 'Use of marine by-products on agricultural

crops', int conf on *Fish By-Products*, Anchorage, Alaska Sea Grant Program, 1990, 187–95.

8. BRINTON W F, 'Composting seafood processing by-products: Solution for the 90s', int conf on *Fish By-Products*, Anchorage, Alaska Sea Grant Program, 1990, 183–6.

9. SCHMIDTSDORFF W, 'Fish meal and fish oil – not only by-products', in Ruiter A, *Fish and Fishery Products*, Wallingford, CAB International, 1995, 347–76.

10. ARASON S, 'Production of fish silage' in Martin A M, *Fisheries Processing: Biotechnological Applications*, London, Chapman & Hall, 1994, 244–72.

11. GILDBERG A and AKSE L, *Quality Demands in Utilisation of Fishery Byproducts – part 2 – Fish Silage* (in Norwegian), Norwegian Institute of Fisheries and Aquaculture, Tromsø, 1992.

12. RAA J, GILDBERG A and STRØM T, 'Silage production – theory and practice' in Ledward D A, Taylor A J and Lawrie R A, *Upgrading Waste for Feeds and Food*, London, Butterworths, 1983, 117–32.

13. FAGBENRO O A and BELLO-OLUSOJI O A, 'Preparation, nutrient composition and digestibility of fermented shrimp head silage', *Food Chem*, 1997 **60**(4) 489–93.

14. DAPKEVICIUS M L E, BATISTA I, NOUT M J R, ROMBOUTS F M and HOUBEN J H, 'Lipid and protein changes during the ensilage of blue whiting (*Micromesistius poutassou* Risso) by acid and biological methods', *Food Chem*, 1998 **63**(1) 97–102.

15. STRØM T and EGGUM B O, 'Nutritional value of fish viscera silage', *J Sci Food Agric*, 1981 **32** 115–20.

16. RAMSØY D E, *Byproducts from Salmon*, Thesis (in Norwegian), Norwegian College of Fisheries, Tromsø, 2000.

17. AKSE L, JOENSON S, EILERTSEN G and BARSTAD H, *Utilisation of Byproducts from Fish Landed Uneviscerated*, Project rep (in Norwegian), Norwegian Institute of Fisheries and Aquaculture, Tromsø, 2001.

18. GILDBERG A, 'Enzymic processing of marine raw materials', *Process Biochem*, 1993 **28** 1–15.

19. BEYNON R J and BOND J S, *Proteolytic Enzymes – a Practical Approach*, Oxford, IRL Press, 1989.

20. BARRETT A J, 'Lysosomal and related proteinases', in Reich E, Rifkin D B and Shaw E, *Proteases and Biological Control*, New York, Cold Spring Harbour Lab, 1975, 467–82.

21. STEFANSSON G and STEINGRIMSDÓTTIR U, 'Application of enzymes for fish processing in Iceland – present and future aspects', in Voigt M N and Botta J R, *Advances in Fisheries Technology and Biotechnology for Increased Profitability*, Lancaster, Technomic Publ Comp, 1990, 237–50.

22. NILSEN K, VIANA M T and RAA J, 'Biotechnology in squid processing: removing skins enzymatically', *Infofish Internat*, 1989(2) 27–8.

23. GILDBERG A and SVENNING R, 'Method and equipment for removal of

connective tissue membranes and parasites from fish liver', Norwegian patent no 951873 (in Norwegian), 1995.

24. HAARD N F, 'Specialty enzymes from marine organisms', *Food Technol*, **52**(7) 64–7.

25. RAA J, 'Biotechnology in aquaculture and fish processing industry: a success story in Norway', in Voigt M N and Botta J R, *Advances in Fisheries Technology and Biotechnology for Increased Profitability*, Lancaster, Technomic Publ Comp, 1990, 509–24.

26. HAARD N F and SIMPSON B K, 'Proteases from aquatic organisms and their uses in the seafood industry', in Martin A M, *Fisheries Processing: Biotechnological Applications*, London, Chapman & Hall, 1994, 132–54.

27. SAISITHI P, 'Traditional fermented fish: fish sauce production', in Martin A M, *Fisheries Processing: Biotechnological Application*, London, Chapman & Hall, 1994, 110–31.

28. IN T, 'Seafood flavourants produced by enzymatic hydrolysis', in Voigt M N and Botta J R, *Advances in Fisheries Technology and Biotechnology for Increased Profitability*, Lancaster, Technomic Publ Comp, 1990, 425–36.

29. CANCRE I, RAVALLEC R, VAN WORMHOUDT A, STENBERG E, GILDBERG A and LE GAL Y, 'Secretagogues and growth factors in fish and crustacean protein hydrolysates', *Marine Biotechnol*, 1999 **1** 489–94.

30. HATATE H, NUMATA Y and KŌCHI M, 'Synergistic effect of sardine myofibril protein hydrolyzates with antioxidants', *Nippon Suisan Gakkaishi*, 1990 **56**(6) 1011.

31. GILDBERG A, BØGWALD J, JOHANSEN A and STENBERG E, 'Isolation of acid peptide fractions from fish protein hydrolysate with strong stimulatory effect on Atlantic salmon *(Salmo salar)* head kidney leucocytes', *Comp Biochem Physiol*, 1996 **114B** 97–101.

32. BØGWALD J, DALMO R A, LEIFSON R M, STENBERG E and GILDBERG A, 'The stimulatory effect of a muscle protein hydrolysate from Atlantic cod, *Gadus morhua* L., head kidney leucocytes', *Fish & Shellfish Immunol*, 1996 **6** 3–16.

33. CORCORAN T H, 'Roman fish sauce', *Classical J*, 1963 **58**(5) 204–10.

34. AMANO K, 'The influence of fermentation on the nutritive value of fish with special reference to fermented fish products of South-East Asia', in Heen E and Kreuzer R, *Fish in Nutrition*, London, Fishing News Books, 1962, 180–97.

35. BEDDOWS C G and ARDESHIR A G, 'The production of soluble fish protein solution for use in fish sauce manufacture. I. The use of added enzymes', *J Food Technol*, 1979 **14** 603–12.

36. RAKSAKULTHAI N, LEE Y Z and HAARD N F, 'Effect of enzyme supplements on the production of fish sauce from male capelin *(Mallotus villosus)*, *Can Inst Food Sci Technol J*, 1986 **19**(1) 28–33.

37. GILDBERG A, 'Utilisation of male Arctic capelin and Atlantic cod intestines for fish sauce production – evaluation of fermentation conditions', *Bioresource Technol*, 2001 **76** 119–23.

38. MABESA R C, CARPIO E V and MABESA L B, 'An accelerated process for fish sauce (patis) production', in Reilly P J A, Parry R W H and Barile L E, *Post-Harvest Technology, Preservation and Quality of Fish in Southeast Asia*, Manila, Echanis Press, 1990, 45–9.

39. BEDDOWS C G and ARDESHIR A G, 'The production of soluble fish protein solution for use in fish sauce manufacture. II The use of acids at ambient temperature', *J Food Technol*, 1979 **14** 613–23.

40. GILDBERG A, HERMES J E and OREJANA F M, 'Acceleration of autolysis during fish sauce fermentation by adding acid and reducing the salt content', *J Sci Food Agric*, 1984 **35** 1363–9.

41. GILDBERG A and THONGTHAI C, 'The effect of reduced salt content and addition of halophilic lactic acid bacteria on quality and composition of fish sauce made from sprat', *J Aquatic Food Product Technol*, 2001 **10**(1) 77–88.

42. RAKSAKULTHAI N and HAARD N F, 'Fish sauce from capelin *(Mallotus villosus)*: contribution of cathepsin C to the fermentation', *ASEAN Food J*, 1992 **7**(3) 147–51.

43. GILDBERG A, 'Aspartic proteinases in fishes and aquatic invertebrates', *Comp Biochem Physiol*, 1988 **91B**(3) 425–35.

44. KROGDAHL Å, 'Fish silage as a protein source for poultry. I. Experiments with layer-type chicks and hens', *Acta Agric Scand*, 1985 **35** 3–23.

45. BALOGUN A M, FASAKIN E A and OWOLANKE D, 'Evaluation of fish silage/ soybean meal blends as protein feedstuff for *Clarias gariepinus* (Burchell, 1822) fingerlings', *J Appl Anim Res*, 1997 **11** 129–36.

46. LOPEZ C S, 'Microbial ensilage of trash fish for animal feeds', in Reilly P J A, Parry R W H and Barile L E, *Post-Harvest Technology, Preservation and Quality of Fish in Southeast Asia*, Manila, Echanis Press, 1990, 189–92.

47. ZUBERI R, FATIMA R, SHAMSHAD S I and QUADRI R B, 'Preparation of fish silage by microbial fermentation', *FAO Fish Rep*, 1992 No 470 Suppl 155–61.

48. BELLO R, CARDILLO E and MARTINEZ R, 'Estudio sobre la elaboración de ensilado microbiano a partir de pescado eviscerado', *Arch Latinoameric de Nutr*, 1993 **43**(3) 221–7.

49. NOEL H S, 'By-catches for protein', *World Fishing*, 1974 March 77.

50. KRISTINSSON H G and RASCO B A, 'Fish protein hydrolysates: production, biochemical, and functional properties', *Crit Rev Food Sci Nutr*, 2000 **40**(1) 43–81.

51. SUGIYAMA K, EGAWA M, ONZUKA H and OBA K, 'Characteristics of sardine muscle hydrolysates prepared by various enzymic treatments', *Nippon Suisan Gakkaishi*, 1991 **57**(3) 475–9.

52. ALMÅS K A, 'Utilization of marine biomass for production of microbial growth media and biochemicals' in Voigt M N and Botta J R, *Advances in Fisheries Technology and Biotechnology for Increased Profitability*, Lancaster, Technomic Publ Comp, 1990, 361–72.

53. GOLDHOR S H, CURREN R A, SOLSTAD O, LEVIN R E and NICHOLS D, 'Hydrolysis and fermentation of fishery by-products: costs and benefits of some processing variables', int conf on *Fish By-Products*, Anchorage, Alaska Sea Grant College Program, 1990, 203–8.

54. UCHIDA Y, HUKUHARA H, SHIRAKAWA Y and SHOJI Y, 'Bio-fish flour', int conf on *Fish By-Products*, Anchorage, Alaska Sea Grant College Program, 1990, 95–9.

55. SIMPSON B K, GAGNE N and SIMPSON M V, 'Bioprocessing of chitin and chitosan', in Martin A M, *Fisheries Processing: Biotechnological Applications*, London, Chapman & Hall, 1994, 155–73.

56. STRØM T and RAA J, 'Marine biotechnology in Norway', *J Mar Biotechnol*, 1993 **1** 3–7.

57. SKAUGERUD Ø and SARGENT G, 'Chitin and chitosan: crustacean biopolymers with potential', int conf on *Fish By-Products*, Anchorage, Alaska Sea Grant College Program, 1990, 61–9.

58. WACHTER R and STENBERG E, 'HYDAGEN CMF in cosmetic applications. Efficacy in different *in-vitro* and *in-vivo* measurements', *Adv Chitin Sci*, 1996 **1** 381–8.

59. HÖRNER V, PITTERMANN W and WACHTER R, 'Efficiency of high molecular weight chitosan in skin care application', in Domard A, Roberts G A F and Vårum K M, *Advances in Chitin Science Vol II*, Lyon, Jacques André Publ, 1997, 671–7.

60. GORBACH V I, KRASIKOVA I N, LUKYANOV P A, LOENKO Y N, SOLOVEVA T F, OVODOV Y S, DEEV V V and PIMENOV A A, 'New glycolipids (chitooligosaccharide derivatives) posessing immunostimulating and antitumor avtivities', *Carbohydr Res*, 1994 **260**(1) 73–82.

61. HOFFMAN J, JOHANSEN A, STEIRO K, GILDBERG A, STENBERG A and BØGWALD J, 'Chitooligosaccharides stimulate Atlantic salmon *Salmo salar* L., head kidney leukocytes to enhanced superoxide anion production *in vitro*', *Comp Biochem Physiol*, 1997 **118B** 105–15.

62. SIMPSON B K, AWAFO V and RAMASWAMY H, 'Biotechnological approaches for the production of value added substances from by-products of seafood harvesting', in *4th International Food Convention – Proc Tech Sess*, Mysore, Jwalamukhi Job Press, 1999, 151–7.

63. BABEL W, 'Gelatine a versatile biopolymer', *Chemie in unserer Zeit*, 1998 (06), information from Deutsche Gelatine-Fabriken Stoess AG, 1–10.

64. CHOI S S and REGENSTEIN J M, 'Physicochemical and sensory characteristics of fish gelatin', *J Food Sci*, 2000 **65**(2) 194–9.

65. DE CLERCQ M, 'Photographic gelatin production', *J Imag Sci Technol*, 1995 **39**(4) 367–72.

66. GUDMUNDSSON M and HAFSTEINSSON H, 'Gelatin from cod skins as affected by chemical treatments', *J Food Sci*, 1997 **62**(1) 37–39(47).

67. ARNESEN J A and GILDBERG A, 'Preparation and characterisation of gelatine from the skin of harp seal *(Phoca groendlandica)*, *Bioresource Technol*, 2002, 82 191–4.

68. NORLAND R E, 'Fish gelatin', in Voigt M N and Botta J R, *Advances in Fisheries Technology and Biotechnology for Increased Profitability*, Lancaster, Technomic Publ Comp, 1990, 325–33.

69. RAVALLEC-PLE R, GILMARTIN L, VAN WOURMHOUDT A and LE GAL Y, 'Influence of the hydrolysis process on the biological activities of the protein hydrolysates from cod *(Gadus morhua)*. *J Sci Food Agric*, 2000 **80**(15) 2176–80.

70. MATSUMURA N, FUJII M, YASUHIKO T, SUGITA K and SHIMIZU T, 'Angiotensin I-converting enzyme inhibitory peptides derived from bonito bowels autolysate' *Biosci Biotech Biochem*, 1993 **57**(5) 695–7.

71. BYUN H G and KIM S K, 'Purification and characterization of angiotensin I converting enzyme (ACE) inhibitory peptides from Alaska pollack *(Theragra chalcogramma)* skin, *Process Biochem*, 2001 **36** 1155–62.

72. RAA J, 'The use of immunostimulatory substances in fish and shellfish farming', *Rev Fish Sci*, 1996 **4**(3) 229–88.

73. GILDBERG A and STENBERG E, 'A new process for advanced utilisation of shrimp waste', *Process Biochem*, 2001 **36** 809–12.

74. KIM S K, KIM Y T, BYUN H G, NAM K S, JOO D S and SHAHIDI F, 'Isolation and characterization of antioxidative peptides from gelatin hydrolysate of Alaska pollack skin', *J Agric Food Chem*, 2001 **49** 1984–9.

75. KOLKOVSKI S, TANDLER A and IZQUIERDO M S, 'Effects of live food and dietary digestive enzymes on the efficiency of microdiets for seabass *(Dicentrarchus labrax)* larvae, *Aquaculture*, 1997 **148** 313–22.

76. KOLKOVSKI S, 'Digestive enzymes in fish larvae and juveniles – implications and applications to formulated diets', *Aquaculture*, 2001 **200**(1–2) 181–201.

77. SAKHAROV I Y, GLYANZEV S P, LITVIN F E and SAVVINA T V, 'Potent debriding ability of collagenolytic protease isolated from the hepatopancreas of the king crab *Paralithodes camtschatica*, *Arch Dermatol Res*, 1993 **285** 32–5.

78. HELLGREN L, KARLSTAM B, MOHR V and VINCENT J, 'Krill enzymes – a new concept for efficient debridement of necrotic ulcers', *Internat J Dermatol*, 1991 **30**(2) 102–3.

79. MEKKES J R, LE POOLE I C, DAS P K, BOS J D and WESTERHOF W, 'Efficient debridment of necrotic wounds using proteolytic enzymes derived from Antarctic krill: a double blind, placebo-controlled study in standardized animal wound model', *Wound Rep Regeneration*, 1998 **6**(1) 50–7.

80. BARLOW S M, YOUNG F V K and DUTHIE I F, 'Nutritional recommendations for n-3 polyunsaturated fatty-acids and the challenge to the food-industry', *Proc Nutr Soc*, 1990 **49**(1) 13–21.

81. DYERBERG J and BANG H O, 'Haemostatic function and platelet polyunsaturated fatty acids in Eskimos', *Lancet*, 1979 **2** 433–5.

82. ELVEVOLL E O, MOEN P, OLSEN R L and BROX J, 'Some possible effects of dietary monounsaturated fatty acids on cardiovascular disease', *Atheriosclerosis*, 1990 **81** 71–4.

83. ELVEVOLL E O and JAMES D, 'The emerging importance of dietary lipids, quantity and quality, in the global disease burden: the potential of aquatic resources', *Nutr Health*, 2000 **15** 7–19.

84. TOCHER D R, MOURENTE G and SARGENT J R, 'The use of silages prepared from fish neural tissues as enrichers for rotifers *(Brachionus plicatilis)* and *Artemia* in the nutrition of larval marine fish', *Aquaculture*, 1997 **148** 213–31.

85. HAARD N F and SIMPSON B K, *Seafood Enzymes*, New York, Marcel Dekker, 2000.

86. NORRIS E R and ELAM D W, 'Preparation and properties of crystalline salmon pepsin', *J Biol Chem*, 1940 **134** 443–54.

87. KARLSEN S, HOUGH E and OLSEN R L, 'Structure and proposed amino-acid sequence of pepsin from Atlantic cod *(Gadus morhua)*, *Acta Cryst*, 1998 **D54** 32–46.

88. SVENNING R, STENBERG E, GILDBERG A and NILSEN K, 'Biotechnological descaling of fish', *Infofish Internat*, 1993 (6) 30–1.

89. GILDBERG A, 'Recovery of proteinases and protein hydrolysates from fish viscera', *Bioresource Technol*, 1992 **39** 271—6.

90. TAVARES J F, 'Recovery of milk coagulating enzymes from tuna waste', in Holló J, *Food Industries and Environment*, Amsterdam, Elsevier, 1984, 265–75.

91. HAARD N F, 'Atlantic cod gastric protease. Characterization with casein and milk substrate and influence of sepharose immobilization on salt activation, temperature characteristics and milk clotting reaction', *J Food Sci*, 1986 **51**(2) 313–6(326).

92. GUERARD F and LE GAL Y, 'Dogfish pepsin as a rennet substitute', in Miyachi S, Karube I and Ishida Y, *Current Topics in Marine Biotechnology*, Tokyo, Fuji Technol Press, 1989, 357–60.

93. OPSHAUG K, 'Procedure for rapid enzymatic ripening of herring' Norwegian patent no 148207 (in Norwegian), 1983.

94. SIMPSON B K and HAARD N F, 'Trypsin from Greenland cod as a food-processing aid', *J Appl Biochem*, 1984 **6** 135–43.

95. LOPEZ A C, SIMPSON B K and HAARD N F, 'Extraction of carotenoprotein from shrimp process wastes with the aid of trypsin from Atlantic cod', *J Food Sci*, 1987 **52**(2) 503–4(506).

96. MURATA Y, HAYASHI T, WATANABE E and TOYAMA K, 'Preparation of skipjack spermary extract by enzymolysis', *Nippon Suisan Gakkaishi*, 1991 **57**(6) 1127–32.

97. OLSEN R L, JOHANSEN A and MYRNES B, 'Recovery of enzymes from shrimp waste, *Process Biochem*, 1990 **25**(2) 67–8.

98. OLSEN R L, ØVERBØ K and MYRNES B, 'Alkaline phosphatase from the hepatopancreas of shrimp *(Pandalus borealis)*: a dimeric enzyme with catalytic active subunits. *Comp Biochem Physiol*, 1991 **99B**(4) 755–61.

99. NILSEN I W, ØVERBØ, K and OLSEN R L, 'Thermolabile alkaline phosphatase from northern shrimp *(Pandalus borealis)*: protein and cDNA sequence

analyses', *Comp Biochem Physiol*, 2001 **129B** 853–61.

100. NILSEN I W, SANDSDALEN E and STENBERG E, 'A method of removing nucleic acid contamination in amplification reactions. International patent application no. WO 99/07887, 1997.

101. PRAGER E M and JOLLÈS P, 'Animal lysozymes c and g: an overview', in Jollès P, *Lysozymes: Model Enzymes in Biochemistry and Biology*, Basel, Birkhäuser Verlag, 1996, 9–31.

102. ITO Y, YOSHIKAWA A, HOTANI T, FUKUDA S, SUGIMURA K and IMOTO T, 'Amino acid sequences of lysozymes newly purified from invertebrates imply wide distribution of a novel class in the lysozyme family, *Eur J Biochem*, 1999 **259** 456–61.

103. NILSEN I W, ØVERBØ K, SANDSDALEN E, SANDAKER E, SLETTEN K and MYRNES B, 'Protein purification and gene isolation of chlamysin, a cold active lysozyme-like enzyme with antibacterial activity', *FEBS Lett*, 1999 **464** 153–8.

104 NILSEN I W and MYRNES B, 'The gene of chlamysin, a marine invertebrate-type lysozyme, is organized similar to vertebrate but different from invertebrate chicken-type lysozyme genes', *Gene*, 2001 **269** 27–32.

105. LOVE R M, *The Chemical Biology of Fishes*, London, Academic Press, 1970.

23

Species identification in processed seafoods

C.G. Sotelo and R.I. Pérez-Martín, Instituto de Investigaciones Marinas, Vigo

23.1 Introduction: the importance of species identification

Some decades ago, most European countries consumed seafood products coming from, almost exclusively, their own fishing fleet in the nearby waters. This fact restricted the consumption of fish to a limited number of fish species, that were all well known by the agents involved in the whole chain of capture and distribution of seafood, from fishermen to consumers. Most landings consisted of an important number of whole specimens, even not eviscerated, from one or several species that were easy to recognise because of the presence of all morphological characteristics.

From that time to the present the situation has changed substantially driven by a number of factors. One was the development of high-seas fishing vessels with the ability to sail and fish all over the world; another was the tremendous improvement in processing and storing, both on board and on land. Another was the establishment of a modern fishing industry in some developing countries, like South American, African or Asian countries where only an artisanal fleet, not so many years ago, was operating and landing fish to supply local markets. The increase and improvement of worldwide aerial, maritime and terrestrial communications, and the market globalisation, led to an increase in the number of fish species, both fresh and processed, being present in our markets.

For example, European hake (*Merluccius merluccius*) is a highly priced and appreciated white fish species in southern Europe, belonging to the order Gadiformes. A few years ago, practically all the hake consumed in Europe came from nearby waters and most of it belonged to the species *Merluccius merluccius*. Nowadays, European markets are supplied with fresh or frozen hake coming from other countries and waters, like Argentina and Chile (*M. hubsi*, and

M. gayi) or South Africa (*M. capensis* and *M. productus*) and with other products labelled as hake, such as fish fingers or hake fillets, for instance, which have been produced using cheaper grenadier species (*Macruronus* spp.).

This extended consumption of new species has been accompanied with a continuous increase in total catches. FAO data (2000) indicates that the actual global capture from fisheries (1998 data) is around 90 million tonnes per year, and that total capture had a five-fold increase from 1950 to 1990, increasing also the number of species with commercial value. Fish consumption has followed the same path, human fish consumption has increased from 40 million tonnes in 1970 up to 86 million tonnes in 1998, with an increase of as much as 31% in fish consumption between 1990 and 1997. However, at the same time, decrease in the capture of most appreciated species, like cod, sole, or European hake, has been reported, due mainly to overexploitation of resources. That means that they are being replaced by alternative species.

The above-mentioned factors have played a decisive role in the increase in the fish supply all around Europe, specially of new species rather than traditional ones. Consumption habits have undergone important changes and a bigger proportion of convenience seafood products than in the past can be found in developed countries. This processed fish can be produced in European or third-world countries, from European species or species totally unknown to European consumers. Since the processing of fish implies the removal of significant morphological characteristics, recognition of species is no longer possible and therefore it is necessary to employ alternative methods to identify them.

Evidently, the price of each fish species or derived product is a function of several variables like intrinsic quality, consumer acceptance or demand (variable from country to country), resource scarcity or abundance, etc. In any case, consumers should be informed about the goods they can buy so that a purchase decision can be made by them based on sufficient and reliable information. This also concerns the industry of fish processing because it relies every day on preprocessed starting material (with variable prices depending on the quality and species) and, as the industry is responsible for the labelling of its seafood products, it also needs to know the fish species involved.

During the last decade, the European Union has been issuing food labelling normatives, including for fish and seafood products, that are more demanding about the information that should be included on the label. In order to control compliance with these regulations, the administration should have adequate tools to verify, among other requisites, the authenticity of the species indicated on the label.

23.2   The problem of species identification in seafood products

The substitution of highly priced species, or those more appreciated by consumers for others of less commercial value is an old problem, and it could be said that this fraudulent behaviour is a reflection of the human condition (Dennis and Ashurst, 1996). However, today more species are being processed, making species identification difficult and giving more opportunities for fraudulent species substitution.

Identification of a particular fish species relies on the recognition of a number of taxonomical morphological characteristics, like skin pattern, body shape and size, shape and number of fins, eyes or even internal organs. Even when all morphological characteristics are intact, and when dealing with very closely related species, identification of some fish species is rather difficult and only a taxonomist can perform it correctly. When these morphological identifiers are removed, for instance, when fish is processed (beheaded, eviscerated, skinned, filleted, etc.) or deeply transformed (smoking, cooking or canning) species identification is very difficult or even impossible. Because of that, it seems evident that there is a need for alternative methods of species identification.

Atlantic cod (*Gadus morhua*) has been prized since antiquity and is an example of a highly appreciated product that now has important resource problems. Atlantic cod was over-fished and the populations have been dramatically exhausted. However, the number of products on the market labelled as cod, far from decreasing are increasing. In our own experience, other gadoid species are being used for preparation of products labelled as containing cod.

Pleuronectiformes or flatfish are an order with a high number of species. Sole (*Solea solea*), turbot (*Scophthalmus maximus*) or European plaice (*Pleuronectes platessa*) are the ones most widely known and appreciated in Europe. Products labelled as one of these species can be found in the European market containing other cheaper flatfish species that at best belong to the same family.

Tuna is a migratory fish species with a very high demand in markets all around the world. The commercial value of this group of species is very high and in most parts of the world these species are appreciated and valued differently. Europe has issued rules concerning labelling of albacore (*Thunnus alalunga*) and yellowfin (*T. albacares*), however these species may be substituted by cheaper tunas, like bigeye (*T. obesus*) or skipjack (*Katsuwonus pelamis*).

Finally, protected species, like sturgeons or whales, may be illegally found in food markets, and evidence of this deception is difficult to demonstrate if there is reliance only on morphological characteristics (Baker and Palumbi, 1994; DeSalle and Birstein, 1996).

23.3   The use of biomolecules as species markers

The need for classification of animals dates from very early times, and it has been regarded as the science of taxonomy since the 4th century BC with Aristotle's classification of living organisms (Gosling, 1994). The concept of just what constitutes a species has changed over time, and the most quoted definition of species is the Mayr's (1970) isolation species concept stating that 'species are groups of interbreeding natural populations that are reproductively isolated from other such groups'. Classification is usually achieved by means of the so-called taxonomic characteristics. As quoted by Ayala (1983) a taxonomic characteristic is 'any attribute of a member of a taxon by which it differs or may differ from a member of a different taxon'. Diagnostic characteristics are those taxonomic characteristics that uniquely specify a particular taxon (Ayala, 1983).

Proteins have been widely used as species markers, not only for identification of animals for human consumption, but for taxonomic purposes (Ayala, 1983). Early works on the development of species identification techniques showed that separation of water-soluble proteins by either starch or agarose electrophoresis could be used for this purpose (Cowie, 1968; Mackie, 1969). Later on, electrophoretic techniques were improved with the development of isoelectric focusing, and highly resolved water-soluble protein patterns were obtained which permitted the differentiation of genetically close-related fish species (Mackie, 1980). The IEF patterns became a method of reference for fish species identification (AOAC, 1990) and many studies were published showing its applicability to solve the problem of biochemical identification of fish species in seafood products (Rehbein, 1990; Rehbein et al., 1995; Mackie, 1996). Different IEF protein patterns can be obtained for the same species, thus making identification difficult, and some of the complicating factors involved were described, like the existence of polymorphic bands related to population differences, type of muscle fibre being analysed (Rehbein and Kündiger, 1984), or frozen storage (Rehbein, 1990). The use of highly resolutive techniques, like 2-D electrophoresis (Piñeiro et al., 1998; Piñeiro et al., 1999), seems to result in quite complex patterns making the process of identification a time-consuming task. The value of other techniques, like Matrix Assisted Laser Desorption/Ionization-Time of Flight Mass Spectrometry(MALDI-TOF-MS) analysis of 2D separated fish sarcoplasmic proteins (Piñeiro et al., 2001) for species identification in routine analysis is rather questionable because of the high cost of the equipment required and the complexity of the analysis itself.

The use of native IEF for the identification of processed fish is somewhat more complicated, although it has been successfully used for mildly treated products, like smoked seafood products (Sotelo et al., 1992) by analysing a few heat-stable proteins. In the case of other products, denatured proteins were solubilised by means of urea or SDS and analysed using urea-IEF (Rehbein et al., 1999a) or SDS-PAGE (Scobbie and Mackie, 1988; Piñeiro et al., 1999). However, the level of species resolution is low and these methods were applied to distinguish species belonging to different families or groups. Identifying closely related fish species, like processed tuna or smoked salmon, has proven to be very difficult using these techniques (Mackie et al., 1992; Mackie et al., 2000).

Capillary zone electrophoresis is another electrophoretic technique which has been used for species identification (LeBlanc et al., 1994; Gallardo et al., 1995). The technique shows some advantages compared with other electrophoretic techniques: low cost of reagents, high resolution power and efficiency, small sample requirement, no need for harmful solvents, spectrophotometric detection of separated proteins, short analysis time (10–20 min.) and potential automation of sample load. However, it has the same restrictions as other protein techniques, like the restriction of its utility only for the authentication of fresh or frozen seafood products.

Species identification has also been performed using other approaches, like the analysis of chemical composition by Near and Mid Infrared Spectroscopy (NIR and MIR). These techniques have been employed as a means of differentiation and quantitation of lamb and beef (McElhinney et al., 1999). Although it is a non-destructive technique and may allow on-line detection of species, its use for fish identification seems uncertain. The level of differentiation is achieved through differences in fat content and composition, protein composition and water content. However, the usual problem in the authentication of fish species is the differentiation of species within a family, and fish families present almost identical proximate composition.

The use of lipid analysis for species identification had very little impact, mainly because lipid composition, especially triglyceride composition, depends greatly on feed and therefore it is quite variable (Dotson, 1976). However, results of fatty acid composition of phospholipids showed ability to discriminate closely related fish species, like tuna (Medina et al., 1997). The application of this technique to canned tuna is demonstrated in the work although its application to routine analysis has not been proven.

Water-soluble sarcoplasmic protein profiles, after separation using reverse-phase high-performance liquid chromatography, were used for the identification of raw fish species. Among the advantages of using this method is the possibility of species quantitation, relying on the online detection of separated proteins using different detector systems, like UV detection (Ashoor and Knox, 1985; Armstrong et al., 1992; Piñeiro et al., 1997). Other advantages of the HPLC methods are the speed, the use of a relatively common piece of equipment in most analytical laboratories and the reproducibility of the profiles. The disadvantages are its restricted use to fresh or frozen seafood products, the need for internal and external (reference material) standards, and the problems inherent in comparison of specific profiles.

Immunological techniques for species identification rely on the recognition of diagnostic proteins (antigens) by antibodies raised against epitopes located in those diagnostic proteins. Antigen-antibody reaction reveals species identity. Initially, meat antibodies which recognise specific antigens present in different meat species, were developed and different assay techniques, like agar gel immunodiffusion (Hsieh et al., 1996) and enzyme-linked immunosorbent assay (ELISA), were employed for this purpose (Whitaker et al., 1983, Andrews et al., 1992; Kang'ethe and Lindqvist, 1987; Dominguez et al., 1997). These techniques were also employed for fish species identification (An et al., 1990; Verrez-Bagnis, 1993; Taylor et al., 1994; Huang et al., 1995; Carrera et al., 1996). One of the problems is that the development of the assay is usually conducted for raw material, but most of the time there is a need for identification of species when thermal treatment is employed in the processing of a particular product. Chen *et al.* (1998) developed monoclonal antibodies to porcine thermal stable muscle proteins, the technique allowed the identification of pork in products which had been heated. However, it is not proven if they would resist sterilisation treatment (Chen *et al.*, 1998). Immunological detection of species has the advantage of enabling the development of colorimetric kits for detection, like immunosticks, which permit the rapid identification of species in a product (Carrera *et al.*, 1997) requiring only equipment for the preparation of soluble extracts of proteins from the product. In the case of fish species, and different from meat species, authentication is aimed to differentiate close-related species, i.e., species of the same genus. Therefore, the main problem with the development of immunological techniques for fish is to find a suitable and specific antigen that shows no cross-reaction with the related species.

DNA molecules can be located in the nucleus and in the mitochondria of eukaryotic organisms. There are segments of DNA, called genes, which possess the information required for the synthesis of proteins within the cell. The total amount of DNA in a cell can vary depending on several biological conditions like stage of cellular cycle, type of tissue, age, etc. The size of the nuclear genome of bony fishes is about 0.3–4.0 billion base pairs (Park and Moran, 1994). Genomic DNA encompasses coding and non-coding sequences of DNA, in fact only 1% of the total genomic mammalian DNA regulates or codes for essential proteins (Park and Moran, 1994). This fact illustrates that it is possible to find much more species or population information using DNA-based methods than one can find with protein-based techniques.

Therefore, DNA sequence information has been employed for species identification, especially after the great development of molecular biology techniques driven by the implementation of Polymerase Chain Reaction (PCR) (Saiki *et al.*, 1988). PCR allows the amplification and analysis of selected fragments of DNA in a relatively short time (two hours). The analysis of the sequence contained in the amplified product can be performed using a myriad of techniques; probably the widest used is sequencing. DNA sequences from both nuclear and mitochondrial DNA have been used for species identification purposes. Compared with the use of proteins as taxonomical markers, DNA markers have the following advantages:

- the analyst can select different DNA regions for analysis with different mutation rates and inter and intraspecific variability depending on the level of resolution required (individuals, population, species, genera, family, etc.)
- DNA is more stable than proteins to different industrial processes.

Mitochondrial DNA (mtDNA) has been the most widely used as a target molecule for species identification. The reasons for choosing mitochondrial genes to carry out these types of studies are the haploid character of the molecule, its size (fish have mtDNA of around 15–20 Kb), a generally higher rate of mutation accumulation, meaning more genetic variability (there are exceptions to this and differences among genes), and no introns within the mtDNA.

In summary, protein analysis has been widely employed in the past for species identification and has been useful in the case of fresh, refrigerated and frozen seafood. One of the advantages of the technique, especially IEF, is that it is relatively cheap, allows quick processing time, and it does not require a highly trained staff, or sophisticated and expensive equipment. However, protein analysis has two major drawbacks: it is useless for heated seafood products, because of protein denaturation, and it has low resolution power in some cases when close-related species need to be differentiated.

23.4   The use of DNA for species identification: DNA integrity and the effect of processing

The first step for conducting species identification based on DNA analysis is the extraction and amplification of DNA. The integrity of DNA plays an important role and restricts the choice of analytical technique to be used. Although DNA is a very stable molecule, there are a number of factors involved in its degradation, like activity of nucleases (Martínez et al., 1997; Cerda and Koppen, 1998), oxidation, acidic treatment and thermal degradation (Bossier, 1999). DNA is degraded to fragments smaller than 500 bp by temperatures higher than 100°C, depending on the temperature and time the thermal treatment is applied (Chikuni et al., 1990; Unseld et al., 1995; Ram et al., 1996; Quinteiro et al., 1998). DNA extracted from sterilised products has an average size of around 300 bp (Ebbehøj and Thomson, 1991). Mitochondrial DNA might be the choice for species identification when dealing with heated products because of the higher number of copies per gram of tissue, compared with nuclear DNA, and the possible higher tolerance to heat of mtDNA than that showed by nuclear DNA (Borgo et al., 1996).

23.5   Polymerase Chain Reaction (PCR) techniques

The Polymerase Chain Reaction (PCR) developed by Mullis and colleagues in the 1980s (Mullis and Faloona, 1987; Mullis, 1990) permitted the synthesis of specific DNA fragments in vitro. There have been a myriad of published articles explaining the fundamentals of the reaction and its possible application to different fields (Bej et al., 1991).

As mentioned above, different mitochondrial DNA fragments have often been the target for PCR, since the level of genetic variability is different for different mitochondrial genes. A good primer design is of paramount importance for successful PCR and all derived techniques, since the lack of perfect match between template DNA and 3′ end of the primer might lead to a PCR failure (Bej et al., 1991). Table 23.1 shows different primers and fragments used for fish

Table 23.1 Primer sets most used for fish species identification, identification technique used. Forensically Informative Nucleotide Sequencing (FINS), Restriction Length Polymorphisms (RFLP)

Name of primer set	Primer sequence	Size*/Technique	Species	Reference
CytbH/CytbL (cytb)	5'-CCCTCAAATGATATTTGTCCTA- 3' 5'-CCATCCAACATCTCAGCATGATGAAA-3'	307 bp FINS 307 bp RFLP 307 bp RFLP	Several Flatfishes Several	Bartlett and Davidson, 1992 Céspedes et al., 1998 Cocolin et al., 2000
Tuna334F/Tuna395R (cytb)	5'-TAGGGATCCTYCTHTCIGCAGTMCCMTAYGT-3' 5'-GGTCTCAGGAAGTGGAAKGCRAAGAAYCGG-3'	60 bp RFLP	Tunas	Ram et al., 1996
TunaFOR/TunaREV (cytb)	5'-GGGAATTCCTMTACAAAGAAACMTGAAACA-3' 5'-DAGGGATCCTCAGAANGAYATYTGTCCTCA-3'	59 bp RFLP	Tunas	Ram et al., 1996
59-3/59-5 (cytb)	5'-GCTGGTACCTCTACAAAGAAACATGAAACA-3' 5'-AAACTGCAGCCCCTCAGAATGATATTTGTCCTCA-3'	59 bp	Tunas	Unseld et al., 1995
L14735/H15149ad (cytb)	5'-AAAAACCACCGTTGTTATTCAACTA-3' 5'-GCNCCTCARAATGAYATTTGTCCTCA-3'	419 bp RFLP 419 bp RFLP 419 bp FINS/RFLP 419 bp RFLP	Several Salmons Flatfishes Sturgeons	Wolf et al., 2000 Russell et al., 2000 Sotelo et al., 2001 Wolf et al., 1999
MERFPD1/GADRPD1 (Control region)	5'-TCAACCCATAATACWCATTCC-3' 5'-ATGGACCTGAAGCTAGGCA-3'	157 bp FINS/RFLP	Hakes	Quinteiro et al., 2001
16S-1/16S-2 (16sRNA)	5'-CACCACAACACATACCCC-3' 5'-CGTTAAACCCATAGTCACAG-3'	911 bp RFLP 911 bp RFLP	Salmon Flatfishes	Carrera et al., 1999 Céspedes et al., 2000

* excluding primers

species identification in seafood products. Cytochrome b has been most often used for species identification, although other gene fragments like DNA coding for ribosomal RNA, Cytochrome oxidase, ATPase and control region have also been employed for this purpose.

23.6   Methods not requiring a previous knowledge of the sequence

There are two main methods to perform an amplification of DNA using a polymerase, in the first type non-specific primers are employed, and that means that it is not necessary to know portions of template sequence. This type of methodology may result in a specific profile of bands. The second type of method requires that at least two short fragments of the target sequence are known; usually the amplification is specific meaning that a particular DNA fragment will be obtained. The amplified DNA is then analysed using other secondary analysis methods.

23.6.1   Random Amplification of Polymorphic DNA (RAPD)

The technique involves the use of single and short arbitrary primers (9 to 10 mers) which anneal randomly in several targets of genomic DNA (Dinesh et al., 1993). Primers are not designed to amplify a specific fragment; on the contrary, they scattered anneal throughout the genome (Grosberg et al., 1996). This is the reason why it is not necessary to have prior sequence data for addressing the study in a particular fish species group. The technique is very simple and specific patterns are generated after amplification, separation using electrophoresis (agarose, polyacrylamide), and detection (ethidium bromide or silver staining). These patterns can be used for species or population identification. However, it has some disadvantages: its high susceptibility to changes of DNA quality and/or quantity, the thermalcycler used for the amplification, cycling conditions, polymerase, separation and detection conditions, and intraspecies variability. All these parameters require standardisation in order to obtain reproducible patterns (Martínez and Daníelsdóttir, 2000; Grosberg et al., 1996). Despite these problems some authors have described RAPD methods which were reproducible (Dinesh et al., 1995) or have been used for processed seafood or meat species identification (Martínez, 1997). The identification process using this technique involves the determination of genetic similarity. This similarity index is based on the number of shared bands between the reference and the unknown species (Martínez and Malmheden Yman, 1998) and may be used to construct dendograms that may help to identify species (Crossland et al., 1993).

23.6.2   Single Strand Conformation Polymorphisms (SSCPs)

SSCP is an electrophoretic technique that allows the rapid detection of mutations in a specific DNA fragment. The shorter the DNA fragment, in a specific range, the higher is the resolution obtained with this technique. Usually optimal sizes range around 100–200 bp. The technique relies on the fact that single strand DNA fragments will form folding conformers that migrate through an electrophoretic field under non-denaturing conditions in a sequence-specific way (Law *et al.*, 1996) (Fig. 23.1). In the highest resolution, differences in a single nucleotide are reflected in differences in electrophoretic mobility (Hayashi, 1991). Since denatured DNA is analysed, PCR fragments should be denatured prior to electrophoresis, although electrophoretic conditions are non-denaturing to allow intramolecular interactions to take place. Differences in

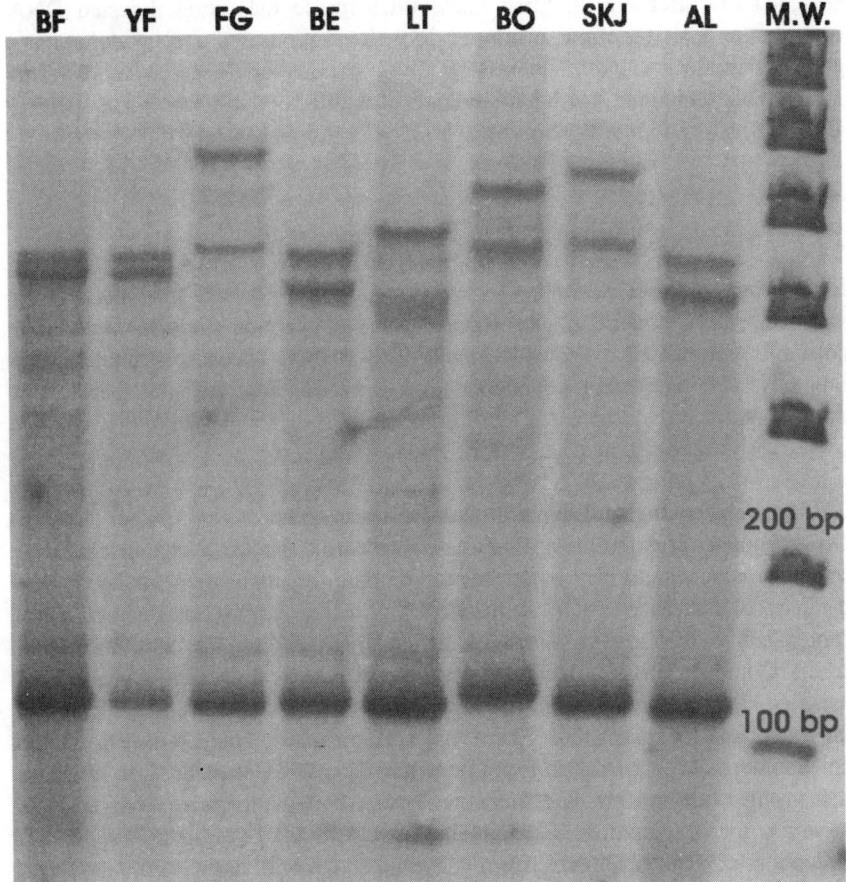

Fig. 23.1   59-3/59-5 SSCP patterns for eight tuna species. BF (bluefin tuna, *Thunnus thynnus*), YF (yellowfin tuna, *T. albacares*), FG (frigate tuna, *Auxis rochei*), BE (bigeye tuna, *T obesus*), LT (little tuna, *Euthynnus aleteratus*), BO (bonito sarda, *Sarda sarda*), SKJ (skipjack tuna, *Katsuwonus pelamis*), AL (albacore, *T. alalunga*).

mobility of single strands of certain fragments are species specific and several papers have demonstrated the usefulness of the technique for routine analysis of species (Rehbein *et al.*, 1997; Rehbein *et al.*, 1998; Rehbein *et al.*, 1999b).

23.6.3   Amplification Fragment Length Polymorphisms (AFLPs)

This is a more sophisticated version of a random amplification, it is also more reproducible, although it has the same restrictions regarding quality of DNA as RAPDs have. Genomic DNA is digested with one or two restriction enzymes, this leaves fragmented DNA with one or two types of sequence overhangs. Selective amplification is performed after ligating specific DNA adapters complementary to the overhangs and using primers designed for these extremes. Primers only amplify a subset of potential annealing sites because they are designed to differ one to three nucleotides inside the target digested DNA fragments. The reaction products are then separated on a denaturing polyacrylamide gel and visualised by silver staining or fluorescence (Bossier, 1999). This technique has been employed for differentiating sturgeon species, allowing also the identification of hybrids (Congiu *et al.*, 2001).

23.7   Methods using sequence information

Identification of species can be performed using the sequence information of a particular DNA fragment. Development of this type of study is more time consuming compared to the above-mentioned methods, because a great deal of time has to be invested in sequencing several specimens of different species, but in the long term they represent a group of methodologies that are quite reliable.

23.7.1   Sequencing and genetic distance measurement

DNA sequencing is the most direct way to identify species if the correct DNA sequence is used, and if it is possible to compare the unknown sequence with a database of reference species sequences. The technique has been named FINS (Forensically Informative Nucleotide Sequencing) (Bartlett and Davidson, 1992). Commercial fish species groups comprise several species which can be genetically closely related, and it is advisable to study as many fish species as those potentially marketed. There have been many studies aimed at the development of identification techniques based on DNA sequences, in different fish groups and species, and there have been a number of research projects aiming at the development of such techniques, particularly those involved in the collection of fish specimens from all over the world. However, there is a notorious lack of effort in collecting all sequence results and presenting them in a useful way for solving the problem of species identification.

One of the crucial parts in the development of an identification technique, based on sequence, is the selection of the target sequence. It should contain enough information to allow the differentiation of all the commercial fish species under study (i.e., gadoids); this means a high interspecific variability, however, it should have as low as possible intraspecific variability (Quinteiro *et al.*, 1998). The length of the sequenced fragment also plays an important role. Fragment size will determine the application of the method to fresh, frozen, mildly heated or thermally treated products. Canned products should be analysed using DNA fragments no longer than 200 bp (Quinteiro *et al.*, 1998).

The use of DNA sequencing also has several advantages with respect to other methods; sequencing permits the classification of an unknown sample in the group of species to which it belongs (Bartlett and Davidson, 1992; Sotelo *et al.*, 2001); and even more interestingly it also allows the detection if a new species is being put on the market or even if unusual species are being sold under a particular label (Cipriano and Palumbi, 1999). It is also possible to overcome problems associated with the existence of intraspecific variability, since the whole sequence information is used to perform the identification.

Once an unknown sequence is obtained, the common method for identifying the species is to calculate a genetic distance among the unknown sequence and a set of reference sequences. Genetic distances can then be used to build a pairwise matrix of distance where the unknown sequence will show the lowest distance with the phylogenetic group to which it belongs (Quinteiro *et al.*, 1998).

Automatic sequencing has caused a revolution in sequencing techniques because radioactive isotope labelling is no longer needed. Labelling of DNA strands is made with fluorochromes which are then separated by electrophoresis, either in a gel format or in a capillary. Automatic sequencing has speeded the whole process of sequencing and many sequencing services are commercially offered in most countries. Figure 23.2 shows a typical identification analysis using this technique.

23.7.2   Restriction Fragment Length Polymorphism of PCR fragments (PCR-RFLP)

This method uses partial sequence information contained in a particular PCR fragment. This sequence information is assayed by using restriction enzymes which recognise a very short sequence within the fragment, usually the restriction enzymes used are four, or six nucleotides cutters, meaning that they would recognise a four or six nucleotide sequence and cut at different points before, within or past the recognised sequence (Fig. 23.3). The technique is very simple, relatively cheap (depending on the restriction enzyme needed), and requires the amplification of a particular DNA fragment and its digestion with one or several restriction enzymes; the restriction fragments are separated by means of electrophoresis. The technique has been used for the identification of several fish species in processed seafood (Chow *et al.*, 1993; Borgo *et al.*, 1996; Ram *et al.*, 1996; Quinteiro *et al.*, 1998; Céspedes *et al.*, 1998; Carrera *et al.*, 1999; Cocolin *et al.*, 2000; Russell *et al.*, 2000; Hold *et al.*, 2001a; Hold *et al.*, 2001b; Sotelo *et al.*, 2001; Quinteiro *et al.*, 2001). However, as in most of the

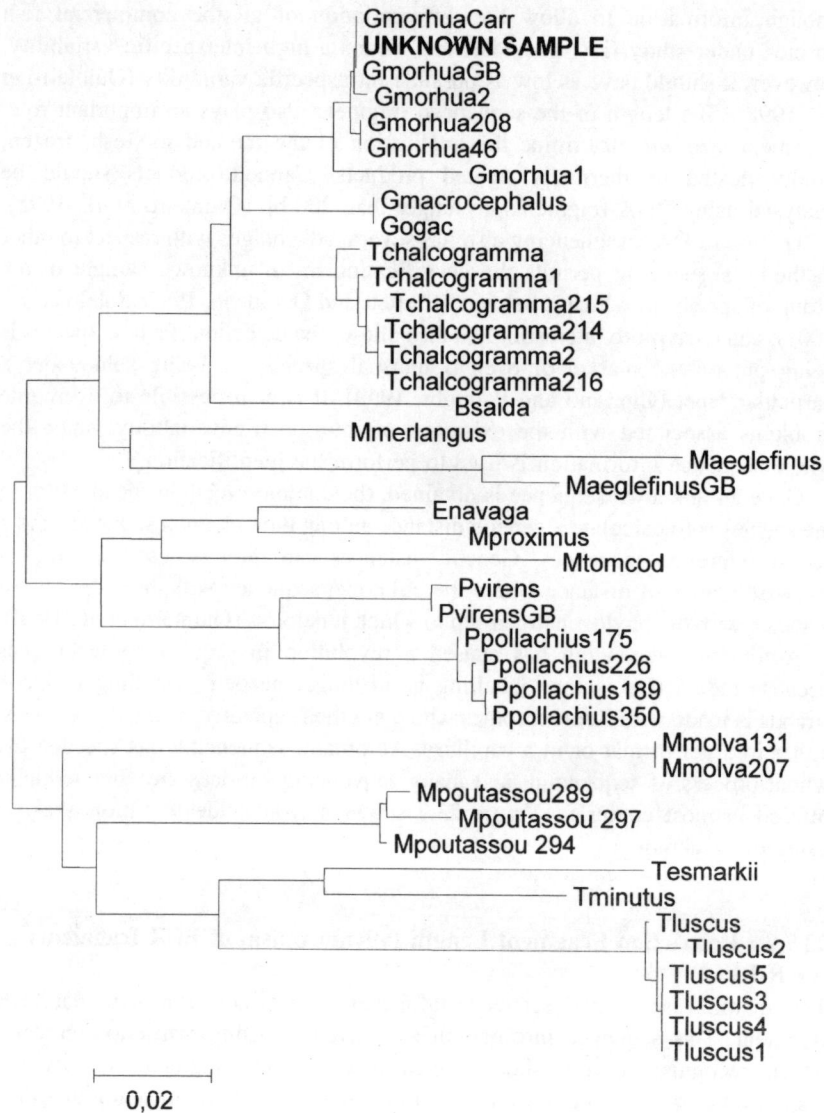

Fig. 23.2 FINS analysis of gadoid species. Unknown sample cytochrome b sequence fragment (marked red) is compared by a genetic distance measurement method with a whole set of reference sequences (same cytochrome b fragment) from gadoid species. Distances were used to produce the phylogenetic tree shown, and the unknown sequence clusters with the Gadus morhua sequences, indicating that this species is present in Unknown sample.

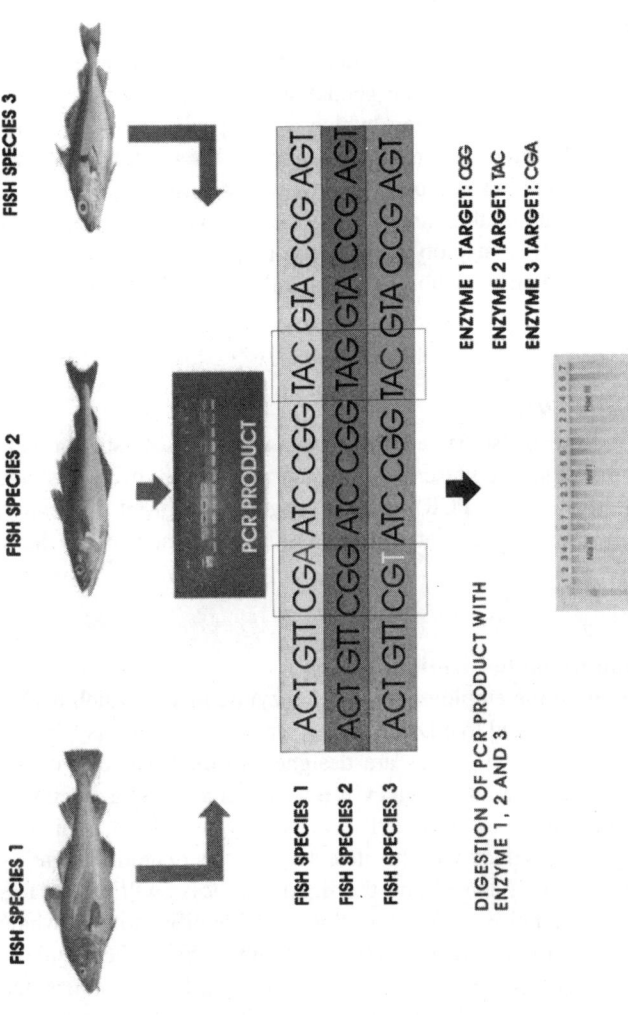

Fig. 23.3 Basis for the identification of fish species by PCR-RFLP.

techniques described here, its usefulness is questionable when a limited number of species is studied when developing the technique (Céspedes *et al.*, 2000). The main problem is that there is no total guarantee that a previously unstudied species will produce a specific RFLP pattern, this means that it could give exactly the same pattern as some of the other species studied. Ideally, the range of species sequences used for selecting the diagnostic restriction enzymes must cover the range of potential species which might be used in a particular product. Unfortunately, the common situation is the substitution of one valuable species by a closely related (same family or same genus) species of a lower price. The analyst usually has to answer the question 'which is the species included in this product?'. If the method being used is designed for differentiating only two species, as has been observed in some published works, for instance *Solea solea* and *Microchirus azevia*, one of the answers may be 'this is not a sole', if the result is that the unknown RFLP haplotype is different from sole. If the pattern is the same as sole, the honest answer should be 'this might be sole but there is no guarantee that it is a sole'.

23.7.3   Species specific PCR
Authentication of species can also be performed by means of PCR using primers targeted for a particular sequence present exclusively in a species of interest and not in others. The presence of the PCR product is regarded as proof of identity. This technique has been employed for differentiating caviar species (DeSalle and Birstein, 1996).

23.7.4   OLA (Oligonucleotide Ligation Assay)
This technique is based on the employment of an enzyme, ligase, which is able to link two DNA probes attached to a DNA strand. One of the probes is labelled with a fluorescent dye. The two probes are designed so they would anneal a DNA sequence, like a single strand from a PCR product, containing a diagnostic site or sequence. If the probes do not match exactly they would not be ligated and therefore no ligated product will be detected. If the diagnostic site or sequence is found, probes will anneal and the ligation process will take place. The technique has been employed for detection and identification of whale species in marketed food products (Cipriano and Palumbi, 1999). These authors were able to identify whale species in 113 out of 130 fresh, frozen and canned products (86.9%). The technique showed no false positives in the survey and only two false negatives out of 130 samples. Although it allows a fast, accurate and relatively cheap method of species identification, there is a need for the appropriate equipment for detecting and sizing the fluorescent ligated product and usually this involves expensive equipment.

23.7.5   DNA probes

Hybridisation of single strand DNA (ssDNA) from a particular species with ssDNA from an unknown has served as the basis for development of several identification methods. Total genomic DNA has been used as a probe using a slot/blot assay (Wintero *et al.*, 1990; Chikuni *et al.*, 1990), however this methodology did not work well with closely related species due to sequence homology producing cross-reactivity. Species specific satellite DNA oligonucleotide probes with no cross-reactivity were tested for the identification of five animal species in a wide range of commercial products (Hunt *et al.*, 1997), furthermore, the use of the probes permitted the semiquantitation of species down to a level of 2.5%. The use of DNA probes for fish species identification has not been proved yet, however, it is expected that specific probes for most important commercial species will soon appear in the literature or even on the market.

23.8   Future trends: rapid methods

The objective of developing methods for species identification is to provide industry, control laboratories, administration or even consumers with methodologies which can be easily implemented in their laboratories. Therefore the methodologies tend to be simplified as much as possible and, if the meat case path is followed, there will be kits on the market for fish species identification as there are already for meat. Some of the kits for meat were based on antibodies recognising specific meat antigens. Examples of these kits are the 'BioKits for Animal Speciation Testing' from Tepnel Biosystems Limited (UK), 'Cooked Meat Species identification kit' from Cortecs Diagnostics, and 'DTEK^{TM} Immunostick' from ELISA Technologies Inc (USA). These methods are easy to perform and no special equipment is needed. No commercial method for fish species identification is available so far, mainly because the development of such methods is more complicated because of the number of fish species which are being commercialised. Also, the level of genetic relationship among fish species to be discriminated is very high, compared to the meat case. Usually the problem is to distinguish fish species within the same genus or, at best, the same family.

The use of rapid DNA methods may facilitate the development of rapid kits for identification. Commercial kits for identification of animal species based on DNA are already available, some examples are 'DNAnimal BOS Ident' from Genescan (Scil Technology holding GmbH) and 'SureFood ID-Animal' from Congen Biotechnologie GmbH (Germany). The methodology behind these kits is based on specific DNA probe hybridisation and detection of hybridised probes with ELISA detection (Fig. 23.4). Again, no commercial kit was found to identify commercial fish species. An alternative to sequencing methods, which is starting to show analytical applications every day, is the use of high-density oligonucleotide probe arrays, commonly known as DNAchips. A DNA array, an

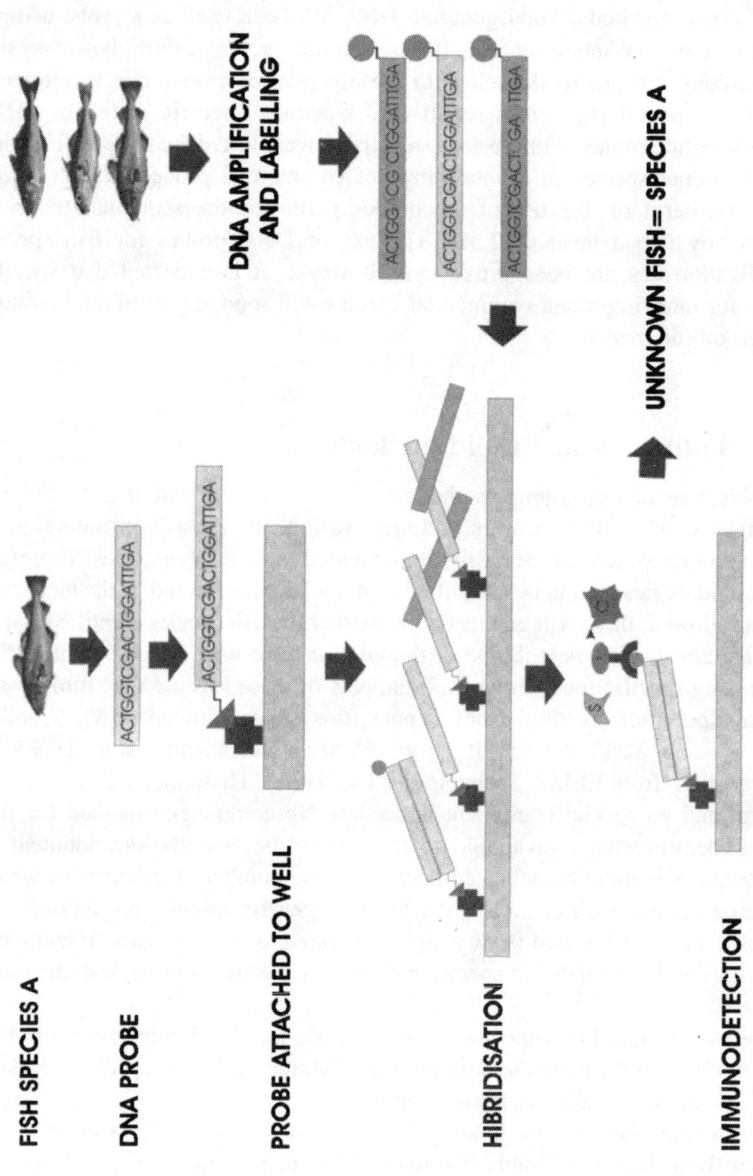

Fig. 23.4 Basis for the identification of fish species using DNA probes, coupled with an antibody detection system.

elevated number of oligonucleotide probes, can be used to determine the identity of a target sequence by detecting specific nucleotides (Chee *et al.*, 1996). For instance 16000 bp sequence can be analysed by using 66000 probes (Chee *et al.*, 1996). Probes are chemically attached to a solid substrate, the target can be labelled with a fluorescent reporter and it is incubated with the array. If the target has a part of the sequence complementary to the probes in the array, a hybridisation will take place. Hybridisation strength will depend on the level of match between probes and target DNA. A system for measuring fluorescence scans the array and detects where the hybridisation has occurred. Software processes the image obtained and correlates this with particular probes being present at particular sites (Lipshutz *et al.*, 1995). The specific pattern is recognised by the software and gives an answer about the species being present in the unknown sample.

23.9   Sources of further information and advice

23.9.1   Web sites related to the problem of species identification
Fishbase http://www.fishbase.org/search.cfm
Fishbase is one of the largest databases on the internet with biological information on fish (family, key features, importance to fisheries, photographs, morphological data, captures, etc.). The database is accessed by a search engine with two main types of entrance: common name and scientific name. It is also possible to search fish families, and a variety of diverse topics can be found for many species like chromosome number, level of ploidy, link to genebank, capture area, etc.

Regulatory Fish Encyclopaedia http://vm.cfsan.fda.gov/~frf/rfe0.html
This is one of the main references for fish identification in seafood products. The regulatory fish encyclopedia presents IEF data of many fish species, with special market interest for the US, there are also photographs showing the species, IEF patterns, common, scientific and market names, entries, etc.

http://www.fisheries.vims.edu/fishgenetics/
This web site on molecular identification of fish species presents some literature and examples, with protocols and figures, of fish species identification using DNA-based methods.

http://www.amonline.net.au/fishes/
This web site presents an online key for morphological fish species identification.

23.10   References

AN H., KLEIN P.A., KAO K.-J., MARSHALL M.R., OTWELL W.S., WEI C. 1990. Development of monoclonal antibody for rock shrimp identification using enzyme-linked immunosorbent assay. *J. Agric. Food Chem.* 38: 2094–100.

ANDREWS C.D., BERGER R.G., MAGEAU R.P., SCHWAB B., JOHNSTON R.W. 1992. Detection of beef, sheep, deer, and horse in cooked meat products by enzyme-linked immunosorbent assay. *J. AOAC Inst.* 75(3): 572–6.

AOAC 1990. Official methods of analysis. 15th edn., p. 883–9. Association of Official Analytical Chemists. Washington, DC.

ARMSTRONG S.G., LEACH D.N., WYLLIE S.G. 1992. The use of HPLC protein profiles in fish species identification. *Food Chem.* 44: 147–55.

ASHOOR S.H., KNOX M.J. 1985. Identification of fish species by high-performance liquid chromatography. *J. Chromat.* 324: 199–202.

AYALA F.J. 1983. Enzymes as taxonomic characters. Systematics Association Special Volume No. 24, *Protein polymorphism: Adaptive and Taxonomic significance*, edited by G.S. Oxford and D. Rollinson, 1983, Academic Press, London and New York.

BAKER C.S., PALUMBI S.R. 1994. Which whales are hunted? A molecular genetic approach to monitoring whaling. *Science* 265: 1538–9.

BARTLETT S.E., DAVIDSON W.S. 1992. FINS (Forensically Informative Nucleotide Sequencing): a procedure for identifying the animal origin of biological specimens. *Biotechniques* 12(3): 408–11.

BEJ A.K., MAHBUBANI M.H., ATLAS R.M. 1991. Amplification of nucleic acids by polymerase chain reaction (PCR) and other methods and their applications. *Crit. Rev. Biochem. Mol. Biol.* 26(3/4): 301–34.

BORGO R., SOUTY-GROSSET C., BOUCHON D., GOMOT L. 1996. PCR-RFLP analysis of mitochondrial DNA for identification of snail meat species. *J. Food Sci.* 61(1): 1–4.

BOSSIER P. 1999. Authentication of seafood products by DNA patterns. *J. Food Sci.* 64(2): 189–93.

CARRERA E., GARCÍA T., CÉSPEDES A., GONZÁLEZ I., SANZ B., HERNÁNDEZ P.E., MARTÍN R. 1997. Immunostick colorimetric ELISA assay for the identification of smoked salmon, trout and bream. *J. Sci. Food Agric.* 74: 547–50.

CARRERA E., MARTÍN R., GARCÍA T., GONZÁLEZ I., SANZ B., HERNÁNDEZ P.E. 1996. Development of an enzyme-linked immunosorbent assay for the identification of smoked salmon (*Salmo salar*), trout (*Oncorhynchus mykiss*) and bream (*Brama raii*). *J. Food Protect.* 59: 521–4.

CARRERA E., GARCÍA T., CÉSPEDES A., GONZÁLEZ I., FERNÁNDEZ A., HERNÁNDEZ P.E., MARTÍN R. 1999. Salmon and Trout analysis by PCR-RFLP for identity authentication. *J. Food Sci.* 64(3): 410–13.

CERDA H., KOPPEN G. 1998. DNA degradation in chilled fresh chicken studied with the neutral comet assay. *Z. Lebensm. Unters. Forsch.* A 207:22–25.

CÉSPEDES A., GARCÍA T., CARRERA E., GONZÁLEZ I., SANZ B., HERNÁNDEZ P.E., MARTÉN R. 1998. Identification of flatfish species using polymerase chain reaction (PCR) amplification and restriction analysis of the cytochrome b gene. *J. Food Sci.* 63: 206–9.

CÉSPEDES A., GARCÍA T., CARRERA E., GONZÁLEZ I., FERNÁNDEZ A., ASENSIO L., HERNÁNDEZ P.E., MARTÍN R. 2000. Genetic differentiation between sole (*Solea solea*) and Greenland halibut (*Reinhardtius hippoglossoides*) by PCR-RFLP analysis of a 12S rRNA gene fragment. *J. Sci. Food Agric.* 80: 29–32.

CHEE M., YANG R., HUBBELL E., BERNO A., HUANG X.C., STERN D., WINKLER J., LOCKHART D.J., MORRIS M.S., FODOR S.P.A. 1996. Accessing genetic information with high-density DNA arrays. *Science* 274:610–4.

CHEN F.-CH., PEGGY-HSIEH Y.-H., BRIDGMAN R.C. 1998. Monoclonal antibodies to porcine thermal stable muscle protein for detection of pork in raw and cooked meats. *J. Food Sci.* 63(2): 201–5.

CHIKUNI K., OZUTSUMI K., KOISHIKAWA T., KATO S. 1990. Species identification of cooked meats by DNA hybridization assay. *Meat Sci.* 27: 119–28.

CHOW S., CLARKE M.E., WALSH P.J. 1993. PCR-RFLP analysis of thirteen western Atlantic snappers (subfamily Lutjaninae): A simple method for species and stock identification. *Fish Bull.* 91: 619–27.

CIPRIANO F., PALUMBI S. R. 1999. Rapid genotyping techniques for identification of species and stock identity in fresh, frozen, cooked and canned whale products. Report to the Scientific Committee, International Whaling Commission SC/51/09.

COCOLIN L., D'AGARO E., MANZANO M., LANARI D., COMI G. 2000. Rapid PCR-RFLP method for the identification of marine fish fillets (Seabass, Seabream, Umbrine, Dentex). *J. Food Sci.* 65(8): 1315–7.

CONGIU L., DUPANLOUP I., PATARNELLO T., FONTANA F., ROSSI R., ARLATI G., ZANE L. 2001. Identification of interspecific hybrids by AFLP: the case of sturgeons. *Mol. Ecol.* 10(9): 2355–9.

COWIE W. 1968. Identification of fish species by thin slab polyacrylamide gel electrophoresis. *J. Sci. Food Agric.* 19: 226–9.

CROSSLAND S., COATES D., GRAHAME J., MILL P.J. 1993. Use of random amplified polymorphic DNAs (RAPDs) in separating two sibling species of Littorina. *Mar. Ecol. Progress. Ser.* 96: 301–5.

DENNIS M.J., ASHURST P.R. 1996. An introduction to food authentication. In *Food Authentication*, Ashurst P.R., Dennis M.J. (eds), Blackie Academic and Professional, London (UK).

DESALLE R., BIRSTEIN V.J. 1996. PCR identification of black caviar. *Nature* 381: 197–8.

DINESH K.R., CHAN W.K., LIM T.M., PHANG V.P.E. 1995. RAPD markers in fish: an evaluation of resolution and reproducibility. *As. Pac. J. Mol. Biol. Biotechnol.* 3: 112–18.

DINESH K.R., LIM T.M., CHUA K.L., CHAN W.K., PHANG V.P.E. 1993. RAPD analysis: an efficient method of DNA fingerprinting of fishes. *Zool. Sci.* 10: 849–54.

DOMINGUEZ E., PÉREZ D., PUYOL P., CALVŎ M. 1997. Use of immunological techniques for detecting species substitution in raw and smoked fish. *Z. Lebensm. Unters. Forsch.* 204: 279-281.

DOTSON R.C. 1976. In *Physiological ecology of tunas.* Sharp D., Dizon A.E. (eds), Academic Press, New York (USA).

EBBEHØJ, K.F., THOMSON P.D. 1991. Species differentiation of treated meat products by DNA hybridization. *Meat Sci.* 30: 221–34.

FAO. 2000. *The state of world fisheries and aquaculture.*

GALLARDO J.M., SOTELO C.G., PIÑEIRO C., PÉREZ-MARTÍN R.I. 1995. Use of capillary zone electrophoresis for fish species identification. Differentiation of flatfish species. *J. Agric. Food Chem.* 43: 1238–44.

GOSLING E.M. 1994. Speciation and wide-scale genetic differentiation. In *Genetics and evolution of Aquatic Organisms.* Beaumont A.R. (ed). Chapman & Hall. London (UK).

GROSBERG R.K., LEVITAN D.R., CAMERON B.B. 1996. Characterization of genetic structure and genealogies using RAPD-PCR markers: a random primer for the novice and nervous. In *Molecular Zoology. Advances, strategies and protocols'.* Ferraris J.D., Palumbi S.R. (eds), Wiley-Liss, Inc.

HAYASHI H. 1991. PCR-SSCP: a simple and sensitive method for detection of mutations in the genomic DNA. *PCR Meth. Applic.* 1: 34–8.

HOLD G.L., RUSSELL V.J., PRYDE S.E., REHBEIN H., QUINTEIRO J., REY-MÉNDEZ M., SOTELO C.G., PÉREZ-MARTÍN R.I., SANTOS A.T., ROSA C. 2001a. Validation of a PCR-RFLP based method for the identification of salmon species in food products. *Eur. J. Food Res. Tech.* 212: 385–9.

HOLD G.L., RUSSELL V.J., PRYDE S.E., REHBEIN H., QUINTEIRO J., VIDAL R., REY-MENDEZ M., SOTELO C.G., PÉREZ-MARTÍN R.I., SANTOS A.T., ROSA C. 2000b. The development of a DNA based method aimed at identifying the fish species present in food products. *J. Agric. Food Chem.* 49(3): 1175–9.

HSIEH Y.H.-P., JOHNSON M.A., WETZSTEIN C.J., GREEN N.R. 1996. Detection of species adulteration in pork products using agar-gel immunodiffusion and enzyme-linked immunosorbent assay. *J. Food Qual.* 19: 1–13.

HUANG T., MARSHALL M.R., KAO K., OTWELL W.S., WEI C. 1995. Development of monoclonal antibodies for red snapper (*Lutjanus campechanus*) identification using enzyme-linked immunosorbent assay. *J. Agric. Food Chem.* 43: 2301–7.

HUNT D.J., PARKES H.C., LUMLEY I.D. 1997. Identification of the species of origin of raw and cooked meat products using oligonucleotide probes. *Food Chem.* 60(3): 437–42.

KANG'ETHE E., LINDQVIST K.J. 1987. Thermostable muscle antigens suitable for use in enzyme immunoassays for identification of meat from various species. *J. Sci. Food Agric.* 39: 179–84.

LAW J.C., FACHER E.A., DEKA A. 1996. Nonradioactive single-strand conformation polymorphism analysis with application for mutation detection in a mixed population of cells. *Anal. Biochem.* 236: 373–5.

LEBLANC E.L., SINGH S., LEBLANC R.J. 1994. Capillary zone electrophoresis of fish muscle sarcoplasmic proteins. *J. Food Sci.* 59(6): 1267–70.

LIPSHUTZ R.J., MORRIS D., CHEE M., HUBBELL E., KOZAL M.J., SHAH N., SHEN N., YANG R., FODOR S.P.A. 1995. Using oligonucleotide probe arrays to access genetic diversity. *Biotechniques* 19(3): 442–7.

MACKIE I.M. 1969. Identification of fish species by a modified polyacrylamide disc electrophoresis technique. *J. Assoc. Publ. Anal.* 83–7.

MACKIE I.M. 1980. A review of some recent applications of electrophoresis and iso-electricfocusing in the identification of species of fish in fish and fish products. In: *Advances in fish science and technology*. Connell J.J. (ed) Fishing New Books, Farnham (UK).

MACKIE I.M. 1996. Authenticity of fish. In *Food Authentication*. Ashurst P.R., Dennis M.J. (eds), Blackie Academic & Professional, London (UK).

MACKIE I.M., CHALMERS M., REECE P., SCOBBIE A.E., RITCHIE A.H. 1992. The application of electrophoretic techniques to the identification of species of canned tuna and bonito. In *Pelagic fish. The resource and its exploitation*. Burt J.R., Hardy R., Whittle K.J. (eds). Fishing News Books, Oxford (UK).

MACKIE I.M., CRAIG A., ETIENNE M., JEROME M., FLEURENCE J., JESSEN F., SMELT A., KRUIJT A., MALMHEDEN YAMN I., FERM M., MARTÍNEZ I., PÉREZ-MARTÍN R.I., PIÑEIRO C., REHBEIN H., KÜNDIGER R. 2000. Species identification of smoked and gravad fish products by sodium dodecylsulphate polyacrylamide gel electrophoresis, urea isoelectric focusing and native isoelectric focusing: a collaborative study. *Food Chem.* 71: 1–7.

MARTÍNEZ I. 1997. DNA typing of fish products for species identification. In *Seafood from producer to consumer, integrated approach to quality*. Luten J.B., Børrensen T. and Oehlenshläger J. (eds), pp. 497–506. Developments in Food Science 38, Elsevier, Amsterdam.

MARTÍNEZ I., DANÍELSDÓTTIR A.K. 2000. Identification of marine mammal species in food products. *J. Sci. Food Agric.* 80: 527–33.

MARTÍNEZ I., MALMHEDEN YMAN I. 1998. Species identification in meat products by RAPD analysis. *Food Res. Int.* 31(6-7): 459–66.

MARTÍNEZ, I., JAKOBSEN T., CARECHE, M. 1997 Post-mortem degradation in the Northern shrimp *Pandalus borealis*. 27th WEFTA Meeting. October 19–22, 1997. Madrid, Spain.

MAYR E. 1970. *Populations, species and evolution*. Harvard University Press, Cambridge, Ma (USA).

MCELHINNEY J., DOWNEY G., O'DONNELL. 1999. Quantitation of lamb content in mixtures with raw minced beef using visible, near and mid-infrared spectroscopy. *J. Food Sci.* 64(4): 587–91.

MEDINA I., AUBOURG S.P., PÉREZ-MARTÍN R.I. 1997. Species differentiation by multivariate analysis of phospholipids from canned Atlantic tuna. *J. Agric. Food Chem.* 45: 2495–9.

MULLIS K.B. 1990. The unusual origin of the polymerase chain reaction. *Scient. Am.* 262: 56–65.

MULLIS K.B., FALOONA F.A. 1987. Specific synthesis of DNA *in vitro* via a polymerase catalysed chain reaction. *Meth. Enzymol.* 155: 335–50.

PARK L.K., MORAN P. 1994. Developments in molecular genetic techniques in fisheries. *Rev. Fish Biol. Fish.* 4: 272–99.

PIÑEIRO C., BARROS-VELÁZQUEZ J., PÉREZ-MARTÍN R.I., MARTÍNEZ I., JACOBSEN T., REHBEIN H., KÜNDIGER R., MENDES R., ETIENNE M., JEROME M., CRAIG A., MACKIE I.M., JESSEN F. 1999. Development of a sodium dodecyl sulfate polyacrylamide gel electrophoresis reference method for the analysis and identification of fish species in raw and heat-processed samples: a collaborative study. *Electrophoresis* 20: 1425–32.

PINEIRO C., BARROS-VELÁZQUEZ J., SOTELO C.G., GALLARDO J.M. 1999. The use of two-dimensional electrophoresis for the identification of commercial flat fish species. *Z. Lebens. Unters. Forsch.* A 208: 342–8.

PIÑEIRO C., SOTELO C.G., MEDINA I., GALLARDO J.M., PÉREZ-MARTÍN R.I. 1997. Reversed-phase HPLC as a method for the identification of gadoid fish species. *Z. Lebensm. Unters. Forsch.* A 204: 411–16.

PIÑEIRO C., VÁZQUEZ J., MARINA A.I., BARROS-VELÁZQUEZ B., GALLARDO J.M. 2001. Characterization and partial sequencing of species-specific sarcoplasmic polypeptides from commercial hake species by mass spectrometry following two-dimensional electrophoresis. *Electrophoresis* 22: 1545-52.

PIÑEIRO C., VELÁZQUEZ J.B., SOTELO C.G., PÉREZ-MARTÍN R.I. AND GALLARDO J.M. 1998. Two-Dimensional electrophoretic study of the water-soluble protein fraction in white muscle of Gadoid fish species. *J. Agr. Food Chem.* 46: 3991–7.

QUINTEIRO J., SOTELO C.G., REHBEIN H., PRYDE S.E., MEDINA I., PÉREZ-MARTÍN R.I., REY-MÉNDEZ M., MACKIE I.M. 1998. Use of mtDNA direct polymerase chain reaction (PCR) sequencing and PCR-restriction fragment length polymorphism methodologies in species identification of canned tuna. *J. Agric. Food Chem.* 46: 1662–9.

QUINTEIRO J., VIDAL R., IZQUIERDO M., SOTELO C.G., CHAPELA M.J., PEREZ-MARTÍN R.I., REHBEIN H., HOLD G., RUSSELL V.J., PRYDE S.E., ROSA C., SANTOS A.T., REY-MÉNDEZ M. 2001. Identification of hake species (*Merluccius genus*) using sequencing and PCR-RFLP analysis of mitochondrial DNA control region sequences. J. Agr. *Food Chem.* 49: 5108–14.

RAM J.L., RAM M.L., BAIDOUN F.F. 1996. Authentication of canned tuna and bonito by sequence and restriction site analysis of polymerase chain reaction products of mitochondrial DNA. *J. Agric. Food Chem.* 44: 2460–7.

REHBEIN H. 1990. Electrophoretic techniques for species identification of fishery products. *Z. Lebensm. Unters. Forsch.* 191: 1–10.

REHBEIN H., ETIENNE M., JEROME M., HATTULA T., KNUDSEN L.B., JESSEN F., LUTEN J., BOUQUET W., MACKIE I.M., RICHIE A.H., MARTIN R., MENDES R. 1995. Influence of variation in technology on the reliability of the isoelectric focusing method of fish species identification. *Food Chem.* 52: 193–7.

REHBEIN H., KÜNDIGER R. 1984. Comparison of the isoelectric focusing patterns of the sarcoplasmic proteins from red and white muscle of various fish species. *Arch. Fisch. Wiss.* 35: 7–16.

REHBEIN H., KÜNDIGER R., MALMHEDEN YMAN I., FERM M., ETIENNE M., JEROME M., CRAIG A., MACKIE I.M., JESSEN F., MARTÍNEZ I., MENDES R., SMELT A., LUTEN J., PIÑEIRO C., PÉREZ-MARTÍN R.I. 1999a. Species identification of cooked fish by urea isoelectric focusing and sodium dodecylsulfate polyacrylamide gel electrophoresis: a collaborative study. *Food Chem.* 67: 333–9.

REHBEIN H., MACKIE I.M., PRYDE S., GONZÁLEZ-SOTELO C., MEDINA I., PÉREZ-MARTÍN R.I., QUINTEIRO J., REY-MÉNDEZ M. 1999b. Fish species identification in canned tuna by PCR-SSCP: validation by a collaborative study and investigation of intra-species variability of the DNA patterns. *Food Chem.* 64: 263–8.

REHBEIN H., MACKIE I.M., PRYDE S., GONZÁLEZ-SOTELO C., PÉREZ-MARTÍN R.I., QUINTEIRO J., REY-MÉNDEZ M. 1998. Comparison of different methods to produce single-strand DNA for identification of canned tuna by single strand conformation polymorphism analysis. *Electrophoresis* 19: 1381–4.

RUSSELL V.J., HOLD G.L., PRYDE S.E., REHBEIN H., QUINTEIRO J., REY-MÉNDEZ M., SOTELO C.G., PÉREZ-MARTÍN R.I., SANTOS A.T., ROSA C. 2000. Use of restriction fragment length polymorphism to distinguish between Salmon species. *J. Agric. Food Chem.* 48: 2184–8.

SAIKI R.K., GELFAND D.H., STOFFEL S., SCHARF S.J., HIGUCHI R., HORN G.T., MULLIS K.B., ERLICH H.A. 1988. Primer-directed enzymatic amplification of DNA with a thermostable DNA polymerase. *Science* 239: 487–90.

SCOBBIE A.E., MACKIE I.M. 1988. The use of Sodium Dodecyl Sulphate polyacrylamide gel electrophoresis in fish species identification – a procedure suitable for cooked and raw fish. *J. Sci. Food Agric.* 44: 343–351.

SOTELO C.G., PIÑEIRO, C. GALLARDO J.M. AND PÉREZ-MARTÍN R.I. 1992. Identification of fish species in smoked fish products by electrophoresis and isoelectric focusing. *Z. Lebens. Unters. Forsch.* 195: 224–7.

SOTELO C.G., CALO-MATA P., CHAPELA M.J., PéREZ-MARTíN R.I., REHBEIN H., HOLD G.L., RUSSELL V., PRYDE S., QUINTEIRO J., IZQUIERDO M., REY-MéNDEZ M., ROSA C. SANTOS A.T. 2001. Identification of flatfish species using DNA based techniques. *J. Agric. Food Chem.* 49: 4562–9.

TAYLOR W.J., PATEL N.P., JONES J.L. 1994. Antibody-based methods for assesing seafood authenticity. *Food Agric. Immunol.* 6: 305–14.

UNSELD M., BEYERMANN B., BRANDT P., HIESEL R. 1995. Identification of the species of origin of highly processed meat products by mitochondrial DNA sequences. *PCR Meth. Applic.* 4: 241–3.

VERREZ-BAGNIS V. 1993. The performance of ELISA and Dot-blot methods for the detection of crab flesh in heated and sterilized surimi-based products. *J. Sci. Food Agric.* 63: 445–9.

WHITAKER R.G., SPENCER T.L., COPLAND J.W. 1983. An enzyme-linked immunosorbent assay for species identification of raw meat. *J. Sci. Food Agric.* 34: 1143–8.

WINTERO A.K., THOMSEN P.D., DAVIES W. 1990. A comparison of DNA-hybridization, immunodiffusion, countercurrent immunoelectrophoresis and isoelectricfocusing for detecting the admixture of pork to beef. *Meat Sci.* 27: 75–85.

WOLF C., BURGENER M., HÜBNER P., LÜTHY J. 2000. PCR-RFLP analysis of mitochondrial DNA: differentiation of fish species. *Lebensm.-Wiss. u.-Technol.* 33: 144–50.

WOLF C., HÜBNER P., LÜTHY J. 1999. Differentiation of sturgeon species by PCR-RFLP. *Food Res. Int.* 32: 699–705.

24

Multivariate spectrometric methods for determining quality attributes

B. M. Jørgensen, Danish Institute for Fisheries Research, Lyngby

24.1 Introduction to multivariate spectroscopic methods

Spectroscopy is a methodology which possesses several advantages for food analysis, e.g. determination of composition and other quality-related properties. The spectroscopic measurements are often fast compared with alternatives like wet chemical methods, thereby providing the possibility of taking many samples and of sampling at short time intervals. Furthermore, several of the spectroscopic techniques can be operated directly on the specimen in question, i.e. without sample preparation at all, or with only a few steps like mincing. Therefore, the methods have potential use for *at-line* or even *on-line* measurements in food production lines.

A large part of the electromagnetic spectrum has been subjected to spectroscopy from the low-energy, high-wavelength, microwaves though far-infrared, mid-infrared, near-infrared, visible, ultraviolet, and to the high-energy X-ray and (γ-ray. Other types of waves, e.g. ultrasound, have also successfully been applied. As they interact with different molecules and molecular bonds to a great extent, they cover a large number of different applications or they may supplement each other, e.g. in characterising a set of food samples.

Common to spectroscopic methods is that they can provide a multivariate result, i.e. a set of values, a vector, is obtained instead of just one figure for each sample. This property is extremely advantageous, because it allows detection of and often correction for interference. The price to pay is that some more or less advanced mathematical procedure is required for obtaining the final result, which for example can be a quality attribute. Spectroscopy thus becomes spectrometry. In the following, three of the spectrometric methods proven very useful for assessment of food quality-related properties are discussed. The types of information provided and the mathematics used for its extraction from the measurement data are different, and together they present a comprehensive picture of the possibilities in spectrometry applied to food.

24.2   Near-infrared (NIR) spectroscopy

The near-infrared (NIR) wavelength region from 780 to 2500 nm has proven extremely useful for food research. Spectroscopic determination of the macro-constituents water, protein, fat and carbohydrate has been published for almost all types of food, including fish and fishery products. Where absorption of light in the visible and ultraviolet region is mostly due to electronic transitions, absorption of infrared light causes transition between vibrational states. The so-called group frequencies in the mid-infrared correspond mostly to transitions between the ground vibrational state and the nearest neighbour states (ground tones) involving one bond at a time, whereas the near-infrared region contains transitions involving several bonds (combination tones) or transitions to higher-lying states (overtones). These transitions are only allowed for anharmonic oscillations and their intensities are much smaller than the ground tones. In practice, only bonds involving the light hydrogen atom contribute substantially to transitions induced by near-infrared light. Such bonds include those abundant in food: C-H (fat and protein), O-H (water and carbohydrate) and N-H (protein).

Two other factors add to the versatility of NIR spectroscopy. The sensitivity is relatively low due to the small transition intensities, and the precise energy difference between the vibrational states, i.e. the wavelength where light is absorbed, is very sensitive to the surroundings of the chemical bonds in question. The low sensitivity is advantageous because measurements can often be made directly on the surface of (or through) biological material without the necessity of a time-consuming sample preparation. This fact opens the possibility of on-line or at-line, non-destructive measurements. A result of the susceptibility of the transition energies to perturbation is that NIR spectra contain information beyond that of the concentration of the substance absorbing light. A compound like NaCl, which does not absorb as it has no covalent bond, may in some circumstances be quantified due to its interaction with water causing a shift in the O-H combination tones and overtones. And perhaps more important, interactions between absorbing bonds in the macro-constituents themselves and between these bonds and solutes give rise to spectral changes that contain quality-related information, e.g. about structure/texture.

The price to pay is that NIR spectra of rather complicated systems like foods contain a large number of absorption bands that overlap to an extent that the single bands are not distinguishable. This is illustrated in Fig. 24.1 showing an example of an NIR spectrum of thawed whole cod. The two most dominating bands can be recognised as stemming from water (the first overtone around 1450 nm and a combination tone around 1910 nm) but each broad band is the result of several overlapping bands. Obviously there are other spectral

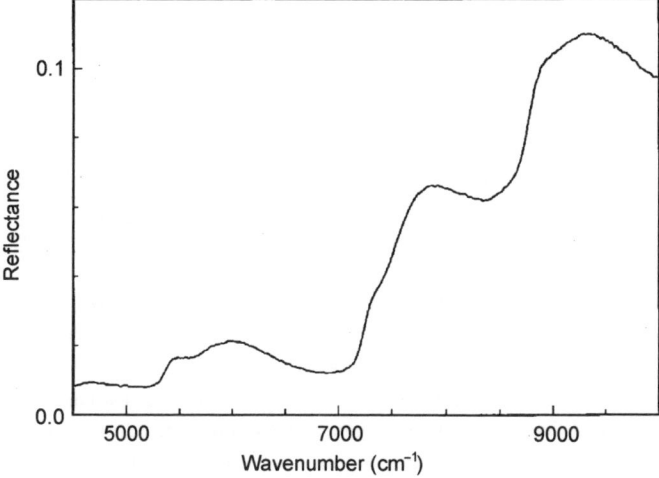

Fig. 24.1 Near-infrared (NIR) diffuse reflectance spectrum of intact cod muscle (fillet). The spectrum was taken in the interval from 4500 to 10000 cm^{-1} (2222 to 1000 nm) with a 12 cm^{-1} data point spacing. The instrument was a Bran&Luebbe InfraProver II Fourier transform spectrometer. More details can be found in Ref. 1 and 43.

structures, probably caused by N-H and C-H bonds in the muscle protein, but not easily interpretable. Measurement at a single wavelength thus is not likely to be of any use neither for quantitative determination of macro-constituents nor for assessment of structure-related quality attributes. This is why NIR spectroscopy when applied to food analysis from 'the early days' has relied on multivariate data analysis. Measurements in the whole wavelength range are used and a 'calibration model' is made based on the functional relationship:

$$y = f(\bar{x}) \hspace{4cm} 24.1$$

where y is the entity sought for (e.g. a concentration, a structural property, or a class assignment) and $\bar{x}$ is a vector of measurements at the different wavelengths, i.e. the spectrum recorded.

Developing the calibration model involves finding the parameters that give the best fit of this equation in a least squares sense. In order to do so, one examines how the difference in x-values between members of a set of samples, the calibration set, affects the difference in y-values of the same samples. That is, when making a calibration one has to determine the y-values by another method, for example a 'classical' chemical method. Obviously, as can also be deduced from the susceptibility of NIR spectra to perturbation as discussed above, a large part of the variation in the spectral values (x-values) may not have a counterpart in y. A moment of thought will convince the reader that if we used univariate measurements, i.e. measurement at a single wavelength only, for determination of y, such variations would introduce an error. But because of the large number of measuring points at different wavelengths, the 'irrelevant' variation may be corrected for by the calibration model, if the variation in question is included in the samples of the calibration set. Sometimes the calibration model takes sufficient care of the correction itself, sometimes it is advantageous, or even necessary, to make a so-called 'pre-processing' of the x-variables. This is applicable when a transformation of the x-variables results in a new set of variables whose variation between samples better correlates to the variation in y:

$$\bar{z} = g(\bar{x}); \quad y = f(\bar{z}) \tag{24.2}$$

where the z-vector contains the measured values after pre-processing with the function g.

Before discussing this point further, it is appropriate to summarise the different measurement modes used in NIR spectroscopy. One is the transmission measurement, well-known to everybody that has ever used a UV/VIS spectrometer. Light is led through the sample and the transmittance, T, is defined as the ratio between the intensity of transmitted light and the intensity of incident light. For dilute transparent solutions, the familiar Lambert-Beer law applies, so that the absorbance, A, which is the base-10 logarithm to $1/T$, is proportional to the concentration of the absorbing substance(s). This is also valid in the near-infrared wavelength region, where the sample may behave as if transparent although it is not so with regard to visible light. (The low sensitivity of NIR spectroscopy is due to the small absorption coefficients.) Thus a usable calibration 24.1 or 24.2 can often be based on a *linear* function f of the absorbance.

Another measurement mode which is popular in NIR spectroscopy is that of diffuse reflectance, R. This quantity is defined as the ratio between intensities of the diffuse reflected light from a sample and the light reflected by a reference, e.g. a white ceramic surface. Only the light interacting with the sample molecules in a way that spectral changes take place, i.e. not mirrored light, ought to be measured. With this technique, even thick samples can be measured, but it should be kept in mind that they are only probed to a certain depth.[1] In the majority of applications where reflectance measurements provide useful spectra, this technique is well suited for at-line determinations.

The theory behind diffuse reflectance is far more complicated and not as well explored as is the case for transmittance. Under the conditions of infinite sample thickness, i.e. larger than the penetration depth, the Kubelka-Munck equation[2–5] has some validity. It has also been found that $\log(1/R)$, an analogue to absorbance and often named absorbance among NIR spectroscopists, either directly or after an appropriate pre-processing step (Eqn 24.2) to a good approximation is linearly dependent of the concentration of the absorbing compound.[5] But in other cases – probably when structural effects are predominant – the use of R directly provides the best calibration result.

The diffuse reflectance is also dependent on the size of the absorbing particles in the sample. Heterogeneous samples like fish muscle thus give rise to complicated spectra where both the structural heterogeneity, the concentration of absorbing compounds and their interactions contribute. Sometimes the spectra have individual baselines or the measurements appear as having been multiplied with a wavelength-dependent factor. In the first case, one may choose to pre-process the spectra by differentiation. An offset is removed by taking first derivatives, and a linear baseline is absent from the second derivative spectrum. But this is gained by the expense of increased noise, although the use of an algorithm that combines differentiation with smoothing partly avoids this increase. Also, differentiation enhances narrow bands compared with wide ones,[6] causing an intrinsic re-weighting of the data.

The pre-processing technique, originally called 'multiplicative scatter correction'[7-8] from its versatility to spectra of powder where the particle size influences the reflectance, can deal with the second type of distortion. Due to its general applicability it is also commonly referred to as 'multiplicative signal correction'. With this technique, each spectrum is linearly regressed on the average spectrum of the *calibration* samples, and the residuals used as data:

$$g: \quad \bar{z}_i = \bar{x}_i - (a_i \bar{x}_{ave} + b_i) \qquad\qquad 24.3$$

where a_i and b_i are the parameters of the regression for sample i. An underlying assumption when using the average of the calibration spectra is that this has the most representative general shape of the type of samples in question. Thus it is important that the set of calibration samples includes samples with all types of interactions found in the 'unknown' samples to be analysed in the future.

Other pre-processing functions may of course be used and often there is no theoretical guidance as to which one to choose. It has to be based on experience (trial-and-error). The aim is to remove as much as possible of the irrelevant between sample variation in the spectral data, i.e. variation not correlated to the quantity, y, that is to be determined. A new class of pre-processing methods are designed to accomplish this directly by removing components orthogonal to y from the set of spectral data.[9-12] These methods look very promising but the various algorithms presently available still have some shortcomings.[12] Within a short time, however, this type of pre-processing is likely to be of great use to NIR spectroscopy.

As already discussed, the use of NIR spectroscopy in food research is dependent on a calibration. The calibration model can be based on linear methods like partial least squares (PLS) regression, or by non-linear pattern recognition methods like the so-called neural networks. Partial least squares regression takes advantage of the rather high degree of covariance between neighbour measuring points (wavelengths). This property results in the existence of latent factors, i.e. underlying spectral structures that are common to all samples, and whose contributions to the total spectrum are correlated to the quantity that is to be determined from the spectral measurements. If only one such factor exists, its shape should be recognisable as a spectrum of the component(s) involved. However, several factors are often necessary in order to cope with interferences. Non-linearity between the spectral signal and the variable y can also to some extent be modelled by extra factors.

The linear model, f, can be expressed as:

$$y_i = b_0 + b_1 x_{i1} + b_2 x_{i2} + \ldots + b_n x_{in} \qquad\qquad 24.4$$

where $(x_{i1}, x_{i2}, \ldots x_{in})$ is the vector of spectral values (measurements at different wavelengths) for sample i and $(b_1, b_2, \ldots b_n)$ is the vector of coefficients common to all samples. When a pre-processing has taken place, z is substituted for x in Eqn 24.4. The b-coefficients are determined from the PLS regression where only factors correlating well with y are included giving rise to some noise reduction and making the calibration more robust. In many cases, only some parts of the NIR spectrum contain useful information and excluding other parts may improve the calibration. The parts to keep can be identified by calculating which of the b-coefficients in Eqn 24.4 are significantly different from zero. For this task a jack-knifing procedure has been published.[13–14] The b-coefficients are presumed to be t-distributed and their variances are estimated from the variation in sub-model b-coefficients determined during cross-validation of the full PLS regression model.

If the function f in Eqn 24.1 is far from linear and no pre-processing can be found to overcome this, one may try a prediction model based on a so called neural network (NN). This technique has a lot of variants and is highly flexible, but almost always functions as a 'black box' when the model is validated, so no insight in the spectral components behind the model is provided. The number of samples and the time necessary for an NN calibration increase rapidly as a function of the number of variables; thus a reduction in this number may be advantageous. An interesting approach that has been tried with some success in meat research is to use PLS regression as a pre-processing step.[15] The first few factors are extracted from the data set and their contribution to the data, the score values, are used as input to the NN algorithm. This procedure results in a huge reduction in the number of variables without losing much relevant information as the factors well correlated to y are preserved.

Applications of NIR spectroscopy within fish research are few compared with those in other branches of food research. The technique has been mostly used for determination of macro constituents. However, measurements of physical properties like texture attributes and water holding capacity have also been reported. A list of examples is shown in Table 24.1.

Table 24.1   Recent examples of the use of NIR spectroscopy in fish research

Application	Authors	Reference number
Process control of heated fish and shellfish	Uddin *et al.*	44
Quality attributes for shrimp	Brodersen and Bremner	45
Freshness of cod and salmon	Nilsen	46
Sensory attributes for cod, saithe, rainbow trout, herring and flounder	Warm *et al.*	47
Quality attributes for frozen/thawed red hake	Pink *et al.*	48
Quality attributes for frozen stored cod	Bechmann and Jørgensen	43
Fat and water content of salmon	Wold and Isaksson	49

24.3   Fluorescence spectroscopy

Where a low sensitivity and specificity are characteristics of NIR spectroscopy, quite the opposite can be said about fluorescence spectroscopy. Only few compounds present in foods emit fluorescenct light, but when they do the intensity is relatively high so that a small amount of substance is sufficient for providing enough light to be measured. It is also possible in some cases to add fluorescent compounds that react with substances that otherwise would not emit fluorescence. In that way indirect measurements can be made. Fluorescence appears when a substance absorbs light resulting in an excitation of an electronic state in the molecule and this is followed by a transition back to another state, most often of a lower energy yet higher than that of the original state which normally is the ground state. Thus, the emitted light is of lower energy, i.e. has a higher wavelength than the incident light. Univariate measurements have been made for many years, where monochromatic light is absorbed and one specific wavelength (or band) of emitted light selected by a monochromator or a filter is measured. The wavelength where the incident light is absorbed, as the wavelength of the emitted light, is characteristic for the molecule. Fluorescent compounds of special relevance for muscle-based food are aromatic amino acids like tryptophane and tyrosine and coenzymes like the reduced form of nicotinamide adenine dinucleotide (NADH). It is however not possible to avoid overlapping of either the emitted or absorbed wavelength ranges, so there will be some interference even if only few substances contribute to the fluorescence emission.

A multivariate treatment of the fluorescence spectra, on the other hand, provides a very efficient way of spectral profile deconvolution, i.e. mathematically separating contributions from the various components. This is accomplished by scanning in both the absorbance and emission modes: Emission spectra are taken at a number of wavelengths of the incident light. The result is a so called 'fluorescence landscape' like the one shown in Fig. 24.2. For low concentrations, the data will be approximately tri-linear as the intensity of the emitted light is proportional to the amount of emitting fluorophore and also to the intensity of the incident light:

$$I_i(\lambda_e, \lambda_a) \propto \sum_{j=1}^{N} c_{ij}\phi_j(\lambda_e)\epsilon_j(\lambda_a) \qquad\qquad 24.5$$

where $I_i(\lambda_e, \lambda_a)$ is the intensity of emitted light at the measurement wavelength λ_e after excitation at the wavelength λ_a, c_{ij} is the concentration of fluorophore j in sample i, $\phi_j(\lambda_e)$ is the emission profile, $\epsilon_j(\lambda_a)$ is the excitation profile, and the summation is over the N fluorescent compounds present.

In more concentrated systems there may be some severe non-linearity due to absorbance of the emitted light by other components (the inner filter effect). This can give rise to distortions, but often such interferences contain information themselves, e.g. about the amount of absorbing component. Moreover, the fluorescence intensity may be quenched, i.e. the transition probability of the

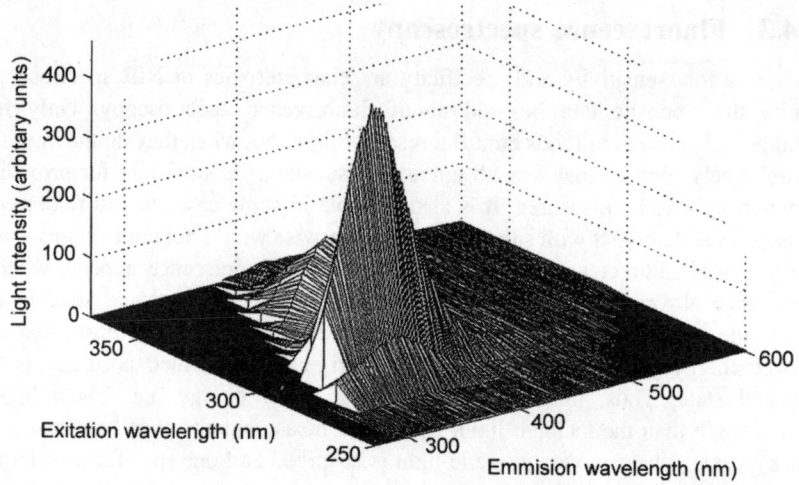

Fig. 24.2 Fluorescence landscape from an aqueous extract of cod muscle. The excitation wavelength range was from 250 to 370 nm in steps of 10 nm and the emission wavelength range was from 270 to 600 nm in steps of 1 nm. The instrument was a Perkin Elmer LS50B spectrofluorometer. By the courtesy of Mrs. Charlotte M. Andersen, Danish Institute for Fisheries Research.

transition that causes fluorescence may be lowered by other substances in the sample. In that case, the light detected is a *decreasing* function of the number of quenching molecules.

For tri-linear systems that have been measured in a way providing tri-linear data (Eqn 24.5), a very efficient multivariate algorithm exists, namely parallel factor analysis (PARAFAC).[16–19] The input to this algorithm is a three-directional tensor with one slab for each sample containing the fluorescence landscape. The PARAFAC algorithm has the property of providing real profiles, in this case emission and excitation* spectra, when its underlying assumptions are met. That is opposite to bi-linear decomposition algorithms like principal component analysis (PCA) and partial least squares (PLS) regression which result in linear combinations of the real shapes only. So when PARAFAC is working properly, a sort of 'mathematical chromatography'[20–21] can be obtained. It should be remembered that one of the dimensions, indexed by i in Eqn 24.5, is constituted of the samples. These must be representative in the sense that a common set of components are present, although in varying amounts. When this prerequisite is met together with an approximate proportionality between amount of substance and absorbed light and emitted

---

* The excitation spectrum resembles the absorbance spectrum with the exception that it is flat at wavelengths where absorbed light does not induce fluorescence. This is because absorption is not measured *per se*; only indirectly through its ability to induce emitted fluorescence, which is the quantity measured

light, PARAFAC is efficient in providing interpretable spectra. In situations where quenching or inner filter effects take place, a non-negativity constraint on the PARAFAC scores can be incorporated to stabilise the algorithm so it provides useful profiles.

The useful wavelength ranges of the excitation and the emitted light often overlap. A large fraction of the incident light is scattered (unchanged wavelength), and some of it will be detected resulting in large diagonal elements in the fluorescence landscape. As this signal has nothing to do with fluorescence, a band of data around the diagonal is normally excluded from the calculations by setting the matrix elements to 'missing values'. The part of the data matrix corresponding to lower wavelengths of the emitted than of the excitation light also consists of missing values. When implementing PARAFAC it is therefore important to use an algorithm that handles missing values efficiently without causing undue distortions of the resulting profiles.

One slight complication inherent in the PARAFAC technique is that of arbitrary scaling. As is obvious from Eqn 24.5, the three vectors c, ϕ and ϵ can be multiplied with arbitrary constants as long as the product of these is one. And the constants need not be the same for all components, $j = 1$ to N. The PARAFAC model provides three sets of vectors. One is named 'scores' and consists of vectors s_j with elements proportional to the concentrations c_j. The other two are named 'loadings' and consist of vectors $l1_j$ and $l2_j$ proportional to the emission and excitation spectra $\phi_j(\lambda_e)$ and $\epsilon_j(\lambda_a)$, respectively. The actual scaling is algorithm-dependent; often the loadings vectors are scaled to a length of unity. Where the shapes and not the absolute values of the spectra are of interest – they are merely used for providing guidance in identifying the compounds – one would like to extract quantitative information from the score vectors. Relative comparisons between samples of the concentration of a given substance are immediately possible by using such score values, as a scaling factor is common to all samples. The scaling factors may therefore be determined by including standards in the sample set. A series of samples with different amounts of internal standard added are to be preferred for pure standard solutions due to the sample matrix effects mentioned. The use of a standard requires, of course, that the fluorophore has been identified, but a previous guess can also be supported or contradicted by an internal standard. A correct standard adds to the score value and a wrong standard creates a new component (i.e. increases the number N of components by one).

Within fish research, fluorescence with 'classical' data analysis has been used for detecting bones that exhibits a characteristic emission easily distinguishable from that of the fish fillet.[22] A few applications to determination of quality attributes have also been published (Table 24.2). Examples of the use of tri-linear decomposition (PARAFAC) has appeared in the literature concerning other types of food (e.g. Reference 20), and reports on its use within fish research are on the way.

Table 24.2   Recent examples of the use of fluorescence spectroscopy in fish research

Application	Authors	Reference number
Lipid-protein interaction in mackerel	Saeed *et al.*	39
Quality changes during frozen storage of sardine	Aubourg *et al.*	50
Quality assessment of canned sardine	Aubourg and Medina	51
Stability of cod and bovine trypsin	Amin Amiza and Owusu Apenten	52

24.4   Nuclear magnetic resonance (NMR) spectroscopy

The applications of nuclear magnetic resonance (NMR) methods to food research may be divided into two main groups after the type of equipment used. One type makes use of high-resolution measurements by large and expensive instruments, operated by expert personnel. With that sort of equipment, nuclei like carbon, phosphorus and sodium can be monitored, and so can hydrogen in a way where atoms bonded in different positions in a molecule can be distinguished. The other type uses small bench-top instruments with a magnetic field strength suitable for hydrogen (protons) only and without the resolution power obtained with the high-field instruments. Applications of the second type most often deal with time-dependent phenomena related to protons in water and/ or lipid, namely the rate of relaxation of the nuclear spin magnetisation after an electromagnetic pulse.

Nowadays, almost all NMR instruments operate in the time-domain, i.e. they create a signal that is a function of time after a certain pulse – or sequence of pulses – has been applied to the sample. The primary signal in the time-domain actually consists of two components that may be considered as its real (Re) and imaginary (Im) part. For data analysis in the time-domain, either the amplitude, $\sqrt{(Re2 + Im2)}$, or even better the phase rotated signal (Im$\approx$0) is chosen to assure that the (small) noise component maintains its symmetrical (normal) distribution around zero.[23–24] Otherwise, the noise becomes all positive due to the square terms, inducing a slight baseline shift. For data analysis in the frequency-domain, a Fourier transform of the time-domain signal is calculated. Hereby one obtains an NMR spectrum which in the high-resolution case contains a number of peaks due to small differences in the so-called chemical shift. With low-resolution data only a single broad peak appears, limiting the usefulness of the spectrum. The time-domain data, however, contains very useful information about the mobility of the molecules in different environments in the food product.

In this chapter, the focus will be on low-field data as they are within reach of most fish research laboratories and the methodology also applicable at sight in the fisheries industry. Some relevant uses of high-field NMR measurements will be briefly listed to complete the picture, though.

24.4.1   High-resolution methods

High-resolution spectra (frequency-domain data) of nuclei like ^{1}H, ^{13}C and ^{31}P have been used for measuring the composition of food taking advantage of the ability to identify the various substances. By combined use of ^{1}H and ^{13}C NMR spectra an impressively detailed picture of the lipid composition of fish material was obtained.[25–28] Among the information extracted were the phospholipid content and distribution, the number, positions and stereo chemistry of unsaturated fatty acid double bonds and the position of various fatty acyl chains in triglyceride and phospholipid. This information was deduced from the spectra by classical means, i.e. by making use of the small differences in chemical shift due to position for identification and the peak integrals for quantification. There is, however, no doubt that some type of multivariate correlation of the spectra to process parameters or quality attributes would be possible, in the first place without identifying what chemical changes have taken place. A pattern recognition algorithm would probably be a good choice for that purpose as variation along the x-axis (chemical shift) is likely to be important.

^{31}P NMR spectra contain separated peaks from one, two or three phosphor nucleotides of which the adenine-containing ones, AMP, ADP and ATP, and the degradation product IMP, are the most abundant. Moreover, creatine phosphate and 'inorganic phosphate' can be identified and quantified. As these compounds change in relative amounts during the period of storage of the fish, the spectra also correlate to quality attributes, especially those related to *rigor mortis*, i.e. texture parameters. The exact chemical shifts of the nucleotide phosphates is pH-dependent, so the spectra also contain information about this quantity. As is the case for high-resolution ^{1}H and ^{13}C NMR spectra, the potential in a multivariate treatment of ^{31}P NMR spectra has not as yet been thoroughly explored.

A few attempts to apply ^{23}Na NMR relaxation curves (time-domain data) to the diffusion of salt in salted fish products have been made and this application should be obvious. The data analysis possibilities are analogous to those of low-field data discussed in the following section. Hitherto, the exponential fit (hard modelling) and the inverse Laplace transform approaches have been used, but apparently not the slicing technique (see later).

A collection of high-field applications is referenced in Table 24.3.

Table 24.3   Recent examples of the use of high-field NMR spectroscopy in fish research

Application	Authors	Reference number
Classification of wild and farmed salmon	Aursand *et al.*	53
Selection of thermal processing conditions for canned tuna	Medina *et al.*	54
Lipid oxidation during thermal processing of salmon	Medina *et al.*	55
Post mortem changes in carp muscle	Yokoyama *et al.*	56
Post mortem changes in oyster	Yokoyama *et al.*	57

24.4.2 Low-resolution methods

The resolution of the proton time-domain signal obtained with low-field instruments is not sufficient for extracting information on the position of protons in macromolecules like protein. In fact, it may be difficult to record contributions from these protons at all because of their fast relaxation compared with the time scale of the instrument. The dominating compound giving rise to magnetic relaxation is water due to its high amount in the tissue and its suitable relaxation rate. In systems with a high lipid content, the relaxation rates of lipid protons may be measurable too. The relaxation rate for lipid protons is much smaller than that for water, and one straight-forward application is to determine the lipid to water ratio from the initial signal intensity and the intensity after (almost) full relaxation of the water protons.

About 80 percent of the muscle of lean fish is water. It is therefore perhaps not surprising that the state of water – its mobility, distribution within the heterogeneous food system and binding to proteins – plays an important role for the perception of sensory quality. It also governs, or at least influences, biochemical reactions that take place in the muscle *post mortem*, thereby affecting other quality changes too. In a heterogeneous system like muscle, there may be a lot of different interactions between water and other constituents, and water movement may also be restricted by physical barriers, i.e. intracellular and intercellular membranes. It seems, however, that a fairly rapid water exchange takes place resulting in a few distinguishable 'sorts' of water. Such sorts are named water pools. Examples are extracellular water, bulk cytoplasmic water, and water bound to the muscle fibres.

NMR is an excellent tool for measuring water pools of different water mobility. This is because the rate of magnetic relaxation after application of a suitable electromagnetic pulse is characterised by some rate constants that are dependent of the water mobility. And the single amplitudes of which the signal is composed, are proportional to the number of nuclei relaxing, i.e. quantitatively represent the size of the pools. The reciprocal rate constants are named relaxation times, of which there are two of different origin: the spin-lattice relaxation time, *T1*, and the spin-spin, or transverse, relaxation time, *T2*. The transverse relaxation time is the easiest to determine, and *T2* is an increasing function of mobility. In the following, the treatment of data from such a spin-spin relaxation experiment involving a so called Carr-Purcell-Meiboom-Gill (CPMG) electromagnetic pulse sequence[29–30] will be discussed. This pulse sequence is preferable as the full relaxation curve is obtained in a single experiment and the contribution from inhomogeneity of the magnetic field is avoided.

The NMR transverse relaxation signal of sample *i*, $R_i(t)$, sampled at the even echoes after a CPMG pulse sequence, is composed of a sum of mono-exponentials:

$$R_i(t) = \sum_{j=1}^{N} a_{ij} e^{-k_j t}$$

24.6

plus a noise component that with modern instruments is small. The parameters in this expression are the amplitudes a_{ij}, the relaxation rate constants k_j, which are the reciprocal transverse relaxation times, $T2 = 1/k$, and the number N of distinguishable components. The variable t is time.

There are several ways to handle Eqn 24.6. Perhaps the reader recognises that it can be considered as a Laplace transform of the product of a continuous function $a_i(k)$ multiplied with a finite sum of Dirac delta functions, $\delta(k - k_j)$. If one was able to calculate the inverse Laplace transform of R_i, the amplitude distribution, $a_i(k) \cdot \delta(k - k_j)$, would immediately be obtained. This problem is, however, numerically ill-conditioned, and a smoothing function has to be introduced so that the solution becomes a distribution with peaks of a certain width. Even then, reaching convergence may be problematic, but a useful procedure has been published together with an associated computer program.[31–33] A few uses of this method for solving food-related problems have appeared in the literature.

The second approach, which is the 'classical' one, is to least squares fit the parameters in Eqn 24.6 directly. As is immediately apparent, the number of parameters for each sample, i, is $2N+1$, namely the N a_{ij}, the N k_j and N itself. This fitting is not a simple task, because the parameters are strongly correlated and several local minima in the objective function, the sum of squared residuals, are likely to exist. By increasing the number N, an apparent better fit will always be achieved, but an overfit situation may be reached for even small N-values. Determining the right N is difficult and in much published work a value seemingly is presumed without a firm basis in the experimental results. But for systems where N has been established and good initial guesses of the set of k_j can be made, the direct fit may converge rapidly to a meaningful result.

The third possibility is that of soft modelling by multivariate methods. Unlike the two former methods which can operate on each sample independently, the multivariate approach is dependent on a set of samples that have the NMR relaxation component in common. The measured signal, R, shown in Eqn 24.6, is seen to be perfectly bi-linear, and can be expressed as

$$R_i(t) = \sum_{j=1}^{N} s_{ij} l_j(t)$$ 24.7

where s_j is the j'th score vector indexed by i, and l_j is the j'th loading vector indexed by t. This opens up for an efficient data analysis with the well-known 'generic' chemometric methods PCA and PLS regression. For predictive purposes – for example using a PLS regression model for predicting quality-related properties – the bilinear property is fully sufficient, but as it operates with 'latent variables' the relaxation curve parameters themselves are not determined. And these parameters, the rate constants k_j and the water pool sizes a_{ij} are of particular interest for characterising the samples.

Fortunately, it is possible to turn the bi-linear system to a tri-linear one. When doing so, the powerful PARAFAC method mentioned in connection with fluorescence can be applied, giving as result the 'pure' mono-exponential components. And from these curves it is straightforward to calculate the parameters. Hereby one gets a more detailed picture of the water pools and their mobility in the set of samples. The trick is to take advantage of the special property of the exponential function shown in Eqn 24.8:

$$\exp(x + u) = \exp(x) \cdot \exp(u) \qquad\qquad 24.8$$

This can be accomplished by 'slicing' the data matrix, whose elements are $X_{it} = R_i(t)$, into two or more 'slabs' as illustrated in Fig. 24.3. The 'lag' variable, u, then spans the third dimension, and the data matrix becomes a three-directional tensor with elements $X_{itu} = R_i(t + u)$. Due to the property shown in Eqn 24.8, the elements in X fulfill:

$$R_i(t + u) = \sum_{j=1}^{N} a_{ij} e^{-k_j t} e^{-k_j u} \qquad\qquad 24.9$$

which is tri-linear. A PARAFAC model can thus be calculated resulting in a set of scores and two sets of loadings (Eqn 24.10), and because of the uniqueness of the PARAFAC solution, these parameters are those that were aimed for.

$$R_i(t + u) = \sum_{j=1}^{N} s_{ij} l1_j(t) l2_j(u) \qquad\qquad 24.10$$

The scores s_{ij} are proportional to the amplitudes a_{ij}, and the loadings $l1_j(t)$ and $l2_j(u)$ are proportional to the 'pure' mono-exponential relaxation curves in the time and the lag directions. The rate constants can be easily calculated as slopes of the weighted mono-logarithmic plots of $l1_j(t)$ against t

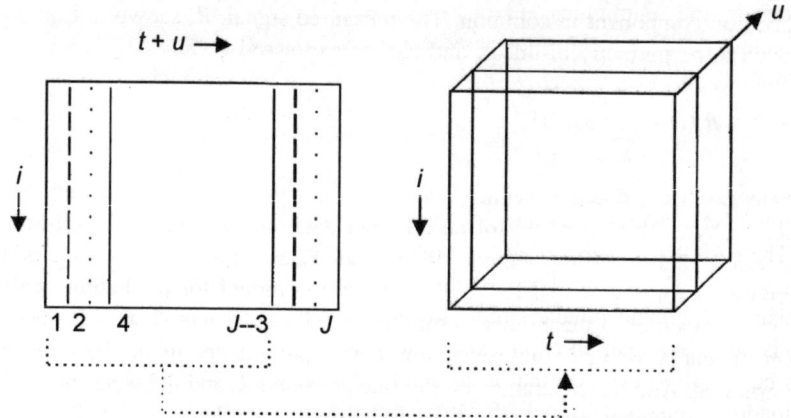

Fig. 24.3 Illustration of a slicing procedure for NMR relaxation profiles. The data matrix, made up of J columns, is sliced into three slabs. Columns 1 to J-3 are in the first slab, columns 2 to J-2 in the second slab and columns 4 to J in the third slab. This corresponds to u-values of 0, 1 and 3, respectively.

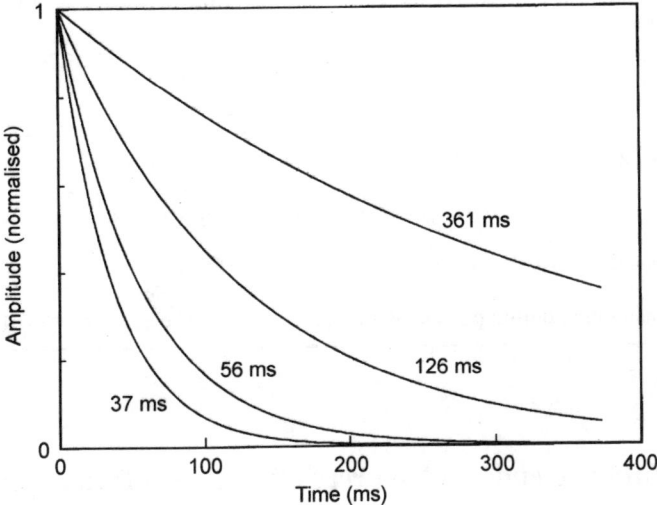

Fig. 24.4 Mono-exponential nuclear magnetic resonance (NMR) transverse relaxation curves. The curves are normalised to an amplitude of one. The $T2$ relaxation times in this example are 37 ms, 56 ms, 126 ms and 361 ms.

$$\ln\{l1_j(t)\} = -k_j t + \ln(q_j) \qquad\qquad 24.11$$

where q_j is the proportionality constant, the magnitude of which is dependent on the PARAFAC algorithm used.

After having determined the set of k_j values from Eqn 24.11, the mono-exponential relaxation curves can be constructed (Fig. 24.4) and the amplitudes calculated by simple multi-linear regression using Eqn 24.6. As the absolute values of the amplitudes are proportional to the number of relaxing nuclei (protons), i.e. dependent of the sample size, it may be preferable to operate with normalised amplitudes, A_{ij}, defined as

$$A_{ij} = a_{ij} / \sum_{j=1}^{N} a_{ij} \qquad\qquad 24.12$$

which then are direct measures of the relative pool sizes.

The PARAFAC method also requires that the correct value of the number of components is chosen. There are several ways for assisting in making this decision. One conceptually simple means to gradually increase N until loading $l1_N$ is not any longer an exponential. Also, if the data set contains duplicates, it can be split into two sets that in principle are equal except for the noise.[35] The $l1$-loadings obtained from PARAFAC models of each subset should therefore also be equal as long as noise-components are not included.

The tri-linearising technique looks very promising, but reports on applications have just started to appear in the literature. A few relevant references are collected in Table 24.4.

Table 24.4 Recent examples of the use of low-field NMR relaxation measurements in fish research

Application	Authors	Reference number
Water pools and water mobility in frozen and chill stored cod	Jensen *et al.*	58
Oil and water in salmon and water holding capacity in cod	Jepsen *et al.*	59
Texture changes in frozen cod mince	Steen and Lambelet	60
Protein denaturation during processing of cod	Lambelet *et al.*	61

24.5   Future trends and sources of further information and advice

The methods discussed in this chapter are but only a few of the versatile fast instrumental methods available. Others like Raman spectroscopy[36–38] and electron spin resonance (ESR) spectroscopy[39] have also been successfully applied. The possibilities are not limited to techniques based on electromagnetic waves; the use of e.g. sound waves[40] provides valuable information too. All methods that supply measurement results as a vector or a matrix benefit from the use of multivariate data analysis for which several efficient 'generic' procedures exist. No doubt an increasing number of papers using such methods for extracting reliable information from measurement data will appear in the near future. Although other very efficient detection principles based on micro sensors and the even smaller nano technology gain an increasing importance in the field of quality control and quality monitoring, spectrometry still plays a role. But for spectroscopy to remain 'competitive', an optimal use of the results obtained is necessary. Nowadays, multivariate data analysis provides the tools for accomplishing this task.

For more information about the topics covered in this chapter, an internet search is recommended. The number of pages dealing with these matters is still accelerating. To those preferring books, the monographs on multivariate methods by Malinowski,[41] Martens and Næs[8] and Martens and Martens[14] are highly recommended. Multi-way methods and algorithms for those are excellently treated in the Ph.D. dissertation by Bro.[35] A thorough treatment of NIR spectroscopy is given by Osborne *et al.*,[5] and the newly published book edited by Gunasekaran[42] contains papers on other spectroscopic methods including 2D image analysis and NMR imaging.

24.6 References

1 JØRGENSEN B M and JENSEN H S, 'Can near-infrared spectrometry be used to measure quality attributes in frozen cod?', in Luten J B, Børresen T and Oehlenschläger J, 'Seafood from producer to consumer, Integrated approach to quality', *Dev Food Sci*, 1997 **38**, Amsterdam, Elsevier, 491–6.

2 KUBELKA P, 'New contributions to the optics of intensely light-scattering materials, Part 1', *J Opt Soc Am*, 1948 **38** 448–57.

3 KUBELKA P and MUNK F, 'Ein Beitrag zur Optik der Farbanstriche', *Z Tech Phys*, 1931 **12** 593–604.

4 BULL C R, 'A model of the reflectance of near-infrared radiation', *J Mod Opt*, 1990 **37** 1955–64.

5 OSBORNE B G, FEARN T and HINDLE P H, *Practical NIR spectroscopy with applications in food and beverage analysis*, 2 ed, Harlow, Longman, 1993.

6 JØRGENSEN B M, 'Quantification of overlapping VIS-absorption bands by Fourier domain fitting of a generalized band shape function' *Appl Spectrosc*, 1990 **44** 313–17.

7 MARTENS H, JENSEN, S A and GELADI P, 'Multivariate linearity transformations for near-infrared reflectance spectroscopy', *Proc Nordic Symp Appl Stat*, Stavanger, Stokkland Forlag, 1983.

8 MARTENS H and NÆS T, *Multivariate calibration*, Chichester, Wiley & Sons, 1989.

9 WOLD S, ANTTI H, LINDGREN F and ÖHMAN J, 'Orthogonal signal correction of near-infrared spectra', *Chemom Intell Lab Syst*, 1998 **44** 175–85.

10 ANDERSSON C, 'Direct orthogonalization', *Chemom Intell Lab Syst*, 1999 **47** 51–63.

11 FEARN T, 'On orthogonal signal correction' *Chemom Intell Lab Syst*, 2000 **50** 47–52.

12 WESTERHUIS J A, DE JONG S and SMILDE A K, 'Direct orthogonal signal correction' *Chemom Intell Lab Syst*, 2001 **56** 13–25.

13 MARTENS H and MARTENS M, 'Modified Jack-knife estimation of parameter uncertainty in bilinear modelling by partial least squares regression (PLSR)', *Food Qual Prefer*, **11** 5–16.

14 MARTENS H and MARTENS M, *Multivariate Analysis of Quality. An Introduction*, Chichester, Wiley & Sons, 2001.

15 BORGGAARD C (1995), 'Modelling non-linear data using neural networks regression in connection with PLS or PCA' in Andrews D L and Davies A M C, *Frontiers in analytical spectroscopy*, Cambridge, The Royal Society of Chemistry, 209–17.

16 HARSHMAN R A, 'Foundations of the PARAFAC procedure: model and conditions for an explanatory multi-mode factor analysis', *UCLA Working Papers in phonetics*, 1970 **16** 1–84.

17 CARROLL J D and CHANG J, 'Analysis of individual differences in multidimensional scaling via an N-way generalization of Eckart-Young decomposition', *Psychometrika*, 1970 **35** 283–319.

18   HARSHMAN R A and LUNDY M E, 'PARAFAC: Parallel factor analysis', *Comput Stat Data Anal*, 1994 **18** 39–72.

19   BRO R, 'PARAFAC. Tutorial and applications', *Chemom Intell Lab Syst*, 1997 **38** 149–71.

20   MUNCK L, NØRGAARD L, ENGELSEN S B, BRO R and ANDERSSON C A 'Chemometrics in food science – a demonstration of the feasibility of a highly exploratory, inductive evaluation strategy of fundamental scientific significance', *Chemom Intell Lab Syst*, 1998 **44** 31–60.

21   BAUNSGAARD D, ANDERSSON C A, ARNDAL A and MUNCK L, 'Multi-way chemometrics for mathematical separation of fluorescent colorants and colour precursors from spectrofluorimetry of beet sugar and beet sugar thick juice as validated by HPLC analysis', *Food Chem*, 2000 **70** 113–21.

22   HUSS H H, SIGSGAARD P and JENSEN S AA, 'Fluorescence of fish bones', *J Food Prot*, 1985 **48** 393–6.

23   VAN DER WEERD L, VERGELDT F J, DE JAGER P A and VAN AS H, 'Evaluation of algorithms for analysis of NMR relaxation decay curves', *Mag Reson Imag*, 2000 **18** 1151–7.

24   PEDERSEN H T, *Low-field nuclear magnetic resonance and chemometrics applied in food science*, Dissertation, The Royal Veterinary and Agricultural University, Frederiksberg, Denmark, 2001.

25   AURSAND M, RAINUZZO J R and GRASDALEN H 'Quantitative high resolution ^{13}C and ^{1}H nuclear magnetic resonance of omega-3 fatty acids from white muscle of Atlantic salmon *(Salmo salar)*', *J Am Oil Chem Soc*, 1994 **71** 971–81.

26   AURSAND M, JØRGENSEN L and GRASDALEN H, 'Positional distribution of ω-3 fatty acids in marine lipid triacylglycerols by high-resolution ^{13}C nuclear magnetic resonance spectroscopy', *J Am Oil Chem Soc*, 1995 **72** 293–7.

27   SACCHI R, MEDINA I, AUBOURG S P, ADDEO F and PAOLILLO L, 'Proton nuclear magnetic resonance rapid and structure-specific determination of ω-3 polyunsaturated fatty acids in fish lipids', *J Am Oil Chem Soc*, 1993 **70** 225–28.

28   MEDINA I and SACCHI R, 'Acyl stereospecific analysis of tuna phospholipids via high resolution ^{13}C-NMR spectroscopy', *Chem Phys Lipids*, 1994 **70** 53–61.

29   CARR H Y and PURCELL E M, 'Effects of diffusion on free precession in nuclear magnetic resonance experiments', *Am J Physiol*, 1954 **94** 630–8.

30   MEIBOOM S and GILL D, 'Modified spin-echo method for measuring nuclear times', *Rev Sci Instr*, 1958 **29** 688–91.

31   PROVENCHER S W, 'A Fourier method for the analysis of exponential decay curves', *Biophys J*, 1976 **16** 27–41.

32   PROVENCHER S W, 'An eigenfunction expansion method for the analysis of exponential decay curves', *J Chem Phys*, 1976 **64** 2772–7.

33   PROVENCHER S W, 'A constrained regularisation method for inverting data represented by linear algebraic or integral equations' *Comp Phys Commun*, 1982 **27** 213–27.

34   RUAN R R and CHEN P L, *Water in foods and biological materials. A nuclear magnetic resonance approach*, Lancaster, Technomic Pub. Co., 1998.

35   BRO R, *Multi-way analysis in the food industry. Models algorithms and applications*, Dissertation, University of Amsterdam, Amsterdam, The Netherlands, 1998.

36   CARECHE M and LI-CHAN E C Y, 'Structural changes in cod myosin after modification with formaldehyde or frozen storage', *J Food Sci*, 1997 **62** 717–23.

37. CARECHE M, HERRERO A M, RODRIGUEZ-CASADO A, MAZO M L and DEL CARMONA P, 'Structural changes of hake (*Merluccius merluccius* L.) fillets: effects of freezing and frozen storage', *J Agric Food Chem*, 1999 **47** 952–9.

38   OGAWA M, NAKAMURA S, HORIMOTO Y, HAEJUNG AN TSUCHIYA T and NAKAI S, 'Raman spectroscopic study of changes in fish actomyosin during setting', *J Agric Food Chem*, 1999 **47** 3309–18.

39   SAEED S, FAWTHROP S A and HOWELL N K, 'Electron spin resonance (ESR) study on free radical transfer in fish lipid-protein interaction', *J Sci Food Agric*, 1999 **79** 1809–16.

40   COUPLAND J and MCCLEMENTS D J (2001) 'Ultrasonics' in Ref. 42, 217–41.

41   MALINOWSKI E, *Factor analysis in chemistry*, New York, Wiley & Sons, 2nd ed, 1991.

42   GUNASEKARAN S, *Nondestructive food evaluation. Techniques to analyze properties and quality*, New York, Marcel Dekker, 2001.

43   BECHMANN I E and JØRGENSEN B M, 'Rapid assessment of quality parameters for frozen cod using near-infrared spectroscopy', *J Food Sci Technol* 1998 **31** 648–52.

44   UDDIN M, ISHIZAKI S, OKAZAKI E and TANAKA M, 'Near-infrared reflectance spectroscopy for determining end-point temperature of heated fish and shellfish meats', *J Sci Food Agric*, 2002 **82** 286–92.

45   BRODERSEN K and BREMNER H A, 'Exploration of the use of NIR reflectance spectroscopy to distinguish and measure attributes of conditioned and cooked shrimp (*Pandalus borealis*)', *Food Sci Technol*, 2001 **34** 533–41.

46   NILSEN H A, 'Freshness measured by near-infrared technology', *Food Technol Int*, 2001 107–9.

47   WARM K, MARTENS H and NIELSEN J, 'Sensory quality criteria for five fish species predicted from near-infrared (NIR) reflectance measurement', *J Food Qual*, 2001 **24** 389–404.

48   PINK J, NACZK M and PINK D, 'Evaluation of the quality of frozen minced red hake: use of Fourier transform near-infrared spectroscopy', *J Agric Food Chem*, 1999 **47** 4280–84.

49   WOLD J P and ISAKSSON T, 'Non-destructive determination of fat and moisture in whole Atlantic salmon by near-infrared diffuse spectrometry', *J Food Sci*, 1997 **62** 734–6.

50   AUBOURG S P, SOTELO C G and PEREZ-MARTIN R, 'Assessment of quality

changes in frozen sardine (*Sardina pilchardus*) by fluorescence detection', *J Am Oil Chem Soc*, 1998 **75** 575–80.

51   AUBOURG S and MEDINA I, 'Quality differences assessment in canned sardine (*Sardina pilchardus*) by fluorescence detection', *J Agric Food Chem*, 1997 **45** 3617–21.

52   AMIN AMIZA M and OWUSU APENTEN R K, 'Urea and heat unfolding of cold-adapted Atlantic cod (*Gadus morhua*) trypsin and bovine trypsin', *J Sci Food Agric*, 1996 **70** 1–10.

53   AURSAND M, MABON F and MARTIN G, 'Characterization of farmed and wild salmon *(Salmo salar)* by a combined use of compositional and isotopic analyses', *J Am Oil Chem Soc*, 2000 **77** 659–66.

54   MEDINA I, SACCHI R and AUBOURG S, 'Application of ^{13}C NMR to the selection of the thermal processing conditions of canned fatty fish', *Eur Food Res Technol*, 2000 **210** 176–8.

55   MEDINA I, SACCHI R, GIUDICIANNI I and AUBOURG S, 'Oxidation in fish lipids during thermal stress as studied by ^{13}C nuclear magnetic resonance spectroscopy', *J Am Oil Chem Soc*, 1998 **75** 147–54.

56   YOKOYAMA Y, AZUMA Y, SAKAGUCHI M, KAWAI F and KANAMORI M, '^{31}P NMR study of bioenergetic changes in carp muscle with cold-CO_2 anesthesia and non-destructive evaluation of freshness', *Fisheries Sci*, 1996 **62** 267–71.

57   YOKOYAMA Y, AZUMA Y, SAKAGUCHI M, KAWAI F and KANAMORI M, 'Non-destructive phosphorus-31 nuclear magnetic resonance study of postmortem changes in oyster tissues', *Fisheries Sci*, 1996 **62** 416–20.

58   JENSEN K N, GULDAGER H S and JØRGENSEN B M, 'Three-way modelling of NMR relaxation profiles from thawed cod muscle', *J Aquat Food Prod Technol*, 2002, in press.

59   JEPSEN S M, PEDERSEN H T and ENGELSEN S B, 'Application of chemometrics to low-field ^{1}H NMR relaxation data of intact fish flesh', *J Sci Food Agric*, 1999 **79** 1793–802.

60   STEEN C and LAMBELET P, 'Texture changes in frozen cod mince measured by low-field nuclear magnetic resonance spectroscopy', *J Sci Food Agric*, 1997 **75** 268–72.

61   LAMBELET P, RENEVEY F, KAABI C and RAEMY A, 'Low-field nuclear magnetic resonance relaxation study of stored or processed cod', *J Agric Food Chem*, 1995 **43** 1462–6.

Index